A CHRONOLOGY OF THE HISTORY OF SCIENCE, 1450–1900

GARLAND REFERENCE LIBRARY
OF THE HUMANITIES
(Vol. 714)

By the Same Author

A Historical Catalogue of Scientists and Scientific Books: From the Earliest Times to the Close of the Nineteenth Century. Garland Publishing, Inc., 1984.

A Historical Catalogue of Scientific Periodicals, 1665–1900: With a Survey of Their Development. Garland Publishing, Inc., 1985.

A CHRONOLOGY OF
THE HISTORY OF SCIENCE,
1450–1900

Robert Mortimer Gascoigne

GARLAND PUBLISHING, INC. • NEW YORK & LONDON
1987

Library of Congress Cataloging-in-Publication Data

Gascoigne, Robert Mortimer, 1918–
A Chronology of the History of Science, 1450–1900.

(Garland Reference Library of the Humanities;
vol. 714)
Includes index.
1. Science—Chronology. 2. Scientists—Biography.
3. Science—History. I. Title. II. Series: Garland
Reference Library of the Humanities; v. 714.

Q125.G39   1987   509   86-33503
ISBN 0-8240-9106-X (alk. paper)

Printed on acid-free, 250-year-life paper
Manufactured in the United States of America

CONTENTS

PART 2. THE SOCIAL DIMENSION

Arrangement by Countries or Regions

INTRODUCTION

Chronology may be the most rudimentary form of history but it has its uses. The present example of the genre, extending from the printing of Aristotle's works in 1472 (entry 1) to the establishment of some American scientific societies in the 1890s (entries 3073–3076), is intended to be of use in various ways to anyone interested in the history of science. The chronological arrangement is by decades, which allows much more flexibility than an annual arrangement, and in general the entries are longer and more discursive than is often the case in chronologies. The book was compiled mainly by means of a biographical approach which made possible the division into two parts, presenting the social aspect of the history of science as well as the more usual cognitive aspect. Part 2, dealing with the social aspect, consists largely of brief career sketches (rather than biographies) of scientists, especially in relation to scientific and educational institutions. There are cross-references linking the two parts and the index enables the book to be used also as a biographical dictionary. The conception of the book is, I believe, unique: I am not aware of anything similar having been published previously.

Two quite different kinds of history of science are included: the cognitive dimension of the subject, which is the traditional approach, and the social dimension which is a fairly new development. The term "cognitive" is used to denote the fundamental kind of history of science which deals with discoveries and new ideas and theories, and in general with the growth of scientific knowledge. On the other hand the social history of science (as the term is used here) takes cognitive history for granted and is concerned with persons—with the many scientists who, over the centuries, have built up the great edifice of scientific knowledge. And concerned with them not just as separate individuals but as members of the scientific community which began to take shape after the introduction of printing in the fifteenth century, and through the following centuries gradually increased in cohesion and internal organization as well as in size.

(It should be mentioned that there are other, very different, types of social history of science—for example the study of the effects of great advances in scientific understanding, such as Newtonianism and Darwinism, on general culture and society at large, or the investigation of the influence, real or supposed, of "external" social factors on scientific thought. These other types of social history are barely touched upon in the present work.)

The social history of the scientific community, which is the subject of Part 2 of the book, has as its basic principle of organization the division into different national settings. Early modern science began chiefly in Italy but from the mid-seventeenth century Italian science was in a state of decline, while rapid growth was taking place in France, Britain, and the Netherlands. In Britain there was a brilliant period in the 1650s and 1660s but soon afterwards British science went into a century-long eclipse, and through the eighteenth century and until about the 1830s France had the leadership. Thereafter it was Germany's turn, and the great period of German science extended into the early twentieth century. Meanwhile the scientific community was putting down its roots in the two potential superstates that flanked Europe—Russia and the United States.

In the past (to say nothing of the present) the activities of the scientific community in any particular country were profoundly affected by the national culture and the social and political characteristics of the country. The relations of the community to its national setting were chiefly mediated by institutions of several kinds, especially the universities and other teaching bodies and the state-supported academies and other organizations for research. To a large extent it was the various kinds of institutions that gave men (hardly any women!) the chance to become scientists and shaped their careers. The career sketches in Part 2, together with the entries dealing with institutions, are presented as basic data for the social history of the scientific community, paralleling the basic data for the cognitive history of science in Part 1. The interconnections between the two parts, effected by means of the cross-references, may contribute to the attempt to relate the cognitive to the social history of science, a topic which is attracting interest today.

To turn now to the methods employed in the construction of the book: though a variety of sources has been used its character has been shaped by the biographical approach. The primary aim has been to include in Part 1 the most important work of the most important scientists in the period covered. To keep the size within reasonable limits the number of scientists was restricted to one thousand. The first task, then, was to select the thousand most important scientists of the period. This was done by going through my *Historical Catalogue of Scientists and Scientific Books* (New York, 1984), which contains brief information about 12,300 scientists in the period, and using as the basis of selection the degree of importance accorded to each of them in various biographical dictionaries and encyclopaedias and in histories of the individual sciences. Much use has, of course, been made of the *Dictionary of Scientific Biography*, but for the purpose it had to be used in conjunction with other biographical sources and histories of the sciences.* In general, scientists who made theoretical advances were preferred before those who confined themselves to experimental, observational, or descriptive work; many important cases of the latter class have been included, however.

Though as much use as possible was made of relevant reference books and histories of the sciences, the list of a thousand scientists finally arrived at has inevitably been shaped to a considerable extent by my own subjective judgement. Everyone would, of course, produce a different list. The differences, however, would be largely confined to cases close to the borderline between inclusion and exclusion, i.e., to the least important (though most numerous) cases; the more important a scientist the more agreement there would be that he should be included. There is a very wide range of significance among scientists and it is likely that, as in other fields of human achievement, the rule of "the higher, the fewer" can be represented by a distribution curve of the inverse-square type. Thus a relatively small number of scientists on the top portion of the curve have contributed much more to the advancement of scientific knowledge than very large numbers of scientists lower down on the curve. The intention in compiling the above-mentioned list was to try to identify the top thousand on the distribution curve.

Part 1 of the chronology was then constructed by selecting the most important work of each of the scientists in the list (though, as explained later, not all of them were included in Part 1). This second stage in the selection process was made on the same basis as the first stage—i.e., with reference to biographies and histories of the individual sciences—and generally in conjunction with it. Here too opinions will inevitably differ about what should, or should not, be included—but again there would be a large measure of agreement. Part 1 also contains much additional mate-

---

*Though the *DSB* includes all the most important figures in the history of science it also includes many minor figures—some very minor. Thus no conclusion can be drawn about the degree of significance of a particular scientist simply from the fact that he is included in the *DSB*. Furthermore, the editors of the *DSB* do not claim to have provided a fully comprehensive coverage of scientists of medium importance. The list of a thousand scientists selected for the present work includes twenty-one who are not in the *DSB*.

rial: in particular there are many entries concerning significant advances made, or significant books written, by persons who did nothing else of importance or whose other achievements were outside the subject range of the present work. Such entries are headed by keywords or book titles—not by personal names—and the persons concerned are not included in Part 2. (Titles of books or articles are also used as entry headings in the case of works written jointly by two or more persons.)

Part 2 consists of career sketches of the thousand scientists together with entries relating to institutions. Each individual is placed in the chronological arrangement according to the decade in which he began his career (generally about twenty to thirty years after his date of birth but with exceptions in various special cases). Here and elsewhere in the chronology the emphasis is on flexibility, and entries within decades are not necessarily in chronological order. The time limit of 1900 has been observed fairly strictly in Part 1 but, where necessary, the career sketches in Part 2 continue into the early twentieth century. Some of the persons in Part 2 do not appear in Part 1 because they were selected for their institutional, organizational, or pedagogical importance rather than for contributions to knowledge.

The institutions included in Part 2 nearly all fall into one or other of the following categories: (a) universities and other teaching bodies; (b) state-supported academies and private societies (preeminently the Royal Society of London) concerned with the cultivation of science generally; (c) societies, generally national, for the individual sciences; and (d) special institutions for research in particular sciences, namely botanical gardens, natural history museums, astronomical observatories, and geological surveys.

The teaching institutions in category (a) are prominent in the text but generally do not have separate entries; they are dealt with in the Appendix (pp. 537–544). Institutions in categories (b) and (c) invariably published periodicals and the chief criterion for their inclusion was the importance of their periodical, information on this score being derived from my *Historical Catalogue of Scientific Periodicals, 1665–1900* (New York, 1985). (These two categories contain numerous famous institutions but category (b) also includes a multitude of small, provincial societies, and category (c) many small societies for natural history and other popular subjects which belonged to the extensive "amateur fringe" of the scientific community. These were generally excluded.) In category (d) only the earliest or the most famous botanical gardens were included (the lead of the pioneers was followed, from the seventeenth century onward, by the establishment of gardens in many of the main cities of Europe) while the other kinds of institutions—which also became quite numerous in the later periods—were assessed mainly by the extent to which they figured in the relevant career sketches and subject histories.

Finally, it may be worth mentioning that I propose to use information from the present work, including statistical analyses of the data, in a forthcoming book on the social history of the scientific community.

R.M.G.
School of History and Philosophy of Science
University of New South Wales
Sydney

PART 1

THE COGNITIVE DIMENSION

Arrangement by Subjects

1.01 SCIENCE IN GENERAL

Including natural philosophy and philosophy of science.

The term 'physics' (or its equivalent in other languages) occurs in some of the titles listed in this section and it should be remembered that before the mid-nineteenth century the term usually meant science generally (including the natural history sciences) or, more specifically, the non-mathematical sciences.

1470s

1. ARISTOTLE (ca. 384-322 B.C.) Latin translations of his works appeared in print from 1472 onward and there were soon numerous editions of the various individual works and of the *Opera*. (The printing of the Greek texts came later) Besides the vast importance, positive and negative, of his writings for natural philosophy, several of his individual works were significant for particular sciences--especially zoology (1410), botany (1276), earth science (998), and mechanics (490).

1500s

2. *MARGARITA philosophica*. Freiburg, 1503. By G. Reisch. The first significant encyclopaedia of modern times. It included a considerable amount of arithmetic, geometry, and astronomy. There were at least sixteen editions before 1600.

1550s

3. CARDANO, G. (1831) *De subtilitate*. Nuremberg, 1550. (6th ed., 1560) A wide-ranging encyclopaedia of science, technology, alchemy, magic, etc. A supplement with the title *De rerum varietate* was published at Basel in 1557 (5th ed., 1581).

4. SCALIGER, J.C. (1976) *Exotericarum exercitationum*. Paris, 1557. A comprehensive criticism of Cardano's *De subtilitate*. Scaliger's work attracted much attention and was held in high regard by many, including Francis Bacon.

1560s

5. *DE COMMUNIBUS omnium rerum naturalium principiis et affectionibus*. Rome, 1562. (Many later editions) A treatise on Aristotelian natural philosophy which was used as a textbook in the Jesuit schools and was widely read generally (it was cited by Galileo). The author was B. Pereira, a Jesuit and a teacher of various subjects in Rome.

1570s

6. CESALPINO, A. (1847) *Quaestionum peripateticarum.* Venice, 1571.
   Aristotelian natural philosophy by an important botanist and
   physician.

7. *PHILOSOPHIAE naturalis adversus Aristotelem.* Rome, 1574. One
   of the first works concerned with reviving the Greek atomic
   theory. The author was S. Basso, a French physician.

1580s

8. PORTA, G. della. (1861) *Magia naturalis.* Naples, 1589. A highly
   popular compendium of natural magic which went through many
   editions and was translated into all the chief European languages.
   It contained much miscellaneous lore, a proportion of which can
   be regarded as a rudimentary form of experimental science, the
   section on optics being the most significant.

1590s

9. ZABARELLA, J. (1853) Best known for his writings on logic and
   methodology but he also wrote an important and influential work
   on natural philosophy--*De rebus naturalibus*, Venice, 1590.

1610s

10. FLUDD, R. (2230) During the 1610s and 1620s he was one of the
    most outstanding exponents of occultism, the Hermetic philos-
    ophy, the microcosm-macrocosm analogy, and related doctrines.
    His writings attracted much attention throughout Europe and
    caused much controversy, being attacked by such figures as
    Kepler, Mersenne, and Gassendi.

1620s

11. BACON, F. (2229) The most important spokesman of the scientific
    movement of the early seventeenth century and a major influence
    in the shaping of the scientific tradition. His ideas were
    especially influential in England in the second half of the
    century. He was, as he himself said, "the herald of the new
    age."
         Bacon's most important books relevant to science appeared
    in the 1620s--*Novum organum* (1620), *De augmentis scientiarum*
    (1623; the much enlarged Latin version of his earlier *Advance-
    ment of learning*), *Sylva sylvarum* (1627), and the visionary
    New Atlantis (1627), the last two being posthumous.

12. GALILEI, G. (1863) *Il saggiatore* ["The assayer"]. Rome, 1623.
    (Published under the auspices of the Accademia dei Lincei)
    "This book has been justly called Galileo's scientific manifesto
    ... [It] marked a crucial point in the history of [his] thought.
    Before, he had spoken as the experimental scientist; later he
    was to speak as a theoretical scientist. In this work he speaks
    as a philosopher of science." (S. Drake) In particular the book
    contained his famous distinction between primary and secondary

qualities, and his no less famous declaration that the book of
nature is written in mathematical characters and can be deciphered only by those who understand mathematics.   (See also 16)

13. CAMPANELLA, T. (1868) Best known for his anti-traditional philo-
    sophical writings which covered a range of other topics as well
    as natural philosophy and appeared mostly in the 1620s and 1630s.
    In his *Apologia pro Galilaeo* (Frankfurt, 1622) he supported
    Galileo as regards the Copernican theory and also on wider issues,
    especially the question of the relations between science and
    theology.

14. MERSENNE, M. (1989) As well as playing a highly important role
    in the social history of science (see 1989) he was also of con-
    siderable importance for its cognitive history. He wrote many
    books in the period 1623–44 (with some posthumous items); they
    are rather unsystematic, most of them ranging over a number of
    subjects. The fields with which he was chiefly concerned were:
    (a) Natural philosophy, where he was one of the leading advocates
    of the new mechanistic conception of nature and a resolute opp-
    onent of occultism and Hermeticism; **(b)** Mathematics, in which he
    dealt with many topics but made his chief contribution in the
    theory of numbers; **(c)** Mechanics, which--as the leading science
    of the age--he cultivated in all its aspects, making some sig-
    nificant discoveries; **(d)** Acoustics, the field in which he made
    his chief contribution to physics; and **(e)** Optics.

15. GASSENDI, P. (1990) A major figure in the scientific movement of
    the mid-seventeenth century because of his espousal of Epicurean
    atomism as a replacement for the Aristotelian natural philosophy.
    His philosophical writings were very influential and in the
    seventeenth century he was much more highly regarded as a phil-
    osopher than he is today. His brand of atomism competed with
    the Cartesian natural philosophy for many years (and as an
    ingredient of Newtonianism eventually superseded it). He was
    also an assiduous astronomical observer from the mid-1620s
    onward and was interested in most of the sciences. In mechanics
    his chief achievement was the correct enunciation of the prin-
    ciple of inertia.
       Gassendi developed his philosophy in several works from
    the early 1620s onward but some of his main writings did not
    appear until the publication of his *Opera omnia* (Lyons, 1658)
    three years after his death. This publication seems to have
    been an important vehicle of his influence.

1630s

16. GALILEI, G. (See also 12) *Lettera a Madama Cristina di Lorena.*
    Strasbourg, 1636. Galileo's views on the proper relations
    between science and theology. The work was written in 1615 and
    circulated in manuscript but was not published until after his
    condemnation and not in Italy.

17. DESCARTES, R. (1991) A central figure in the Scientific Revol-
    ution of the seventeenth century, and one of the most famous of
    all philosophers. Author of an ambitious system of natural
    philosophy based on the new world-view emerging from the Coper-

nican Revolution--a philosophy which had an immense appeal to
his age and seemed to be as systematic and all-embracing as
Aristotle's had been.

His works were published from 1637 onward, beginning
with the celebrated *Discours de la méthode*. The intimate
connection between philosophy, mathematics, and science in his
thought is illustrated by the fact that the *Discours* had app-
ended to it, as examples of his method, three essays, of which
one was of fundamental importance to mathematics (see 135), another was a solid contribution to optics (see 583), and the
third an ambitious attempt to explain the weather (which, we
now know, was not possible in his age).

In time all the sciences were affected in some degree by
the pervasive influence of his mechanistic world-view (set forth
chiefly in his *Principia philosophiae* of 1644). This was especially the case for mechanics, the leading science of the age.
And in physiology his mechanistic appoach led in time to a
veritable revolution (see 1713).

1640s

18. *CIRCULUS Pisanus*. Udine, 1643. A revival of the atomic theory.
    By C.G. de Bérigard, professor of philosophy at Pisa and later
    at Padua.

19. *DEMOCRITUS reviviscens*. Pavia, 1646. A revival of the atomic
    theory. By J.C. Magnenus, professor of medicine and philosophy
    at Pavia. The book was widely read and often cited (e.g. by
    Boyle).

1650s

20. CHARLETON, W. (2246) *Physiologia Epicuro-Gassendo-Charltoniana;
    or, A fabrick of science natural, upon the hypothesis of atoms*.
    London, 1654. ('Physiologia' here means natural philosophy)
    A translation and amplification of a part of one of Gassendi's
    works dating from 1649. Charleton's book was widely read in
    England where it was effective in disseminating Gassendi's
    atomism.

1660s

21. BOYLE, R. (2251) **(a)** *Certain physiological essays*. London, 1661.
    (2nd ed., 1669) (The word 'physiological' here is equivalent
    to 'scientific') **(b)** *The origine of formes and qualities (acc-
    ording to the corpuscular philosophy)*. Oxford, 1666. Both
    works expounded a corpuscularian philosophy which Boyle claimed
    to be able to establish experimentally, in contrast to the
    atomism of Gassendi and the particle theory of Descartes. It
    can be regarded as an experimentalist's working hypothesis, as
    against the rationalist philosophies of Gassendi and Descartes.

1670s

22. ROHAULT, J. (2002) *Traité de physique*. Paris, 1671. (Many later
    editions. Latin trans. by T. Bonet in 1674 and another by S.
    Clarke in 1697; several later editions of both. English trans.,
    1723) A very influential exposition of Cartesian natural phil-
    osophy.

1680s
---

23. FONTENELLE, B. de. (2027) *Entretiens sur la pluralité des mondes.*
    Paris, 1686. (Many later editions and translations) A brilliant
    popularization, not only of Copernican astronomy, but of the
    mechanical philosophy (in its Cartesian formulation) and the
    new science generally.

1690s
---

24. RAY, J. (2254) *The wisdom of God manifested in the works of
    Creation.* London, 1691. (Three enlarged editions in his life-
    time and many reprintings up to the early nineteenth century)
    The paradigm work of the English tradition of natural theology.
    (See also 1023)

1700s
---

25. *LEXICON technicum; or, An universal English dictionary of arts
    and sciences.* 2 vols. London, 1704-10. (5th ed., 1736) The
    first general scientific and technical encyclopaedia. By John
    Harris, F.R.S. and author of works on astronomy and various
    other subjects.

1710s
---

26. NATURAL THEOLOGY. **(a)** *Physico-theology; or, A demonstration of
    the being and attributes of God from his works of creation.*
    London, 1713. (11th ed., 1749. Trans. into French, German,
    and Swedish) By William Derham, a clergyman and F.R.S. who
    published numerous papers in the *Philosophical Transactions*
    on a variety of subjects.
         (b) *Astro-theology; or, A demonstration of the being
    and attributes of God from a survey of the heavens.* London,
    1715. (7th ed., 1755. Trans. into German) By William Derham.
         (c) *Het regt gebruick der wereltbeschouwingen.* Amsterdam,
    1714. (Five editions and later reprints. English trans.: *The
    religious philosopher; or, The right way of contemplating the
    works of the Creator*, 3 vols, London, 1718; 4th ed., 1730.
    French trans., 1725 and later eds. German trans., 1732 and
    later eds) The theological and philosophical argument of the
    book was based on a comprehensive review of the scientific
    knowledge of the time. The author was Bernard Nieuwentijt, a
    Dutch physician and an able mathematician.

1740s
---

27. *SCIENZA della natura.* 2 vols. Naples, 1748-49. (An abridgment
    appeared in 1753 and a much enlarged edition in 1767-70) A com-
    prehensive and well arranged (and well illustrated) presentation
    of the full extent of the scientific knowledge of the time. By
    G.M. Della Torre.

1750s
---

28. BOSCOVICH, R.G. (1916) *Philosophiae naturalis theoria.* Vienna,
    1758. Presented a system of natural philosophy which embodied
    influences from both Newton and Leibniz and included ideas which,
    much later, came to be fundamental in physics. It is thought

that Faraday was influenced by Boscovich's ideas in formulating his concept of the field in 1844.

29. *COLLECTION académique.* 13 vols. Paris, 1755–59. A massive coll-
    ection of articles in the non-mathematical sciences, selected
    from the chief periodicals since their beginning and translated
    into French. The initiator and first editor was J. Berryat;
    later editors included Buffon, Daubenton, and other leading
    French scientists.

1760s

30. EULER, L. (2537) *Lettres à une princesse d'Allemagne sur divers
    sujets de physique et de philosophie.* 3 vols. St. Petersburg,
    1768–72. Originated from lessons on mechanics, physics, and
    astronomy which Euler gave to a niece of Frederick the Great
    during his Berlin period. Admirably clear and attractively
    written, the book became immensely popular, running to some
    forty editions in seven languages.

1770s

31. SENEBIER, J. (2901) *L'art d'observer.* 2 vols. Geneva, 1775.
    An essay on the experimental method which anticipated much that
    was later expressed by Claude Bernard.

1780s

32. *PHYSIKALISCHES Wörterbuch.* 4 vols. Leipzig, 1787–91. By J.S.T.
    Gehler. See also 35.

1800s

33. *REPERTORIUM commentationum a societatibus litterariis editarum.*
    16 vols. Göttingen, 1801–21. A bibliography, classified by
    subject (and with author indexes), of the papers in the period-
    ical publications of academies and other learned societies from
    their beginnings in the seventeenth century until 1800. By J.D.
    Reuss, librarian at the University of Göttingen.

1810s

34. *ENCYCLOPAEDIA metropolitana; or, Universal dictionary of knowledge.*
    29 vols. London, 1817–45. Arranged by subjects. The natural
    sciences were dealt with in Vols I–VIII, the articles being
    written by some of the leading British scientists of the time.

1820s

35. *PHYSIKALISCHES Wörterbuch. Neu bearbeitet....* 23 vols. Leipzig,
    1825–45. A greatly enlarged edition by H.W. Brandes et al. of
    the work of J.S.T. Gehler (32).

1830s

36. HERSCHEL, J. (2346) *A preliminary discourse on the study of
    natural philosophy.* London, 1830. (Many reprintings) A dist-
    inguished work on the philosophy of science in the Baconian

tradition. "... contains much the best and most comprehensive formulation up to its date of the methods of scientific investigation, and strongly stimulated or influenced the labors in this field of his contemporaries Whewell and J.S. Mill." (R.M. Blake et al.) (See also 43)

37. BABBAGE, C. (2345) *Reflections on the decline of science in England, and on some of its causes.* London, 1830. A forthright criticism of the composition and management of the Royal Society (cf. 2366). The book was a major factor leading to the formation of the British Association in the following year.

38. *BRIDGEWATER treatises on the power, wisdom and goodness of God as manifested in the Creation.* Eight separate treatises published in the period 1833-36 under a bequest from the eighth Earl of Bridgewater. The authors were: T. Chalmers, J. Kidd. W. Whewell, C. Bell, P.M. Roget, W. Buckland, W. Kirby, and W. Prout.

39. AMPÈRE, A.M. (2115) *Essai sur la philosophie des sciences; ou, Exposition analytique d'une classification naturelle de toutes les connaissances humaines.* Paris, 1834.

40. *ON THE CONNEXION of the physical sciences.* London, 1834. (10th ed., 1877) By Mary Somerville. A widely praised work which, together with her earlier book (434), won its authoress honorary membership of the Royal Astronomical Society and several other scientific societies. The successive editions were revised and it retained its value for over forty years.

1840s
---

41. WHEWELL, W. (2344) *The philosophy of the inductive sciences.* 2 vols. London, 1840. (3rd ed., much enlarged, 1858-60) Unlike the work of Herschel (36) and J.S. Mill's *System of logic* (published in the following year) it departed from the tradition of British empiricism and adopted a Kantian approach.

42. *DAS BUCH der Natur, die Lehren der Physik, Chemie, Mineralogie, Geologie, Physiologie, Botanik und Zoologie umfassend.* Brunswick, 1846. (22 eds up to 1884. English trans., 1851 and later) By F.K.L. Schroeder.

43. HERSCHEL, J. (See also 36) *A manual of scientific enquiry. Prepared for the use of Her Majesty's Navy, and adapted for travellers in general.* London, The Admiralty, 1849. (5th ed., 1886) Edited by Herschel. (See also 46)

1850s
---

44. *THEORIE der Induction.* Jena, 1854. By E.F. Apelt. A notable work in the philosophy of science. Being in the Kantian tradition it had some affinities with Whewell's work (41) and, like it, made use of case studies in the history of science.

45. *ETUDES et expériences sur les sciences d'observation.* Paris, 1855. A successful essay in *haute vulgarisation* by the physicist, J. Babinet.

46. HERSCHEL, J. (See also 43) *Essays from the Edinburgh and Quarterly Reviews, with addresses and other pieces.* London, 1857. (See also 52)

47. *A CYCLOPAEDIA of the physical sciences*. London, 1857. (3rd ed.,
    1868) By the astronomer, J.P. Nichol.

1860s

48. POGGENDORFF, J.C. (2598) *Biographisch-literarisches Handwörter-
    buch zur Geschichte der exakten Wissenschaften*. 2 vols. Leipzig,
    1863. A well known and widely used work giving brief biographies
    and extended bibliographies of 8,400 scientists of all periods
    (not including biologists). The work had a current as well as
    a historical value and after Poggendorff's death it was continued
    on his plan in many successive volumes for contemporary scien-
    tists.

49. LIEBIG, J. von. (2603) *Induction und Deduction*. Munich, 1865.
    A significant work in the philosophy of science.

50. HELMHOLTZ,H. von. (2631) *Populäre wissenschaftliche Vorträge*.
    3 vols. Brunswick, 1865-76. (3rd ed., 1884. English trans.:
    Popular lectures on scientific subjects, 1873-81 and later eds)

51. *DICTIONNAIRE générale des sciences théoriques et appliquées*.
    2 vols. Paris, 1865-69. By A. Privat-Deschanel and A. Focillon.

52. HERSCHEL, J. (See also 46) *Familiar lectures on scientific sub-
    jects*. London, 1866. (Several reprintings)

53. *ESSAI sur les institutions scientifiques de la Grande-Bretagne et
    de l'Irlande*. Brussels, 1867. *L'Espagne scientifique*. Brussels,
    1868. Two works by the Belgian astronomer, N.E. Mailly.

54. ROYAL SOCIETY OF LONDON. *Catalogue of scientific papers* [1800-
    1900]. 19 vols published in four series between 1867 and 1925.
    Contained alphabetical sequences of authors' names with the
    references to their periodical articles listed under their names.

1870s

55. HUXLEY, T.H. (2403) Published numerous essays and addresses from
    1870 onward. His *Collected essays* (9 vols. London, 1893-94)
    contained his own selection and arrangement of them.

56. TYNDALL, J. (2411) *Fragments of science for unscientific people.
    A series of detached essays, lectures and reviews*. London, 1871.
    (8th ed., 1892)

57. *SCIENTIFIC London*. London, 1874. By B.H. Becker.

58. *THE PRINCIPLES of science. A treatise on logic and scientific
    method*. 2 vols. London, 1874. (2nd ed., 1877) By W.S. Jevons.

59. *ATOMISTIK und Kriticismus. Ein Beitrag zur erkentniss-theoretischen
    Grundlegung der Physik*. Brunswick, 1878. By K. Lasswitz.

60. *ENCYKLOPÄDIE der Naturwissenschaften*. 42 vols. Breslau, 1879-
    1902. Edited by G. Jäger, A. Kenngott, et al.

1880s

61. *WISSENSCHAFTLICHEN Vereine und Gesellschaften Deutschlands im
    neunzehnten Jahrhundert. Bibliographie ihrer Veröffentlichungen*.
    2 vols in 3. Berlin, 1883/87-1917. By L. Müller.

62. *THE COMMON sense of the exact sciences*. London, 1885. (3rd ed., 1892) A well-known discussion of the philosophy of science by the mathematician, W.K. Clifford.

63. BERTHELOT, M. (2176) *Science et philosophie*. Paris, 1886. A collection of essays.

64. DU BOIS-REYMOND, E.H. (2648) *Reden*. 2 vols. Leipzig, 1886-87. A collection of numerous addresses given mostly in the 1870s and 1880s on a wide variety of issues in the cultural relations of science.

1890s

65. *ÜBER DIE Grundlagen der Erkenntnis in den exacten Wissenschaften*. Tübingen, 1890. By the mathematician, P.D. Du Bois-Reymond.

66. PEARSON, K. (2467) *The grammar of science*. London, 1892. (2nd ed., 1900) A very influential account of the philosophy of science.

67. MACH, E. (2960) *Populärwissenschaftliche Vorlesen*. Leipzig, 1896. (5th ed., 1923. English trans.: *Popular scientific lectures*, 1895; 5th ed., 1943)

1.02 MATHEMATICS

This section is confined as far as possible to pure mathematics;
applied mathematics is included in Mechanics, Astronomy, Physics
and occasionally some other sections.

1460s

68. REGIOMONTANUS, J. (2933) *De triangulis omnimodis.* (Written about
    1464 but not printed until 1533) A treatise on plane and spher-
    ical trigonometry. Very influential after it was printed because
    it was the first systematization of the subject as a branch of
    mathematics independent of astronomy.

1470s

69. *ARTE dell'abbaco.* Treviso, 1478. (Anonymous) The first arith-
    metic of known date. It was soon followed by many others, espec-
    ially in Italy and Germany.

1480s

70. CHUQUET, N. (1963) *Triparty en la science des nombres*--a section
    of a longer work written in 1484 and 'published' in manuscript
    form by a firm of copyists. It is a treatise on algebra of out-
    standing originality but because it was never printed (until
    1880) its influence was very restricted.

71. EUCLID (fl. ca. 295 B.C.) The first printed edition of his
    Elements, issued in 1482 and many times subsequently, was of
    one of the medieval Latin translations from the Arabic. It was
    the first book (of known date) to contain diagrams. The first
    Renaissance translation from the Greek to be printed appeared
    in 1505 and was significantly different from the medieval trans-
    lations. Later translations from the Greek, notably the elab-
    orate one by Commandino (1843) in 1572, contained commentaries
    by various ancient mathematicians and sometimes Euclid's *Optics*
    and other minor works. The Greek text was first printed in 1533.
    Translations into the vernaculars were as follows: Italian, from
    1543; German, from 1562; French, from 1564; English, from 1570;
    Spanish, from 1576.

1490s

72. PACIOLI, L. (1823) *Summa de arithmetica, geometria et proportion-
    alita.* Venice, 1494. (In Italian. Various later editions, in
    whole or part) An extensive treatise on theoretical and pract-
    ical arithmetic, with some algebra and geometry. Widely used
    and very influential.   (See also 75)

13

1500s

73. FERRO, S. (1826) No writings of his are known, either in manu-
    script or printed. However the accounts of his work by Tartaglia,
    Cardano, and others make it clear that he was an important algeb-
    raist. His chief discoveries seem to have been made in the
    first or second decade of the sixteenth century.

74. BOUVELLES, C. (1964) *Geometricae introductionis*. Paris, 1503.
    (French trans., 1542 and several later eds) Included a notable
    section on stellated polygons as well as several attempts to
    solve the long-standing problem of the quadrature of the circle.

75. PACIOLI, L. (See also 72) *Divina proportione*. Venice, 1509. (In
    Italian) A collection of various writings, the most important
    being those on the golden section (or divine proportion) and on
    the geometry of the regular polyhedra.

1510s

76. BOUVELLES, C. (See also 74) **(a)** *Liber de XII numeris*. Paris,
    1510. On perfect numbers. **(b)** *Géométrie en françoys*. Paris,
    1511. Probably the first book on geometry to be printed in
    French.

77. ORTEGA, J. de. (2921) *Tractado subtilismo d'aritmetica y de
    geometria*. Barcelona, 1512. (Eight editions known, including
    a translation into French) A well-known work.

78. LAX, G. (2920) **(a)** *Arithmetica speculativa*. Paris, 1515.
    (b) *Proportiones*. Paris, 1515.

79. RIES, A. (2479) "The most influential of the German writers in
    the movement to replace the old computation by means of counters
    by the more modern written computation." (D.E. Smith) From 1518
    onward he wrote a number of commercial arithmetics (which can be
    regarded as improved and enlarged versions of the one work).
    They went through more than a hundred editions up to the mid-
    seventeenth century.

1520s

80. WERNER, J. (2478) Published in 1522 a collection of tracts on
    mathematics and astronomy. It included the first original study
    of conic sections in modern times.

81. TUNSTALL, C. (2215) *De arte supputandi*. London, 1522. (Seven
    later editions known) A highly regarded treatise on arithmetic.

82. DÜRER, A. (2477) *Underweysung der Messung*. Nuremberg, 1525.
    A treatise on geometry and its applications in the arts. The
    first outstanding work on mathematics to be printed in German.

83. RUDOLFF, C. (2935) **(a)** *Behend und hübsch Rechnung...*. Strasbourg,
    1525. The first significant work on algebra in German. The
    title commonly used for it was *Coss*, the old German word for
    algebra. **(b)** *Künstliche Rechnung...*. Vienna, 1526. (Eleven
    editions known) A popular arithmetic.

1530s

---

84. KÖBEL, J. (2476) **(a)** *Astrolabii declaratio.* 1532. **(b)** *Geometrei.* 1535. **(c)** *Zwey Rechenbüchlin.* 1537. These and his other popular books on mathematics and astronomy were reprinted repeatedly up to about 1600 and were important in disseminating a knowledge of mathematics, including Hindu-Arabic numerals and methods of using them.

85. APIAN, P. (2481) *Instrumentum sinuum.* Nuremberg, 1534. Tables of sines for every minute, with the radius divided decimally. The first such tables to be printed. Useful especially for astronomers.

86. APOLLONIUS OF PERGA (ca. 262-190 B.C.) Only the first four of the eight books of his celebrated *Conics* were preserved in Greek and known in Europe before 1661. (For the fortunes of the other books see 150) Their first appearance in print was in a rather poor Latin translation in 1537. It was superseded in 1566 by a much better one by Commandino who also added some of the ancient commentaries.

87. CARDANO, G. (1831) *Practica arithmetica.* Milan, 1539. His first mathematical work--brillant though not very original. See also 90.

1540s

---

88. RECORDE, R. (2218) *The ground of artes.* London, 1543. (Many later editions, the last in 1699) The first arithmetic of any importance to be published in English. It was the most popular of all his works. (See also 95)

89. STIFEL, M. (2485) The most important German algebraist of the sixteen century. Both the following works included algebra as well as arithmetic. **(a)** *Arithmetica integra.* Nuremberg, 1544. **(b)** *Deutsche arithmetica.* Nuremberg, 1545.

90. CARDANO, G. (See also 87) *Ars magna.* Nuremberg, 1545. The greatest of his numerous works, full of new ideas. It was the first Latin treatise devoted solely to algebra.

91. FERRARI, L. (1834) A protégé and collaborator of Cardano. He left no publications in his own name but his important results on the cubic and quartic equations were included in Cardano's *Ars magna.*

92. TARTAGLIA, N. (1828) *Quesiti ed invenzioni diverse.* Venice, 1546. Included important work in algebra, arithmetic, and geometry, as well as applications of mathematics and mechanics to a variety of practical concerns. (See also 98)

93. PELETIER, J. (1970) *L'arithmétique.* Poitiers, 1549. (Many later editions) A combination of a commercial arithmetic and a theoretical treatise. (See also 97)

94. ARCHIMEDES (ca. 287-212 B.C.) The first printed texts appeared in 1503 and were of two of his works in Moerbeke's thirteenth-century Latin translation. These, together with the other works

in Moerbeke's translation, were published as the *Opera* by Tartaglia in 1543. More important was another publication of the
Opera in the same year: it comprised the Latin translation of
Jacopo da Cremona, made about 1450, and the Greek text. Thereafter a knowledge of Archimedes' achievements spread rapidly.
Another important Latin translation was that of Commandino,
published in 1558.

1550s

95. RECORDE. R. (See also 88) **(a)** *The pathway to knowledge*. London,
    1551. (Two later editions known) An adaptation of the first
    four books of Euclid's *Elements*. **(b)** *The whetstone of witte*.
    London, 1557. A sequel to his first book (88), containing ad-
    vanced arithmetic and some algebra.

96. BUTEO, J. (1975) **(a)** *Opera geometrica*. Lyons, 1554. **(b)** *De
    quadratura circuli*. Lyons, 1559. **(c)** *Logistica*. Lyons, 1559.
    His most important book.

97. PELETIER, J. (See also 93) **(a)** *L'algèbre*. Lyons, 1554. **(b)** *In
    Euclidis "Elementa geometrica" demonstrationum*. Lyons, 1557.
    Both works were at an advanced level and contained some origin-
    al features.

98. TARTAGLIA, N. (See also 92) *Il general trattato di numeri et
    misure*. 6 parts. Venice, 1556-60. The best Italian treatise
    on arithmetic in the sixteenth century. Contained an extensive
    account of commercial arithmetic and its economic and social
    setting.

1570s

99. DEE, J. (2220) His *Mathematicall praeface* to Billingsley's trans-
    lation of Euclid (in 1570--the first into English) was widely
    read in England and was an influential survey of the various
    kinds of mathematics and their practical applications.

100. DIGGES, L. (2219. One of the earliest mathematical practitioners
     of importance in England) **(a)** *A geometrical practise named
     Pantometria*. London, 1571. A manual of practical mathematics,
     especially notable for its section on surveying. **(b)** *Prognos-
     ticon everlastinge*. London, 1576. Both works were posthumous
     and were published by his son, Thomas Digges, who added to the
     latter an important account of Copernican astronomy; see 336)

101. BOMBELLI, R. (1844) *L'algebra*. Bologna, 1572. A systematic
     treatise including important contributions of his own. It was
     the first work to spread a knowledge of Diophantus and was
     highly regarded by later mathematicians such as Stevin and
     Leibniz.

102. CLAVIUS, C. (1854) *Euclidis "Elementorum" libri XV*. Rome, 1574.
     (Many later editions) An extensive commentary containing, in
     additions to his own notes, a large amount of material collected
     from earlier commentators and editors.   (See also 105)

103. MAUROLICO, F. (1835) *Opuscula mathematica*. Venice, 1575. A coll-
     ection of eight treatises, one of them on astronomy, one on music,
     and the others containing some significant researches in arith-
     metic and geometry.

104. VIÈTE, F. (1979) *Canon mathematicus*. Paris, 1579. On plane and
     spherical trigonometry.   (See also 108)

1580s

105. CLAVIUS. C. (See also 102) *Epitome arithmeticae practicae*. Rome,
     1583. (Several later editions. Italian trans., 1586) An
     important textbook.   (See also 114)

106. FINK, T. (2825) *Geometriae rotundi*. Basel, 1583. An influen-
     tial textbook of trigonometry which had a number of original
     features.

107. STEVIN, S. (2760) **(a)** *De thiende*. Leiden, 1585. (English trans.,
     1608 and 1619. French trans., 1608) A small book that was
     influential in introducing decimal fractions. **(b)** *L'arithmé-
     tique*. Leiden, 1585. A general treatment, including algebra
     to which he made some valuable contributions.

1590s

108. VIÈTE, F. (See also 104) **(a)** *In artem analyticem isagoge*. Tours,
     1591. The most important of his many works on algebra. In it
     he introduced the use of letters for both known and unknown
     quantities; it is thus the first work on symbolic algebra. He
     also introduced many technical terms, some of which are still
     used. **(b)** *Supplementum geometriae*. Tours, 1593.
     (See also 118)

109. ROOMEN, A. van. (2808) **(a)** *Ideae mathematicae*, Antwerp, 1593.
     On regular polygons. **(b)** *In Archimedis circuli dimensionem*.
     Geneva, 1597.   (See also 116)

110. RHETICUS, G. (2487) *Opus Palatinum de triangulis*. Heidelberg,
     1596. (Posthumous) An immense set of tables of trigonometrical
     functions whose computation required vast labour. A revised
     edition was published in 1613 by B. Pitiscus.

111. CEULEN, L. van. (2759) *Van den circkel*. Delft, 1596. Includes
     a computation of the number π to twenty decimal places using
     Archimedes' method and a decimal notation. He later reached
     thirty-five decimal places.

1600s

112. PITISCUS, B. (2502) *Trigonometriae; sive, De dimensione triang-
     ulorum*. Augsburg, 1600. (3rd ed., 1612. English trans., 1614
     and later. Partial French trans., 1619) A major treatise on
     trigonometry (and the first use of the word).

113. GHETALDI, M. (1869) Between 1603 and 1613 he published several
     books on the mathematics of Archimedes and Apollonius which
     were partly analyses and commentaries, partly extensions of
     their work and, in the case of Apollonius, partly an attempt
     to reconstruct the missing books.

114. CLAVIUS, C. (See also 105) **(a)** *Geometria practica*. Rome, 1604.
     (b) *Algebra*. Rome, 1608. **(c)** *Triangula sphaerica*. Mainz, 1611.
     Outstanding textbooks.

115. VALERIO, L. (1862) **(a)** *De centro gravitatis solidorum.* Rome,
       1604. **(b)** *Quadratura parabolae.* Rome, 1606. Both works
       were firmly in the Archimedean tradition.

116. ROOMEN, A. van. (See also 109) *Canon triangulorum sphaericorum.*
       Mainz, 1609. A significant work on spherical trigonometry.

1610s

117. CATALDI, P.A. (1859) *Trattato del modo brevissimo di trovar la
       radice quadra delli numeri.* Bologna, 1613. A notable work on
       the determination of square roots by means of infinite series
       and unlimited continued fractions.

118. VIÈTE, F. (See also 108) *De aequationum recognitione et emenda-
       tione.* Paris, 1615. (Posthumous) His main work on the theory
       of equations, giving general methods for solving equations of
       the second, third, and fourth degrees.   (See also 138)

119. NAPIER, J. (2221) **(a)** *Mirifici logarithmorum canonis descriptio.*
       Edinburgh, 1614. (English trans., 1616) Announcement of the
       discovery of logarithms. **(b)** *Mirifici logarithmorum canonis
       constructio.* Edinburgh, 1619. (Posthumous) A fuller account.

120. BRIGGS, H. (2228) Known in the history of mathematics as the
       person who developed and publicized the use of logarithms. He
       worked in close conjunction with Napier until the latter's
       death in 1617. Brigg's first work was *Logarithmorum chilias
       prima* (London, 1617; English trans., 1630?) which contained the
       first table of common or decimal logarithms. For his later
       works see 123 and 130.

1620s

121. GUNTER, E. (2233. A notable figure in the tradition of the English
       mathematical practitioners) His chief books were published
       between 1620 and 1624, and his collected works appeared in 1624
       (two years before his death) and ran to at least five later
       editions.

122. DIOPHANTUS OF ALEXANDRIA (fl. A.D. 250) *Arithmeticorum libri sex.*
       Paris, 1621. The first edition of the Greek text with a Latin
       translation. (An earlier Latin translation had appeared in
       1575) It was edited by C.G. Bachet de Méziriac whose copious
       notes included some original contributions of his own to the
       theory of numbers. The work was reprinted in 1670 with addition
       of notes by Fermat.

.... MERSENNE, M. See 14.

123. BRIGGS, H. (See also 120) *Arithmetica logarithmica.* London,
       1624. Included 30,000 logarithms together with a general dis-
       cussion of their nature and use.   (See also 130)

124. GIRARD, A. (2765) **(a)** *Tables des sinus, tangentes et sécantes.*
       The Hague, 1626. Included a short treatise on trigonometry.
       (b) *Invention nouvelle en l'algèbre.* Amsterdam, 1629. Both
       works included many original features.

1630s

125. FAULHABER, J. (2505) Author of numerous mathematical works,
     chiefly textbooks and semi-popular writings. His chief import-
     ance lay in his dissemination, in the early 1630s, of a know-
     ledge of logarithms in Germany.

126. HARRIOT, T. (2227) *Artis analyticae praxis*. London, 1631.
     (Posthumous) A work on algebra of outstanding importance,
     especially for the theory of equations.

127. OUGHTRED, W. (2234) Author of numerous mathematical textbooks,
     the best of which was his *Clavis mathematicae* (London, 1631.
     2nd ed., 1648; four reprintings. English trans., 1647 and
     1694). It was praised by such notable figures as Newton, Wallis,
     and Boyle. Oughtred was also the inventor of the slide rule.

128. MYDORGE, C. (1992) *Conicorum operis* .... (Also issued as *De
     sectionibus conicis*) Paris, 1631. An enlargement and refine-
     ment of Apollonius' famous work on conic sections.

129. CAVALIERI, B. (1880) (**a**) *Directorium* ... *in quo trigonometriae
     logarithmicae fundamenta ac regulae demonstrantur*. Bologna,
     1632. In this and some later works he disseminated a know-
     ledge of logarithms in Italy. (**b**) *Geometria indivisibilibus
     continuorum nova quaedam ratione promota*. Bologna, 1635. His
     chief work. It introduced his method of indivisibles (a major
     step towards the infinitesimal calculus).

130. BRIGGS, H. (See also 123) *Trigonometria Britannica*. Gouda, 1633.
     (English trans., 1658) On the application of logarithms to
     plane and spherical trigonometry, with appropriate tables.

131. ROBERVAL, G.P. (1995) Though most of his work was not published
     it came to be known in Paris, apparently through his teaching
     at the Collège Royal from 1634 onward and in other informal
     ways. He was best known for his discoveries in the field of
     higher plane curves and in the geometry of infinitesimals
     (which was to prove a significant step in the direction of the
     calculus).

132. GULDIN, P. (2940) *Centrobaryca; seu, De centro gravitatis*. 4
     vols. Vienna, 1635-41. As well as dealing with centres of
     gravity in the Archimedean style the work also discussed var-
     ious geometrical topics, especially solids of revolution.

133. DESARGUES, G. (1993) In the period 1636-48 he published several
     works on geometry and perspective, with applications in stone-
     masonry and architecture. His geometrical ideas were original
     and profound but they attracted little attention at the time
     (though appreciated by Pascal, Mersenne, and Descartes), partly
     because his books were published in very limited editions, but
     chiefly because his kind of approach to geometry was quite
     overshadowed by the great success of Descartes' algebraic app-
     roach. His work was forgotten until the mid-nineteenth century
     when its importance for the theory of projective geometry was
     at last realized.

134. FERMAT, P. de. (1994) From 1636 onward he was in correspondence

with Mersenne and Roberval, and later with others such as
Descartes, Carcavi, Pascal, and Huygens, and this correspondence continued until the early 1660s, not long before his death.
It was only through the dissemination of his letters and the
manuscripts he sometimes sent with them that his work became
known to the circle of French mathematicians of the time. Fourteen years after his death much of his work was published by
his son (*Varia opera mathematica*, Toulouse, 1679) but it was
not until the twentieth century that the full extent of his
achievement became known.

Fermat is now regarded as one of the greatest mathematicians of his age. His system of analytical geometry (which
was completed by 1636) was very similar to that of Descartes,
and his researches on the quadrature of curves and on maxima
and minima were fundamental. This work had some influence in
his own time but his achievements in the theory of numbers, for
which he is now best known, had hardly any influence until the
nineteenth century.

135. DESCARTES, R. (1991) *La géometrie* [Appendix to his *Discours de
     la méthode*]. Leiden, 1637. The establishment of analytical
     geometry--often said to be the greatest single step ever taken
     in the history of mathematics. The work became best known in
     the form of the Latin edition--*Geometria*--which was brought out,
     with a commentary, by F. van Schooten (2768) in 1649. The sec-
     ond Latin edition in two volumes (1659-61), with a much enlarged
     commentary and other material, became a standard work.

1640s

136. PASCAL, B. (2003) **(a)** *Essay pour les coniques*. Paris, 1640.
     (Only a few copies were published and it was not reprinted
     until 1779) Presented the basic ideas of projective geometry
     but, like the work of Desargues in the same field (133), it
     was neglected until the nineteenth century. **(b)** *Lettre ... sur
     le sujet de la machine nouvellement inventée ... pour faire
     toutes sortes d'opérations d'arithmétique*. Paris, 1645. A pam-
     phlet describing his calculating machine.   (See also 147)

137. TORRICELLI, E. (1884) *Opera geometrica*. Florence, 1644. A
     highly regarded work which, among many other things, applied
     and extended Cavalieri's method of indivisibles. It also inclu-
     ded a section on mechanics--*De motu gravium*--in which Torricelli
     extended and supplemented Galileo's study of free fall and the
     motion of projectiles, and added some pioneering work in hydro-
     dynamics, notably the law of efflux velocity named after him.

138. VIÈTE, F. (See also 118) *Opera mathematica*. Leiden, 1646. (Ed-
     ited by F. van Schooten) Some of Viète's many works--which
     were published between 1591 and his death in 1603--had only a
     limited circulation and did not become generally available
     until this publication of his collected works.

139. SAINT VINCENT, G. (2809) *Opus geometricum quadraturae circuli et
     sectionum coni*. Antwerp, 1647. A large treatise containing
     many original approaches in geometry, notably in connection
     with the method of indivisibles developed by Cavalieri.

1650s

140. MENGOLI, P. (1886) **(a)** *Novae quadraturae arithmeticae*. Bologna, 1650. **(b)** *Geometriae speciosae elementa*. Bologna, 1659. "His significance to the history of science lies in the transitional position of his mathematics, midway between Cavalieri's method of indivisibles and Newton's fluxions and Leibniz' differentials." (*DSB*)

141. HUYGENS, C. (2772) **(a)** *Theoremata de quadratura hyperboles, ellipsis et circuli....* Leiden, 1651.
(b) *De circuli magnitudine inventa*. Leiden, 1654.
(c) *Tractatus de ratiociniis in aleae ludo*. Leiden, 1657. The first treatise on the theory of probability (which had been initiated earlier in correspondence between Pascal and Fermat). (See also 156)

142. TACQUET, A. (2812) Author of some original works of no great importance; best known for his *Elementa geometriae* (Antwerp, 1654; many later editions and translations, up to 1805), one of the most famous mathematical textbooks. It was based closely on Euclid and also included selections from Archimedes.

143. WALLIS, J. (2242) **(a)** *De sectionibus conicis*. Oxford, 1655. An original treatment in terms of analytical geometry. **(b)** *Arithmetica infinitorum*. Oxford, 1656. His chief work, incorporating his analytical approach. It had an important influence on Newton. **(c)** *Tractatus duo ... De cycloide, et ... De cissoide*. Oxford, 1659. Also used his analytical approach. (See also 165)

144. FRENICLE DE BESSY, B. (1996) A brillant mathematician who devoted himself chiefly to problems in the theory of numbers, his chief work being *Solutio duorum problematum circa numeros cubos et quadratos* (Paris, 1657).

145. SCHOOTEN, F. van. (2768) *Exercitationes mathematicae*. Leiden, 1657. A collection of original contributions in geometry, algebra, and arithmetic. He was best known for his Latin edition, with commentary, of Descartes' *Géometrie* (135).

146. SLUSE, R.F. de. (2813) *Mesolabum*. Liège, 1659. (2nd ed., enl., 1668) Included work on the cycloid, the cubature of solids, solutions of third- and fourth-degree equations, the derivation of geometric means, etc.

147. PASCAL, B. (See also 136) *Lettres de A. Dettonville contenant quelques-unes de ses inventions de géometrie*. Paris, 1659. An unsystematic and unfinished account of his important work on the cycloid and the method of indivisibles. (See also 151)

1660s

148. ANGELI, S. degli. (1893) In his chief works, published in the 1660s, he developed and applied the method of indivisibles that had been introduced by Cavalieri and taken up by Torricelli.

149. VIVIANI, V. (1892) An accomplished mathematician who devoted his talents mainly to commentaries of various kinds on the works of

the ancient Greek mathematicians. He published several books
in this field from 1659 to 1702.

150. APOLLONIUS OF PERGA (ca. 262-190 B.C.) The first four books of
his famous *Conics*, which had been preserved in Greek, had been
printed in the sixteenth century (see 86). Books V-VII had
been translated into Arabic in the ninth century and a copy of
the Arabic text, which had been acquired by the Medici of Florence, became available to Borelli who arranged for it to be
translated into Latin and published in 1661. An improved translation was published in 1710 by Halley, together with a recon-
struction of Book VIII.

151. PASCAL, B. (See also 147) *Traité du triangle arithmétique*. Paris,
1665. (Completed in 1654 but not published until three years
after his death) The beginning of the calculus of probabilities
(cf. 141c)

152. GREGORY, J. (2264) **(a)** *Vera circuli et hyperbolae quadratura*.
Padua, 1667. **(b)** *Geometriae pars universalis*. Padua, 1668.

153. MERCATOR, N. (2248) *Logarithmotechnia*. London, 1668. On a new
method of constructing logarithms from first principles.

154. WREN, C. (2263) "The generation of an hyperbolic cylindroid...."
—a paper in the *Philosophical Transactions*, 1669.
See also 158.

1670s

155. NEWTON, I. (2265) From about 1670 much of his most important
work in mathematics circulated widely in manuscript and was not
published until much later--some of it not until after his death
--when it was already well known. (He was rebuked by Wallis
for not publishing discoveries that would have brought credit
to his country)   (See also 170)

156. HUYGENS, C. (See also 141) Some of his best work in pure mathematics was included in his *Horologium oscillatorium* (515),
notably his theory of evolutes. From the 1670s up to his death
in 1695 he also published various mathematical articles in the
Journal des Sçavans, *Acta Eruditorum*, and other periodicals.

157. BARROW, I. (2262) **(a)** *Lectiones geometricae*. London, 1670.
(English trans., 1735) **(b)** *Lectiones opticae et geometricae*.
London, 1674.   (See also 161)

158. WREN, C. (See also 154) "On finding a straight line equal to
that of a cycloid and to the parts thereof"--*Philosophical
Transactions*, 1673.

159. LA HIRE, P. de. (2021) *Nouvelle méthode en géometrie pour les
sections des superficies coniques et cylindriques*. Paris, 1673.
A treatise on conic sections strongly influenced by the ideas
of Desargues and using the projective approach. (See also 166)

160. CEVA, G. (1902) *De lineis rectis se invicem secantibus*. Milan,
1678. His first and most important book. He continued to publish books of some importance on pure and applied mathematics
until 1711.

1680s

161. BARROW, I. (See also 157) *Lectiones mathematicae*. London, 1683.
     (Posthumous; the lectures were given in 1664-66. English trans.
     entitled *The usefulness of mathematical learning*, 1734)

162. LEIBNIZ, G.W. (2522) He was developing his mathematical ideas
     from the mid-1670s but it was not until the *Acta Eruditorum* was
     founded in 1682 that he began to publish them. Thereafter a
     large number of his papers on pure and applied mathematics
     appeared in the journal until approximately 1700. The cele-
     brated papers on the differential and integral calculus were
     published in 1684-86.

163. TSCHIRNHAUS, E.W. (2524) During the 1680s and 1690s he published
     many mathematical papers in the *Acta Eruditorum* on such topics
     as the study of curves, including caustics and catacaustics,
     maxima and minima, the theory of equations, etc. He was one of
     the first to give serious attention to Leibniz' papers on the
     'new calculus.'

164. BERNOULLI, Jakob (I). (2891) In the period 1683-1701 he published
     a large number of articles in the *Acta Eruditorum* on various
     mathematical topics. His chief achievement was to develop and
     apply the Leibnizian calculus, and generally make it known and
     accepted. He also made important contributions to algebra, the
     calculus of variations, mechanics, the theory of series, and
     the theory of probability. His book on the latter topic, *Ars
     conjectandi*, was published posthumously in 1713.

165. WALLIS, J. (See also 143) *Treatise of algebra, both historical
     and practical*. London, 1685. A comprehensive exposition of
     algebra, together with a history of the subject.

166. LA HIRE, P. de. (See also 159) *Sectiones conicae*. Paris, 1685.
     A treatise along the same lines as his earlier work on the sub-
     ject (159) but more mature and much more extensive.

1690s

167. ROLLE, M. (2028) **(a)** *Traité d'algèbra*. Paris, 1690. **(b)** *Méthode
     pour resoudre les équations indéterminées de l'algèbre*. Paris,
     1699.

168. BERNOULLI, Johann (I). (2892) From 1691 until the 1730s he pub-
     lished a large number of mathematical papers (in the *Acta Eru-
     ditorum* up to about 1700 and theafter in the periodicals of the
     Paris, Berlin, and St. Petersburg Academies). His main work
     was in the development and application of the Leibnizian calc-
     ulus—especially the integral calculus and its use with diff-
     erential equations, as well as the exponential calculus—but
     he also wrote on a wide range of other topics. In addition he
     made some important contributions to mechanics (see 526).

.... VARIGNON, P. See 521.

169. L'HOSPITAL, G.F.A. de. (2035) *Analyse des infiniment petits pour
     l'intelligence des lignes courbes*. Paris, 1696. (Three later
     editions) The first treatise on the differential calculus; it

had a big influence in France and elsewhere. L'Hospital (who had studied with Johann Bernoulli about 1691) was an important figure in the early development of the Leibnizian calculus.

1700s

170. NEWTON, I. (See also 155) *Arithmetica universalis*. Cambridge, 1707. (Rev. ed., London, 1722. English trans., 1720) On algebra and the theory of equations (lecture notes from 1673-83).

171. *ANALYSE démontrée*. Paris, 1708. (2nd ed., 1736-38, with comments by Varignon) The first textbook on the infinitesimal calculus. By C.R. Reyneau, an associate of Malebranche.

1710s

172. RICCATI, J.F. (1912) Published numerous articles in several periodicals over the period 1710-50 on several branches of mathematics, pure and applied, and is best known for his work on differential equations. He was an enthusiastic advocate of Newtonianism.

173. COTES, R. (2288) Published only one article in his short life (in 1714) but in 1722, six years after his death, his mathematical papers were published in book form and revealed his great ability.

174. FAGNANO, G.C. (1913) His periodical articles, which appeared over the period 1714-42, were collected in his *Produzioni matematiche* (2 vols, Pesaro, 1750) His main work was in analytical geometry and integral calculus, including elliptical integrals.

175. TAYLOR, B. (2291) **(a)** *Methodus incrementorum directa et inversa*. London, 1715. Included the well-known theorem named after him for expanding functions into infinite series. **(b)** *Linear perspective*. London, 1715. (Four later editions, three translations, and many expanded versions by others)

176. MOIVRE, A. de. (2283) *The doctrine of chances*. London, 1718. One of the early treatises on probability theory. He continued to make important contributions to the subject (especially his approximation to the binomial distribution in 1733) and to some other branches of mathematics--chiefly trigonometry and complex number theory--in several papers up to 1744. The revised editions of his book, in 1738 and 1756, included his later work.

1720s

177. MACLAURIN, C. (2292) *Geometrica organica*. London, 1720. His most important work. It dealt with the general properties of conic sections and other plane curves, and introduced the method of generating conics that bears his name. It also developed numerous theorems, some of which were in Newton's *Principia* but without proof.   (See also 182)

1730s

178. EULER, L. (2973) The most prolific mathematician in history and

probably the most versatile. From about 1730 until his death in 1783 he published about 550 articles (mostly in the periodicals of the St. Petersburg and Berlin Academies) and numerous books--to say nothing of a large amount of material left unpublished at his death. In his work, as in eighteenth-century mathematics generally, analysis predominated, but he contributed to virtually every field of the mathematics of his age, both pure and applied.   (See also 181, 184, and 190)

179. CLAIRAUT, A.C. (2052) **(a)** *Recherches sur les courbes à double courbure*. Paris, 1731. The first significant analytical treatment of curves in space (using their projections on two perpendicular planes). **(b)** From 1731 until shortly before his death in 1765 he published many papers in the *Mémoires* of the Paris Academy on various branches of mathematics, notably the calculus of variations, differential equations, and the integral cakculus. (See also 180)

1740s
---

180. CLAIRAUT, A.C. (See also 179) **(a)** *Elémens de géometrie*. Paris, 1741. **(b)** *Elémens d'algèbre*. Paris, 1746. Two well-conceived and important textbooks which both went through many editions and several translations. The latter was especially influential.

181. EULER, L. (See also 178) **(a)** *Methodus inveniendi lineas curvas maximi minimive proprietate gaudentes*. Lausanne, 1744. On the calculus of variations. His first major work in pure mathematics. **(b)** *Introductio in analysin infinitorum*. 2 vols. Lausanne, 1748. (German trans., 1788-90. French trans., 1796-97) A classical treatise in which he presented analysis as a science of functions.   (See also 184)

182. MACLAURIN, C. (See also 177) **(a)** *A treatise of fluxions*. Edinburgh, 1742. Written to defend Newton's calculus against its critics (especially the philosopher, George Berkeley), it was the first systematic exposition of the Newtonian methods. **(b)** *A treatise of algebra*. London, 1748. (6th ed., 1796)
        "His influence on the progress of British mathematics was on the whole unfortunate. By himself abandoning the use both of analysis and of the infinitesimal calculus, he induced Newton's countrymen to confine themselves to Newton's methods ...." (W.W.R. Ball)

1750s
---

183. CRAMER, G. (2894) *Introduction à l'analyse des lignes courbes algébriques*. Geneva, 1750. Included what came to be known as Cramer's rule and Cramer's paradox.

184. EULER, L. (See also 181) *Institutiones calculi differentialis*. Berlin, 1755. (German trans., 1790-93) A classical treatise on the differential calculus which served for many years as a prototype for such works.   (See also 190)

185. LAMBERT, J.H. (2542) *Die freye Perspektive*. Zurich, 1759. "Intended for the artist wishing to give a perspective drawing without first having to construct a ground plan, it is nevertheless a masterpiece in descriptive geometry, containing a wealth of geometrical discoveries." (*DSB*)
(See also 188)

1760s

186. LAGRANGE, J.L. (1921) In a large number of papers published
     chiefly in the period 1760-86 (initially in the *Mélanges de
     Turin* and later in the *Mémoires* of the Berlin Academy) he made
     many important contributions, especially in the calculus of
     variations (which he initiated at the beginning of his career),
     prime number theory, the number-theoretic equation, algebra
     (especially the algebraic solvability of equations), differ-
     ential equations, the calculus, and probability theory.
     (See also 197)

187. ALEMBERT, J. d'. (2058) *Opuscules mathématiques*. 8 vols. Paris,
     1761-80. Contained some contributions to pure mathematics but
     most of the contents were essays in applied mathematics dealing
     with problems in mechanics, astronomy, and optics.

188. LAMBERT, J.H. (See also 185) **(a)** *Beyträge zum Gebrauch der Math-
     ematik und deren Anwendung*. 4 vols. Berlin, 1765-72. A coll-
     ection of papers dealing with various topics in pure and applied
     mathematics. It included his best-known paper, proving the
     irrationality of π and e, as well as other important work in
     the theory of numbers, hyperbolic trigonometry, etc.
     (b) *Theorie der Parallel-Linien*--a work written in 1766 but not
     published until 1786, after his death, when it appeared as a
     periodical article. It was neglected until the late nineteenth
     century when it was seen to be a significant investigation of
     the possibility of proving Euclid's axiom of parallel lines,
     and thus a foreshadowing of non-Euclidean geometry.

189. RICCATI, V. (1917) *Institutiones analyticae*. 2 vols. Bologna,
     1765-67. Written with G. Saladini. The first extensive treat-
     ise on the integral calculus. It also introduced the use of
     hyperbolic functions.

190. EULER, L. (See also 184) **(a)** *Institutiones calculi integralis*.
     3 vols. St. Petersburg, 1768-70. (3rd ed., 1824-25) A class-
     ical treatise on the integral calculus which, like his compan-
     ion work on the differential calculus (184), long served as a
     prototype.

1770s

191. LAPLACE, P.S. (2075) From 1771 to the early years of the nine-
     teenth century he published many articles in the *Mémoires* of
     the Paris Academy, chiefly on probability theory but also in
     some other areas, especially integral calculus and differential
     equations. His work on probability was summed up in his treat-
     ise on the subject in 1812 (see 209). He also developed a
     number of mathematical techniques such as generating functions,
     the variation of constants, the gravitation function, and what
     came to be called the Laplace transform.

192. BEZOUT, E. (2064) *Théorie générale des équations algébriques*.
     Paris, 1779. A work on the theory of equations, important
     especially for its method of elimination by means of symmetric
     functions.

1780s

193. *ENCYCLOPÉDIE méthodique ... Mathématiques.* 3 vols and supp.
    1784-92. The mathematical section of this vast, topically-
    arranged encyclopaedia. It was written by d'Alembert and
    included, in addition to pure mathematics, hydrostatics, optics,
    perspective, and astronomy.

194. CONDORCET, M.J.A.N. (2070) *Essai sur l'application de l'analyse
    à la probabilité des décisions rendues à la pluralité des voix.*
    Paris, 1785. An important work in the theory of probability;
    conceived by its author as a step towards "social mathematics."

195. LEGENDRE, A.M. (2076) In four papers published in the *Mémoires*
    of the Paris Academy between 1785 and 1793 he introduced the
    special functions now known as the Legendre polynomials. His
    other papers in the 1780s dealt with the calculus of variations
    (the Legendre conditions), the Legendre transformation, and the
    beginning of his theory of elliptic functions. (See also 196)

1790s

196. LEGENDRE, A.M. (See also 195) (a) *Elémens de géometrie.* Paris,
    1794. (12 editions up to 1823 and thereafter many adaptions.
    English trans., 1819 with many later editions. German trans.,
    1822) A famous textbook in which he rearranged and simplified
    many of the propositions in Euclid's *Elements.* (b) *Essai sur
    la théorie des nombres.* Paris, 1798. (3rd ed., 1830) An
    original work of much significance in the field.
    (See also 208)

197. LAGRANGE, J.L. (See also 186) (a) *Leçons sur le calcul des fonc-
    tions.* Lectures given at the Ecole Normale in 1795 and pub-
    lished in book form in 1806. (b) *Théorie des fonctions analyt-
    iques.* Paris, 1797. A treatise based on his lectures at the
    Ecole Polytechnique. (c) *Traité de la résolution des équations
    numériques de tous les degrés.* Paris, 1798. (3rd ed., 1826)
    Based on his lectures at the Ecole Polytechnique.

198. *A MATHEMATICAL and philosophical dictionary.* 2 vols. London,
    1795-96. (Another ed., 1815) By C. Hutton.

199. CARNOT, L. (2087) *Réflexions sur la métaphysique du calcul
    infinitésimal.* Paris, 1797. (Rev. ed., 1813. Many reprintings.
    Trans. into five languages) An investigation of the underlying
    logic of the calculus and an attempt to make it more rigorous.
    (See also 204)

200. LACROIX, S.F. (2086) *Traité du calcul différential et du calcul
    intégral.* 2 vols. Paris, 1797-98. (2nd ed., 1810-19) An
    outstanding treatise which successfully synthesized the work of
    the eighteenth century in the field. (See also 203)

201. MONGE, G. (2074) *Géometrie descriptive.* Paris, 1799. (7th ed.,
    1847. Trans. into five languages) The first exposition of
    descriptive geometry, of which he was the creator (despite
    various predecessors). In addition to its mathematical signif-
    icance it revolutionized engineering design. (See also 206)

1800s

202. GAUSS, C.F. (2570) **(a)** *Disquisitiones arithmeticae.* Leipzig,
     1801. (French trans., 1807) A work on number theory of great
     brillance and deep and lasting significance. **(b)** Up to the
     1840s he published a large body of important work in pure
     mathematics (in addition to his achievements in several fields
     of applied mathematics), nearly all as periodical articles. He
     contributed to most fields of mathematics and is perhaps best
     known for his method of least squares and his development of
     the theory of surfaces.

203. LACROIX, S.F. (See also 200) *Traité élémentaire du calcul diff-
     érentiel et du calcul intégral.* Paris, 1802. (9th ed., 1881.
     English trans., 1816. German trans., 1830-31) An important
     textbook. The English translation was published by Babbage,
     Herschel, and Peacock as part of their campaign to introduce
     Continental methods into Cambridge mathematics.

204. CARNOT, L. (See also 199) *Géometrie de position.* Paris, 1803.
     One of the works that initiated the revival of projective
     geometry. Unlike his teacher, Monge, he refused to use analyt-
     ical methods and advocated a return to pure geometry.

205. ARGAND DIAGRAM. J.R. Argand (a Parisian bookseller about whom
     very little is known) proposed in 1806 a graphical represent-
     ation of complex numbers and operations upon them--"a critical
     idea for which the time was ripe."

206. MONGE, G. (See also 201) *Application de l'analyse à la géometrie.*
     Paris, 1807. Established the algebraic methods of three-dimen-
     sional geometry.

1810s

207. POISSON, S.D. (2111) From about 1810 until his death in 1840 he
     made many contributions to pure mathematics (as well as to sev-
     eral fields of applied mathematics), especially in differential
     equations, algebraic equations, definite integrals, surfaces,
     the calculus of variations, and the theory of probability.

208. LEGENDRE, A.M. (See also 196) *Exercices de calcul intégral.* 3
     vols. Paris, 1811-17. A synthesis of the results he had
     obtained in his pioneering studies of elliptic integrals since
     the 1780s. It was superseded by his later *Traité* (218).

209. LAPLACE, P.S. (See also 191) **(a)** *Théorie analytique des probab-
     ilités.* Paris, 1812. (3rd ed., 1820) A synthesis of the
     mathematical methods he had been developing since the 1770s
     for the calculation of probabilities, with applications to such
     fields as games of chance, demography, error theory, decision
     theory, etc. A pioneering work; nearly all later developments
     in the field were based on it. **(b)** *Essai philosophique sur les
     probabilités.* Paris, 1814. (5th ed., 1825) A popularization
     which, like his famous popularization of planetary astronomy
     (418a), was a model of French prose.

210. RUFFINI, P. (1934) *Riflessioni intorno alla soluzione delle
     equazioni algebriche generali.* Modena, 1813. Put forward his
     theorem of the impossibility of solving the general algebraic

equation of the fifth (or higher) degree by means of radicals. At the same time the work took an important step towards the algebraic theory of groups.

211. BOLZANO, B. (2945) *Rein analytischer Beweis....* Prague, 1817. One of a number of contemporary works which sought to impart more rigour into the concepts and proofs of analysis. In particular it was the first successful attempt to eliminate the use of infinitesimals from the differential calculus. (See also 254)

1820s

212. BABBAGE, C. (2345) Through the 1820s and 1830s he was engaged in developing a mechanical computational device which would not only calculate various kinds of mathematical tables but also print them (thereby eliminating a major source of error). He managed to obtain governmental support for the project in 1823 and his first model--his "difference engine"--was sufficiently developed by 1837 to be able to produce excellent tables of eight-figure logarithms. By 1834 he had made plans for an "analytical engine", a more sophisticated device which can be regarded as the precursor of today's computers. In 1842 however the government refused further financial support and the whole project lapsed.

213. CAUCHY, A.L. (2123) **(a)** *Cours d'analyse*. Paris, 1821. **(b)** *Résumé des leçons sur le calcul infinitésimal*. Paris, 1823. **(c)** *Leçons sur les applications du calcul infinitésimal à la géométrie*. Paris, 1826-28. These three treatises incorporate many of his chief contributions--above all the clear and rigorous methods he introduced into analysis.
   Cauchy was second only to Euler in the number of his publications and he continued to pour out papers until his death in 1857. He contributed to all branches of pure mathematics--especially calculus, complex functions, and the theory of equations--and made many important advances in applied mathematics.

214. PONCELET, J.V. (2131) *Traité des propriétés projectives des figures*. Metz/Paris, 1822. A pioneering work in projective geometry which had a decisive influence on the subsequent development of the field.

215. BOLYAI, J. (2949) "Appendix scientiam spatii absolute veram exhibens"--an essay setting forth a consistent system of non-Euclidean geometry. It was completed in 1823 and published in 1831 as an appendix to a mathematical treatise by his father, Farkas Bolyai. Its importance was not recognized until the 1860s.

216. BESSEL, F.W. (2578. The great astronomer) In 1817 he had first made use of what are now called Bessel functions, and in 1824, in a paper on planetary perturbations, he gave an account of some of their general properties. Though particular cases of these functions had been investigated earlier by Euler, Lagrange, and others, it was Bessel who gave a general treatment of them.

217. ABEL, N.H. (2855) **(a)** His numerous contributions to the theory of equations included a proof in 1824--using a more rigorous

argument than Ruffini (210)—of the impossibility of solving
the general algebraic equation of the fifth degree by means of
radicals. He also gave the first rigorous proof of the general
binomial theorem. **(b)** From the early 1820s he published many
papers developing a general theory of elliptic functions and in
1828 announced his famous discovery of what came to be called
Abelian integrals.

218. LEGENDRE, A.M. (See also 208) *Traité des fonctions elliptiques.*
3 vols. Paris, 1825–28. Intended to be the final statement of
his work on elliptic functions, with their applications in geometry and mechanics. When he learnt that Abel and Jacobi had
published works on the subject almost at the same time he brought
out three supplements in rapid succession (1828, 1829, 1832).

219. DIRICHLET, G.P.L. (2594) From the late 1820s to the late 1850s
he published numerous important papers on the theory of numbers,
his first paper on analytic number theory appearing in 1837.
Over the same period he also published extensively on analysis,
quadratic forms, and various aspects of applied mathematics.
(See also 253)

220. MÖBIUS, A.F. (2577) *Der barycentrische Calcul.* Leipzig, 1827.
Introduced homogeneous coordinates into analytical geometry.
The book had many other important features, including a treatment of geometrical transformations, especially projective
transformations, and the introduction of the construction now
known as the Möbius net which became important in the development of projective geometry.   (See also 263)

221. PLÜCKER, J. (2592) *Analytisch-geometrische Entwicklungen.* 2 vols.
Essen, 1828–31. An important contribution to the renewal of
analytical geometry. It embodied the proposal that the fundamental geometric element could be the straight line rather than
the point; this led to a detailed formulation of the principle
of duality.   (See also 233)

222. STURM, C.F. (2133) His well-known theorem—a major contribution
to the theory of equations—was first announced in 1829.
(See also 224)

223. JACOBI, C.G.J. (2593) *Fundamenta nova theoriae functionum ellip-
ticarum.* Königsberg, 1829. Presented a theory of elliptic
functions based on four theta functions defined by their series
expansions.   (See also 230)

1830s

224. STURM, C.F. (See also 222) Through the 1830s he made important
contributions to the theory of second-order differential equations, partly in association with Liouville.

225. KUMMER, E.E. (2614) Published numerous papers during the 1830s
on aspects of function theory. The most notable were those
dealing with hypergeometric series.   (See also 236)

226. OSTROGRADSKY, M.V. (2984) From 1830 until his death in 1862 he
produced numerous papers on topics in pure mathematics (calculus of variations, theory of differential equations, algebraic

functions, etc.) and applied mathematics (elasticity, heat con-
duction, dynamics, theory of percussion, etc.).

227. LOBACHEVSKY, N.I. (2981) His system of non-Euclidean geometry
was first published in an obscure Russian journal in 1829-30.
It first appeared in the West in a long article in French in
Crelle's *Journal* in 1837 and three years later Lobachevsky
published a full exposition of it in his *Geometrische Untersuchungen zur Theorie der Parallellinien* (Berlin, 1840). Its
importance was not recognized however until the 1860s.

228. DE MORGAN, A. (2354) **(a)** *Elements of arithmetic*. London, 1830.
Included a philosophical treatment of the ideas of number and
magnitude. **(b)** In 1838 he introduced and defined the term
'mathematical induction', signifying a logical process long
used by mathematicians but hitherto unclarified. (See also 248)

229. STEINER, J. (2590) *Systematische Entwickelung der Abhängigkeit
geometrischer Gestalten von einander*. Berlin, 1832. The major
work of one of the chief founders of projective geometry. His
many later contributions to the subject were surveyed in his
posthumous *Vorlesungen über synthetische Geometrie* (2 vols
Leipzig, 1867)

230. JACOBI, C.G.J. (See also 223) **(a)** His successful treatment of
hyperelliptical functions in 1832 led him to formulate the
theory of Abelian functions of p variables. **(b)** *Canon arith-
meticus*. Berlin, 1839. A notable contribution to the theory
of numbers.   (See also 238)

231. LIOUVILLE, J. (2149) From 1832 to 1857 he published about a
hundred papers on various aspects of analysis. His most significant discovery was of transcendental numbers, in 1844. His
approach to boundary-value problems, later designated as Sturm-
Liouville theory, was to be of importance for mathematical
physics in the twentieth century.   (See also 260)

232. LAMÉ, G. (2132) Introduced in 1833 the use of curvilinear co-
ordinates which he used thereafter in many problems in applied
mathematics. Much of his work was summed up in his *Leçons sur
les coordonnés curvilignes et leurs diverses applications*
(Paris, 1859)

233. PLÜCKER, J. (See also 221) **(a)** *System der analytischen Geometrie*.
Berlin, 1835. Introduced the use of linear functions instead
of the usual coordinate points. Most of the book dealt with
the study and classification of algebraic curves, a topic which
had been neglected since the mid-eighteenth century.
(b) *Theorie der algebraischen Curven*. Bonn, 1839. A sequel to
the foregoing work. It dealt especially with the equations
detailing the singularities on algebraic curves. (See also 245)

234. *DICTIONNAIRE de mathématiques pures et appliquées*. 3 vols. Paris,
1835-40. (2nd ed., 1845) By A. de Montferrier. (cf. 261)

235. CHASLES, M. (2148) *Aperçu historique sur l'origine et le dével-
oppment des méthodes en géometrie*. Paris, 1837. (Later reprints. German trans., 1839) On the basis of his historical
study he asserted the claims of geometry against algebraic

analysis. The book also contained the results of his early
researches in projective geometry, of which he was one of the
main founders.   (See also 256)

1840s

236. KUMMER, E.E. (See also 225) His many papers on the theory of
     numbers during the 1840s and 1850s included notably his devel-
     opment of the concept of ideal numbers during the mid-1840s.
     By means of these he was able to show that Fermat's famous
     theorem is correct for many prime numbers. Other mathematicians
     also made extensive use of them.   (See also 269)

237. WEIERSTRASS, K. (2625) "The father of modern analysis" and one
     of the greatest mathematicians of the nineteenth century. From
     the 1840s he devoted his working life to developing his theory
     of functions. He did not publish very much but from the mid-
     1850s disseminated his ideas through his influential teaching
     at the University of Berlin. In his last years he began pub-
     lishing his lectures and other writings as *Mathematische Werke*
     but died after the second volume appeared (seven volumes were
     published between 1894 and 1927).

238. JACOBI, C.G.J. (See also 230) **(a)** "De formatione et proprietat-
     ibus determinantium"--a paper of 1841 which made a pioneering
     contribution to the theory of determinants. **(b)** In the early
     1840s he was the first to apply function theory to the theory
     of numbers: by the use of elliptic functions he proved one of
     Fermat's assertions. **(c)** See 559.

239. CAYLEY, A. (2384) From the early 1840s to his death in 1895 he
     published nearly a thousand papers and notes on almost every
     part of pure mathematics (and some applied mathematics). He is
     best known for his work on invariants, matrices, n-dimensional
     geometry, linear transformations, and skew surfaces.
     (See also 249)

240. CHEBYSHEV, P.L. (2988) **(a)** From the early 1840s to about 1880 he
     produced many papers on a range of topics--notably theory of
     numbers, probability, theory of integrals, quadratic forms, and
     orthogonal functions. **(b)** *Teoria sravneny* ["Theory of congru-
     ences"]. St. Petersburg, 1849. (3rd ed., 1901. German trans.,
     1888) A treatise on the theory of numbers which included his
     own work in the subject; it was long used as a textbook in
     Russia.

241. HERMITE, C. (2160) Author of numerous papers from the early
     1840s in several fields, especially the theory of functions,
     the arithmetical theory of quadratic forms, and the theory of
     invariants.   (See also 264)

242. HAMILTON, W.R. (2353) Made in 1843 his famous discovery of quat-
     ernions--ordered sets of four numbers obeying special algebraic
     laws--which are useful for manipulating quantities having mag-
     nitude and direction in space. He spent the remaining twenty-
     two years of his life developing the theory and its applications.
     The fruits of his efforts were presented in his *Lectures on
     quaternions* (Dublin, 1853) and the posthumous *Elements of quat-
     ernions* (London, 1866).

243. GRASSMANN, H.G. (2613) *Die lineale Ausdehnungslehre*. Leipzig, 1844. A highly original treatise on geometric analysis. It appeared before its time and was moreover written in a rather obscure fashion; consequently it was neglected almost completely until about 1870. Thereafter it largely inspired the Continental school of vector analysis and continued to exert a seminal influence into the twentieth century.

244. EISENSTEIN, F. (2626) The papers he published from 1844 until his death in 1852 included important contributions to the theory of ternary and quadratic forms, the theory of numbers, and the theory of functions. Some of his papers also helped to open the way to the theory of invariants.

245. PLÜCKER, J. (See also 233) *System der Geometrie des Raumes*. Düsseldorf, 1846. A systematic and elegant treatise on the analytical geometry of space, including his own substantial contributions.

246. GALOIS, E. (2150) His brief but profound writings on the theory of algebraic equations of higher degree dated from the period just before his tragic death in 1832 but were first published in 1846 by J. Liouville in his *Journal de Mathématiques*. His work did not begin to attract much attention until the 1850s and it was not until the late 1860s that its full significance was appreciated.

247. STAUDT, C. (2591) *Geometrie der Lage*. Nuremberg, 1847. A very rigorous approach to projective geometry ("geometry of position"), dispensing with all dependence on metrical entities or relations. (See also 259)

248. DE MORGAN, A. (See also 228) **(a)** *Formal logic*. London, 1847. A pioneering work in symbolic logic. **(b)** *Trigonometry and double algebra*. London, 1849. Treated algebra as based on a collection of meaningless symbols together with the laws for manipulating them.

249. SYLVESTER, J.J. (2369) and CAYLEY, A. (2384). Working in close association they largely created the theory of algebraic forms (quantics) and the theory of invariants from the late 1840s onward.

250. *ENCYCLOPAEDIA of pure mathematics, forming part of the "Encyclopaedia metropolitana."* 1847. (cf. 34)

1850s

251. KRONECKER, L. (2627) His numerous papers, published mostly from the mid-1850s onward, included important contributions to elliptic functions, the theory of equations, and the theory of algebraic numbers. He was also noted for his concern with the philosophical foundations of mathematics and his attempts to unify arithmetic, algebra, and analysis.

252. SMITH, H.J.S. (2407) Author of some distinguished papers on the theory of numbers from the mid-1850s to the late 1860s. He is said to have been the foremost follower of Gauss in this field. (In his later period, until his death in 1883, he wrote mainly on elliptic functions)

253. DIRICHLET, G.P.L. (See also 219) Published in 1850 an important
     paper on the boundary-value problem (now known as Dirichlet's
     problem) concerning partial differential equations used in the
     study of the flow of heat, electricity, etc.   (See also 268)

254. BOLZANO, B. (See also 211) *Paradoxien des Unendlichen*. Leipzig,
     1851 (Posthumous) Though more philosophical than mathematical,
     the book was of some importance in leading towards the theory
     of sets.

255. RIEMANN, B. (2650) **(a)** *Grundlagen für eine allgemeine Theorie
     der Functionen einer veränderlichen complexen Grösse*. Göttin-
     gen, 1851. (2nd ed., 1867) His doctoral thesis. A major
     contribution to the theory of complex functions. His treatment
     was based on geometrical conceptions and led to the idea of the
     Riemann surface. **(b)** The papers he published between the mid-
     1850s and his early death in 1866 were not numerous but they
     were mostly of very high quality. An indication of their influ-
     ence is the large number of concepts, theorems, and methods
     that are named after him. Besides his greatest work on non-
     Euclidean geometry (see 274) and on complex functions (above)
     his papers dealt principally with differential equations, diff-
     erential geometry, foundations of analysis, theory of numbers,
     topology, and trigonometric series. He also wrote on various
     aspects of mathematical physics.

256. CHASLES, M. (See also 235) *Traité de géometrie supérieure*.
     Paris, 1852. (Reprinted 1880. German trans., 1856) Contained
     much original work of his own on projective geometry.
     (See also 271)

257. BETTI, E. (1943) One of the first to take up the work of Galois
     (246). In the early 1850s he developed a detailed and rigorous
     exposition of the theory of equations that Galois had sketched
     out.   (See also 277)

258. BOOLE, G. (2383) **(a)** *An investigation of the laws of thought on
     which are founded the mathematical theories of logic and prob-
     ability*. London, 1854. A famous pioneering work in mathemat-
     ical logic.
            (b) *Treatise on differential equations*. Cambridge,
     1859. (4th ed., 1877) **(c)** *Treatise on the calculus of finite
     differences*. Cambridge, 1860. (3rd ed., 1880) Two widely-
     used textbooks which contained much of his own original work.

259. STAUDT, C. (See also 247) *Beiträge zur Geometrie der Lage*. 3
     vols. Nuremberg, 1856-60. Contained the first general theory
     of imaginary points, lines, and planes in projective geometry.

260. LIOUVILLE, J. (See also 231) Began in 1856 a remarkable series
     of papers in the theory of numbers which he continued until the
     late 1860s.

261. *ENCYCLOPÉDIE de mathématiques élémentaires*. 4 vols. Paris,
     1856-59. By A. de Montferrier. (cf. 234)

262. CHRISTOFFEL, E.B. (2651) His papers, published mostly from the
     late 1850s to the early 1880s, included significant work in
     higher analysis and geometry. He also wrote on applied mathe-
     matics.

263. MÖBIUS, A.F. (See also 220) One of the pioneers of topology. Well known for his discovery in 1858 of the Möbius strip, a one-sided surface.

264. HERMITE, C. (See also 241) *Sur la résolution de l'équation du cinquième degré*. Paris, 1858. The first solution of the general equation of the fifth degree. He achieved it by means of elliptic functions.   (See also 284)

1860s

265. BELTRAMI, E. (1948) Through the 1860s he produced several notable papers on the differential geometry of curves and surfaces. His work was especially significant in making non-Euclidean geometry acceptable to mathematicians generally.

266. ROCHE, E.A. (2161) Around 1860 he put forward an important generalization of Taylor's theorem (175a).

267. HESSE, L.O. (2624) **(a)** *Vorlesungen über analytische Geometrie des Raumes*. Leipzig, 1861. (3rd ed., 1876) **(b)** *Vorlesungen über analytische Geometrie der geraden Linie, des Punktes und des Kreis in der Ebene*. Leipzig, 1865. (4th ed., 1909) Two influential textbooks. They incorporated his own substantial contributions.

268. DIRICHLET, G.P.L. (See also 253) *Vorlesungen über Zahlentheorie*. Brunswick, 1863. (Posthumous; ed. by R. Dedekind)

269. KUMMER, E.E. (See also 236) Discovered in 1863 the fourth-order surface named after him.

270. CREMONA, L. (1946) Set forth his general theory of geometrical transformations in two papers in 1863 and 1866. (The papers were translated into German and published in book form in 1870) Another paper in 1871 completed his theory by extending it from plane curves to space curves.   (See also 281)

271. CHASLES, M. (See also 256) **(a)** *Traité des sections coniques*. Paris, 1865. Like his earlier treatises (235, 256) it incorporated much original work of his own in projective geometry. **(b)** From 1864 he published many papers using his important "principle of correspondence" and "method of characteristics" for the solution of numerous problems.

272. CLEBSCH, R. (2653) *Theorie der Abelschen Funktionen*. Leipzig, 1866. With P. Gordan. A notable contribution to algebraic geometry.   (See also 283)

273. TAIT, P.G. (2408) *An elementary treatise on quaternions*. Oxford, 1867. (3rd ed., 1890. German trans., 1880) Despite the title it included his many important contributions to the subject, increasing from edition to edition.

274. RIEMANN, B. (See also 255) *Über die Hypothesen welche der Geometrie zu Grunde liegen*. Leipzig, 1867. (Posthumous) His *Habilitationsvortrag* of 1854. "One of the highlights in the history of mathematics: young, timid Riemann lecturing to the aged, legendary Gauss, who would not live past the next spring, on consequences of ideas the old man must have recognized as his own and which he had long secretly cultivated." (*DSB*) The

lecture set forth a system of non-Euclidean geometry (independently of Bolyai and Lobachevsky) which in the twentieth century
was to form part of the theory of relativity.

275. GORDAN, P.A. (2671) From 1868 onward for many years he worked in
the theory of invariants, becoming the leading figure in the
field. His chief contribution was made at the beginning of his
researches, in 1868, with his "theorem of finiteness."
(See also 272)

1870s

276. CANTOR, G. (2673) Before developing his famous theory of sets
(from 1874 onward; see 296) he published a series of ten papers
between 1869 and 1873 which made notable contributions to the
theory of numbers and trigonometric series.

277. BETTI, E. (See also 257) His friendship with Riemann inspired
him to write an important article on topology in 1870-71 and
to undertake some significant researches in applied mathematics
(especially on elasticity) in the late 1870s.

278. JORDAN, C. (2185) (a) *Traité des substitutions et des équations
algébriques*. Paris, 1870. He was the first to systematically
develop Galois' theory of groups and his treatise includes a
comprehensive account of the theory with applications to algebraic equations. (b) Through the 1870s he produced a series of
papers containing proofs of his important "finiteness theorems."
(See also 293)

279. LIE, S. (2868) In 1870 he discovered contact transformations
which, with transformation groups, he developed in many later
papers. His work in the field was expounded in his *Theorie der
Transformationsgruppen* (3 vols, Leipzig, 1888-93; written with
his assistant, F. Engel).　　　　　　　　　　　　　　(See also 307)

280. KLEIN, F. (2701) (a) In two papers in 1871 and 1873 he showed
that the non-Euclidean geometries can be subsumed under projective geometry. (b) In his inaugural address as professor at
Erlangen in 1872 he put forward his *Erlanger Programm* for the
unification of mathematics in terms of group theory. The *Programm* was very well received (his address was translated into
six languages) and was influential in guiding the researches of
many mathematicians.　　　　　　　　　　　　　　　(See also 295)

281. CREMONA, L. (See also 270) (a) *Le figure reciproche nella statica
grafica*. Milan, 1872. (English trans., 1890) (b) *Elementi di
geometria proiettiva*. Turin, 1873. (English trans., 1885)
(c) *Elementi di calcolo grafico*. Turin, 1874. Three important
textbooks which included much of his own original work.

282. DEDEKIND, R. (2652) (a) *Stetigkeit und irrationale Zahlen*.
Brunswick, 1872. Contained a purely arithmetical conception of
the continuum, with definitions of rational and irrational
numbers which depended on what came to be called "the Dedekind
cut." (b) "Über die Theorie der ganzen algebraischen Zahlen"--
a paper of 1877 which put forward his 'ideal' theory of collections of numbers.　　　　　　　　　　　　　　　　　(See also 303)

283. CLEBSCH, R. (See also 272) (a) *Theorie der binären algebraischen*

Formen. Leipzig, 1872. (b) *Vorlesungen über Geometrie*. 2 vols. Leipzig, 1876-91. (Posthumous; ed. by K. Lindemann)

284. HERMITE, C. (See also 264) Proved in 1873 that the number *e* is transcendental. His proof was adapted slightly by F. Lindemann nine years later to prove the same for the number π.

285. DARBOUX, G. (2186) *Sur une classe remarquable de courbes et de surfaces algébriques*. Paris, 1873. Developed the analytic and geometric theory of the cyclids (a class of surfaces). (See also 301)

286. POINCARÉ, H. (2194) From the late 1870s until his death in 1912 he published almost 500 papers covering most branches of pure and applied mathematics. His chief contributions were in the fields of automorphic functions, Abelian functions and algebraic geometry, the qualitative theory of differential equations, probability, number theory, and algebraic topology. He also wrote extensively on celestial mechanics (see 487) and on almost every aspect of mathematical physics (with H.A. Lorentz he did much to set the scene for the special theory of relativity).

287. PICARD, E. (2195) From the late 1870s to about 1900 he published many papers on a variety of topics, notably algebraic geometry, quadratic forms, group theory, and differential equations. (Much of this work was included in his two books--see 308) He also did important work in mathematical physics.

288. STIELTJES, T.J. (2793) From the late 1870s until his untimely death in 1894 he published on almost every aspect of analysis. He is best known for his study of continued fractions and for the integral named after him, but he also did important work in number theory, spherical harmonics, and the theory of Riemann's function.

289. FREGE, G. (2699) *Begriffsschrift*. Halle, 1879. A pioneering work in modern mathematical logic.   (See also 298)

290. FROBENIUS, G.F. (2700) Published a major paper in 1879 on group theory, especially the concept of abstract groups. (See also 304)

1880s

291. MARKOV, A.A. (3004) The papers and monographs he published during the 1880s and 1890s were concerned with number theory and function theory. The work on probability (notably Markov chains), for which he is best known, was done after 1900.

292. VOLTERRA, V. (1956) Author of many papers from 1881 to the 1930s on various aspects of pure and applied mathematics. His first major contribution dated from 1883 when he began constructing a general theory of functionals. This led to the development of new areas of analysis, notably the type of integral equations named after him.

293. JORDAN, C. (See also 278) *Cours d'analyse de l'Ecole Polytechnique*. 3 vols. Paris, 1882. (3rd ed., 1909-15) An influential textbook which included much original work of his own (especially in the later editions).

294. PASCH, M. (2672) *Vorlesungen über neuere Geometrie.* Leipzig,
     1882. A pioneering work on the foundations of geometry. It
     initiated the axiomatic method.

295. KLEIN, F. (See also 280) (a) *Über Riemanns Theorie der algebra-*
     ischen Funktionen und ihrer Integrale. Leipzig, 1882. A major
     work on function theory. (b) *Vorlesungen über das Ikosaeder*
     und die Auflösung der Gleichungen vom 5. Grade. Leipzig, 1884.
     Derivation of a theory of the fifth-degree equation from a
     treatment of the icosahedron.   (See also 306)

296. CANTOR, G. (See also 276) *Grundlagen einer allgemeinen Mannig-*
     faltigkeitslehre. Leipzig, 1883. A presentation of the theory
     of sets which he had been developing in a series of papers
     since 1874.   (See also 311)

297. SEGRE, C. (1957) Began in 1883 the distinguished researches in
     geometry that he was to continue to the early 1920s.

298. FREGE, G. (See also 289) *Die Grundlagen der Arithmetik.* Breslau,
     1884. An attempt to construct arithmetic as an extension of
     logic.   (See also 309)

299. PEANO, G. (1955) (a) *Calcolo differenziale e principii di calcolo*
     integrale. 1884. A major work containing many results of his
     own. (b) *Applicazioni geometriche del calcolo infinitesimale.*
     An introduction to his geometrical calculus. (c) *Arithmetices*
     principia, novo methodo exposita. 1889. An exposition of his
     system of mathematical logic.   (See also 310)

300. HILBERT, D. (2727) During the period 1885 to 1893 he made some
     highly important contributions to the theory of invariants.
     (See also 314)

301. DARBOUX, G. (See also 285) *Leçons sur la théorie générale des*
     surfaces et les applications géométriques du calcul infinités-
     imal. 4 vols. Paris, 1887-96. A collection of essays incorp-
     orating most of his earlier researches on the application of
     analysis to curves and surfaces.   (See also 315)

302. RICCI-CURBASTRO, G. (1952) In a series of papers between 1887
     and 1896 he laid the foundations of what he called the absolute
     differential calculus--sometimes termed the Ricci calculus and
     now known as tensor analysis. After 1896 his pupils, notably
     T. Levi-Civita, continued to develop the subject. (It was used
     by Einstein in his general theory of relativity)

303. DEDEKIND, R. (See also 282) *Was sind und was sollen die Zahlen?*
     Brunswick, 1888. A further development of his ideas in the
     theory of numbers.

1890s

304. FROBENIUS, G.F. (See also 290) In collaboration with his student,
     I. Schur, he published several papers during the 1890s develop-
     ing group theory by means of the theory of finite groups of
     linear substitutions.

305. HADAMARD, J. (2206) (a) His important work on analytical func-
     tions and especially the analytical continuation of a Taylor
     series, done during the 1890s, was summarized in his monograph

La série de Taylor et son prolongement analytique (1901).
(b) Some of the results of his work in function theory enabled him in 1896 to solve the famous problem concerning the distribution of the prime numbers.

306. KLEIN, F. (See also 295) **(a)** *Vorlesungen über die Theorie der elliptischen Modulfunktionen.* 2 vols. Leipzig, 1890–92. Written with his student, R. Fricke. **(b)** *Vorlesungen über die Theorie der automorphen Funktionen.* 2 vols. Leipzig, 1897–1902. With R. Fricke.

307. LIE, S. (See also 279) **(a)** *Vorlesungen über Differentialgleichungen mit bekannten infinitesimalen Transformationen.* Leipzig, 1891. With G. Scheffers. **(b)** *Vorlesungen über continuierliche Gruppen.* Leipzig, 1896. With G. Scheffers. Two standard works.

308. PICARD, E. (See also 287) **(a)** *Traité d'analyse.* 3 vols. Paris, 1891–96. (Some later editions) An important textbook which included many of his own results. **(b)** *Théorie des fonctions algébriques de deux variables indépendantes.* 2 vols. Paris, 1897–1906. With G. Simart. An exposition of his earlier work.

309. FREGE, G. (See also 298) *Grundgesetze der Arithmetik.* 2 vols. Jena, 1893–1903. A development and refinement of his two earlier works on mathematical logic.

310. PEANO, G. (See also 299) **(a)** *Lezioni di analisi infinitesimale.* 2 vols. 1893. A distinguished work. **(b)** *Formulario matematico.* 1895. (5th ed., 1908) (Written with several collaborators) A remarkable attempt to present all known theorems in mathematics using the notation of his mathematical logic. It had a major influence on subsequent attempts to restructure mathematics.

311. CANTOR, G. (See also 296) *Beiträge zur Begründung der transfiniten Mengelehre.* 1895–97. (English trans., 1915) An exposition of his ideas of continuity and the infinite, including infinite ordinals and cardinals.

312. MINKOWSKI, H. (2728) *Geometrie der Zahlen.* Leipzig, 1896. (2nd ed., 1910) A pioneering work in the geometrical theory of numbers.

313. VALLÉE-POUSSIN, C. de la. (2822) In 1896--at the same time as Hadamard (305b)--he solved the problem of the distribution of prime numbers.

314. HILBERT, D. (See also 300) **(a)** *Der Zahlbericht.* 1897. A classical exposition of algebraic number theory. **(b)** *Grundlagen der Geometrie.* Leipzig, 1899. (9th ed., 1962. English trans., 1902) A famous work in the axiomatic treatment of geometry.

315. DARBOUX, G. (See also 301) *Leçons sur les systèmes orthogonaux et les coordonnées curvilignes.* Paris, 1898. Further extensions of his earlier geometrical researches.

316. *ENCYKLOPÄDIE der mathematischen Wissenschaften.* 23 vols. 1898–ca. 1935. Published under the auspices of the Göttingen, Leipzig, Munich, and Vienna Academies.

1.03 ASTRONOMY

Including celestial mechanics and astrophysics.

1450s

317. BIANCHINI, G. (1820) Composed a commentary on the Alphonsine Tables in 1458, on the order of the Emperor Frederick III. Presumably this led to his *Tabulae celestium motuum eorumque canones* (not printed until 1495).

318. PEUERBACH, G. (2932) His *Theoricae novae planetarum* was completed in 1454 and printed about 1474 (56 editions are known up to the mid-seventeenth century and many manuscript copies survive, mostly from the late fifteenth century). It was a carefully written textbook of planetary theory based on the Ptolemaic tradition transmitted through the Arabs, and became the standard work on the subject until the time of Tycho Brahe. A more technical work was his *Tabulae eclipsium*, based on the Alphonsine Tables but with substantial improvements. It was completed probably in 1459 but not printed until 1514; however it circulated in manuscript. It also continued in use up to the time of Tycho Brahe. Peuerbach's early death prevented him completing his other major work, *Epitoma Almagesti Ptolemaei*, which was finished by his pupil, Regiomontanus.

1460s

319. REGIOMONTANUS, J. (2933) **(a)** *Epitoma in Almagestum Ptolemaei*-- a work begun by Peuerbach who died before he could finish it; Regiomontanus completed it about 1462 and it was printed at Venice in 1496. It consisted of a condensed translation of Ptolemy's famous work, with revisions and notes. **(b)** *Tabulae directionum*--completed in 1467 and printed on Regiomontanus' own press at Nuremberg in 1475. (Many later reprints) It contained tables of the longitudes of the celestial bodies in relation to the (apparent) daily rotation of the heavens.

1470s

320. SACROBOSCO, Johannes de. (ca. 1200-1250) *Sphaera mundi*. The standard astronomical textbook of the Middle Ages. It was first printed in 1472 and continued in use to the end of the sixteenth century.

1490s

321. AL-FARGHĀNĪ (called ALFRAGANUS). (fl. 863) The medieval translation of his work, entitled *Compilatio astronomica* (or *Elem-*

enta astronomica), first appeared in print in 1493. It was a
clear, comprehensive, and non-mathematical account of Ptolemaic
astronomy which was popular and widely influential.

1510s

322. STÖFFLER, J. (2475) **(a)** *Elucidatio fabricae ususque astrolabi.*
Oppenheim, 1513. (French trans., 1560) **(b)** *Tabulae astronom-
icae*. Tübingen, 1514.   (See also 324)

323. PTOLEMY. (ca. 100–170) The first printed edition of the *Almagest*
appeared in 1515 in the form of the twelfth-century Latin translation from the Arabic. (Earlier and probably more influential
printed accounts of Ptolemaic astronomy were Regiomontanus'
Epitoma and al-Farghānī's *Compilatio*; see above) The first
Latin translation of the *Almagest* from the Greek was published
in 1528 and the Greek text was printed in 1538.

1530s

324. STÖFFLER, J. (See also 322) *In Procli Diadochi ... Sphaeram
mundi ... commentarius*. Tübingen, 1534.

325. NUÑEZ (SALACIENSE), P. (2922) *Tradado da sphera*. Lisbon, 1537.
(A Latin version was published in 1566 and had two later editions) A collection of writings on astronomy, geography, and
navigation.

326. AL-BATTĀNĪ (called ALBATEGNIUS) (fl. 858–929) The medieval translation, *De motu stellarum*, of his great work was first printed
in 1537. In manuscript the work had been well known to Peuerbach, Regiomontanus, and Copernicus.

1540s

327. APIAN, P. (2481) *Astronomicon Caesareum*. Ingolstadt, 1540.
Notable especially for its pioneering observations on comets.

328. REINHOLD, E. (2486) **(a)** *Themata*.... Wittenberg, 1541. (Many
reprintings) A short treatise on spherical astronomy.
(b) *Theoricae novae planetarum*. Wittenberg, 1542. (Rev. ed.,
1553. Many reprintings) A commentary on Peuerbach's book of
the same name. **(c)** *Mathematicae constructionis*. Wittenberg,
1549. (Several reprintings) A commentary on Ptolemy's
Almagest.   (See also 331)

329. RHETICUS, G. (2487) In 1539 he visited Copernicus who had long
been hesitating about publishing his masterpiece but gave
Rheticus permission to write an account of it. This he did in
his *De libris Revolutionum ... Nicolai Copernici ... narratio
prima* (Danzig, 1540; 2nd ed., 1541)—the first printed announcement of Copernicus' work. Rheticus was later the editor of the
De revolutionibus.

330. COPERNICUS, N. (2934) *De revolutionibus orbium coelestium*.
Nuremberg, 1543. The famous book which eventually transformed
astronomy and had a profound influence on the Scientific Revolution of the seventeenth century.

1550s

331. REINHOLD, E. (See also 328) **(a)** *Prutenicae tabulae coelestium motuum.* Tübingen, 1551. (Four later editions) His chief work; it soon became the most widely used set of astronomical tables. They were a rearrangement and improvement of Copernicus' tables. **(b)** *Primus liber tabularum directionum.* Tübingen, 1554. (Two later editions)

332. RECORDE, R. (2218) *The castle of knowledge, a treatise on astronomy and the sphere.* London, 1556. An elementary textbook based on Ptolemy, Sacrobosco, etc. It included a brief, favorable reference to Copernicus.

1560s

333. WILHELM IV, *Landgrave of Hesse.* (2494) Built a private observatory in 1561 and later employed Christoph Rothmann as observer and calculator, and Joost Bürgi as instrument-maker; both were very competent. Wilhelm was acquainted with Tycho Brahe and, like him, was especially concerned to improve the techniques and accuracy of astronomical observation. He began an ambitious project to compile a new star catalogue based on data from his own observatory but was unable to bring it to completion.

1570s

334. ARISTARCHUS (ca. 310-230 B.C.) The first significant Latin translation of his *Treatise on the sizes and distances of the sun and moon* was published by Commandino (1843) in 1572. (An earlier translation had been printed in a heterogenous collection of texts in 1498)

335. BRAHE, T. (2823) *De nova ... stella.* Copenhagen, 1573. His first work. A brief but significant tract discussing the new star of 1572. (In 1602 his much fuller account was published --see 345)   (See also 341)

336. DIGGES, T. (2222) **(a)** *Alae seu scalae mathematicae.* London, 1573. Included significant observations of the new star of 1572. **(b)** "A perfit description of the caelestiall orbes"--an appendix to his father's *Prognosticon everlastinge* of 1576 (see 100b). It contained a translation of parts of Book I of Copernicus' *De revolutionibus* with Digges' own conception of an infinite universe in which the stars are fixed and are at different distances from the earth.

1580s

337. MOLETI, G. (1855) *Tabulae Gregorianae ex Prutenicis deductae.* Venice, 1580. Tables derived from Reinhold's Prutenic Tables and compiled by order of Pope Gregory XIII and the Senate of Venice for the reform of the calendar.

338. GREGORIAN CALENDAR. Introduced in 1582 by Pope Gregory XIII.

339. CLAVIUS, C. (1854) *In Sphaeram Ioannis de Sacro Bosco commentarius.* Rome, 1581. Anti-Copernican.

340. MÄSTLIN, M. (2500) *Epitome astronomiae.* Heidelberg, 1582. (At
     least six later editions) A popular textbook. Despite the
     author's firm attachment to Copernicanism it expounded the
     Ptolemaic system as being easier for students to understand.

341. BRAHE, T. (See also 335) *De mundi aetheri recentioribus phaenom-
     inis.* Uraniborg, 1588. On the comet of 1577. The book also
     introduced his geoheliocentric world-system.   (See also 342)

1590s

342. BRAHE, T. (See also 341) **(a)** *Epistolarum astronomicarum liber.*
     Uraniborg, 1596. A selection from his extensive correspondence,
     beginning with discussions of the comet of 1585 and dealing
     largely with his instruments and techniques of observation.
     An appendix gave an account of his observatory at Uraniborg.
     (b) *Astronomiae instauratae mechanica.* Wandsbeck, 1598. An
     account of his instruments and their use, with good woodcut
     illustrations.   (See also 345)

343. KEPLER, J. (2937) *Mysterium cosmographicum.* Tübingen, 1596.
     The first important astronomical work that was committed to the
     Copernican theory. Starting from a Neo-Platonic vision of
     celestial mathematical harmonies, which underlaid all his work,
     he arrived at the law connecting the relative distances of the
     planets.   (See also 347)

344. VARIABLE STARS. In 1596 David Fabricius was the first to observe
     a variable star (o Ceti). He called it Mira Ceti.

1600s

345. BRAHE, T. (See also 342) *Astronomiae instauratae progymnasmata.*
     Prague, 1602. (Posthumous; edited by Kepler) On the new star
     of 1572.

346. BAYER, J. (2507) *Uranometria.* Augsburg, 1603. A popular guide
     to the heavens in which he introduced a new system of nomencla-
     ture for stars which was much more exact and convenient than
     the one inherited from the Greeks. His nomenclature is still
     used today for stars visible to the naked eye.

347. KEPLER. J. (See also 343) **(a)** *Astronomiae pars optica.* Frank-
     furt, 1604. See 577. The astronomical chapters included
     discussions of parallax, atmospheric refraction, etc.
     (b) *De stella nova.* Prague, 1606. On the new star of 1604.
     (c) *Astronomia nova.* Prague, 1609. The book in which Kepler,
     working from Tycho Brahe's observations, established that the
     orbit of Mars (and likewise the other planets) is elliptical.
     He was the first to point out that the other planets resemble
     the earth in being material bodies.   (See also 349)

1610s

348. GALILEI, G. (1863) **(a)** *Siderius nuncius.* Venice, 1610. An
     account of the first use of the telescope in astronomy and the
     discoveries he made with it. The book caused a sensation
     throughout Europe. **(b)** *Istoria e dimonstrazioni intorno alle*

macchie solari. Rome, 1613 (published under the auspices of the Accademia dei Lincei). On sunspots. In this work he explicitly supported the Copernican system for the first time. **(c)** *Discorso sulle comete*. Florence, 1619. (See also 356)

349. KEPLER, J. (See also 347) **(a)** *Dioptrice*. Augsburg, 1611. See 580. **(b)** *Epitome astronomiae Copernicanae*. 3 parts. Linz, 1618–21. An introductory exposition of Copernican astronomy (including Kepler's own contribution to it). Widely read and very influential. **(c)** *De cometis*. Augsburg, 1619. A discussion of the bright comets of 1607 and 1618. **(d)** *Harmonices mundi*. Linz, 1619. "A great cosmic vision, woven out of science, poetry, philosophy, mysticism...." (M. Caspar) It contained his third law of planetary motion. (See also 354)

350. CHRISTMANN, J. (2506) **(a)** *Theoria lunae ex novis hypothesibus et observationibus demonstrata*. Heidelberg, 1611. **(b)** *Nodus Gordius ex doctrina sinuum explicatus*. Heidelberg, 1612. Both works included original contributions to astronomy, with some trigonometrical methods in the latter.

351. SCHEINER, C. (2512) Constructed a telescope for astronomical observations in 1611 and discovered sunspots (which he thought were small planets circling the sun). His discovery was made public by letter and, though it led to a priority dispute, was independent of Galileo's discovery of the same phenomenon (which was also discovered independently by several others about the same time). Of his several books on astronomy the chief one was *Rosa ursina* (Brescia, 1630) which contained a detailed study of sunspots. He also made an important improvement in the telescope, incorporating a suggestion of Kepler.

352. MAYR, S. (2508) *Mundus Jovialis*. Nuremberg, 1614. A study of the four satellites of Jupiter then known, including tables of their mean periodic motions. Though Galileo was the discoverer of the four satellites (in 1610) Mayr's tables of their mean motions were earlier and more accurate than his.

353. CYSAT, J.B. (2513) *Mathemata astronomica de loco, motu, magnitudine, et causis cometae....* Ingolstadt, 1619. A report of the first telescopic observations of a comet, carried out in much detail over a period of two months, with measurements of position. He also made some other notable observations, including the moons of Saturn and the Orion nebula.

1620s

354. KEPLER, J. (See also 349) *Tabulae Rudolphinae*. Ulm, 1627. Tables for calculating the positions of the planets, the sun, and the moon at any time. Based on Tycho Brahe's observations and incorporating Kepler's own discoveries, they were far more accurate than all previous tables.

1630s

355. VERNIER SCALE. Invented in 1630 by Pierre Vernier, a French military engineer. Its original applications were in astronomy but for technical reasons it did not come into general use until about the end of the century.

356. GALILEI, G. (See also 348) *Dialogo ....sopra i due massimi sistemi del mondo.* Florence, 1632. His celebrated *Dialogue concerning the two chief world-systems.*

357. BLAEU, W.J. (2762) *Institutio astronomica bipartita, de usu globorum coelestium et terrestrium.* Amsterdam, 1634. (Originally in Dutch. At least five reprintings of the Latin edition. French trans., 1642. English trans., 1654) An introduction to the use of the celestial and terrestrial globes that he made for sale. It was in two parts, the first based on the Ptolemaic scheme and the second on the Copernican.

358. WILKINS, J. (2243) **(a)** *Discovery of a new world ... in the moon.* London, 1638. **(b)** *Discourse of a new planet ... our earth....* London, 1640. (The two books were in effect two parts of the one book and were so treated in the two later editions) Important popularizations of the new astronomy and the new science generally. Capably and effectively written and based on a wide knowledge of the relevant literature (cf. 508)

359. GASCOIGNE, W. (2237) Though he never published anything his surviving correspondence and descriptions of his instruments establish that in the late 1630s he had adapted the telescope for use as a measuring instrument in positional astronomy. A few years later he was killed in the Civil War and his improvements to the telescope were re-invented by Auzout and Picard in the late 1660s (see 370).

1640s

360. BOULLIAU, I. (1997) *Astronomia philolaica.* Paris, 1645. A partial acceptance of Kepler's discoveries. Unlike nearly all astronomers at the time he accepted the ellipticity of orbits, but he rejected Kepler's physical approach and proposed a very implausible geometrical hypothesis. This was later made redundant by Ward (363).

361. HEVELIUS, J. (2941) *Selenographia; sive, Lunae descriptio.* Danzig, 1647. A detailed description and map of the moon. Observations of some of the planets were also included, and an account of his excellent instruments.   (See also 371)

1650s

362. RICCIOLI, G. (1887) *Almagestum novum.* Bologna, 1651. (Despite the title he was in favour of Tycho Brahe's system) A bulky collection of significant astronomical observations and discussions. Among much else it included a lunar map (drawn by his pupil, Grimaldi) which introduced the kinds of names of lunar topographical features which are still used.

363. WARD, S. (2244) *Astronomia geometrica.* Oxford, 1656. Proposed an alternative formulation of Kepler's law of areas that was more tractable than Kepler's. Until the advent of Newtonianism it was generally used in planetary computations.

364. HUYGENS, C. (2772) **(a)** *De Saturni luna observatio nova.* The Hague, 1656. Discovery of a satellite of Saturn (later named Titan). **(b)** *Systema Saturnium.* The Hague, 1657. A clarific-

ation of the earlier, rather puzzling, observations of Saturn by Galileo and others, with the explanation that the planet is surrounded by a thin ring inclined at a considerable angle to the plane of the planet's orbit around the sun.

1660s

365. CASSINI, G.D. (1894) During his period in Bologna he published a large number of observations from 1656 onward, many of which were of considerable importance--notably those of the visible features of Jupiter and Mars which established the rotation periods of these planets. His chief publication before he left Italy was *Ephemerides Bononienses Mediceorum siderum* (Bologna, 1668)--tables of the motions of the main satellites of Jupiter (a task that Galileo and others had attempted) which were widely used for many years for determination of longitude. (See also 374)

366. STREETE, T. (2249) *Astronomia Carolina. A new theorie of the celestial motions.* London, 1661. (Many later editions) "One of the most popular expositions of astronomy in the second half of the seventeenth century, it served as a textbook for Newton, Flamsteed, and Halley ... an important vehicle for the dissem-- ination of Kepler's astronomical ideas, which were as yet by no means generally accepted." (*DSB*)

367. HORROCKS, J. (2238) *Venus in sole visa.* Danzig, 1662. (Post-- humous; edited with notes by J. Hevelius) An account dating from 1639 of the first observation of a transit of Venus across the sun (which he had predicted). (See also 375)

368. PETIT, P. (2004) **(a)** *Dissertation sur la nature des comètes.* Paris, 1665. **(b)** "Lettre touchant une nouvelle machine pour mesurer exactement les diamètres des astres"--a paper in the *Journal des Sçavans*, 1667, which gave an account of his filar micrometer (also invented independently by Auzout and Picard).

369. BORELLI, G.A. (1885) **(a)** *Del movimento della cometa....* Pisa, 1665. A short account of a recent comet in which he suggested that the comet might have followed a parabolic path. **(b)** *Theor-- icae Mediceorum planetarum.* Florence, 1666. A work in which he considered the physical causes of the orbital motions of the satellites of Jupiter, and analogously of the planets around the sun. He suggested, among other things, that there is a balance between a centrifugal tendency and a centripetal tend-- ency.

370. PICARD, J. (2005) In 1666, in collaboration with Auzout, he per-- fected the use of a moveable-wire micrometer attached to a tele-- scope, thereby making possible accurate measurements of angular widths or small separations of celestial objects. The final step in converting the telescope into a precision instrument for positional astronomy--in place of the open sights used hitherto--was taken by Picard in 1667 when he fitted telescopic sights to the quadrants and sectors used in making angular measurements. (See also 373)

371. HEVELIUS, J. (See also 361) *Cometographia: Totam naturam comet-- arum ... exhibens.* Danzig, 1668. An extensive account of the comets of the two previous centuries. (See also 377)

372. NEWTON, I. (2265) Constructed the first reflecting telescope in
     1668.   (See also 382)

1670s

.... PARIS OBSERVATORY. See 2022.

.... GREENWICH OBSERVATORY. See 2271.

373. PICARD, J. (See also 370) **(a)** *Mesure de la terre*. Paris, 1671.
     An account of a project, undertaken in 1668-70 on behalf of the
     newly-established Académie des Sciences, to measure an arc of
     meridian in order to obtain an accurate value of the radius of
     the earth. Picard measured the distance from Amiens to a spot
     a few miles south of Paris very accurately by means of triang-
     ulation, using instruments fitted with the telescopic sights
     he had designed. The resulting figure for the earth's radius
     led to major improvements in cartography (and enabled Newton
     to verify his theory of gravitation in 1684).
     (b) *Voyage d'Uraniborg*. Paris, 1680. An account of another
     project undertaken on behalf of the Academy in 1671-72. This
     was a journey to Tycho Brahe's old observatory at Uraniborg in
     order to determine its co-ordinates accurately so that Tycho
     Brahe's observations could be more precisely compared with those
     to be made at the new Paris Observatory.

374. CASSINI, G.D. (2014) From the opening of the Paris Observatory
     in 1672 until the end of the century he carried out a great
     deal of research which he reported to the Academy in numerous
     memoirs. His most notable work was on the satellites and ring
     of Saturn, on new tables of atmospheric refraction, and on the
     development of the geodetic programme that had been initiated
     by Picard (373a)

375. HORROCKS, J. (See also 367) *Opera posthuma*. London, 1672-73.
     He died in 1641 before he could publish any of his important
     work. His manuscripts circulated widely from the late 1650s
     and were finally printed in this publication under the auspices
     of the Royal Society.

376. MONTANARI, G. (1898) *Discorso accademico sopra la sparizione di
     alcune stelle*. Bologna, 1672. An account of variable stars,
     especially Algol (β Persei).

377. HEVELIUS, J. (See also 371) *Machina coelestis*. 2 parts. Danzig,
     1673-79. Part 1 was a detailed account of his instruments and
     techniques; because of their excellence it was of great interest
     to other astronomers. Part 2 was a large collection of observ-
     ational data.   (See also 383)

378. RÖMER, O. (2023) From his observations in 1675 of the eclipses
     of Io, the first satellite of Jupiter, he established the finite
     velocity of light (as against the prevailing view of its instan-
     taneous propagation) and made an approximate estimate of its
     value.

379. HALLEY, E. (2274) *Catalogus stellarum australium*. London, 1678.
     A catalogue of southern stars which he compiled from his observ-
     ations made on the island of St. Helena when he was still a
     student.   (See also 386)

380. RICHER, J. (2015) Was sent by the Paris Academy in 1671 to make observations at Cayenne, in French Guiana (about 5° north of the equator). During the year he spent there he made observations of Mars which, when later compared with corresponding observations made in Paris, yielded a good estimate of the parallax of Mars and hence of the sun, and in turn a good estimate of the distance of the earth from the sun and of the dimensions of the solar system--the first such measurement ever achieved.

  Another set of observations Richer made, in accordance with the instructions given him by the Academy, also had a momentous result. He found that a pendulum which had beaten seconds exactly in Paris had to be shortened somewhat to do the same in Cayenne. The inference was that the force of gravity near the equator is somewhat less than at higher latitudes. From this it could be concluded that the diameter of the earth is greater at the equator than at the poles.

1680s

381. FLAMSTEED, J. (2273) *The doctrine of the sphere*. London, 1680. Included his lunar theory.   (See also 390)

382. NEWTON, I. (See also 372) As well as being of unparalleled importance for terrestrial mechanics, his *Principia* (see 518) also established the foundations of celestial mechanics.

1690s

383. HEVELIUS, J. (See also 377) *Prodromus astronomiae*. Danzig, 1690. (Posthumous) A catalogue of over 1500 stars. Much used by later astronomers.

1700s

384. CASSINI, J. (2039) During the period 1700-40, in addition to his work in geodesy (1029) he published a large number of papers in the *Mémoires* of the Paris Academy on astronomical subjects, as well as some books. They contained a number of substantial contributions (especially his announcement in 1738 of the proper motions of the stars) the value of which was however reduced by his unfortunate theoretical committments. "A timid Copernican but a convinced Cartesian and a fervent disciple of his father, Cassini fought unceasingly to defend the work of his father and to reconcile the facts of observation with the theory of vortices; he also never admitted the value of the theory of gravitation." (*DSB*)

385. GREGORY, D. (2279) *Astronomiae physicae et geometricae elementa*. Oxford, 1702. (English trans., 1715; reprinted 1726) The first textbook of astronomy on Newtonian principles.

386. HALLEY, E. (See also 379) *Astronomiae cometicae synopsis*. Oxford, 1705. Contained calculations of the paths of many comets previously recorded and the famous prediction that the comet of 1682 (and earlier) would return in 1758 (see 402). (See also 387)

1710s

387. HALLEY, E. (See also 386) **(a)** "An account of several nebulae...."
     Philosophical Transactions, 1715. Contained the suggestion
     that the nebulae consist of material spread out over vast areas
     of space and shining with its own light. **(b)** "Considerations
     of the change of the latitudes of some of the principal fixt
     stars", *Philosophical Transactions*, 1718. The discovery of the
     proper motions of stars.

388. MANFREDI, E. (1908) *Ephemerides motuum coelestium*. 2 vols.
     Bologna, 1715. Ephemerides for the period 1715-25; the work
     was followed by a similar one published in 1725 for the period
     1726-50. Manfredi's ephemerides were popular with practical
     astronomers, being well arranged and unusually comprehensive.
     (See also 391)

1720s

389. REFLECTING TELESCOPE. The type of reflecting telescope intro-
     duced by Newton in 1668 was first developed by J. Hadley, a
     skilled optician and F.R.S. In 1721 he completed two excellent
     instruments, one of which he donated to the Royal Society.

390. FLAMSTEED, J. (See also 381) **(a)** *Historia coelestis Britannica*.
     3 vols. London, 1725. The greatest star catalogue since Tycho
     Brahe's and the first comprehensive one based on telescopic
     observation. It contained the places of over 3000 stars, deriv-
     ed from a vast number of observations of hitherto unparalleled
     accuracy. **(b)** *Atlas coelestis*. London, 1729. A star atlas to
     accompany the *Historia*.

391. MANFREDI, E. (See also 388) *De annuis inerrantium stellarum
     aberrationibus*. Bologna, 1729. Discovery of the annual aber-
     ation of fixed stars, a phenomenon that was explained in the
     same year by Bradley.

392. BRADLEY, J. (2297) "Account of a newly discovered motion of the
     fixed stars", *Philosophical Transactions*, 1729. A discovery
     which he correctly interpreted as being due to the 'aberration'
     of light. It provided the first observational evidence of the
     earth's rotation around the sun. Bradley was able to deduce
     from it a reasonably good value for the velocity of light.
     (See also 394)

1740s

393. EULER, L. (2537) **(a)** *Theoria motuum planetarum et cometarum*.
     Berlin, 1744. A major treatise on the calculation of orbits.
     (b) In a series of papers in the periodicals of the Berlin and
     St. Petersburg Academies, beginning in 1747 and continuing
     almost to his death in 1783, he made repeated attempts to calc-
     ulate the perturbations of planetary orbits caused by the mutual
     gravitation of Jupiter and Saturn, as well as of the earth and
     other planets. Though he was unable to solve these intractable
     problems he had some success on subsidiary points.
     (See also 398)

394. BRADLEY, J. (See also 392) "On the apparent motion of the fixed
     stars", *Philosophical Transactions*, 1748. Discovery of the

nutation of the earth's axis, resulting from the changing
direction of the gravitational pull of the moon. (See also 396)

395. ALEMBERT, J. d'. (2058) *Recherches sur la précession des équi-
noxes et sur la nutation de l'axe de la terre.* Paris, 1749.
A rigorous mathematical demonstration that precession and
nutation are due to the action of the moon's gravitational
attraction on the earth's equatorial bulge.   (See also 400)

1750s

396. BRADLEY, J. (See also 394) Some years after his appointment as
astronomer royal he was able to get a sizable grant from the
Admiralty to re-equip Greenwich Observatory. With the new
instruments he carried out an extensive series of highly accurate observations from 1750 until his death in 1762. Publication
of his results was delayed until 1798-1805 but after they had
been reduced by Bessel in 1818 (see 425) they were of fundamental importance for nineteenth-century astronomy.

397. CLAIRAUT, A.C. (2052) *Théorie de la lune.* St. Petersburg, 1752.
Incorporated the first approximate solution of the three-body
problem. His *Tables de la lune* (Paris, 1754) were based upon
it.   (See also 402)

398. EULER, L. (See also 393) *Theoria motus lunae, exhibens omnes
ejus inaequalitates.* Berlin, 1753. His first lunar theory,
utilizing his approximate solution of the three-body problem.
It provided the basis for Tobias Mayer's lunar tables (399).
(See also 409)

399. MAYER, T. (2538) His new tables for the motion of the moon were
first published in the *Commentarii* of the Göttingen Society in
1753. They were based on Eular's lunar theory (398) and made
extensive use of observational data which were skilfully incorporated into the theoretical framework, yielding a set of tables
which were considerably more accurate that any hitherto. Mayer
sent them to the British Admiralty which, after further improvements, published them in 1770 as a important navigational aid
for the determination of longitude.   (See also 411)

400. ALEMBERT, J. d'. (See also 395) *Recherches sur différens points
importans du système du monde.* 3 vols. Paris, 1754-56. Dealt
chiefly with his lunar theory and his work on the three-body
problem which he had presented in memoirs to the Academy from
1747 onward (in competition with Clairaut and Euler). It also
included a set of lunar tables.

401. LACAILLE, N.L. de. (2053) *Astronomiae fundamenta.* Paris, 1757.
Included reduction tables and a catalogue of 400 of the brightest stars, observed and reduced with the greatest care. The
catalogue was far superior to all its predecessors.
(See also 405)

402. HALLEY'S COMET. In 1705 Halley had predicted that the comet of
1682 (and earlier) would return in 1758 or 1759, and had attributed the variations in the periods of its earlier appearances
to perturbations caused by the gravitational attraction of
Jupiter and Saturn. As the time of the predicted return app-

roached, Clairaut set to work to calculate the effects of the
two giant planets. In November 1758 he announced to the Paris
Academy that his calculations indicated that the comet would
pass its perihelion about the middle of April 1759, but due to
some difficulties in the calculations there might be an error
of a month either way. The returning comet was first sighted
in December 1758 and passed its perihelion on 13 March 1759.

   The success, both of Halley's prediction and of Clairaut's detailed calculations, met with general acclaim in intellectual circles everywhere and was a powerful confirmation of
the Newtonian system.

1760s

403. TRANSITS OF VENUS, 1761 and 1769. Halley had pointed out in the
late seventeenth century that the two transits of Venus predicted to occur in 1761 and 1769 could yield valuable astronomical
data. If a transit were observed from a number of stations in
different parts of the earth, and the time the planet took to
cross the face of the sun determined accurately, then it would
be possible to calculate the solar parallax; from the value of
the parallax the distance of the sun from the earth and the
dimensions of the solar system generally could easily be deduced.

   Accordingly, scientific bodies led by the Paris Academy
and the Royal Society of London, with the support of several
governments, organized expeditions to many parts of the earth
to observe the two transits--expeditions which were the first
international co-operative undertakings in the history of
science. The results were rather disappointing: for a number
of reasons (especially certain optical effects) it was not
found possible to get consistent figures for the time of passage
of the planet across the sun. Nevertheless the value of the
solar parallax, if still imprecise, was known much more definitely than previously. (See also 429)

404. LAMBERT, J.H. (2542) *Cosmologische Briefe über die Einrichtung
des Weltbaues.* Augsburg, 1761. Unaware of similar speculations
published by Thomas Wright in 1750 and Immanuel Kant in 1755,
Lambert suggested that the Milky Way might be a visual effect
produced by a lens-shaped conglomeration of thousands of stars.
There might even be, he thought, many such Milky Way systems.
His book, which contained further bold speculations going well
beyond astronomical knowledge, caused something of a sensation
and was translated into French, Russian, and English.

   Lambert's contributions to astronomy were not limited
to such speculations: he made many significant observations,
notably on some long-term fluctuations in the motions of
Saturn and Jupiter, and was the founder of the Berlin Academy's
Astronomisches Jahrbuch (which began in 1774 under the editorship of J.E. Bode).

405. LACAILLE, N.L. de. (See also 401) *Coelum australe stelliferum.*
Paris, 1763. (Posthumous) The first extensive catalogue of
the southern skies, based on his observations at the Cape of
Good Hope. It contained data on nearly 2000 stars as well as
a star map. (He also observed 8000 other stars but his observations were not reduced until the 1840s)

406. LALANDE, J.J. (2066) *Traité d'astronomie*. 2 vols. Paris, 1764. (3rd ed., 1791) An outstanding textbook. (See also 421)

407. LAGRANGE, J.L. (2548) In a series of papers published mostly between 1764 and 1786 (and mostly in the *Mémoires* of the Berlin Academy) he made many contributions to celestial mechanics—chiefly in lunar theory, planetary theory (especially periodic and secular inequalities), the stability of the solar system, and cometary perturbations. Much of this work was occasioned by the prize competitions held periodically by the Paris Academy to stimulate research in celestial mechanics. He won the Academy's prize five times (1764, 1766, 1772, 1774, and 1780).

408. BODE, J.E. (2553) *Anleitung zur Kenntnis des gestirnten Himmels*. Hamburg, 1768. (11th ed., Berlin, 1858) A highly successful popularization. (See also 410)

1770s

409. EULER, L. (See also 398) *Theoria motuum lunae, nova methodo pertractata*. St. Petersburg, 1772. His second lunar theory, worked out on a different basis from his first (398). (The elaborate calculations for it he performed in his head because by this time he was blind)

410. BODE, J.D. (See also 408) Confirmed and publicized in 1772 a suggestion made six years earlier by the Wittenberg professor, J.D. Titius, concerning an arithmetic expression for the relative distances of the planets from the sun. This empirically-derived relationship came to be known as the Titius-Bode law (or often as Bode's law). (See also 414)

411. MAYER, T. (See also 399) *Opera inedita*. Göttingen, 1775. Ed. by G.C. Lichtenberg) Contained papers and treatises on a variety of astronomical topics that were left unpublished at his early death in 1762. The most notable item was a detailed map of the moon which embodied his earlier studies of its surface.

412. LAPLACE, P.S. (2075) From 1775 he published many memoirs on celestial mechanics which were later synthesized in his celebrated *Mécanique céleste* (418b).

1780s

413. GOODRICKE, J. (2321) Discovered the periodicity of variable stars in 1782. He noticed the (previously observed but forgotten) variability of Algol (β Persei) and established that the variation in brightness is periodic, measuring the period very accurately. He suggested that the changes in the star's light were caused by a dark body revolving around it (a suggestion not confirmed until 1880). A few years later he discovered the variability of two other stars, β Lyrae and δ Cephei.

414. BODE, J.E. (See also 410) *Vorstellung der Gestirne*. Berlin, 1782. A widely used star atlas. It contained over 5000 stars. (See also 419)

415. *CATALOGUE des nebuleuses et des amas d'étoiles*. Published in the *Connaissance des Temps* for 1784. The first catalogue of nebulae and star clusters; it included 103 of them. By C. Messier. His numbering system is still used for the well-known objects.

416. HERSCHEL, W. (2314) His discovery of the planet Uranus in 1781
     made him famous but most of his important work was concerned
     with stars. It was published in some seventy papers in the
     Philosophical Transactions between 1780 and 1818, the chief
     features being as follows.
          (a) Over the period 1783–1802 he compiled catalogues
     listing 2500 nebulae and star clusters (in place of the hundred
     or so known hitherto). Finding that many nebulae could be
     resolved by his powerful telescopes (see 2314) into clusters of
     innumerable stars he assumed that all of them could be. Later
     however he discovered that some nebulae consist of clouds of
     "luminous fluid" from which, he thought, stars gradually conden-
     sed. Though he considered that many clusters and nebulae belong
     to our own star system he suspected that some were external
     "island universes."
          (b) In 1784–85 he developed his method of "star-gauging"
     or counting the number of stars visible in different areas of the
     sky, and thereby established the roughly lens-like shape of our
     star system. Such a conclusion had been suggested earlier from
     the nature of the Milky Way (cf. 404) but Herschel was the first
     to establish it from numerical data.
          (c) It had been known since the early eighteenth century
     that certain stars have (very small) 'proper motions'--i.e. that
     they are moving slowly in relation to the general body of stars
     --and since it was now generally assumed that the sun is a star
     there was the possibility that it too might be moving. By a
     very ingenious method Herschel was able to establish in 1783
     that the sun does in fact move.
          (d) He discovered the existence of double stars--i.e.
     pairs of stars very close together--and catalogued over 800 of
     them. Later he was able to show in some cases that they are
     revolving around each other.
          (e) A few variable stars had been known earlier but he
     discovered additional ones and made a study of them.

1790s

---

417. OLBERS, H.W.M. (2559) Published in 1797 a book describing a
     simplified method of computing cometary orbits from a few
     observations, hitherto a very laborious process. His method
     was widely used for a century or more.   (See also 420)

418. LAPLACE, P.S. (See also 412) **(a)** *Exposition du système du monde.*
     2 vols. Paris, 1796. (5th ed., 1824. The 6th ed. was revised
     by Laplace but did not appear unti 1835. English trans., 1809
     and 1830) One of the most successful popularizations in the
     history of science. A finely written, non-mathematical account
     of the solar system, based on his *Mécanique céleste*. The con-
     cluding chapter contained his famous speculation about the
     origin of the solar system--his 'nebular hypothesis.'
          (b) *Traité de mécanique céleste.* Paris, 1799–1805,
     Vols I–IV; 1823–25, Vol. V, in several parts. (2nd ed., 4 vols,
     1829–39. German trans. of Vols I and II, 1800–02. English
     trans., with a valuable commentary by Bowditch, 1829–39) A
     famous classic which summed up all that had been accomplished in
     the field since Newton. It reduced planetary astronomy, in

principle at least, to a set of problems in mechanics, founded on the law of universal gravitation. The treatment was so complete that little could be added to the subject for generations afterwards.

1800s

419. BODE, J.E. (See also 414) *Uranographia*. Berlin, 1801. A star atlas containing over 17,000 stars, the largest hitherto published. It was the first to contain the nebulae, clusters, and double stars discovered by William Herschel.

420. OLBERS, H.W.M. (See also 417) Took a leading part in the discovery of asteroids (or minor planets) in the space between Mars and Jupiter--an apparently vacant space that was then being searched by several astronomers. The first asteroid was discovered in 1801 by the Italian astronomer, G. Piazzi, who named it Ceres. He was prevented from keeping track of it, and it was lost. Olbers re-discovered it in 1802, and in the following year discovered a second one (Pallas). Others were subsequently found, including another (Vesta) by Olbers in 1807. From the unusual characteristics of the orbits of these asteroids he inferred that they were fragments of an original planet which had somehow been shattered. (See also 423)

421. LALANDE, J.J. (See also 406) *Bibliographie astronomique*. Paris, 1803. A massive bibliography which also included a review of the subject over the preceding twenty years.

422. GAUSS, C.F. (2570) *Theoria motus corporum coelestium....* Hamburg, 1809. An outstanding work in celestial mechanics in which he developed his method of calculating the orbit of a planet from only three complete observations of its position. It made use of his famous statistical technique of least squares for the combination of observations. (See also 424)

1810s

423. OLBERS, H.W.M. (See also 420) Discovered in 1815 the comet named after him (as well as three others at various times). He suggested that the tail of a comet is a cloud of matter that had been expelled from its nucleus, and that the tail always points away from the sun because of some kind of repulsion by the sun. (See also 430)

424. GAUSS, C.F. (See also 422) The new observatory at Göttingen was completed in 1816 and thereafter Gauss, as its director, worked steadily for the rest of his life as an observational astronomer (in addition to his researches in numerous other fields of science).

425. BESSEL, F.W. (2578) *Fundamenta astronomiae pro anno 1755 deducta ex observationibus ... James Bradley*. Königsberg, 1818. An elaborate, systematic reduction of the important series of observations made by Bradley in the 1750s (396). Bessel's work constituted a major contribution to practical astronomy, both in providing the first highly accurate star positions and in establishing the theory of astronomical measurement and reduction of observations. (See also 431)

426. ENCKE, J.F. (2579) Established in 1819 that comet sightings of
     1786, 1795, 1805, and 1818 were all of the same comet. Encke's
     comet, as it came to be known, was the second (after Halley's)
     to be recognised as periodic.   (See also 429)

1820s

427. CAPE OBSERVATORY. The Royal Observatory at Cape Town was estab-
     lished in 1820 to make observations in the southern hemisphere
     strictly comparable with those made at Greenwich.

428. HERSCHEL, J. (2346)   During the period 1821-23, in collaboration
     with James South, he continued his father's pioneering work on
     double stars, re-observing the known ones (to detect changes of
     position), discovering new ones, and verifying their orbital
     motion. For their catalogue of 380 double stars, published in
     1824, the two authors were awarded prizes by the Astronomical
     Society and the Paris Academy.   (See also 433)

429. ENCKE, J.F. (See also 426) Re-calculated in 1822-24 the results
     of the observations of the transits of Venus made in the 1760s
     (see 403). By applying Gauss' method of least squares to the
     observations (and using improved determinations of the longi-
     tudes of some of the observing stations) he obtained a figure
     for the solar parallax–and hence of the distance of the earth
     from the sun––which was generally accepted for many years. (It
     was about 2.5% greater than the presently accepted figure)
     (See also 432)

430. OLBERS, H.W.M. (See also 423) Put forward in 1823 the famous
     paradox that bears his name: if the universe is infinite (as
     was generally believed) then the infinite number of stars
     should cover the whole sky, making it uniformly bright. His
     own solution was that space contains some dark matter which
     obscures some of the starlight.

1830s

431. BESSEL, F.W. (See also 425) (a) *Tabulae Regiomontanae*. Königs-
     berg, 1830. Ephemerides for thirty-eight stars over the period
     1750-1850. They provided a reference system for detailed stud-
     ies of the motion of members of the solar system as well as of
     proper motions of stars. They were found to be very useful.
     (b) Announced in 1838 the first definite detection of the par-
     allax of a fixed star (61 Cygni) and consequently the first
     determination of the distance of a star from the earth. His
     achievement in an undertaking that had baffled astronomers for
     centuries brought him high praise.   (See also 439)

432. ENCKE, J.F. (See also 429) (a) Was the first to develop, in 1830,
     a method for computing the orbits of double stars. (b) Discov-
     ered in 1837 a small gap––now called Encke's division––within
     the outermost ring of Saturn.

433. HERSCHEL, J. (See also 428) (a) In the early 1830s he continued
     research along the lines established by his father, especially
     the observation of nebulae, clusters, and double stars. In
     1833 he published a massive revision and extension of his

father's catalogue of nebulae and clusters, and in 1836 his
sixth catalogue of double stars appeared. **(b)** During the period
1834-38 he worked at a private observatory he established at
the Cape of Good Hope, doing for the southern hemisphere what
his father had done for the northern. "Never before or since
did one astronomer in so short a time reap so rich a harvest.
No telescope with anything like the space-penetrating power of
the Herschel instruments had ever before been trained upon the
southern constellations." (R.L. Waterfield) The results of his
four years of intense activity at the Cape were finally published in 1847, after much labour in reducing and arranging the
observations, as *Results of astronomical observations made ...
at the Cape of Good Hope. Being a completion of a telescopic
survey of the whole surface of the visible heavens, commenced
in 1825.* He was the only person who had ever swept the whole
heavens from pole to pole. (See also 445)

434. *THE MECHANISM of the heavens.* London, 1831. By Mary Somerville.
A highly praised work by a brilliant expositor of science (cf.
40). It was used as a textbook at Cambridge and elsewhere for
many years.

435. *MAPPA selenographica.* 4 vols. Berlin, 1834-36. By W. Beer and
J. Mädler. A very detailed and accurate map of the moon. The
accompaning text included measurements of the diameters of many
lunar craters and the relative heights of lunar mountains.

436. STRUVE, W. (2982) **(a)** In a major paper in 1837 he presented his
observations of more than 3000 double stars, with measurements
of their angular separation and details of other features.
(b) In 1838 he succeeded in measuring the parallax of a fixed
star (Vega), shortly after Bessel's similar achievement.

437. ARGELANDER, F. (2596) *Über die eigene Bewegung des Sonnensystems.*
St. Petersburg, 1837. From a study of the proper motions of
nearly 400 stars he confirmed William Herschel's deduction,
hitherto controversial, that the solar system is moving through
space. (See also 441)

438. HANSEN, P.A. (2595) *Fundamenta nova investigationis....* Gotha,
1838. Developed a highly accurate theory of the moon's motion
which became the basis of tables published by the British Admiralty in 1857 and used for the *Nautical Almanac* until 1923.

1840s

439. BESSEL, F.W. (See also 431) His papers published in the early
1840s included the following. **(a)** The detection of small
irregularities in the orbit of Uranus which he suggested were
caused by an unknown planet beyond it.. **(b)** Observations of
minute wave-like motions of Sirius from which he inferred that
the star has a dark companion (which was actually observed in
1862). **(c)** Studies of the orbital periods of each of the main
moons of Jupiter from which he calculated the planet's mass and
volume, concluding that its density is only 1.35 times that of
water.

440. SUNSPOT CYCLE. The eleven-year cycle was discovered in 1843 by

the amateur astronomer, S.H. Schwabe, from statistics of sunspot counts that he had compiled over a period of seventeen
years.

441. ARGELANDER, F. (See also 437) In 1844 he began a study of variable stars (of which only eighteen were known at the time). By
devising a simple means of estimating their changing brightness
he established the subject as a new branch of astronomy.
(See also 450)

442. ADAMS, J.C. (2385) Began in 1843 an investigation of the irregularities in the motion of Uranus and came to the conclusion
that they were caused by the presence of an unknown planet. In
the following year he calculated the orbit and mass of the
hypothetical planet and in 1845 gave copies of his calculations
to the director of the Cambridge Observatory (J. Challis) and
the astronomer royal at Greenwich Observatory (G.B. Airy). No
search was made for the planet however and it was found by a
German astronomer at the position predicted by Le Verrier who
had independently come to the same conclusions as Adams.

443. LE VERRIER, U.J.J. (2151) From the late 1830s he devoted himself
to celestial mechanics, beginning with an extension of the work
of Laplace on the stability of the solar system and going on to
the theory of Mercury's orbit and the perturbation of comets.
In 1845 he began investigating the irregularities in the orbit
of Uranus and concluded that they were due to the perturbing
effect of an unknown planet. He calculated its position in
1846 and asked a number of astronomers to look for it. The new
planet was promptly found by J.G. Galle, of the Berlin Observatory, within a degree of the predicted position. It was later
named Neptune.   (See also 442)

444. PARSONS, W. (*Lord* ROSSE) (2356) During the late 1840s he used his
giant telescope (see 2356) to make observations of nebulae that
had hitherto been impossible. He was able to resolve many of
them into clusters of stars, but many others could not be resolved. Some of them he found to have a spiral structure which
he depicted in excellent drawings. His other observations
included the Crab Nebula, which he named, and the Great Nebula
in Orion, of which he made a detailed study.

445. HERSCHEL, J. (See also 433) *Outlines of astronomy*. London, 1849.
(Many editions and translations) A very popular introduction.

_____

                              1850s
_____

.... FOUCAULT, L. Improvements in the construction of reflecting
telescopes: see 694b.

446. ROCHE, E.A. (2161) **(a)** Calculated mathematically in 1850 the
minimum distance ('the Roche limit') that a satellite can
approach the planet it orbits without being torn apart by tidal
forces. If the satellite and planet are of similar density the
theoretical limit is 2.44 times the radius of the planet. (An
important case is Saturn's ring, the outer edge of which has a
radius 2.3 times that of Saturn) **(b)** Analyzed in 1859 the
effect on a comet's envelope of a hypothetical force emanating

from the sun. He was thus able to give an explanation of the
shape of comets many years before the discovery of the physical
cause. (See also 463)

447. CARRINGTON, R.C. (2386) His systematic and detailed observations
of sunspots, carried out over the period 1853-61, were published
in his *Observations of the spots on the sun* (London, 1863). He
established: **(a)** the direction of the sun's axis very accurately,
(b) that the sun rotates somewhat faster at its equator than
nearer its poles, **(c)** that the statistical distribution of the
sunspots is low near the equator and low also above 35° latitude,
and **(d)** that the zones of spots move towards the equator (where
they disappear) as the eleven-year cycle progresses. In 1859
he was the first to observe a solar flare; it lasted only five
minutes and was very brillant.

448. HELMHOLTZ, H. (2631) Suggested in 1854 that the energy radiated
from the sun is derived from the mutual gravitation of its parts
as a result of shrinkage. Calculations based on reasonable
assumptions appeared to support Helmholtz' theory and it was
generally accepted until the twentieth century (though in 1862
Kelvin calculated that on this theory the life history of the
sun would comprise only fifty million years).

449. MAXWELL, J.C. (2413) Won a major prize at Cambridge in 1859 with
an essay on the mathematical theory of Saturn's rings. He
showed that the rings could not be solid, as had been thought
hitherto, but must consist of swarms of separate particles.
Many years later his deduction was confirmed observationally
(485b and 486b).

450. ARGELANDER, F. (See also 441) *Bonner Sternverzeichnis.* 3 vols.
Bonn, 1859-62. (Often known as the 'Bonner Durchmusterung' or
the 'Bonn Catalogue') A catalogue and atlas of some 325,000
stars, down to a magnitude of 9.5, located over the northern
celestial hemisphere. It surpassed all former catalogues in
completeness and reliability, and immediately became a tool of
first importance for all observatories. It is still used today
for some purposes (it was reprinted in 1950).

1860s

451. ASTRONOMICAL PHOTOGRAPHY. From the middle of the nineteenth century there were numerous experiments in the application of photography to astronomy. One of the main difficulties was that
ordinary telescope lenses were not optically suitable for the
photographically-active component of the light and so did not
give sharp images. In the early 1860s the American experimenter,
L.M. Rutherfurd, devised a telescopic lens system optically
designed for photography—a camera with a telescope as a lens—
which produced usable photographs. Thereafter photography was
progressively adopted by astronomers, every improvement in
photographic technique being exploited (cf. 461, 462, 465, 480).

452. ASTRONOMICAL SPECTROSCOPY. In October 1859, in the course of his
physical researches, Kirchhoff (704) had shown that the chemical
composition of the sun (or, more exactly, its surface or atmosphere) could be deduced from the dark (absorption) lines in the

solar spectrum. This discovery opened the way to astronomical spectroscopy which was soon taken up by Huggins and Secchi, followed by many others.

453. ASTROPHOTOMETER. Invented in 1860 by J.K.F. Zöllner.

454. SCHIAPARELLI, G.V. (1947) **(a)** In 1861, almost at the beginning of his career, he discovered the asteroid, Hesperia.
(b) During the 1860s he studied comets and meteor showers, and firmly established (what had long been suspected) that meteor swarms are the remnants of comets which have disintegrated. (See also 471)

455. SPÖRER, G.F.W. (2628) In the early 1860s he independently made the same discoveries as R.C. Carrington (447). He continued for thirty years to study the regularities in the formation, movements, and distribution of sunspots.

456. HUGGINS. W. (2409) One of the first to apply spectroscopy to astronomy. His chief results were as follows.
(a) In collaboration with W.A. Miller (professor of chemistry at King's College, London, who was already working in chemical spectroscopy) he attached a spectroscope to a telescope and began observing stellar spectra. Their results were published in 1863-64 and showed that, while stars differ considerably in their chemical composition, they contain the same elements as the earth and the sun. The fact that the spectra are absorption spectra they interpreted to mean (following Kirchhoff) that the light from a hot star passes through an absorbing surface layer, or atmosphere, of somewhat lower temperature.
(b) In 1864 Huggins examined the spectra of many nebulae and found that some had spectra similar to those of stars (these, he inferred, could be resolved into star clusters) but others exhibited only a few bright (emission) lines. The latter nebulae, he concluded, were not star clusters but consisted of luminous gas.
(c) A nova appeared in 1866 and he found its spectrum to contain bright hydrogen lines superimposed on a normal spectrum of numerous dark (absorption) lines. He interpreted this to mean that the nova had expelled a shell of hydrogen gas which was at a higher temperature than the star's surface.
(d) Continuing his spectroscopic explorations, Huggins found in 1868 that comets have spectra similar to the spectrum of a candle flame, and inferred that they contain luminescent carbon-containing gases. He later found that meteors emit similar spectra.
(e) Perceiving the possibility of observing the Doppler effect (678) in the spectra of stars, he succeeded in 1868 in detecting a minute displacement of the F hydrogen line in the spectrum of Sirius; this he interpreted to mean that the star is receding from the solar system at high speed. He later found similar displacements in the spectra of some thirty other stars. (The value of the method was not established however until accurate measurements of the Doppler shifts became possible--see 482).   (See also 462)

457. SECCHI, A. (1944) One of the first to apply spectroscopy to
     astronomy. During the period 1863-69 he examined the spectra
     of over 4000 stars and found that they could nearly all be
     classified into four spectral types, ranging from the high-
     temperature white or bluish stars to the low-temperature red
     ones. His classification was generally accepted.
     (See also 469)

458. JANSSEN, P.J.C. (2173) During the mid-1860s he elucidated the
     effect of the earth's atmosphere on the spectrum of solar
     radiation, demonstrating selective absorption by oxygen and
     water vapour. Subsequently he used the results to detect the
     presence of water vapour in the atmosphere of Mars.
     (See also 459 and 465)

459. SOLAR PROMINENCES. The solar eclipse of 1868 was the first to
     be observed spectroscopically. The spectrum of the chromosphere
     and prominences was found to consist of a few bright (emission)
     lines, showing that they consist of gas in which hydrogen was
     identified. After the eclipse P.J.C. Janssen and J.N. Lockyer
     independently devised a method for observing a spectrum of a
     prominence at the edge of the sun's disc in ordinary daylight,
     without need of an eclipse. The method, which was improved in
     the following year by W. Huggins, proved to be very valuable
     for subsequent studies of the chromosphere.

460. LOCKYER, J.N. (2426) In 1868, following his discovery of a
     method for observing solar prominences (459), he found a
     yellow line in the spectrum of the prominences and chromosphere.
     He attributed the line to an unknown element which he named
     helium. His conclusion was not generally accepted until 1895
     when helium was discovered on earth (989).   (See also 466)

1870s

461. DRAPER, H. (3038) Made the first successful photograph of the
     spectrum of a star in 1872 and two years later photographed
     the transit of Venus. He continued developing the techniques
     of astronomical photography and made many photographs of stellar
     spectra and various objects (including comets and the Orion
     nebula) until his early death in 1882. He did much to make
     photography a vital part of astronomy.

462. HUGGINS, W. (See also 456) From the beginning of his spectros-
     copic researches he had attempted to photograph stellar spectra
     but the techniques of photography were not adequate to the task
     until the 1870s. Following Draper's success in 1872 he used
     photography extensively and by 1879 was photographing ultra-
     violet spectra.

463. ROCHE, E.A. (See also 446) Made in 1873 the first rigorous
     mathematical analysis of Laplace's nebular hypothesis (418a)
     and as a result suggested some important additions which made
     the hypothesis more self-consistent.

464. TRANSIT OF VENUS, 1874. The first since that of 1769 (see 403).
     As before, many expeditions were sent to various parts of the

earth to determine the time taken by the planet to cross the
face of the sun. This time however photographic methods were
used (cf. 465).

465. JANSSEN, P.J.C. (See also 458) One of the chief pioneers of
   astronomical photography. In 1874 he took a rapid sequence
   of photographs of the transit of Venus, and from 1876 until
   1903 compiled a remarkable series of photographs of the sun's
   surface over that period.

466. LOCKYER, J.N. (See also 460) *Contributions to solar physics.*
   London, 1874. Included among other things his spectroscopic
   observations of sunspots which he concluded were cooler than
   other parts of the sun's surface. From his application of
   Doppler's principle to their spectra he deduced that they
   contain gases moving at very high velocities.

467. *L'ASTRONOMIE et les observatoires.* 5 vols. Paris, 1874-78. By
   G.A.P. Rayet et al. A survey of the equipment and historical
   development of the chief observatories of the world.

468. VOGEL, H.C. (2675) *Untersuchungen über die Spectra der Planeten.*
   Leipzig, 1874. Investigations of the spectra of the planets
   in order to gain information about their atmospheres.
   (See also 482)

469. SECCHI, A. (See also 457) *Le soleil.* 4 vols. Paris, 1875-77.
   An account of his extensive investigations of the sun, conducted
   at various times over the previous twenty years.

1880s

---

470. NEWCOMB, S. (3037) Nearly all his chief work was published during
   the 1880s and 1890s in the series *Astronomical papers prepared
   for the use of the "American ephemeris and nautical almanac"*
   which he had begun in 1879 with the purpose of providing, as he
   put it, "a systematic determination of the constants of astronomy from the best existing data, a reinvestigation of the
   theories of the celestial motions, and the preparation of
   tables, formulae, and precepts for the construction of ephemerides, and for other applications." Most of the papers and
   tables in the first seven volumes were written by him and they
   constitute one of the chief achievements in celestial mechanics
   in the late nineteenth century.

471. SCHIAPARELLI, G.V. (See also 454) During the oppositions of Mars
   from 1877 to 1890 he studied the topography of the planet as
   thoroughly as his telescopes permitted. His catalogue and map
   of all the visible features included the word *canali* which was
   incorrectly translated into English as 'canals' (instead of
   'channels') and gave rise to widespread and persistent speculation about whether the 'canals' had been made by intelligent
   beings.

472. DARWIN, G.H. (2433) Through the 1880s he published a series of
   papers in applied mathematics dealing with tidal friction as
   an important agent in the development of the earth-moon system
   and satellite systems generally. Though his main conclusions

are no longer accepted his work was significant as a pioneering effort to elucidate the evolution of the solar system (or aspects of it) and to use dynamical analysis for the purpose.

473. BARNARD, E.E. (3065) During the 1880s he discovered many comets and made numerous important observations (one of which proved the particulate nature of Saturn's ring). In 1889 he began a programme of photographing the Milky Way with a large-aperture camera; the photographs he accumulated revealed a wealth of new detail.   (See also 484)

474. PICKERING, E.C. (3043) **(a)** Laid the foundations of exact stellar photometry in 1879 with his invention of the meridian photometer which enabled the brightness of any star to be compared with that of a standard reference star (Polaris). With its use he compiled the first catalogue of photometric magnitudes, containing over 4000 stars; it was published in 1884 as *Harvard photometry*. He continued this cataloguing work for many years.
(b) Made a detailed photometric study of the variable star Algol in 1880 and showed that its fluctuations are due to a dark companion revolving around it. **(c)** In 1886 the Harvard Observatory, under his direction, instituted a major photographic survey of stellar spectra which continued for many years.
(d) Independently of Vogel (482b) he discovered a case of a spectroscopic binary in 1889.

475. LANGLEY, S.P. (3042) Around 1880 he invented the bolometer--a highly sensitive instrument for measuring radiant energy--with which he measured the intensity of solar radiation (as a function of wavelength) far into the infrared region. He also used it to measure the selective absorption of the earth's atmosphere. The bolometer was especially valuable for measurements in the far infrared.

476. *BIBLIOGRAPHIE générale de l'astronomie*. 2 vols in 3. Brussels, 1882-89. By J.C. Houzeau and A. Lancaster.

477. SEELIGER, H. (2702) The first since the Herschels to apply statistical analysis to the distribution of stars, a subject he developed in a series of papers from 1884 to 1909. He arrived at an estimate of the size of our stellar system (i.e. the Milky Way) but his figures were modified by later studies.

478. INTERNATIONAL Conference at Washington for the Fixing of a Prime Meridian and a Universal Day. 1884. The Conference decided on Greenwich as the common prime meridian for the world.

479. *LISTE générale des observatoires et des astronomes, des sociétés et des revues astronomiques*. Brussels, 1886. (3rd ed., 1890) By A. Lancaster.

480. CARTE DU CIEL. At the instigation of the director of the Paris Observatory, E.B. Mouchez, an International Conference for Astrophotography was held in Paris in 1887. It decided on the compilation by photographic means of a large-scale map of the heavens, together with a catalogue of all stars up to the eleventh magnitude (i.e. over two million stars). Twenty-two observatories in several countries took part in the enormous project which, with interruptions, took over fifty years.

481. *NEW GENERAL CATALOGUE of nebulae and clusters of stars.* London,
     1888. By the Danish-Irish astronomer, J.L.E. Dryer. A monu-
     mental work based on John Herschel's catalogue (433a). Two
     supplements (called 'index catalogues') were published in 1895
     and 1909.

482. VOGEL, H.C. (See also 468) **(a)** His accurate measurements, in
     1889, of the Doppler shifts in the spectra of many stars estab-
     lished the value of the method for determining the radial motion
     of stars towards or away from the solar system. (The original
     use of the method by Huggins in 1868 (456e) had given incon-
     clusive results because he had had to use visual measurement of
     the minute shifts. Now however it was possible to make much
     more accurate measurements photographically) **(b)** In the course
     of the foregoing work Vogel showed spectroscopically that Algol
     is a component of a double star system, being periodically
     eclipsed by its dark companion (as had already been shown photo-
     metrically by Pickering--474b). Many other examples of spectros-
     copic binaries were soon discovered, including some double stars
     that are so close together that the individual stars cannot be
     detected telescopically.

483. TISSERAND, F.F. (2187) *Traité de mécanique céleste.* 4 vols.
     Paris, 1889-96. (Vols 1 and 2 reprinted, 1960) A worthy
     successor to the great work of Laplace.

1890s

484. BARNARD, E.E. (See also 473) During the 1890s he built up a vast
     collection of photographs of comets which is still in use today.
     Of his many discoveries the most outstanding was that of Jupit-
     er's fifth satellite, in 1892.

485. DESLANDRES, H. (2196) **(a)** In the early 1890s, independently of
     --and almost at the same time as--the American, G.E. Hale, he
     invented the spectroheliograph which enabled photographs of the
     sun and its atmosphere to be made in monochromatic light of any
     desired wavelength. The invention proved very valuable for
     studies of the sun and in the following years Deslandres made
     extensive use of it. **(b)** In the mid-1890s, by measuring Doppler
     shifts, he was able to show that Saturn's ring rotates as a
     system of individual particles (not as a solid body, as had
     earlier been thought), and that the rotation of Uranus about
     its axis is in a direction different from that of all the other
     planets.

486. KEELER, J.E. (3066) **(a)** From 1890 until his untimely death in
     1900 he photographed and studied the spectra of hundreds of
     nebulae. He found that most of them have a spiral structure
     and was able to show spectroscopically that, like stars, they
     are moving towards or away from the solar system. **(b)** At the
     same time as Deslandres, and in the same way, he proved that
     Saturn's ring rotates as a system of individual particles.

487. POINCARÉ, H. (2194. The great mathematician) *Les méthodes nouv-
     elles de la mécanique céleste.* 3 vols. Paris, 1892-99. Inaug-
     urated a more rigorous treatment of celestial mechanics and

    developed powerful new mathematical techniques for the subject.
The work also included important contributions to the theory
of orbits and the classical three-body problem.

488. CONFÉRENCE Internationale des Étoiles Fondamentales. Paris, 1896.

489. GILL, D. (2432) **(a)** *Cape Photographic Durchmusterung*. 1896–1900.
A catalogue of over 450,000 stars of the southern hemisphere
and the first major work of its kind to be carried our photo-
graphically. The photographs were taken by Gill at the Cape
Observatory and the data were processed by J.C. Kapteyn in
Holland. **(b)** In 1897 Gill obtained a highly accurate value
for the solar parallax by means of heliometric observations,
carried out for several years earlier, of minor planets at or
close to opposition. His value for this important constant
remained in widespread use until 1968.

1.04 MECHANICS

Including rational (or analytical) mechanics and engineering mechanics.
Celestial mechanics is included under Astronomy.

1470s

490. ARISTOTLE. (See entry 1) *Mechanica*. A discussion of problems
in statics by one of Aristotle's early followers. (Mistakenly
attributed to Aristotle himself)

1480s

491. LEONARDO DA VINCI. (1824) His investigations in mechanics and
mechanical technology began in the 1480s and continued until
his death in 1519. However in this field, as in others, his
work was never brought to the stage of publication.

1500s

492. THOMAZ, A. (1965) *Liber de triplici motu*. Paris, 1509. "This
work shows Thomaz to be a mathematician and physicist of considerable ability who understood and organized the teachings of
fourteenth-century English calculators [especially Richard
Swineshead] and Parisian terminists, such as Oresme, making
them available to a wide audience of European scholars in the
sixteenth century." (*DSB*)

1530s

493. TARTAGLIA, N. (1828) *La nuova scienza*. Venice, 1537. The first
work to apply mathematics to gunnery.

1540s

.... ARCHIMEDES. *Opera*; printed 1543-44. See entry 94.

1550s

494. BENEDETTI, G.B. (1845) Published a succession of small works on
various aspects of mathematics, mechanics, and astronomy from
1553 onward. His chief work was his last--*Diversarum specu-
lationum mathematicarum et physicarum liber* (Turin, 1585).
His very original ideas in mechanics, especially in his discussion of falling bodies, make him the chief immediate precursor
of Galileo but his books were not widely read and his influence
upon Galileo and others appears to have been slight.

1570s

495. CARDANO, G. (1831) *Opus novum de proportionibus.* Basel, 1570.
     Included discussions of natural and violent motion, equilibria,
     and the moment of a force. It also described an experimental
     comparison of the resistance of air and water to the motion of
     a projectile.

496. MONTE, G. del. (1860) *Liber mechanicorum.* Pesaro, 1577. A
     major work on statics which, in humanist style, rejected the
     medieval work in the subject and returned to the Greeks, espec-
     ially Archimedes. Through this work and through personal con-
     tact del Monte had a major influence on Galileo.

497. *THÉATRE des instrumens mathématiques et mécaniques.* Lyons, 1579.
     A beautifully illustrated description of many kinds of machines
     and mechanisms as well as instruments. The introduction to the
     book stressed the importance of mathematics. The author was
     Jacques Besson, an instrument maker and teacher of mathematics.

1580s

498. STEVIN, S. (2760) **(a)** *De beghinselen der weeghconst.* Leiden,
     1586. (Latin trans. by W. Snel, 1605–08. French trans., 1634)
     Chiefly on statics in the Archimedean tradition. It included
     his discovery of the law of the inclined plane (or the parallel-
     ogram of forces). **(b)** *De beghinselen des waterwichts.* Leiden,
     1586. "The first systematic treatise on hydrostatics since
     Archimedes."

499. *LE DIVERSE ed artificiose macchine.* Paris, 1588. (Text in both
     Italian and French) A monumental work which, with the aid of
     nearly two hundred fine illustrations, described a variety of
     machines and mechanisms, including many kinds of pumps, mills,
     etc., and featured a preface extolling the value of mathematics.
     The author was A. Ramelli, a military engineer.

1590s

500. *MACHINAE novae.* Venice, [ca. 1595]. An illustrated description
     of various kinds of machinery. By F. Veranzio.

1610s

501. GALILEI, G. (1863) *Discorso ... intorno alle cose che stanno in
     sù l'acqua....* Florence, 1612. A work on hydrostatics in the
     Archimedean tradition.   (See also 506 and 507)

1620s

502. BALDI, B. (1864) *In Mechanica Aristotelis problemata exercit-
     ationes.* Mainz, 1621. (Posthumous; written about 1589) An
     original contribution to mechanics in the form of a commentary
     on the *Mechanica problemata* (incorrectly attributed to
     Aristotle--see 490).

503. CASTELLI, B. (1876) *Della misura dell'acque correnti.* Rome,
     1628. (Many later eds. English trans., 1661. French trans.,

1664) A pioneering work which is now regarded as constituting the beginning of modern hydraulics.

504. *LE MACHINE*. Rome, 1629. An illustrated description of many kinds of machines and mechanisms. By G. Branca.

.... MERSENNE, M. See 14 and 506.

1630s

505. ROBERVAL, G.P. de. (1995) *Traité de méchanique*. Paris, 1636. A small treatise on the composition of forces. He said it was only a sample of a larger work to appear later. However, as in mathematics (see 131), most of his work, which was considerable, was not published but disseminated informally.

506. GALILEO/MERSENNE. **(a)** *Les méchaniques*. Paris, 1634. A free translation by Mersenne, with additions of his own, of an unpublished treatise by Galileo, dating from about 1600. **(b)** *Les nouvelles pensées de Galilée*. Paris, 1639. Translation by Mersenne of Galileo's *Two new sciences* (507).

507. GALILEI, G. (See also 501 and 506) *Discorsi e dimonstrazioni matematiche intorno a due nuove scienze*. Leiden, 1638. (French trans.: see 506b. English trans. by T. Salusbury, 1665) The two new sciences were the engineering science of the strength of materials and the mathematical treatment of motion, the latter being Galileo's greatest achievement and one of the pillars of modern science.

1640s

.... TORRICELLI, E. See 137.

508. WILKINS, J. (2243) *Mathematical magick; or, Wonders that may be performed by mechanical geometry*. London, 1648. An important popularization emphasizing practical applications of mechanical devices, with numerous illustrations and a minimum of theory. It was based chiefly on the Latin writings of several Italian authors and of Mersenne (who was also an important source for Wilkin's popularization of the new astronomy--see 358).

1650s

509. HUYGENS, C. (2772) *Horologium*. The Hague, 1658. An account of his adaptation of the pendulum to the regulation of clocks. (Not to be confused with his later *Horologium oscillatorium*-- 515)   (See also 511)

1660s

.... PASCAL, B. Work on hydrostatics: see 590.

510. BORELLI, G.A. (1885) **(a)** *De vi percussioni*. Bologna, 1667. A detailed treatment of percussion, with which he also included many other problems of motion--of fluids, pendulums, vibrations, etc. **(b)** *De motionibus naturalibus a gravitate pendentibus*. Reggio, 1670. A discussion of *gravitas* which he regarded as the result of an external action exerted upon a body, and not an intrinsic principle of motion; he thus argued against *levitas* as a positive principle. He went on to discuss a number

of related topics such as siphons, pumps, the Torricellian
experiment, etc.

511. HUYGENS, C. (See also 509) "Règles du mouvement dans la rencontre
     des corps", *Journal des Sçavans*, 1669. A summary of his funda-
     mental work on the collision of elastic bodies. He had written
     a treatise on the subject--*De motu corporum ex percussione*--as
     early as 1656 but it was not published in full until after his
     death.   (See also 515)

512. WREN, C. (2263) "Lex collisionis corporum", *Philosophical Trans-
     actions*, 1669. An important discussion of the collision of
     elastic bodies.

513. WALLIS, J. (2242) *Mechanica; sive, De motu tractatus geometricus.*
     3 parts. London, 1669-71. A comprehensive treatise with an
     emphasis on percussion.

1670s

514. MARIOTTE, E. (2016) *Traité de la percussion ou choc des corps.*
     Paris, 1673. (3rd ed., 1684) An experimentally based account
     of the laws of inelastic and elastic impact, and their applic-
     ation to several problems in mechanics. It was the first exten-
     sive treatment of the subject and had a lasting influence.
     (See also 517)

515. HUYGENS, C. (See also 511) *Horologium oscillatorium.* Paris,
     1673. His chief work. "Contained a theory on the mathematics
     of curvatures, as well as complete solutions to such problems
     of dynamics as the derivation of the formula for the time of
     oscillation of the simple pendulum, the oscillation of a body
     about a stationery axis, and the laws of centrifugal force for
     uniform circular motion." (J. Herivel)   (See also 520)

516. HOOKE, R. (2266) **(a)** *Lectures de potentia restutiva or of spring.*
     London, 1678. Statement of the law of elasticity, still known
     as Hooke's law. **(b)** *Lectiones Cutlerianae.* London, 1679. The
     six lectures collected in this volume covered a number of topics
     in the fields of instrumentation, astronomy, and mechanics. In
     one of them Hooke proposed a system of planetary dynamics in
     which gravitational attraction was a central feature. Later in
     1679 he began a correspondence with Newton, in the course of
     which he suggested that the force of gravity is inversely pro-
     portional to the square of the distance. Hooke never took the
     matter any further however.

1680s

517. MARIOTTE, E. (See also 514) *Traité du mouvement des eaux.* Paris,
     1686. A treatise in five parts, mostly dealing with hydrostat-
     ics and hydrodynamics. (Part 1 was concerned with meteorology
     and hydrology) It included a careful experimental investigation
     of Torricelli's law of efflux velocity (137).

518. NEWTON, I. (2265) *Philosophia naturalis principia mathematica.*
     London, 1687. (2nd ed., 1713; 3rd ed., 1726. Several Contin-
     ental reprints and editions in the eighteenth century. English
     trans., 1729. French trans., 1759) The great classic which
     constituted the culmination of the Scientific Revolution of the
     seventeenth century.

.... BERNOULLI, Jakob. See 164.

519. LEIBNIZ, G.W. (2522) Published an article in the *Acta Eruditorum* in 1686 attacking Cartesian dynamics; it led to a controversy (known as the *vis viva* controversy) with several Cartesians which lasted until 1691. Through the 1690s he continued to publish important papers on mechanics.

1690s
---

520. HUYGENS, C. (See also 515) *Discours de la cause de la pesenteur.* (Published as an appendix to his *Traité de la lumière* of 1690) A mechanical explanation of gravity, based on Cartesian vortices. It can be regarded as his answer to Newton's *Principia.* (See also 525)

521. VARIGNON, P. (2029) One of the first French mathematicians to recognize the value of the new calculus. From the early 1690s he published a number of articles in the *Mémoires* of the Paris Academy applying it to problems in mechanics. He also played a part in the framing of the principle of virtual velocities and the principle of the composition of forces. His chief book, *Nouvelle mécanique* (2 vols, Paris, 1725) appeared three years after his death.

522. LA HIRE, P. de. (2021) *Traité de mécanique.* Paris, 1695. A work which combined a theoretical and mathematical approach (though non-Newtonian and non-infinitesimal) with a concern for practical problems. It "marks a significant step toward the elaboration of a modern manual of practical mechanics, suitable for engineers of various disciplines." (*DSB*)

523. GUGLIELMINI, D. (1903) *Della natura de' fiumi, trattato fisico-matematico.* Bologna, 1697. A widely acclaimed treatise dealing with hydrodynamic principles and their application in hydraulic engineering.

1700s
---

524. KEILL, John. (2284) *Introductio ad veram physicam.* Oxford, 1701. (Trans. as *An introduction to natural philosophy*, London, 1720) One of the first works to propagate the Newtonian system. It was based on the course of lectures, with experimental demonstrations, that Keill had been giving at Oxford since 1694.

525. HUYGENS, C. (See also 520) *Opuscula posthuma.* Leiden, 1703. Contained much important work in mechanics (and some in optics) which was not published--or at least not published in full--during his lifetime and which consequently did not have the influence that it might have had.

1710s
---

526. BERNOULLI, Johann. (2892) In addition to his many contributions to pure mathematics (see 168) he also made important advances in mechanics, especially during the 1710s and 1720s. His only book, *Théorie de la manoeuvre des vaissaux* (Basel, 1714), dealt with an application of mechanics which was of great practical importance in the age of sailing ships. He also wrote on the theory of tides and enunciated the principle of virtual displacements.

1720s
---

527. 'sGRAVESANDE, W.J. (2783) *Physices elementa mathematica; sive, Introductio ad philosophiam Newtonianam*. 2 vols. Leiden, 1720-21. (3rd ed., 1742. English trans., 1720-21; five later eds. French trans., 1746) A very influential exposition of Newtonianism (with an emphasis on experimental verification). It was the first such work to appear on the Continent.

528. *RACCOLTA d'autori che trattano del moto dell'acque*. 3 vols. Florence, 1723. (2nd ed., rev. & enl., 9 vols, 1765-74) A collection reflecting the long-standing Italian concern with hydrology and flood control. See also 540.

529. LEUPOLD, J. (2531) *Theatrum machinarum generale*. 9 vols. Leipzig, 1723-27. "The first attempt to produce a systematic treatise on mechanism. It has chapters on cams, on the crank, and on machines for converting circular motion into rectilinear motion ... Its importance lies in the fact that it was one of the first attempts to consider the constituent elements of a whole machine separately and examine their motions and constraints." (A.F. Burstall)

530. BERNOULLI, D. (2895) From 1726 to the late 1770s he published many papers on numerous aspects of mechanics, notably rotating bodies, conservation of *vis viva*, friction, motion of bodies in a resisting medium, and vibrating systems (in this last area his work was especially important). For his treatise on hydrodynamics see 532.

1730s
---

531. EULER, L. (2973) *Mechanica sive motus scientia analytice exposita*. 2 vols. St. Petersburg, 1736. His first major work. A systematic treatise on the kinematics and dynamics of a point-mass which introduced a comprehensive use of analysis into the subject. "For the first time the full power of the calculus was directed against mechanics, and the modern era in that basic science began. Euler was to be surpassed in this direction by his friend Lagrange, but the credit for having taken the decisive step is Euler's." (E.T. Bell)   (See also 535)

532. BERNOULLI, D. (See also 530) *Hydrodynamica*. Strasbourg, 1738. His most famous work. A comprehensive treatise on fluid flow. (One chapter is devoted to "elastic fluids"--i.e. gases--and contains a remarkable anticipation of the kinetic theory of gases)

1740s
---

533. ALEMBERT, J. d'. (2058) **(a)** *Traité de dynamique*. Paris, 1743. The first and most famous of his works, continuing the process of formalizing the still-new science of dynamics. It began with a long philosophical preface reflecting the epistemology and metaphysics of the time. **(b)** *Traité de l'équilibre et du mouvement des fluides*. Paris, 1744. A treatise on fluid mechanics as a companion work to the foregoing which was restricted to rigid bodies. **(c)** *Réflexions sur la cause générale des vents*. Paris, 1747. A work on fluid mechanics

rather than meteorology and significant chiefly for its use of
partial differential equations--their first general use in
mechanics.   (See also 536)

534. MAUPERTUIS, P.L.M. de. (2049) "Les lois du mouvement et du repos
déduites d'un principe métaphysique"--an article in the *Mémoires*
of the Berlin Academy for 1746 enunciating the famous principle
of least action.

535. EULER, L. (See also 531) *Scientia navalis*. 2 vols. St. Petersburg, 1749. Vol. 1 developed a general theory of floating
bodies which was applied to the case of a ship in Vol. 2.
(See also 537)

1750s

536. ALEMBERT, J. d'. (See also 533) *Essai d'une nouvelle théorie de
la résistance des fluides*. Paris, 1752. An important contribution to hydrodynamics.

537. EULER, L. (See also 535) In three major papers published in 1757
in the *Mémoires* of the Berlin Academy he presented a detailed
analytical theory of hydrostatics and hydrodynamics.
(See also 539)

1760s

538. BORDA, J.C. (2065) During the 1760s he published several theoretical and experimental papers on fluid mechanics. His chief
contributions were his analysis of the difficult efflux problem
(previously studied by d'Alembert and others with conflicting
results) and his investigations of the resistance of fluids to
objects of various shapes.

539. EULER, L. (See also 537) *Theoria motus corporum solidorum*.
Rostock/Greifswald, 1765. A new and updated treatment of the
subject of his first book (531).

540. *NUOVA RACCOLTA d'autori che trattano del moto dell'acque*. 7 vols.
Parma, 1766-68. Consisted largely of the writings of Guglielmini. Edited by J. Belgrado. (cf. 528)

1770s

541. COULOMB, C.A. (2077) From 1776 onward he published several papers
on aspects of engineering mechanics, especially structural
design (arches, retaining walls, etc.), strength of materials
(rupture of wooden beams and masonry piers), soil mechanics,
friction, and ergonomics. To a large extent this work laid the
foundations for the development of engineering theory in the
early nineteenth century.

1780s

542. CARNOT, L. (2087) *Essai sur les machines en général*. Dijon,
1783. A revised version published under the title *Principes
fondamentaux de l'équilibre et du mouvement* (Paris, 1803)
attracted much more notice (partly no doubt because by that
time he was a national hero). It was a pioneering treatise on
the abstract theory of machines and had an important long-range
influence (see for example 551).

543. ATWOOD'S MACHINE. A useful device for demonstrating the laws of
     falling bodies. First described in 1784 by G. Atwood in his
     textbook on Newtonian mechanics.

544. LAGRANGE, J.L. (2085) *Mécanique analitique*. Paris, 1788. (2nd
     ed., partly rev. and enl. by himself, 2 vols, 1811-15. 3rd ed.,
     1853-55. 4th ed., 1888-89) A great classic, presenting a
     lucid synthesis of the hundred years of development in mechanics
     since Newton. It was thoroughly analytical in its approach (it
     contained no diagrams) and was based on Lagrange's own calculus
     of variations. There were two main parts, statics and dynamics,
     each of which dealt with solid bodies and fluids separately.

1790s

545. PRONY, G.C. (2088) From 1790 to about 1830 he published many
     books, pamphlets, and periodical articles on analytical mechan-
     ics, engineering mechanics, and practical engineering. His
     textbooks and manuals were very influential.

.... CAVENDISH, H. Measurement of the gravitational constant and the
     mean density and mass of the earth: see 1043.

1800s

546. POINSOT, L. (2112) *Élémens de statique*. Paris, 1803. (12th ed.,
     1877. English trans., 1847) An important textbook which took
     an original approach in applying geometrical methods to the
     basic problems of mechanics. It also introduced the concept
     of the couple.   (See also 552)

1810s

547. POISSON, S.D. (2111) *Traité de mécanique*. 2 vols. Paris, 1811.
     (2nd ed., enl., 1833. English trans., 1842) The standard work
     in the subject for many years. Poisson also made a number of
     original contributions of some significance, notably what came
     to be known in the theory of elasticity as Poisson's ratio.

1820s

548. PONCELET, J.V. (2131) From the early 1820s to the late 1840s he
     published many papers on experimental and engineering mechanics
     and the theory of machines. His improvements of waterwheels
     and turbines were especially important.

549. CAUCHY, A.L. (2123) In several papers between 1822 and 1829 he
     established the foundation and much of the superstructure of
     the modern theory of elasticity.

550. NAVIER, H. (2113) (a) "Lois de l'équilibre et du mouvement des
     corps solides élastiques"--a paper of 1821 containing general
     equations for the equilibrium and vibration of elastic solids.
     In other papers later in the decade he dealt with fluids and
     established what is now called the Navier-Stokes equation of
     linear momentum for a viscous compressible fluid. (b) *Résumé
     des leçons ... sur l'application de la mécanique à l'établisse-
     ment des constructions et des machines*. Paris, 1826. (3rd ed.,

1864) A pioneering treatise on engineering theory. It included his own major contributions, especially in the application of mathematical analysis, the study of the strength of materials, and the development of a criterion for the performance of machines (in which connection he came close to the concept of mechanical work).

551. CORIOLIS, G. (2124) *Du calcul de l'effet des machines*. Paris, 1829. A work inspired by Lazare Carnot's pioneering treatise (542). It effectively introduced the concept of mechanical work (which had been "in the air" for some years) and also the convenient formulation $\frac{1}{2}mv^2$ for what he termed *force vive* (later kinetic energy). (See also 554)

1830s

552. POINSOT, L. (See also 546) *Théorie nouvelle de la rotation des corps*. Paris, 1834. Established geometrically that a rigid body suspended from its centre of gravity must conserve any rotation imparted to it if the rotation is about one of the privileged axes. The principle was later used by Foucault in his invention of the gyroscope.

553. HAMILTON, W.R. (2353) "On a general method in dynamics"—a famous paper of 1834 which (with a supplementary paper in the following year) applied to dynamics the kind of mathematical approach he had already used in optics (660). His approach led to a new formulation of the fundamental equations of dynamics (which in the twentieth century was to prove valuable for quantum mechanics).

554. CORIOLIS, G. (See also 551) "Sur les équations du mouvement relatif des systèmes de corps"--a paper of 1835 which established the existence of an inertial force (later termed the Coriolis force) acting on a body moving in a rotating frame of reference.

1840s

555. STOKES, G.G. (2387) In a series of papers on hydrodynamics in the 1840s he made a study of viscosity, or the internal friction of fluids, and derived the law (subsequently known as Stokes' law) governing the settling velocities of small spherical particles in a fluid medium.

556. DIRICHLET, G.P.L. (2594) From 1840 to 1857 he published numerous significant papers on various aspects of theoretical mechanics.

1850s

557. FOUCAULT, L. (2162) Announced in 1851 his famous pendulum experiment, demonstrating the rotation of the earth about its axis. His invention of the gyroscope the next year was a sequel to it.

558. CHEBYSHEV, P.L. (2988) From the early 1850s, in addition to his publications on various aspects of theoretical mechanics, he was also inventing mechanisms relevant to his theoretical interests--especially hinge-lever gears and linkages for converting rotary motion into rectilinear motion. Some of his work in pure mathematics (240) stemmed from problems in theoretical mechanics.

1860s

559. JACOBI, C.G.J. (2593) *Vorlesungen über Dynamik*. Leipzig, 1866. Posthumous publication by R. Clebsch (a one-time member of Jacobi's school) of lectures given in 1842-43. They included Jacobi's important work on differential equations of the first order and their applications in dynamics, and also the derivation of what came to be called the Hamilton-Jacobi equation.

.... *TREATISE on natural philosophy*. By Thomson and Tait. See 712.

1870s

560. BELTRAMI, E. (1948) *Richerche sulle cinematica dei fluidi*. Bologna, 1875.

.... POINCARÉ, H. See 286.

561. LAMB, H. (2434) *A treatise on the motion of fluids*. Cambridge, 1879. (The second and later editions, much enlarged, were entitled *Hydrodynamics*. 6th ed., 1932) A famous treatise which incorporated his own substantial contributions (increasingly in the successive editions).

1880s

562. LYAPUNOV, A.M. (3005) From 1881 until after 1900 he published papers on various aspects of theoretical mechanics, especially the stability of the motion of mechanical systems and the stability of figures of equilibrium of a rotating liquid.

563. MACH, E. (2960) *Die Mechanik in ihrer Entwickelung historisch-kritisch dargestellt*. Leipzig, 1883. (9th ed., 1933. English trans., 1893 and later eds) A critique of mechanics in its historical development from the standpoint of his radically positivist epistemology. His criticisms of Newton's ideas of absolute space, time, and motion prepared the way for the theory of relativity. (Einstein acknowledged the influence the book had on him)

564. REYNOLDS, O. (2427) **(a)** Presented in 1883 an experimentally-based analysis of the flow of liquids through pipes and channels, demonstrating the two distinct types of fluid flow--streamline and turbulent--and establishing the existence of a critical velocity for the transition between them. **(b)** His paper of 1886 on the theory of lubrication laid the foundation for later work on the subject and had important practical consequences.

1890s

565. ZHUKOVSKY, N.E. (3001) Began in the early 1890s the researches in aerodynamics which in the twentieth century were to earn him the title of "the father of Russian aviation."

566. LOVE, A. (2458) *A treatise on the mathematical theory of elasticity*. 2 vols. Cambridge, 1892-93. (4th ed., 1927. Trans. into several languages) A brilliantly written work and a classic in the field.

1.05 PHYSICS

1450s

567. MARLIANI, G. (1821) His collected works were printed at Pavia
in 1482; they comprised several treatises written in the period
1448-72. Three of the treatises dealt with the topic of heat,
approaching it in a very Aristotelian way, the most notable
features being the distinction Marliani made between the intensity of heat (i.e. temperature) and its extension (or quantity),
and his use of a numerical scale to represent the intensity.
He also wrote on dynamics in the tradition of late medieval
Scholasticism.

1480s

568. PECHAM, John (ca. 1230/35-1292) *Perspectiva communis*. The chief
textbook on optics of the Middle Ages. It was first printed in
1482 and continued in use through the sixteenth century.

1530s

569. WITELO. (fl. 1235-1275) *Perspectiva*. An important medieval work
on optics, first printed in 1535. See also 572.

1540s

570. HARTMANN, G. (2480) Discovered magnetic inclination (or dip)
and was probably the first to measure the magnetic declination
on land. In 1544 he reported both his discovery and his
measurement in a letter to the Archduke of Prussia. His
results may have become known to others even though his letter
long remained unpublished. (Magnetic dip was rediscovered by
Robert Norman--see 575)

1570s

571. FLEISCHER, J. (2498) *De iridibus*. Wittenberg, 1571. A very
capable treatise on the optics of the rainbow.

572. IBN AL-HAYTHAM (called ALHAZEN). (965-ca. 1040) *Opticae thesaurus
Alhazeni ...* [With] *Vitellonis ... libri X. Omnes instaurati
... a Federico Risnero*. Basel, 1572. An important edition by
F. Risner (1981) of the optical works of Ibn al-Haytham and
Witelo (cf. 569). It had a wide influence, especially as it
was the first printed edition of Ibn al-Haytham's *Optics*.

573. HERO OF ALEXANDRIA. (fl. A.D. 62) His *Pneumatics* was first
printed in 1575 as a Latin translation (under the title of
Spiritalium liber) by Commandino (1843).

574. MAUROLICO, F. (1835) *Photismi de lumine et umbra....* Venice,
     [1575?]. (Best known in a Naples edition of 1611) An import-
     ant work on optics, covering such topics as the theory of vision,
     lenses, mirrors, the rainbow, etc.

1580s

575. NORMAN, R. (2224) *The new attractive, containing a short dis-*
     course of the magnes or lodestone. London, 1581. (Often re-
     printed) A treatise on the magnet which included his discovery
     of the deviation of a magnetic needle from the horizontal (cf.
     570).

1600s

576. GILBERT, W. (2223) *De magnete magneticisque corporibus, et de*
     magno magnete tellure. London, 1600. The first comprehensive
     treatment of magnetism which, among other things, introduced
     the idea of the earth as a giant magnet. It also included a
     chapter on amber and other "electrics" and their power of
     attraction which Gilbert was at pains to distinguish from
     magnetic attraction. The work was significant for its experi-
     mental methodology as well as for its content.

577. KEPLER, J. (2937) *Astronomia pars optica.* Frankfurt, 1604. An
     analysis of the process of vision which laid the foundations of
     physiological optics. The book was an off-shoot of his astron-
     omical investigations and part of it was concerned with such
     problems as atmospheric refraction.   (See also 580)

578. RISNER, F. (1981) *Opticae libri quatuor.* Kassel, 1606. (Post-
     humous) Written in conjunction with Ramus (1969). It depended
     basically on Witelo's *Perspectiva* which Risner had edited (see
     572). It was later annotated by Snell.

1610s

579. DOMINIS, M.A. de. (1873) *De radiis visus et lucis.* Venice, 1611.
     Gave an explanation of the recently-invented telescope as well
     as dealing with the traditional topics of optics, especially
     the rainbow.

580. KEPLER, J. (See also 577) *Dioptrice.* Augsburg, 1611. Following
     Galileo's introduction of the telescope into astronomy, Kepler
     extended his earlier optical investigations (577) with a thor-
     ough description of the optics of lenses and its application to
     the telescope. As a result he was able to suggest considerable
     improvements to the new instrument.

1620s

581. SNEL (or SNELL), W. (2763) Author of several works on mathematics
     (chiefly trigonometry) but best known for his discovery of the
     law of refraction of light rays, which was made about 1621. It
     was not published but the manuscript containing his results was
     seen by Isaac Vossius and Huygens.

.... MERSENNE, M. See 14.

1630s

.... VERNIER SCALE. See 355.

582. CASTELLI, B. (1876) The results of his researches on optics and radiant heat were communicated by letter to Galileo and others in the 1630s and eventually published in his posthumous *Alcuni opuscoli filosofici* (Bologna, 1669).

583. DESCARTES, R. (1991) *La dioptrique* [Appendix to his *Discours de la méthode*]. Leiden, 1637. Included the law of refraction (which he discovered independently of Snel), the theory of lenses, and an optical analysis of the human eye.

1640s

584. TORRICELLI, E. (1884) He first announced the performance and explanation of his famous barometric experiment in two letters in June 1644. Copies of them circulated in the Italian scientific community and were also sent to Mersenne in Paris. A few months later Mersenne happened to be in Italy and witnessed a demonstration of the experiment by Torricelli himself; through his information network he rapidly spread a knowledge of it throughout Europe.

As well as constituting the starting point for a major new direction in experimental physics, Torricelli's experiment immediately became the central issue in the current debate about the existence of the vacuum, and thereby played no small part in the Scientific Revolution of the seventeenth century. (Though Torricelli's explanation was eventually accepted as correct it was hotly contested until after the middle of the century)

585. PETIT, P. (2004) *Observation touchant le vuide faite pour la première fois en France*. Paris, 1647. A pamphlet describing a repetition of the Torricellian experiment by Pascal and himself at Rouen in 1646 (see also 586).

586. PASCAL, B. (2003) Attempts to repeat Torricelli's experiment elsewhere were at first unsuccessful because of the unavailability of glass tubes strong enough. In October 1646 Pascal, in collaboration with Petit (585), was the first to repeat the experiment in France, and after further investigation of the phenomenon he published in October 1647 a pamphlet, *Expériences nouvelles touchant le vide*, which precipitated a vigorous controversy. In September 1648 he carried out his famous experiment on the Puy de Dôme and forthwith published a second pamphlet, *Récit de la grande expérience de l'équilibre des liqueurs*, in which he accepted the existence of the vacuum and the concept of the weight (or rather pressure) of the air. Controversy continued but this explanation was gradually accepted. (See also 590)

1650s

587. GUERICKE, O. von. (2516) About 1650 his interest in the current philosophical question concerning the vacuum led him to devise an air pump with which he discovered the elasticity of air.

This discovery led him to several important experiments dealing
with air pressure and the nature of the vacuum. (His famous
demonstration with 'the Magdeburg hemispheres' was first performed in 1657) His findings were made public through reports
of them by several people, especially his friend Gaspar Schott
(2518) in two of his books in 1657 and 1664, and also by some
correspondence of Guericke himself. They were further described
in his book, *Experimenta nova ... de vacuo spatio*, of 1672.

1660s

588. RENALDINI, C. (1888) As a member of the Accademia del Cimento
     he took part in the experiments on heat and cold and thermometry
     which were published in the Academy's *Saggi* (1891). His part
     in this work was not published in his own name until much later
     in his *Philosophia naturalis* (Padua, 1694). Apparently he was
     the originator of the idea of using the freezing and boiling
     points of water as fixed points for the thermometric scale.

589. BOYLE, R. (2251) **(a)** *New experiments physico-mechanical touching
     the spring of the air and its effects*. Oxford, 1660. A work
     of great significance in the rapidly developing field of pneu-
     matics. Following Guericke's invention of the air pump, Boyle
     got his young assistant, Robert Hooke, to design and construct
     a much better one with which he carried out a series of experi-
     ments of fundamental importance. In an appendix to the second
     edition (1662) of the book Boyle announced the law that bears
     his name. **(b)** *Experiments and considerations touching colours*.
     London, 1664. Contained many significant observations, includ-
     ing some on the colours of thin films. Boyle contended that
     colours had no separate existence of their own and were due to
     changes in the light reflected from the surface of bodies.
     (c) *Hydrostatical paradoxes*. Oxford, 1666. "Both a penetrating
     critique of Pascal's work on hydrostatics [590], full of acute
     observations upon Pascal's experimental method, and a presen-
     tation of a series of important and ingenious experiments upon
     fluid pressure." (*DSB*)

590. PASCAL, B. (See also 586) *Traités de l'équilibre des liqueurs
     et de la pesanteur de la masse de l'air*. Paris, 1663. (Written
     in 1654 or earlier but not published until a year after his
     death) A major contribution to hydrostatics--especially as
     regards the concept of pressure and what is now known as
     Pascal's principle. The second part, on the effects of the
     weight of the air, assimilated aerostatics to hydrostatics
     but was largely a systematization and development of his two
     earlier pamphlets (586). In the latter respect the book was
     rather out of date because of the rapid progress that had been
     made in pneumatics.

591. GRIMALDI, F.M. (1889) *Physico-mathesis de lumine, coloribus, et
     iride*. Bologna, 1665. A thorough study of light, including
     an account of his discovery of diffraction. He proved that the
     phenomenon could not be explained by reflection or refraction.

592. BARTHOLIN, E. (2828) *Experimenta crystalli Islandici*. Copenhagen,
     1669. Discovery of double refraction in Icelandic spar.

1670s

593. NEWTON, I. (2265) Papers on his famous theory of light and colours, *Philosophical Transactions*, 1672-76. (See also 601)

.... BARROW, I. *Lectiones opticae*. See 157b.

594. "STEAM DIGESTER" In the late 1670s, while working with Robert Boyle, Denis Papin invented the pressure cooker (and a safety valve for it)--an invention which doubtless arose out of the contemporary work in pneumatics.

595. MARIOTTE, E. (2016) (a) *De la nature de l'air*. Paris, 1679. Concerned chiefly with the weight and elasticity of air, and its solubility in water. It included an account of the volume -pressure law, earlier discovered by Boyle, and made use of it in an ingenious attempt to estimate the height of the atmosphere. It also discussed the barometer as a meteorological instrument. (b) *Du chaud et du froid*. Paris, 1679.

1680s

596. MARIOTTE, E. (See also 595) *De la nature des couleurs*. Paris, 1681. A review of experiment and theory in the subject which rejected Newton's ideas and came to no definite conclusions.

597. HYGROMETER. The first hygrometer was invented in 1687 by G. Amontons. He subsequently invented or improved various other instruments and devices.

1690s

598. HUYGENS, C. (2772) *Traité de la lumière*. Leiden, 1690. An exposition of his wave theory of light, including elegant explanations of reflection and refraction.

599. HARTSOEKER, N. (2779) *Essai de dioptrique*. Paris, 1694. A comprehensive treatise.

1700s

600. SAUVEUR, J. (2030) His pioneering researches in acoustics (the term is due to him) were published in several papers in the *Mémoires* of the Paris Academy from 1700 until his death in 1716. He investigated vibrations of strings and frequencies of notes, using beats (which he interpreted correctly) to determine frequency differences. With organ pipes he was able to measure frequencies in cycles per second. He also explained the nature of harmonics as higher frequencies superimposed on a basic frequency, and pointed out that all musical tones include harmonics.

601. NEWTON, I. (See also 593) *Opticks*. London, 1704. (4th ed., 1730. Latin trans., 1706. French trans., 1720) The great classic of experimental science which exercised a profound influence through much of the eighteenth century.

602. HAUKSBEE, F. (2289) The first to conduct a programme of experiments on electricity. During the period 1705-13, in his capacity as curator of experiments for the Royal Society, he devised

many striking experiments on static electricity and electro-
luminescence (quite inexplicable at the time) which became well
known and aroused much interest in the field. His demonstration
that glass is a convenient material for generating frictional
electricity was especially useful. He also made a significant
experimental investigation of capillarity. His *Physico-mechanical experiments* (London, 1709; 2nd ed., 1719) was translated
into Italian, Dutch, and French, and widely read.

1710s

603. THERMOMETRY. The first accurate and reliable thermometers were
produced by the Amsterdam instrument-maker, Daniel Fahrenheit.
In 1714 he introduced the use of mercury and by 1717 had settled
on the scale that bears his name.

1720s

604. MUSSCHENBROEK, P. van. (2784) His highly successful lecture
course on experimental physics was published in successive
versions, beginning with an *Epitome* in 1726 and continuing
through progressively expanded editions to the final *Introductio ad philosophiam naturalem* of 1762. The work was widely
read and various editions were translated into the main European languages. (For his part in the discovery of the Leyden
jar see 610)

605. BOUGUER, P. (2048) *Essai d'optique sur la gradation de la
lumière*. Paris, 1729. A pioneering work on photometry in
which he showed how to compare the apparent brightness of
celestial objects with a standard candle flame. He also formulated a law regarding the amount of light absorbed from a
beam in its passage through a transparent medium.
(See also 618)

1730s

606. GRAY, S. (2285) In a paper in the *Philosophical Transactions* of
1732 he described his experiments in communicating electrification from one body to another and his consequent discovery of
the distinction between conductors and insulators.

607. DUFAY, C.F. de C. (2047) Stimulated by Gray's paper of 1732, he
embarked on a series of methodical and careful experiments
which resulted in eight important memoirs presented to the
Paris Academy in 1733-37. In these he greatly clarified the
investigation of electricity, establishing that: **(a)** All bodies
can, in the right circumstances, be electrified by friction
(excepting metals and those too soft or fluid to rub); **(b)** All
bodies without exception can be electrified by contact with one
already electrified, provided they are insulated; **(c)** Two bodies
electrified by contact with one previously electrified will
repel each other; **(d)** There are two kinds of electricity--
"vitreous" and "resinous"; **(e)** Two bodies electrified by one
kind of electricity repel each other, while two electrified by
the two different kinds attract each other.

608. SMITH, R. (2294) *A compleat system of opticks*. Cambridge, 1738.
(Trans. into Dutch, German, and French. Abridged ed., 1778)

Probably the most influential textbook of optics in the eighteenth century. Smith gave his full support to Newton's particle theory of light, and the popularity of his book probably contributed much to the supremacy of Newton's theory throughout the century.

1740s

609. NOLLET, J.A. (2054) **(a)** *Leçons de physique expérimentale*. 6 vols. Paris, 1743-45. (Many reprints) Based on his very popular lecture course. **(b)** In 1745 he put forward a general theory of electricity which had the misfortune to appear just before the discovery of the Leyden jar. It was superseded later by Franklin's theory.

610. THE LEYDEN JAR. The discovery of the electric capacitor, in the form of a glass vessel containing water, was made independently and almost simultaneously in 1745/46 by E.J. von Kleist in Germany and by workers in Musschenbroek's laboratory at Leiden. It was Musschenbroek who made the discovery widely known, initially with a letter to his French acquaintances who reported this "expérience nouvelle mais terrible" to the Paris Academy.
    "[It] was a true discovery, the finding of something opposed to expectation, a piece of serendipity on the part of two unpracticed operators. The importance of this discovery, on which the development of modern electrostatics hinged, was recognized immediately." (J.L. Heilbron)

611. FRANKLIN, B. (3016) The accounts of his electrical experiments, carried out from 1746 onward, were communicated to his English correspondent, Collinson (2298), who in 1751 published them in London on Franklin's behalf as *Experiments and observations on electricity made at Philadelphia in America*, with supplements in 1753 and 1754 containing his later experiments. (5th ed. of the whole, 1774. Trans. into French, German, and Italian)
    His single fluid theory of electricity was widely adopted, and several terms and concepts that he introduced are still used, e.g. 'positive' and 'negative', conservation of charge, transfer of charge, etc. His analysis of the distribution of charge in the Leyden jar was particularly elegant. With the general public everywhere he became famous for his demonstration that lightning is an electrical phenomenon and for his invention of the lightning conductor.

612. EULER, L. (2537) "Nova theoria lucis et colorum"--a tract in his *Opuscula varii argumenti* (Berlin, 1746). In contrast to the dominant corpuscularian ideas of the time, his theory attempted to explain light and its properties in terms of oscillations in a universal ether. It attracted little interest.
    (See also 622)

1750s

613. BECCARIA, G. (1918) **(a)** *Dell'elettricismo artificiale e naturale*. Turin, 1753. A well-presented account of Franklin's new theory with additional experiments of his own. The book was influential in disseminating Franklin's theory, especially as an extract from it was published in French. **(b)** *Dell'elettricismo lettere*. Bologna, 1758. A series of articles in the form of

letters, some of which dealt with developments of Franklin's
ideas while others described numerous experiments relating to
atmospheric electricity.

614. KLINGENSTIERNA, S. (2835) Proved theoretically in 1754 the
     invalidity of Newton's conclusion that it would be impossible
     to make an achromatic lens. He communicated this result to
     the London optician and instrument-maker, John Dollond, who in
     1758 was successful in making achromatic lenses. In 1760
     Klingenstierna published a comprehensive theory of achromatic
     and aplanatic lens systems, referring to Dollond's experimental
     results.

615. COOLING BY EVAPORATION. The production of cold (or absorption of
     heat) by the evaporation of liquids was strikingly demonstrated
     by W. Cullen in 1756 by evaporating a variety of liquids
     (including "nitrous aether") in a receptacle evacuated by an
     air pump.

616. AEPINUS, F.U.T. (2543) *Tentamen theoriae electricitatis et
     magnetismi*. St. Petersburg, 1759. An original and important
     work which, discarding the current ideas of electrical atmos-
     pheres (but retaining the essentials of Franklin's single fluid
     theory), explained the phenomena of electrostatics by deducing
     them, in Newtonian style, from forces of attraction and repul-
     sion acting at a distance. It was also the first systematic
     attempt to apply mathematics to the theory of electricity.
          The book had a small circulation and was mathematical
     and difficult. It did not have much influence until a French
     epitome of it was published by R.J. Haüy in 1787.

1760s

617. LAMBERT, J.H. (2542) *Photometria; sive, De mensura et gradibus
     luminis, colorum et umbrae*. Augsburg, 1760. An important
     work in photometry. It established what were later called
     Lambert's law of light absorption and Lambert's cosine law of
     brightness.   (See also 624)

618. BOUGUER, P. (See also 605) *Traité d'optique sur la gradation de
     la lumière*. Paris, 1760. A considerable expansion of his
     earlier work, describing various kinds of photometers and
     tackling some difficult theoretical topics--successfully on
     the question of visual range through an obscuring atmosphere.

619. *DICTIONNAIRE de physique*. 3 vols. Avignon, 1761. (At least
     eight editions) By A.H. Paulian.

620. BLACK, J. (2302) Established the phenomenon of latent heat in
     1761. At about the same time he realized the existence of
     specific heat but did not establish it until 1764, partly in
     conjunction with James Watt who was then engaged in his famous
     work on the steam engine. Black never published either of
     these discoveries in writing but he did include them in his
     lecture course which was very popular and well known.

621. PRIESTLEY, J. (2307) *The history and present state of electricity,
     with original experiments*. London, 1767. (3rd ed., 1775. Trans.
     into French, German, and Dutch) A review of recent work in the

field (rather than a history in the modern sense). The book
was very successful and the later editions contained much new
material, including his own experiments. The most notable of
them was one which suggested that the attraction between
electric charges might follow an inverse-square law.

1770s

622. EULER, L. (See also 612) *Dioptrica*. 3 vols. Berlin, 1770-71.
Established the principles for calculating the properties of
optical systems and for designing optical instruments.

623. WILCKE, J.C. (2842) Independently discovered the phenomenon of
latent heat in 1772 and, unlike Black (620), published his
discovery.   (See also 626)

624. LAMBERT, J.H. (See also 617) **(a)** *Hygrometrie*. 2 parts. Augsburg, 1774-75. A work resulting from his meteorological
investigations and dealing chiefly with measurement of the
humidity of the air. **(b)** *Pyrometrie*. Berlin, 1779. (Posthumous) On the measurement of heat.

625. *A COMPLETE treatise on electricity in theory and practice*.
London, 1777. (3rd ed., 3 vols, 1786-95) A well-written and
widely read work. By T. Cavallo, F.R.S., an Italian-born
instrument-maker living in London.

1780s

626. WILCKE, J.C. (See also 623) He probably discovered the phenomenon
of specific heat independently before 1780 but in that year he
read a second-hand account of Black's ideas on the subject (620).
In the following year he published his measurements of the
specific heats of a number of metals and other substances.

627. COULOMB, C.A. (2077) His fundamental researches in electricity
and magnetism were published in the *Mémoires* of the Paris
Academy as follows.

(a) Two papers, in 1780 and 1784, in which he established the theory of torsion in thin threads and described his
celebrated torsion balance which provided the instrumental
basis for his subsequent work.

(b) A series of seven famous papers in 1785-91 in which
he established that Newton's law of inverse squares holds also
for electrical and magnetic attraction and repulsion, and that
the degree of attraction or repulsion depends on the amount of
electric charge or magnetic pole strength ("electric mass" and
"magnetic density" in his terms).

He also studied the surface distribution of charge on
conducting bodies, and the leakage of charge. In place of the
existing fluid theory of magnetism he proposed a theory based
on molecular polarization.

628. *DICTIONNAIRE de physique*. 5 vols. Paris, 1781-82. By J.A.
Sigaud de Lafond.

.... "MEMOIRE sur la chaleur." By Lavoisier and Laplace. See 830.

629. CHLADNI, E. (2560) From the 1780s onward he studied experimentally the vibrations of strings, plates, and rods. "Chladni's

figures", formed by spreading sand on plates and then vibrating
them, were intriguing and sometimes spectacular, and attracted
a great deal of interest; in 1808 he gave a demonstration of
them in Paris at the Institut National.

Chladni discovered longitudinal vibrations in rods and
used them to determine the speed of sound in solids. He also
determined the speed of sound in gases other than air by filling
organ pipes with the gas under study and measuring the resulting
pitch.

630. *NEW EXPERIMENTS on electricity*. London, 1789. By Abraham Bennet,
F.R.S., a country clergyman. It included accounts of the sensitive gold-leaf electrometer and the induction machine he had
invented, as well as numerous experimental results.

1790s

631. THOMPSON, B. (2322) From about 1790 his investigations of heat
led him to reject the accepted theory of a calorific fluid and
espouse the vibratory theory. Of the many experiments he used
in order to judge between the two theories the best known is
his account in 1798 of the process of boring cannon with a
blunt drill--a process which generated unlimited amounts of heat.

632. THEORY OF HEAT EXCHANGES. A Swiss *savant*, Pierre Prevost, suggested in 1791 that, irrespective of temperature, a body is constantly giving off heat and at the same time receiving it from
neighbouring bodies. Even if the temperatures of bodies are
the same, heat is still being exchanged between them (in equal
amounts). The idea was later found to be a fertile one (cf.
703)

633. *ENCYCLOPÉDIE méthodique ... Physique*. 4 vols & atlas. Paris,
1793-1822. The physics section of this enormous, topically
arranged encyclopedia. It was written by G. Monge.

634. VOLTA, A. (1929) When he learnt of Galvani's discoveries in 1791
(see 1735) Volta was initially sceptical but, finding that he
could confirm them, he set out to investigate the phenomena.
He found that the frogs' legs constituted a highly sensitive
electroscope and gradually came to the conclusion by 1793 that
the electricity they revealed was not "animal electricity", as
Galvani thought, but arose from the contact of two dissimilar
metals (Galvani's brass hooks on an iron railing). This phenomenon of contact electricity he proceeded to investigate,
determining the "electromotive force" (his term) of various
combinations of metals and moist conductors.

Between 1796 and 1800 his work was interrupted by the
French invasion of Italy, but by the latter year he had invented
his famous pile, consisting of pairs of zinc and silver discs
with pieces of moist cardboard between the pairs. He announced
his discoveries in a letter to the president of the Royal Society of London in 1800.

1800s

635. HERSCHEL, W. (2314. The famous astronomer) In the course of some
studies of the sun using "various combinations of differently
coloured darkening glasses" he discovered in 1800 that with

some combinations light came through his telescope with very
little heat, while with other combinations heat came through
with very little light. He experimented with this invisible
(infrared) radiant heat--from terrestrial sources as well as
solar--and established that it was reflected and refracted
like light.

636. YOUNG, T. (2330) In a paper read to the Royal Society in 1800
he upheld the wave theory of light, in opposition to the
accepted corpuscular theory, and in the following year discovered and explained the phenomenon of interference. Though the
importance of this discovery was recognized his advocacy of the
wave theory was rejected and even ridiculed (by Henry Brougham
in some notorious articles in the *Edinburgh Review*).

637. RITTER, J.W. (2571) Following Herschel's discovery of invisible
light beyond the red end of the spectrum (635) he looked for
something similar at the other end, and in 1801 discovered that
silver chloride is blackened by invisible light just beyond
the violet.

638. WOLLASTON, W.H. (2336) Established in 1801 that frictional and
galvanic electricity are identical. In the same year he
discovered--independently of Ritter--the existence of invisible
light beyond the violet.

639. LAPLACE, P.S. (2075) During the period 1805-15, after the completion of his great work in celestial mechanics and while he
was at the height of his fame and influence, he developed a
programme of research in physics and used his patronage to
direct the work of promising young men (especially Biot, Arago,
Malus, and Poisson) into areas connected with it (cf. 2110).
The Laplacian programme envisaged a mathematical theory of
forces between particles which would be applicable to physical
phenomena generally (and to chemical reactions as well) and
would do for the several branches of physics what Newtonianism
had done for celestial mechanics. Laplace initiated the programme with the publication in 1805 of treatments of refraction
and capillarity, and his protégés took up analogous researches.

After 1815 however the Laplacian school disintegrated,
partly because of the waning of Laplace's influence and partly
because several new developments in physics (and chemistry)
tended to undermine its chief assumptions. Nevertheless it may
well have had a beneficial influence on the development of
physics through its emphasis on mathematization and unification
--two things which physics needed at the time.

640. MALUS, E.L. (2116) While experimenting on the phenomenon of
double refraction he accidently discovered in 1808 that reflected
light is polarized. Hitherto polarization (his term) had been
known only as an isolated effect accompaning double refraction
in some crystals but Malus now showed it to be a general property of light. He proposed an explanation of it in terms of
the generally-accepted corpuscular theory (and his discovery
seemed at the time to refute the opposing wave theory).

641. BREWSTER, D. (2335) From about 1812, following the discovery by
     Malus of the polarization of light by reflection, he carried
     out a long series of experiments on polarization, polarized
     light, and the optical properties of crystals. His chief
     discovery, in 1815, was the law named after him--that the
     tangent of the angle of polarization is equal to the refractive
     index of the reflecting material. He also discovered that light
     reflected from metallic surfaces is elliptically polarized. He
     interpreted all his findings in terms of the corpuscular theory
     of light which he firmly supported.

642. BIOT, J.B. (2114) From 1812 he carried out a large number of
     investigations on polarized light and its properties. His
     main discoveries were as follows.
         (a) In 1812 he found that plates cut from some crystals
     of quartz rotated the plane of polarization of a beam of polar-
     ized light to the right, while plates cut from other crystals
     rotated it to the left.
         (b) In the same year he discovered the phenomenon of
     rotatory dispersion, i.e. that the degree of rotation of the
     plane of polarization depends on the colour of the light, the
     angle of rotation being inversely proportional to the square
     of the wavelength (though at the time he did not use the con-
     cept of wavelength, being a supporter of the corpuscular theory).
         (c) He found unexpectedly in 1815 that liquids (includ-
     ing solutions) can possess the property of rotating the plane
     of polarization. He realized that this meant that it was the
     molecules of the liquids (or dissolved solids) that affect the
     polarized light.
         (d) The foregoing discovery led him later to develop
     the technique of polarimetry and to demonstrate its usefulness
     for chemistry (and especially for the sugar industry).

643. POISSON, S.D. (2111) The chief disciple of Laplace and a strong
     supporter of the Laplacian programme for the mathematization of
     physics (see 639). His first major contribution to mathematical
     physics was his comprehensive treatment of electrostatics in
     1812.   (See also 666)

644. FRAUNHOFER, J. (2580) In 1814, in the course of a technically-
     oriented study of methods of determining very accurate refrac-
     tive indexes of different kinds of optical glass, he made use
     of the two bright lines in flame spectra as a source of mono-
     chromatic light. He went on to compare flame spectra with the
     solar spectrum which he found to contain many dark lines. He
     plotted the relative positions of many of the 574 lines that he
     detected, labelling the chief ones with capital letters (which
     are still used), and noticed that some of the dark lines in the
     solar spectrum seemed to correspond to the bright lines in flame
     spectra (especially the two yellow lines). Spectra of the light
     from the moon and planets he found to have the same pattern as
     the solar spectrum, but several stars he examined each had their
     own individual pattern, different from the sun and from each
     other.   (See also 653)

645. ARAGO, F. (2117) Some early experiments in optics (in the course
of which he invented the polarimeter in 1811) led him to lose
confidence in the accepted corpuscular theory and in 1815 he
joined forces with Fresnel. They collaborated in a series of
papers advocating the wave theory, Fresnel providing the key
concepts and the mathematical analyses, with Arago doing much
of the experimental work. They were also at pains to rebut the
criticisms of the adherents of the corpuscular theory, especially Laplace, Biot, and Poisson. Arago was the chief publicist
for the wave theory and in 1838 he suggested an experiment to
decide between the two theories; it was later performed successfully by Fizeau and Foucault (683a).

646. FRESNEL, A.J. (2126) In a series of papers (some in collaboration
with Arago) from 1815 to the early 1820s he adduced powerful
experimental and theoretical support for the wave theory of
light. He first investigated diffraction and discovered the
phenomenon of interference (unaware that it had been discovered
earlier by Young). In further investigations he applied mathematical analysis to the wave theory in a way that had not been
done before, thereby removing a number of objections and winning
serious consideration for it. The phenomenon of polarization
presented a serious challenge but Fresnel met this with the
bold proposal (also suggested half-heartedly by Young) that the
vibrations constituting light are transverse, rather than longitudinal, to the direction of propagation of the light. Though
it raised apparent difficulties about the nature of the ether
this proposal was found to be in accord with the experimental
findings, especially in the important case of double refraction.
         Not long after Fresnel's death in 1827 the wave theory
became generally accepted.

647. DULONG, P.L. (2125) In collaboration with A.T. Petit he carried
out some important researches on heat during the period 1815-20.
They first conducted a critical analysis of thermometric scales
(mercury and air thermometers), then a close examination of
Newton's law of cooling, and in 1819 put forward what came to
be known as the law of Dulong and Petit, namely that the product
of the specific heat of an element and its atomic weight is
(approximately) a constant--a finding of considerable value to
chemistry at the time (see 871).

648. "DU ZÉRO ABSOLU de la chaleur et du calorique spécifique"--a paper
of 1819 by N. Clément and C.B. Desormes in which they determined
the ratio of the specific heat of a gas at constant pressure to
its specific heat at constant volume, and made an estimate of
absolute zero.

1820s

649. OERSTED, H.C. (2851) His famous discovery in 1820 of the magnetic
effect of an electric current (specifically, that a magnetic
needle aligns itself across a wire carrying a current) inaugurated the electromagnetic era and led directly to the fundamental
researches of Ampère and Faraday.

650. AMPÈRE, A.M. (2115) On hearing of Oersted's discovery he immediately began an intensive investigation of the relations between

electricity and magnetism. Three weeks later he announced to
the Paris Academy his discovery that two parallel wires each
carrying a current attract each other if the currents are in
the same direction and repel each other if the currents are in
opposite directions. He also found that a solenoid coil
through which a current is passing has the same properties as
a bar magnet. These phenomena, he considered, indicated that
magnetism is the result of electricity in motion--a sharp
break with the accepted opinion that electricity and magnetism
were unrelated.

      After much subsequent experimentation Ampère developed
an elaborate mathematical theory, published in full in 1827,
which explained all the electromagnetic effects then known in
terms of an inverse-square force acting at a distance.

651. BIOT, J.B. (See also 642) When he heard of Oersted's discovery
    in 1820 he immediately set to work, in collaboration with Félix
    Savart, to explore the phenomenon. They found that the magnetic
    force created by a current flowing through a wire acts at right
    angles to a perpendicular to the wire, and that its intensity
    varies inversely with the distance from the wire. This relation-
    ship, known as the Biot-Savart law, was important for the
    subsequent development of electromagnetic theory.

652. POGGENDORFF, J.C. (2598) Following Oersted's discovery he invent-
    ed the first rudimentary form of the galvanometer in 1820. (A
    similar device was invented at the same time by J.S.C. Schwei-
    gger) Poggendorff later improved his instrument and it was
    used by Gauss and others.

653. FRAUNHOFER, J. (See also 644) Following Fresnel's work on inter-
    ference effects (646) he showed in two papers in 1821 and 1823
    that spectra could be produced by means of a diffraction grating
    consisting of a large number of closely aligned parallel wires,
    or alternatively of lines ruled on glass with a diamond. He
    discovered the mathematical relation between the wavelength of
    the diffracted light and the distance apart of the wires (or
    lines) and this enabled him to make the first determinations
    of wavelength. His measurements of the main spectral lines
    which he had discovered earlier (644) were of impressive accuracy.

      The great importance of this and his earlier work was
    not appreciated until the subject was taken up by Bunsen and
    Kirchhoff in the late 1850s (704).

654. THERMOELECTRICITY. The German physicist, Thomas Seebeck, discov-
    ered in 1821 that an electric current will flow in a circuit
    containing two separate bimetallic junctions when they are held
    at different temperatures. Though the phenomenon was not under-
    stood for a long time it was important in showing that heat
    could be converted into electricity.   (See also 671)

655. FOURIER, J. (2101) *Théorie analytique de la chaleur*. Paris,
    1822. A classical work on the conduction of heat in solids
    using as a mathematical technique the kind of infinite series
    now known as Fourier series. It was a landmark in mathematical
    physics, bringing to bear on the study of heat the kind of
    mathematical analysis that had hitherto been applied only to
    problems in mechanics (celestial and terrestrial). Fourier

analysis was subsequently applied in other fields of physics (cf. 679a) and has become one of the main tools of mathematical physics.

656. THE CRITICAL STATE. Discovered in 1822 by C. Cagniard de la Tour. He heated various liquids in sealed tubes and observed that at particular temperatures and pressures liquid and vapour became identical. The pnenomenon was further investigated by Andrews (928).

657. CARNOT, S. (2134) *Réflexions sur la puissance motrice du feu.* Paris, 1824. A work of great originality and extraordinary depth. It had the aim of improving the efficiency of steam engines, and presented a very abstract theory of heat engines in general. It introduced the important concept of cyclic operations and the principle of reversibility, and showed that the efficiency of an ideal, perfectly reversible engine depends only on the temperature range within which the cycle operates.
    With the important exception of an article by Clapeyron (672), the book was ignored almost completely until the advent of the principle of conservation of energy about 1850. It was then realized that Carnot's ideas were of the greatest significance for the new science of thermodynamics.

658. OHM, G.S. (2597) His famous law relating electric current, potential difference, and resistance was announced in two papers in 1826 followed by a book in 1827. It brought much-needed order into the hitherto very confused collection of facts and ideas relating to the electric current and electric circuits.

659. STURM, C.F. (2133) In 1826, in collaboration with D. Colladon (later a leading engineer), he made the first accurate measure of the speed of sound in water, and in the following year wrote a prize-winning essay, including many experimental results, on the compressibility of liquids.

660. HAMILTON, W.R. (2353) "Theory of systems of rays"--a paper of 1827 which, with three supplements published in 1830-32, introduced a major new development in mathematical optics. A significant feature of the theory was that it lent itself to both a corpuscular and an undulatory conception of light. The theory incidentally predicted the quite unexpected phenomenon of conical refraction in biaxial crystals, a prediction that was confirmed experimentally soon afterwards.

661. THE POTENTIAL FUNCTION. In an extraordinary essay, published as a pamphlet in 1828, the self-taught mathematician, George Green, introduced some valuable mathematical techniques into the theory of electricity and magnetism--especially the potential function (which had earlier been used in the analytical dynamics of Lagrange). His essay remained virtually unknown until it was discovered and reprinted by William Thomson (Kelvin) in 1850. Green's use of the potential function was later incorporated into Maxwell's great theory.

662. THE NICOL PRISM. Invented in 1828 by the mineralogist, William Nicol. It proved to be of great value for the study of the polarization of light and also made possible the polarizing microscope (see 1268).

663. THE THERMOPILE. Invented in 1829 by L. Nobili (professor of
     physics at the Florence Museum), on the basis of the Seebeck
     effect (654). For its development and use see 667.

<hr>
1830s
<hr>

664. HENRY, J. (3021) His first researches in electricity, from the
     late 1820s to 1832, had the following main results: **(a)** A great
     improvement in the construction of electromagnets (which hitherto
     had been quite primitive); **(b)** The discovery of electromagnetic
     induction probably a year before Faraday--but Faraday was the
     first to publish it (in 1831); **(c)** The discovery of self-induc-
     tion, which he published in 1832. (Faraday made the same
     discovery and published it in 1835)

665. FARADAY, M. (2358) **(a)** Made his celebrated discovery of electro-
     magnetic induction in 1831 and constructed what was in effect
     the first dynamo. **(b)** In 1837 he was working on electrostatic
     induction and made the discovery of specific inductive capacity
     (or dielectric constant). From that time onward he was develop-
     ing a general theory of electricity incorporating the idea of
     lines of electric force, analogous to his earlier idea of lines
     of magnetic force.   (See also 687)

666. POISSON, S.D. (See also 643) **(a)** *Nouvelle théorie de l'action
     capillaire*. Paris, 1831. **(b)** *Théorie mathématique de la chal-
     eur*. Paris, 1835. Both works included contributions he had
     made over a period of twenty years.

667. MELLONI, M. (1939) The thermopile, which had been invented by
     Nobili (663), was much improved in 1831 by Melloni in conjunc-
     tion with Nobili. By means of it Melloni carried out a series
     of investigations during the 1830s on radiant heat and its
     properties, comparing and contrasting them with those of light.
     He established that, like light, radiant heat could be reflected,
     refracted, and polarized, and could give rise to interference
     effects. On the other hand he could find little correlation in
     the absorption or transmission of the two kinds of radiation by
     solids and liquids.

668. GAUSS, C.F. (2570) In connection with his investigations of
     terrestrial magnetism (1050) he introduced in 1833 a measure-
     ment of magnetism which was expressed in absolute units. He
     pointed out that, since all forces may be measured by the
     motions they produce, only three absolute units are needed--
     namely units of length, mass, and time. This would have the
     advantage that all measurements expressed in terms of these
     three units could be compared with each other.
          His introduction of absolute units for the measurement
     of terrestrial magnetism was the first time such units had been
     used outside mechanics (and later inspired his colleague, W.E.
     Weber, to use them in the measurement of electricity).

669. LENZ, E. (2987) His law describing the direction of the induced
     current in electromagnetic induction was announced in 1834. In
     subsequent papers in the 1830s and 1840s he made many contrib-
     utions to the knowledge of electromagnetic, electrothermal, and
     electrochemical phenomena.

670. WHEATSTONE, C. (2370) In 1834 he made an (unsuccessful) attempt
to measure the speed of electricity through a conductor by
means of spark gaps and the ingenious use of a rapidly rotating
mirror. His rotating mirror technique was later used by Fizeau
and Foucault to measure the speed of light. In his later work
in electricity he invented the rheostat and introduced what
came to be known as the Wheatstone bridge.

671. THERMOELECTRICITY. An intriguing sequel to Seebeck's discovery
(654) was that of the Parisian watchmaker, J.C.A. Peltier, who
found in 1834 that with junctions of some pairs of metals--
especially bismuth/antimony--an electric current could actually
produce a lowering of temperature. Other workers later succeeded in freezing water by the Peltier effect.

672. CLAPEYRON, B.P.E. (2135) "Mémoire sur la puissance motrice de
la chaleur", 1834. An exposition and mathematical treatment
of the (non-mathematical) work of Sadi Carnot (657) which had
hitherto been ignored. Clapeyron bypassed what was incorrect
in Carnot's theory (i.e. the assumption of the conservation of
heat, or rather caloric) and gave a new presentation of its
profound insights. It was through this paper of Clapeyron's
(which was translated into English and German) that Carnot's
ideas were eventually appreciated.

673. FORBES, J.D. (2371) Following the discovery of the refraction of
radiant heat by Melloni (667) he showed in 1834 that it could
also be polarized. This fact added much support to the view
that radiant heat was closely related to light (a view which
at that time Melloni was opposing).

674. THE GALVANOMETER. Early forms of the galvanometer were devised
in the years following Oersted's discovery of the magnetic
effect of a current, notably by Poggendorff (652). A major
advance was the invention in 1836 of the moving-coil galvanometer by the English "electrician", W. Sturgeon.

1840s

675. REGNAULT, H.V. (2152) Though he made no outstanding theoretical
contributions he was a highly skilled experimenter who from
about 1840 onward compiled an extensive body of accurate data
on the thermal properties of many substances, especially gases.
He also designed apparatus for various physical measurements.
His results established that the gas laws are only approximately
true for real gases.

676. STOKES, G.G. (2387) The acceptance of the wave theory of light
raised the problem of the mechanical properties of the luminiferous ether. In a series of papers in the 1840s Stokes elaborated one of the best-known mathematical theories of the ether.
He postulated that it behaved like an elastic solid with respect
to the vibrations constituting light, but like a fluid with
respect to the motion of the earth and other planets through it.
He proposed, furthermore, that the earth and planets dragged
along with them the ether that was close to their surfaces
while the ether further away remained at rest.
(See also 696)

677. GAUSS, C.F. (See also 668) *Dioptrische Untersuchungen*. Göttingen,
     1841. Included a mathematical analysis of the path of a beam
     of light through a system of lenses, and a demonstration that
     any such system is equivalent to a suitably chosen single lens.

678. DOPPLER, J.C. (2953) Announced in 1842 his discovery of the
     effect (later famous under the name of the Doppler effect) on
     the observed frequency of light or sound waves that is produced
     by the motion of the source or of the observer. (For its applic-
     ation in astronomy see 456e)

679. THOMSON, W. (2389) **(a)** At the beginning of his career he was much
     influenced by Fourier's classic (655) and in 1842 constructed a
     theory of electricity which used a mathematical model similar to
     Fourier's, thereby expressing a formal analogy between the flux
     of electricity in a conductor and the flux of heat in a rod.
     (b) On becoming acquainted with Clapeyron's account (672) of
     Carnot's work he pointed out in 1848 the possibility of con-
     structing an absolute scale of temperature, independent of the
     properties of any thermometric substance. Later work, in con-
     junction with Joule, showed that the gas scale (especially
     using hydrogen) is practically coincident with the absolute
     scale.   (See also 692)

680. MAYER, J.R. (2630) In a paper of 1842 he concluded that heat and
     motion are interconvertible, being two different manifestations
     of one indestructible force (the first statement of the principle
     of conservation of energy). He inferred, furthermore, that
     there should be a numerical constant expressing the mechanical
     equivalent of heat, and went on to deduce a value for this
     constant from published data relating to the heat evolved in
     the compression of a gas (i.e. the specific heats of air at
     constant pressure and constant volume).

681. JOULE, J.P. (2388) Found in 1840 that the heat produced in a
     conductor by an electric current is proportional to the product
     of the resistance of the conductor and the square of the current.
     His further researches on voltaic electricity and the interrel-
     ations of chemical, electrical, and thermal effects led him to
     investigate the thermal effect of mechanical action.
             In 1843, in a famous series of experiments, he arrived
     at a figure for the mechanical equivalent of heat. His announce-
     ment of it at a meeting of the British Association in that year
     aroused little or no interest. In subsequent experiments he
     derived the equivalent in other ways and improved the accuracy
     of the figure. His full statement, in 1847, of the principle
     of conservation of "force" gained recognition through the
     response of William Thomson who immediately saw its importance.

682. COLDING, L.A. (2864) In a paper read before the Danish Society
     of Sciences in 1843 (though not published until 1856) he put
     forward the idea of the interconvertibility of heat and work.
     From the subsequent elaboration of his experiments on the
     conversion of work into heat he arrived in 1847 at a numerical
     value for the mechanical equivalent of heat. Further experi-
     ments during the early 1850s considerably improved the accuracy
     of his figure.

683. FIZEAU, H. (2163) **(a)** From 1844 to 1847 he carried out a number of experiments in collaboration with Foucault which provided additional evidence for the wave theory of light--in particular an ingenious demonstration that light travels faster in air than in water (as predicted by the wave theory).
(b) In 1849, in a similarly ingenious experiment, he made the first non-astronomical measurement of the velocity of light.
(c) Unaware of the earlier work of Doppler (678) he formulated in 1848 what is now known as the Doppler effect, and showed in principle that it could be used to determine the motion of stars towards or away from the earth.   (See also 693)

684. FOUCAULT, L. (2162) For his work in collaboration with Fizeau see 683a.   (See also 694)

685. NEUMANN, F.E. (2599) In two papers in 1845 and 1848 he extended Ampère's theory of electrodynamics to include electromagnetic induction, and derived the potential function corresponding to Ampère's force.

686. KIRCHHOF, G.R. (2633) In 1845-46, while still a student, he formulated the laws--later named after him and of considerable practical importance--which govern the distribution of voltage and current in networks of conductors.   (See also 704)

687. FARADAY, M. (See also 665) **(a)** He had long suspected that light and magnetism were related, and in 1845 he was able to show that the plane of polarization of light is rotated in a strong magnetic field, the angle of rotation being proportional to the strength of the field. **(b)** His powerful electromagnet enabled him to verify another of his conjectures later in the same year, with the discovery of diamagnetism. He advanced a theory to explain the phenomena of diamagnetism and paramagnetism in terms of his notion of lines of magnetic force.
        This theory of magnetism, together with his theory of electricity (665b)--undeveloped and non-mathematical though these theories were--transmitted to the future the profound idea that the medium is the site of electromagnetic action--the germ of the field theory.

688. WEBER, W.E. (2600) Put forward in 1846 a general mathematical law (derived from Ampère's theory) expressing the force between moving charges. He elaborated the law--which had wide ramifications--in later papers and it gained much support (at least on the Continent) as a general theory of electrodynamics. Eventually however it was superseded by Maxwell's theory. (See also 699)

689. GROVE, W.R. (2372) *On the correlation of physical forces*. London, 1846. (6th ed., 1874. The successive editions were extensively revised in the light of the new developments that were appearing) The first edition was a notable early statement of the principle of the conservation of energy. In the later editions the work became one of the chief popularizations of the new principle.

690. HELMHOLTZ, H. von. (2631) *Über die Erhaltung der Kraft*. Berlin, 1847. A statement of the principle of conservation of "force" which he arrived at from considerations of mechanistic physiol-

ogy, especially animal heat. He also invoked physical considerations, including the recent work of Joule, and argued for the
dynamical theory of heat.   (See also 717)

1850s
---

691. CLAUSIUS, R. (2632) **(a)** In a famous paper of 1850 he (i) accepted
Joule's equivalent of heat and work, (ii) rejected the caloric
theory and its assumption of the conservation of heat, but (iii)
nevertheless accepted the part of Carnot's theory which did not
depend on that assumption. From these considerations he arrived
at the view that there are *two* separate laws which are fundamental to the science of thermodynamics. (See also 710)
(b) In 1857 he gave the first extended and systematic treatment
of a kinetic theory of gases. (The idea that the properties of
gases could be explained in terms of the motion of their molecules went back to at least the early eighteenth century (cf.
532) but hitherto had never been able to compete with the
Newtonian 'static' theory) He developed the theory considerably,
introducing the concepts of average molecular velocity, and
vibratory and rotational motion as well as translational. In
another paper in the following year he developed the idea of
'mean free path.'

692. THOMSON, W. (See also 679) **(a)** "On the dynamical theory of heat",
1851. A famous paper in which he independently came to the same
conclusion as Clausius (691a) concerning the reconciliation of
the work of Joule and Carnot, and the requirement for two separate laws of thermodynamics (a term he coined three years
later). Like Clausius he also discarded the caloric theory of
heat in favour of the dynamical (or kinetic) theory. Though
he did not try to base the laws of thermodynamics on that theory
he regarded it as supporting them.
     (b) In the early 1850s he developed the concept to
which Rankine (697) in 1853 gave the name 'energy.' Like
Rankine he saw its fundamental importance and wide applicability.
     (c) In 1854, when he first became involved in the plan-
ning for the Atlantic cable, he was asked for an explanation of
the delay in an electric current passing through a very long
cable. This enquiry prompted a major paper in 1855 in which,
referring to his work of 1842 (679a), he pointed out that, just
as the time for the flow of heat through a rod depends only on
the properties of the rod, so the time for the flow of electricity through a cable depends only on the properties of the cable
—its resistance and capacitance. This consideration proved to
be fundamental for the design of the Atlantic cable.
(See also 712)

693. FIZEAU, H. (See also 683) The adoption of the wave theory of
light brought with it a general belief in the existence of a
luminiferous ether. It was assumed that outside material
bodies or media (such as air or water) the ether was at rest.
As early as 1818, however, Fresnel had predicted that the ether
inside material media partook in some degree of their motion.
In 1851 Fizeau devised an experiment to test this prediction,
making use of a very sensitive method depending on interference

fringes. He found that the velocity of light in a moving
column of water is greater downstream than upstream, and he
took this to mean that the water drags the ether along with it,
as Fresnel had predicted. The matter was re-investigated in
1886 by Michelson and Morley (741a) who obtained the same result.

694. FOUCAULT, L. (See also 684) **(a)** Devised in 1851 his famous experimental demonstration of the earth's rotation by means of a
very long pendulum with a heavy bob. This led him in the following year to the invention of the gyroscope. **(b)** During the
1850s he invented many instruments and devices for use in pure
and applied physics, as well as in astronomy (e.g. a regulator
to keep a telescope pointed constantly at a star). Especially
important for astronomy were his method for silvering glass to
make mirrors for reflecting telescopes and his simple but very
accurate method for testing and correcting the shapes of mirrors
and lenses.   (See also 707)

695. RÜHMKORFF COIL. Invented in 1851 by the instrument-maker,
H.D. Rühmkorff.

696. STOKES, G.G. (See also 676) Elucidated the phenomenon of fluor-
escence (his term) in 1852 and used it as a means for studying
the ultraviolet spectrum (before photographic methods were
available).

697. RANKINE, W.J.M. (2410) Around 1850 he was, as an engineer, concerned with calculating the efficiency of steam engines and in
the early 1850s his publications included some of the thermodynamic ideas which were then being formulated by Thomson and
Clausius. In 1853 he introduced the concept of energy and
emphasized its fundamental importance as a primary agent in
nature and its unifying role in physical theory.

698. ÅNGSTRÖM, A.J. (2863) Concluded in 1853 that an incandescent gas
emits spectral lines of the same wavelength as those it absorbs
when cold. He also considered that a flame spectrum is essentially a reversal (or negative) of a part of the solar spectrum.

699. WEBER, W.E. (See also 688) In 1855, in collaboration with R.H.A.
Kohlrausch, he determined the ratio of the electromagnetic and
electrostatic units of charge--a constant that was then of no
special importance (but was later to become prominent in
Maxwell's theory).

700. MAXWELL, J.C. (2413) "On Faraday's lines of force"--a paper of
1855/56 which constituted the beginning of his researches on
electromagnetism.   (See also 705a)

701. THE GEISSLER PUMP. A mercury pump (based on the use of the
Torricellian vacuum), which produced a much better vacuum than
piston pumps, was invented in the late 1850s by J.H.C. Geissler,
instrument-maker at the University of Bonn. It later had a
far-reaching effect on research in physics, especially in the
field of discharge phenomena in rarefied gases (cf. 702, 714,
etc.). In 1865 Geissler's pump was improved and made much
easier to use by H.J.P. Sprengel.

702. PLÜCKER, J. (2592) Discovered in 1858 what he called "the beauti-
ful and mysterious green glow" produced by electrical discharges

in tubes evacuated by means of the Geissler pump. He observed
that the glow was deflected by a magnet.

703. STEWART, B. (2412) In 1858, as a result of his experiments with
      infrared radiation, he extended Prevost's theory of exchanges
      (632) by showing that, at each wavelength, the amount of radia-
      tion emitted by a body equals the amount it absorbs. His work
      was however overshadowed by the famous researches of Kirchhoff
      who reached a similar conclusion from experiments with optical
      spectra the next year (704b).

704. KIRCHHOFF, G.R. (See also 686) (a) In 1859, in a joint paper
      with his Heidelberg colleague, Bunsen, he established the method
      of spectrum analysis, building on the foundations laid long
      before by Fraunhofer (644), whose observations he explained.
      Bunsen went on to exploit the potentialities of the new method
      in chemistry while Kirchhoff made use of it to study the comp-
      osition of the sun. (b) In consequence of his demonstration
      of the relation between emission and absorption spectra he
      concluded that the ratio of the emissive and absorptive powers
      of a body, at each wavelength, is the same for all bodies at
      the same temperature ('Kirchhoff's radiation law').

1860s

705. MAXWELL, J.C. (See also 700 and 706) (a) "Illustrations of the
      dynamical theory of gases"--a paper of 1860 which was inspired
      by Clausius' pioneering investigations (691b). It contained
      the derivation of a statistical formula (which he likened to
      the Gaussian formula for the distribution of errors in measure-
      ments) for the distribution of the velocities of gas molecules.
      (b) In a second paper of 1860 he gave another derivation of his
      velocity distribution function and used it to explain various
      properties of gases such as viscosity, diffusion, and heat
      conduction. (c) The foregoing work was refined and extended
      in a further paper in 1867.

706. MAXWELL, J.C. (See also 700 and 705) (a) "On physical lines of
      force", 1861/62. His second paper on the electromagnetic field.
      It contained the historic conclusion that "light consists in
      the transverse undulations of the same medium which is the cause
      of electric and magnetic phenomena." (b) "A dynamical theory
      of the electromagnetic field", 1865. His fourth paper on the
      subject and the first full presentation of his theory.
      (c) "Note on the electromagnetic theory of light", 1868. An
      important sequel which included a simplification of his
      equations.   (See also 720)

.... ANDREWS, T. On the continuity of the gaseous and liquid states.
      See 928.

707. FOUCAULT, L. (See also 694) In 1862, as a sequel to his early
      work with Fizeau (683a), but with improved apparatus, he
      carried out a very accurate determination of the velocity of
      light.

708. ÅNGSTRÖM, A.J. (See also 698) (a) From his studies of the solar
      spectrum in the early 1860s he concluded that hydrogen is

present in the sun and probably a number of other elements as
well. **(b)** *Recherches sur le spectre solaire*. Uppsala, 1868.
Included an atlas of the solar spectrum giving wavelengths of
about a thousand of the lines, the wavelengths being determined
by means of a diffraction grating. The work was highly regarded
and for many years constituted the standard for wavelength
determinations. The unit in which he expressed his measurements
later became known as the angstrom unit.

709. TYNDALL, J. (2411) **(a)** *Heat considered as a mode of motion*.
London, 1863. (9th ed., 1892) An influential popularization
of the recently accepted dynamical theory of heat. **(b)** In a
series of investigations through the 1860s he studied the
absorption (and radiation) of radiant heat by gases. He found
that elementary gases such as oxygen, nitrogen, and hydrogen
absorb (and radiate) hardly any heat, but more complex gases do,
the amount increasing with the complexity of their molecules.
Water vapour was an especially important case because of its
implications for meteorology and he studied it intensively.
(See also 719)

710. CLAUSIUS, R. (See also 691a) In a celebrated paper in 1865 he
introduced the term 'entropy' and developed its meaning--thus
the famous statement of the second law of thermodynamics: "The
entropy of the universe tends to a maximum." He accepted
Rankine's (and Thomson's) concept of energy (692b) in his
statement of the first law: "The energy of the universe remains
constant."

711. KUNDT, A.A. (2676) Developed in 1865-68 a device--'Kundt's tube'
--with which he made accurate measurements of the speed of
sound in air and other gases.   (See also 726)

712. *TREATISE on natural philosophy*. Oxford, 1867. (2nd ed., 2 vols,
1879-83) By W. Thomson and P.G. Tait. Though limited to kinetics and dynamics (instead of the much wider scope originally
planned) it was a highly influential text. It made much use of
the concept of energy, establishing it as part of the concept-
ual structure of mechanics.

713. BOLTZMANN, L. (2961) In his first major paper, in 1868, he
extended Maxwell's seminal application of statistics to the
kinetic theory of gases (705) and derived a new formula for
the distribution of velocity among gas molecules.
(See also 716)

714. HITTORF, J.W. (2634) Following up the work of Plücker on discharges in Geissler tubes (702), he described in 1869 the
changes accompaning lowering of the gas pressure and various
other phenomena. He established that the glow is produced by
rays originating at the cathode and travelling in straight lines.

1870s

715. KOHLRAUSCH, F. (2677) **(a)** See 944. **(b)** *Leitfaden der praktischen
Physik*. Leipzig, 1870. The first important laboratory manual
for physics. It ran to sixteen editions and was translated into
several languages. The first edition of the English translation,
Introduction to physical measurements, appeared in 1873.

716. BOLTZMANN, L. (See also 713) In a series of papers in the 1870s
     he further extended Maxwell's velocity distribution law, arriv-
     ing at a general statement for the distribution of energy among
     the various parts of a system, and deriving the theorem of the
     equipartition of energy. This work laid the foundations of
     statistical mechanics and interpreted the second law of thermo-
     dynamics in terms of the statistics of molecular motions.
     (See also 738)

717. HELMHOLTZ, H. von. (See also 690) Between 1871 and 1876 he played
     an important role in developments concerning the mathematical
     theory of electrodynamics. He rejected Weber's law (688),
     which was then generally accepted on the Continent, and in an
     extended controversy weakened the support for it and prepared
     the way for its replacement by Maxwell's theory. His own
     theorizing and experimentation gradually led him to an accept-
     ance of Maxwell's theory and many Continental physicists follow-
     ed his lead--as did his most outstanding student, Hertz, who in
     the late 1880s provided powerful experimental support for the
     theory (see 743).

718. STRUTT, J.W. (*Lord* RAYLEIGH) (2437) **(a)** Found in 1871 that the
     degree of scattering of light by very fine particles is a
     function (the inverse fourth power) of the wavelength of the
     light. This relationship ('Rayleigh scattering') provided a
     solution to the old problem of the blue colour of the sky.
     (b) *The theory of sound*. 2 vols. London, 1877-78. (2nd ed.,
     1894-96 and later reprints) A comprehensive and authoritative
     treatise which long remained the leading work in the field.

719. TYNDALL, J. (See also 709) His study in the early 1870s of the
     scattering of light by fine dust particles--in which he eluci-
     dated what is now known as the Tyndall effect as well as
     Rayleigh scattering--led him on, in view of Pasteur's work, to
     a realization of the importance of airborne bacteria and thence
     to an effective intervention in the controversy about spontan-
     eous generation (see 1810).

720. MAXWELL, J.C. (See also 706) *A treatise on electricity and mag-
     netism*. 2 vols. Oxford, 1873. A loosely-arranged work on
     his electromagnetic theory which strove to extend and go beyond
     it. (The systematic presentation which he intended to write
     never appeared because of his untimely death) The work contain-
     ed his prediction, confirmed experimentally by Lebedev in 1899,
     that light exerts pressure.

721. GIBBS, J.W. (3044) In his first papers on thermodynamics in 1873
     he virtually rescued Clausius' concept of entropy from neglect
     and confusion, and gave it its due place in a new and compre-
     hensive statement of thermodynamic principles. His chief paper,
     appearing in 1876, was centered on the properties of equilibrium
     states of simple systems (rather than the changes in heat and
     work in various processes which had been the concern of thermo-
     dynamics hitherto) and greatly enlarged the range of phenomena
     to which the thermodynamic approach could be applied. His chief
     application was to chemical equilibrium, thereby laying the
     foundations of chemical thermodynamics. It was not however
     until about 1890 that his work attracted much attention (largely
     because of his abstract treatment and austere style)--cf. 966.

722. WAALS, J.D. van der. (2794) In his doctoral dissertation of 1873 he put forward an equation for the gas laws ('the van der Waals equation') which contained terms relating to the volumes of the molecules themselves and the attractive forces between them. Pressure-volume relations (for particular values of temperature) derived from the equation agreed well with the experimental findings of Andrews (928) and others. His work gave an explanation in molecular terms of certain observed properties of gases and vapours, especially the critical temperature.

723. HEAVISIDE, O. (2438) For many years from 1873 onward he published numerous papers on the theory of cable telegraphy. His main contributions to the subject had far-reaching theoretical significance, as well as great practical value, and played an important part in developing and applying Maxwell's electromagnetic theory.

724. ROWLAND, H.A. (3052) Established in 1875 that a moving electric charge produces a magnetic field as if it were a current flowing through a conductor.   (See also 735)

725. THE KERR EFFECT. Two different phenomena discovered by the Scottish experimentalist, John Kerr, in 1875 and 1876 respectively, are both often called the Kerr effect. **(a)** When a transparent substance is placed in a strong electric field it becomes optically anisotropic and doubly refracting. **(b)** When plane polarized light is reflected from the (polished) pole of an electromagnet it becomes elliptically polarized. These discoveries were first treated theoretically by G.F. Fitzgerald in 1880.

726. KUNDT, A.A. (See also 711) In joint investigations in 1875-76 with his colleague, E.G. Warburg, he provided important experimental confirmation of the kinetic theory of gases. In particular they established that the ratio of the specific heats, at constant pressure and volume, of monatomic gases (such as mercury) has the value 5/3, as predicted by the theory.

727. CROOKES, W. (2416) In the late 1870s his success in producing a very high vacuum led him to a series of researches on electrical discharges through rarefied gases. He observed the dark space around the cathode (later called the Crookes dark space) and studied the properties of what he called "radiant matter" (i.e. cathode rays) which he thought to consist of negatively charged corpuscles. He found that it travelled in straight lines, was deflected by a magnet, and induced phosphorescence in various materials.

728. GOLDSTEIN, E. (2704) From 1876 onward for many years he studied cathode rays (his term) and numerous phenomena associated with them. In his initial investigations he independently made some of the same findings as were made by Crookes (727). In contrast to Crookes he regarded the rays as analogous to light, even though he had found that they were deflected by a magnet. (See also 742)

729. CAILLETET, L.P. (2174) During 1877-78 he succeeded in liquefying the six gases (oxygen, hydrogen, etc.) which had hitherto been considered "permanent." His success was due to his experimental skill together with his realization--following the introduction

of the concept of critical temperature by Andrews (928)--that
very low temperatures were needed as well as high pressures.

730. DEWAR, J. (2428) From 1877 until after the end of the century he
carried out a long series of researches in cryogenics, in the
course of which he invented the vacuum-jacketed flask named
after him. He devised methods for producing large quantities
of liquid air and oxygen, and after many attempts succeeded in
liquefying and solidifying hydrogen in 1898.
        Dewar's chief interest was in the physical properties
and chemical reactivities of substances at very low temperature:
he found for example that liquid oxygen is magnetic and, in
collaboration with J.A. Fleming, measured the resistance and
other electrical and magnetic properties of metals at decreasing
temperatures; their results indicated that resistance would
gradually disappear as absolute zero is approached.

.... POINCARÉ, J. See 286.

731. THE BROWNIAN MOVEMENT. This phenomenon, discovered in 1827 by
the botanist Robert Brown (see 1358), long resisted explanation
but in 1877 William Ramsay suggested that it was caused by
"thermal molecular motion in the liquid environment." The
theory of the phenomenon continued to interest physicists for
many years.

732. STEFAN, J. (2956) Discovered experimentally in 1879 that the
amount of heat radiated by a surface is proportional to the
fourth power of the absolute temperature (cf. Boltzmann--738).

1880s
----

733. CURIE, P. (2208) In 1880, while working with his brother, Jacques
(who was an assistant in the mineralogical laboratory at the
Sorbonne), he discovered piezoelectricity in crystals that have
no centre of symmetry. This led him later to some important
work on the laws of symmetry.   (See also 753)

734. HYSTERESIS. Discovered in 1881 by the English professor of engineering, J.A. Ewing, who later investigated it in detail. The
phenomenon was also discovered independently at about the same
time by A. Righi and by E.G. Warburg.

735. ROWLAND, H.A. (See also 724) In the early 1880s he developed a
method for making concave diffraction gratings of unprecedented
accuracy. Such gratings (which he made at cost for many other
physicists) were very valuable for spectroscopy, and using one
of them he remapped the solar spectrum with far greater accuracy
than previously.

736. EÖTVÖS, R. von. (2967) (a) His investigations of the surface
tension of liquids in the early 1880s led to his discovery in
1886 of the law relating surface tension to temperature.
(b) From 1888 until his death in 1919 he published numerous
papers dealing with the design and uses of an extremely sensitive torsion balance. Its ability to measure the gravitational
field at the earth's surface very accurately made it an invaluable instrument for geophysical exploration. By means of it
Eötvös was able to demonstrate the equivalence of gravitational
and inertial mass (a cardinal point for the theory of relativity).

737. *PHYSIKALISCH-chemische Tabellen*. Berlin, 1883. By H.H. Landolt and R. Börnstein. Successive editions, progressively enlarged, have continued to the present.

738. BOLTZMANN, L. (See also 716) In 1884 he ingeniously deduced-- from the second law of thermodynamics and Maxwell's theory of electromagnetism--the law relating radiated energy to absolute temperature that had earlier been discovered experimentally by Stefan (732). The 'Stefan-Boltzmann law' was later to provide a point of departure for Wien's work on radiation (754).

739. "ON THE TRANSFER of energy in the electromagnetic field"--a paper of 1884 by J.H. Poynting (professor of physics at Birmingham). In it he derived from Maxwell's theory a simple expression ('Poynting's vector') for the flow of energy in an electro- magnetic field.

740. THE BALMER SERIES. The spectrum of hydrogen in the visible region exhibits a prominent series of lines whose wavelengths can be correlated with remarkable accuracy by a simple numerical form- ula discovered in 1885 by the Swiss schoolteacher, J.J. Balmer. He predicted that similar series would be found in the ultra- violet and infrared regions--as was later found to be the case. Other formulae for spectral series were later found, notably by J.R. Rydberg.

741. MICHELSON, A.A. (3053) (a) In 1886, in collaboration with his colleague, E.W. Morley, he confirmed the notable result that Fizeau (693) had obtained in his "ether-drag" experiment. (b) In the following year, again with Morley, he carried out an improved version of an experiment (which he had first attempt- ed in 1881) to test the hypothesis that the earth is moving through the luminiferous ether. The experiment made use of his interferometer which provided an extremely sensitive method of detecting the postulated change in the velocity of light. No change was found--one of the most significant negative results in the history of science.

742. GOLDSTEIN, E. (See also 728) In the course of his investigations of electrical discharges through gases at low pressure he dis- covered in 1886 what he called *Kanalstrahlen* (canal rays, later named positive rays) which emerged from channels or holes in the cathode in a direction opposite to the cathode rays.

743. HERTZ, H.R. (2730) In a series of experiments in 1886-88 he gener- ated electromagnetic (radio) waves by means of an induction coil and spark gap, and devised a simple detector for them. He found that stationary waves were produced by reflection from the walls of his laboratory and was able to measure their wavelength (about 30 cm); from this value and the calculated frequency of his oscillator he deduced the velocity of propagation of the radiation and found it to be the same as that of light. He also showed that the radiation could be reflected, refracted, and polarized.
       These findings provided overwhelming evidence for Maxwell's theory of electrodynamics. In subsequent theoretical papers Hertz reorganized and greatly clarified Maxwell's theory.

744. THE PHOTOELECTRIC EFFECT. Discovered accidently by Hertz in 1887 in the course of his researches on electromagnetic radiation

which left him no time to follow up the discovery. It was soon investigated further by several other physicists, notably W.L.F. Hallwachs at Strasbourg and A. Righi at Padua.

745. DESLANDRES, H. (2196) In the late 1880s he studied emission spectra of molecules such as nitrogen, cyanogen, and water, and derived two laws characterizing the bands in their spectra. His laws later proved useful in the study of molecular spectra.

746. LODGE, O.J. (2440) In some experiments in 1887–88 with the electrical oscillations produced by the discharge of a Leyden jar he came close to making the discovery that was made at that time by Hertz (743). He subsequently devised the coherer, a detector of radio waves that was widely used for several years before it was superseded. This and his other work in the field during the 1890s played a major part in the genesis of radio-telegraphy.   (See also 752)

747. FITZGERALD, G.F. (2439) Put forward in 1889 an explanation of the negative result of the Michelson-Morley experiment (741b). He suggested—on the basis of certain ideas at the time about the electrical nature of matter—that a body in motion is shorter (along its line of motion) than when it is at rest, and that such a contraction had affected the instruments used in the experiment. The same idea was proposed in 1895 by H.A. Lorentz and was later used by Einstein in the special theory of relativity.

1890s

748. BOYS, C.V. (2441) (a) *Soap-bubbles and the forces which mould them.* London, 1890. A popular book based on lectures given at the Royal Institution and embodying original contributions to the knowledge of surface tension and the properties of thin films. (b) Boys introduced the use of fused quartz into experimental physics and in the early 1890s repeated Cavendish's measurement of the gravitational constant (1043) using a fibre of fused quartz in the torsion balance. His measurement, the accuracy of which has hardly been surpassed, is regarded as one of the classics of experimental physics.

749. *HANDBUCH der Physik.* 5 vols. Breslau, 1891–96. Edited by A.A. Winkelmann. Contained articles by leading German physicists.

750. LORENTZ, H.A. (2795) Following the work of Hertz (743) he took up Maxwell's theory and, in a series of publications between 1892 and 1904, developed it into his 'electron theory' in which the relationship between matter and the ether was explained in terms of the connection between electrons and the electromagnetic field. The discovery of the Zeeman effect by his assistant, Pieter Zeeman, in 1896 (761) provided strong support for his theory.

751. LENARD, P. (2732) In 1892, during a series of experiments on cathode rays, he devised a tube with a thin aluminium window which transmitted the rays outside the tube, thus enabling them to be studied separately. Later, at about the same time as J.J. Thomson and others, he showed that the rays consist of streams of negatively-charged particles (electrons). From an

investigation of the permeability of matter to the rays he suggested that the atoms of matter must be mostly empty space and he advanced some pertinent speculations about atomic structure.

752. LODGE, O.J. (See also 746) The negative result of the famous Michelson and Morley experiment (741b) was generally interpreted as showing that the luminiferous ether in the vicinity of a moving body was carried along with it. In 1893 however Lodge proved, by means of an ingenious experiment, that this is not so. The apparent contradiction did much to discredit the whole idea of the ether.

753. CURIE, P. (See also 733) (a) In the early 1890s, from a study of the effects of temperature on the magnetic properties of substances, he clarified the relations between diamagnetism, paramagnetism, and ferromagnetism, and discovered that paramagnetism is inversely proportional to the absolute temperature.
(b) In 1898 he dropped his own research to join in the investigation that his wife, Marie, had begun early that year into the phenomenon discovered by Becquerel (759); his physical approach was a valuable complement to her chemical approach. Later in the same year they announced the discovery of two new, very radioactive, elements--polonium and radium. Subsequently Pierre showed that radioactivity decreases exponentially at a rate that is unaffected by any agent, and he also measured the (large) amount of heat evolved by radium.

754. WIEN, W. (2734) (a) Working from the Stefan-Boltzmann law (738) he deduced in 1893 that the wavelength at which a black body radiates maximum energy is inversely proportional to the temperature of the body. This 'displacement law' was soon afterwards confirmed experimentally by others.
(b) Wien went on in 1896 to deduce his energy distribution law according to which the energy density at a particular wavelength varies inversely as the fifth power of the wavelength. When tested experimentally in several laboratories this law was found to hold at short wavelengths but to be badly in error at higher wavelengths. (It was this result that in 1900 led Planck to propose the quantum theory as a new approach to the problem)

755. PASCHEN, F. (2735) (a) Discovered experimentally (and independently) Wien's displacement law (754a) in 1894 and Wien's energy distribution law (754b) in 1895. (b) In collaboration with Carl Runge in 1895 he determined very accurately the optical spectrum of the newly-discovered element, helium, and established its spectral line series.

756. LARMOR, J. (2460) His 'electron theory'--comparable to that of H.A. Lorentz (750)--was presented in three papers in 1894-98 and elaborated in his widely-read book *Aether and matter* in 1900.

757. DRUDE, P.K.L. (2733) (a) *Physik des Äthers auf elektro-magnetischer Grundlage*. Stuttgart, 1894. One of the first German treatises based on Maxwell's theory. (b) His work on the physical optics of crystals, done during the 1890s, was included in his *Lehrbuch der Optik* of 1900 which also supported Maxwell's theory.

758. RÖNTGEN, W.C. (2703) While experimenting with a Crookes (cathode
     ray) tube in 1895 he accidently discovered a new kind of radia-
     tion emitted by the glass wall of the tube from the spot where
     the cathode rays impinged on it. He made an intensive study of
     the rays which he named X-rays because of their uncertain nature:
     since they could not be reflected or refracted he thought they
     were not related to light.

759. BECQUEREL, H. (2198) In 1896, while testing a (wrong) hypothesis
     suggested to him by Röntgen's recent discovery of X-rays, he
     discovered that uranium and its compounds emit penetrating
     radiation. The discovery did not seem very significant at the
     time and he did not follow it up. As a result of the Curies'
     work (753) however he returned to the subject and in 1899 demon-
     strated that the radiation from a sample of radium (lent to him
     by the Curies) contained the same kind of "corpuscles" that
     J.J. Thomson had recently identified in cathode rays.

760. RUBENS, H. (2751) In the course of a series of researches explor-
     ing the far infrared region he discovered the phenomenon of
     Reststrahlen in 1896. (His subsequent experiments with them
     yielded results which in 1900 had an important bearing on the
     origin of Planck's quantum theory)

761. ZEEMAN, P. (2804) Following up an experiment of Faraday's, he
     began in 1896 to investigate the effect of a strong magnetic
     field on light and discovered what soon came to be known as
     the Zeeman effect. His discovery provided important evidence
     for Lorentz' theory that light is caused by the oscillation of
     electrons (750). In turn, Lorentz' theory enabled him to calc-
     ulate from his data the charge-to-mass ratio of the oscillating
     particles and to show that they are negatively charged.

762. THOMSON, J.J. (2459) From his investigations of cathode rays he
     established in 1897 that they consist of negatively charged
     "corpuscles" (electrons) of much smaller mass than atoms.
     Cathode rays, furthermore, appeared to be composed of the
     same kind of corpuscles irrespective of the nature of the gas
     in the discharge tube or the metal of the cathode. He concluded
     that the corpuscles were sub-atomic constituents of all kinds
     of matter--a conclusion that was strengthened during the next
     few years when it was found that they could be produced in a
     variety of other ways. He went on to propose the first theory
     of atomic structure.

763. PLANCK, M. (2731) *Vorlesungen über Thermodynamik*. Leipzig, 1897.
     (11th ed., 1966. Trans. into five languages) A masterly treatise.

764. CHARGE ON THE ELECTRON. Measured by students of J.J. Thomson at
     the Cavendish Laboratory in 1898.

765. LEBEDEV, P.N. (3012) By means of ingenious and skilful experiments
     he succeeded in 1899 in not only detecting the pressure of light
     on solid bodies but also measuring it.

In a lecture in 1900 Lord Kelvin said that he saw only two clouds in
the firmament of physics--the failure to explain the earth's motion
through the ether, and the difficulties associated with the equipart-
ition of energy. Both clouds were soon to bring cyclonic disturbances.

1.06 CHEMISTRY

Including medical chemistry and biochemistry.

<hr>

1470s

766. *VON DEN ausgebrannten Wassern.* Augsburg, 1477. (Often reprinted) A distillation book. By M. Puff von Schrick, professor of medicine at Vienna.

1500s

<hr>

767. *LIBER de arte distillandi de simplicibus.* Strasbourg, 1500. (Various later editions. German trans., 1512? English trans., 1527) For many years the most important distillation book, especially because of its numerous illustrations. By H. Brunschwig, an apothecary and surgeon in Strasbourg.

1520s

<hr>

.... PARACELSUS. See 2482.

768. *PROBIERBÜCHLEIN.* Magdeburg, 1524. (The earliest dated edition; there are several others, some undated) An anonymous manual of assaying and metallurgy.

769. *COELUM philosophorum, seu de secretis naturae liber.* Fribourg, 1525. (Over twenty editions and translations) A very popular distillation book, especially significant in that it adopted the laboratory techniques of the alchemical tradition for medical purposes, thus foreshadowing the iatrochemical movement. By P. Ulstadt, a teacher of medicine at the Fribourg Academy.

1540s

<hr>

770. *DE LA pirotechnia.* Venice, 1540. An original account, with many woodcut illustrations, of numerous metallurgical and chemical processes. The first work of its kind. By V. Biringuccio, metallurgist, engineer, and architect.

771. *PLICTHO de larte de tentori.* Venice, 1548. The first book on dyeing. By G. Rosetti.

1550s

<hr>

772. *SECRETI* [supposedly of Alexis of Piedmont]. Venice, 1555. One of the most extraordinary successes in the technical literature of early modern times: by the end of the seventeenth century more then ninety editions had appeared, in virtually every European language. It was a compilation of recipes for medicinal preparations, dyes, pigments, cosmetics, etc. The author

was G. Ruscelli, a minor literary figure involved in the printing and publishing trade in Naples.

.... AGRICOLA, G. *De re metallica*. See 1007.

1560s
---

773. BODENSTEIN, A. von. (2885) During the 1560s and 1570s he published many of the works of Paracelsus which had hitherto
remained in manuscript.

774. DORN, G. (2886) An early follower of Paracelsus, some of whose
writings he edited. His own books, which were numerous and
chiefly concerned with alchemy and occult philosophy, appeared
in the period 1566–84. They also helped to disseminate the
Paracelsian doctrines, especially his dictionary of Paracelsian
terms (which had several editions in the original Latin as well
as some translations).

1570s
---

775. SEVERINUS, P. (2824) *Idea medicinae philosophicae*. Basel, 1571.
"The first major synthesis of the Paracelsian corpus." Widely
discussed and very influential.

776. DUCHESNE, J. (1982) *Ad Iacobo Auberti ... brevis responsio*.
Lyons, 1575. (Many reprints. Trans. into French, German, and
English) A strong defence of chemical medicines and the
Paracelsian doctrines. It attracted much attention.

777. *BESCHREIBUNG allerfürnemisten mineralischen Erzt und Berckwerks-
arten*. Prague, 1574. "The first manual of analytical and
metallurgical chemistry." Editions and translations of it
extended to the mid–eighteenth century. The author was Lazarus
Ercker, an assayer and manager of mints in Saxony, Brunswick,
and finally Prague.

1590s
---

778. LIBAVIUS, A. (2503) *Alchemia*. Frankfurt, 1597. (2nd ed., enl.,
1606) A treatise of iatrochemistry rather than alchemy, it is
regarded as the first textbook of chemistry and the most important book in the subject in the early seventeenth century. It
was distinguished by its clarity and system (in marked contrast
to most other contemporary books in the field) and by the excellence of its illustrations.

1600s
---

779. GUIBERT, N. (1983) *Alchymia ratione et experientia ... impugnata
et expugnata*. Strasbourg, 1603. A vigorous attack on alchemy
by an ex–alchemist. Among other things it argued strongly that
metals are distinct chemical species, not transmutable into each
other and not analogous to biological species.

780. DUCHESNE, J. (See also 776) **(a)** *De priscorum philosophorum verae
medicinae materia*. St. Gervais, 1603. (Many later editions in
various languages) A book which initiated a vigorous controversy in Paris in 1603–04 between the adherents of the traditional Galenic medicine and Duchesne and other iatrochemists.

(b) *Pharmacopoea dogmaticorum restituta.* Paris, 1607. (Twenty-five editions by 1650) A collection of chemical remedies.

781. CROLL, O. (2938) *Basilica chymica.* Frankfurt, 1609. (At least eighteen later editions as well as translations into French, English, and German) The chief work to disseminate a knowledge of Paracelsian iatrochemistry and to make it comparatively respectable.

1610s

782. BEGUIN, J. (1986) *Tyrocinium chymicum.* Paris, 1610. (Many later editions and translations) A clear and straightforward description of the preparation of many chemical remedies. Immensely popular. The first in a long line of important French textbooks of chemistry.

783. *L'ARTE vetraria.* Florence, 1612. (Several later editions and translations) The first book on glassmaking. By A. Neri.

784. SENNERT, D. (2509) (a) *Epitome naturalis scientiae.* Wittenberg, 1618. (b) *De chymicorum cum Aristotelicis et Galenicis consensu ac dissensu.* Wittenberg, 1619. He was best known for his theoretical ideas which were very eclectic, attempting to reconcile the theories of the Paracelsians, atomists, Galenists, and Aristotelians.

1620s

785. SALA, A. (2510) Author of many works on iatrochemistry, the main ones being written in the 1620s. His attempts to clarify the concepts of analysis and synthesis are noteworthy.

1630s

786. HARTMANN, J. (2504) *Praxis chymiatrica.* Leipzig, 1633. (Posthumous. Many reprints) A textbook of iatrochemistry.

787. DAVISON, W. (1999) *Philosophia pyrotechnica.* 4 parts. Paris, 1633-35. An iatrochemical treatise with a heavy emphasis on Neo-Platonic and Paracelsian theory. In 1651, after he had become professor of chemistry at the Jardin du Roi, he brought out a much revised and less philosophical version in French: *Les élémens de la philosophie du feu, ou chimie.*

1640s

788. *EL ARTE de los metales.* Madrid, 1640. (Many editions and translations extending to the mid-eighteenth century) A very competent and full account of metallurgical procedures, especially for the extraction of gold, silver, and copper. By A.A. Barba, a Spanish priest who served in Peru for many years.

789. GLAUBER, J.R. (2767) Between 1646 and his death in 1670 he wrote a large number of books (in German but often with Latin titles), many of them very popular. Various collections of his writings were translated into Latin, English, and French. His best book was his *Furni novi philosophici* (5 vols, Amsterdam, 1646-49). It contained most of his many chemical discoveries and gave details of his improved furnaces and other laboratory equipment. His efficient methods for preparing the mineral acids were especially important.

790. HELMONT, J.B. van. (2810) *Ortus medicinae*. Amsterdam, 1648.
     (Posthumous. His collected works, containing much chemistry.
     Many later editions and translations) The famous chemical
     philosopher. "His work, more than that of any other, was
     responsible for the tremendous influence of chemistry in the
     decades following ... With van Helmont there was a new emphasis
     on observational and experimental data, coupled with an interest
     in the development of new instruments and in the application of
     quantification to chemical and medical problems." (A.G. Debus)

1650s

791. ASHMOLE, E. (2250) *Theatrum chemicum Britannicum*. London, 1652.
     (In English) Contained annotated extracts from the writings
     of English alchemists.

1660s

792. SYLVIUS, F. (2770) His very influential iatrochemical doctrine
     (see 1717), speculative though it was, proved quite beneficial
     to chemistry in several ways, especially by giving it a greater
     significance in the eyes of many in the medical profession.

793. LE FEBVRE, N. (2008) *Traicté de la chymie*. Paris, 1660. (Many
     later editions. English trans., 1662 and later. German trans.,
     1672 and later) A notable textbook, firstly because it trans-
     mitted the fruits of recent German work in chemistry to French
     (and English) readers, and secondly because it constituted the
     culmination of the Paracelsian and Helmontian tradition in the
     seventeenth-century sequence of French iatrochemical textbooks
     (its later successors were, in marked contrast, in the Cartes-
     ian style).

794. BOYLE, R. (2251) *The sceptical chymist*. London, 1661. A polem-
     ical work opposing the iatrochemists' doctrine of the three
     principles (salt, sulphur, and mercury), as well as the four
     elements of the Aristotelians, and advocating his own corpus-
     cular theory (cf. 21).
        Up to his death in 1691 Boyle wrote a large number of
     books and tracts on chemistry, all replete with experiments.
     Though the factual material in them served as a quarry for
     later chemists their theoretical conclusions were slight--
     inevitably, for it was not until a century later that chemistry
     began to acquire a theoretical foundation.   (See also 798)

795. GLASER, C. (2017) *Traité de la chymie*. Paris, 1663. (Many
     later editions and translations) A straightforward and compet-
     ent account of chemical procedures and the preparation of drugs.
     Unlike the better-known textbooks of Le Febvre (793) and Lemery
     (799) it did not indulge in speculative theory.

796. TACHENIUS, O. (1895) (a) *Hippocrates chimicus*. Venice, 1666.
     (b) *Antiquissimae Hippocraticae medicinae clavis*. Venice, 1669.
     Both works were primarily concerned with theoretical (and polem-
     ical) issues, especially the philosophy of acid and alkali as
     universal principles in both the living and non-living realms--
     a doctrine also expounded at that time by Sylvius and others.
     The two books also included much practical chemistry and reveal
     a virtual industrialization of drug production in Venice.

797. BECHER, J.J. (2520) *Physica subterranea*. Frankfurt, 1669. An
     iatrochemical treatise, very theoretical in the Paracelsian
     and Helmontian style. Its chief significance was in the influ-
     ence it had a generation later on G.E. Stahl (803) who developed
     some of its inchoate ideas into the phlogiston theory.

1670s

798. PHOSPHORUS. Discovered about 1674 by a German alchemist, H.
     Brand. He kept the method of preparation secret but enough
     information leaked out to enable others to rediscover it,
     notably Kunckel and Boyle. The latter made a thorough investi-
     gation of its properties which he published, together with his
     method of making it, in a tract of 1680.

799. LEMERY, N. (2024) *Cours de chymie*. Paris, 1675. (There were
     eleven editions apparently revised by the author and over
     twenty more later, as well as translations into several lang-
     uages) A book based on his very popular lecture course and
     owing its immense success both to his clear and attractive
     presentation and to his engaging (but wholly speculative) explan-
     ations of chemical phenomena in terms of the new mechanical
     philosophy. Though in the latter respect he broke away sharply
     from the Paracelsian-Helmontian theorizing of his predecessors,
     notably Le Febvre (793), he nevertheless retained much of the
     practical content and organization of their books.

800. KUNCKEL, J. (2519) **(a)** *Oeffentliche Zuschrift von dem Phosphoro
     mirabili*. Leipzig, 1678. A brief description of phosphorus
     and its properties (cf. 798). **(b)** *Ars vitraria experimentalis*.
     Frankfurt, 1679. (In German. At least four later editions)
     A collection of essays on glassmaking, some by Kunckel himself
     and others translated by him.

1690s

801. HOFFMANN, F. (2526) During the 1690s and following decades he
     wrote a large number of pamphlets and books describing his
     chemical researches. The most important of them dealt with
     methods of qualitative analysis, especially of mineral waters.

802. HOMBERG, W. (2031) During the period 1692-1714 he published over
     seventy papers in the *Mémoires* of the Paris Academy. Hitherto
     chemical research in the Academy had been at a low level but it
     was now given a big impetus, Homberg's example and inspiration
     --as both an experimentalist and a mechanical philosopher--being
     more important than his actual discoveries (which however were
     not negligible, especially his quantitative experiments on the
     neutralization of acids and bases).

1700s

803. STAHL, G.E. (2525) From about 1700 to 1730 he published a large
     number of books in which, along with much experimentation, he
     developed his famous phlogiston theory--"the first comprehensive
     explanation of the phenomena of combustion and of such related
     biological processes as respiration, fermentation, and decay...
     it dominated chemical thought for almost a century and provided
     a rational basis for much useful experimental work." (C.W. Beck)

804. FREIND, J. (2290) *Praelectiones chymicae*. London, 1709. (English trans., 1712) Lectures on chemistry which he had given at
   the (old) Ashmolean Museum, Oxford, in 1704. They constituted
   one of the earliest attempts to use Newtonian principles (especially attraction) to explain chemical phenomena. "Newton's
   chemical queries [in the 1717 edition of the *Opticks*] resemble
   parts of Freind's work so closely that it is impossible to
   suppose that they are independent." (J.R. Partington)

1710s

805. GEOFFROY, E.F. (2040) Published a paper in the Paris Academy's
   Mémoires in 1718 in which he announced a generalization he had
   reached from his study of the multifarious chemical reactions
   described by earlier chemists--namely that in displacement
   reactions of the type   A + BC → AB + C   substance B has a
   greater *rapport* (the word he used) with substance A than it has
   with substance C. He presented a table comprising several
   columns showing the order of displacement of a number of well-
   known substances. This was the first of what were later to be
   called affinity tables but Geoffroy, whose theorizing was in
   Cartesian style, did not use such a word as 'affinity.'

1720s

806. *NOUVEAU cours de chymie, suivant les principes de Newton et
   Sthall* [sic]. Paris, 1723. (Rev. ed., 1737) An anonymous
   work, generally ascribed to the anatomist, J.B. Senac, but
   possibly by E.F. Geoffroy. It was presumably of some significance in spreading the doctrines of Newton and Stahl among
   French chemists. Neither doctrine, however, was widely accepted
   in France until the middle of the century, and Stahl's ideas
   were disseminated chiefly through the teaching of Rouelle (2059).

807. HALES, S. (2295) Besides its fundamental importance for plant
   physiology his *Vegetable statics* (1327) was also of long-range
   importance for chemistry. It contained a chapter describing a
   series of chemical investigations (arising out of some botanical
   observations) in which Hales thermally decomposed a wide range
   of substances and collected the "air" evolved in an apparatus
   that was the forerunner of the pneumatic trough. He thus demonstrated that "air" (actually various gases) was "fixed" in many
   substances--mineral as well as animal and vegetable--and this
   discovery provided the starting-point for the pneumatic chemistry of the second half of the century. In France it was
   made known through the teaching of Rouelle (2059).

1730s

808. BOERHAAVE, H. (2782) *Elementa chemiae*. Leiden, 1732. (Ten
   editions up to 1759. Trans. into German, French, and English)
   A comprehensive and judicious presentation of the state of
   contemporary chemical knowledge by a famous teacher. It was
   the most popular textbook in the field for some decades but
   lost support later because it did not include the phlogiston
   theory.

809. JUNCKER, J. (2532) *Conspectus chemiae theoretico-practicae.* 2 vols. Halle, 1730-38. (Two later editions. German trans., 1749-53. French trans., 1757) An important textbook which gave a more coherent and systematic account of Stahl's chemical theories, notably the phlogiston theory, than could be found in Stahl's own writings. It consequently did much to disseminate the Stahlian doctrines.

810. BRANDT, G. (2833) **(a)** Published the first thorough study of arsenic and its compounds in 1733. **(b)** *Dissertatio de semi-metallis.* 1735. By 'semi-metals' he meant substances similar to metals in appearance and density but not malleable. He says: "As there are six kinds of metals [Au, Ag, Cu, Fe, Pb, Sn] so ... there are also six kinds of semi-metals [Hg, Bi, Zn, Sb, As, Co]." The book included the first recognition of cobalt as a distinct substance.

811. CRAMER, J.A. (2534) *Elementa artis docimasticae.* Leiden, 1739. (Trans. into English, German, and French) The first textbook of assaying and chemical analysis. It included the first description of the use of the blowpipe.

1740s

812. MARGGRAF, A.S. (2533) A skilful laboratory worker who published many significant findings in a long series of papers in the periodical of the Berlin Academy between 1740 and 1780. He is best known for his contributions to the methods of chemical analysis, a field which was to be of the greatest importance for the subsequent development of chemistry. One of his discoveries which was of vast economic potential—and was made possible by his new techniques—was his finding that beet produces a sugar identical with that from sugarcane.

1750s

813. MACQUER, P.J. (2060) His *Elémens de chymie théorique* (1749) and the companion *Elémens de chymie pratique* (1751) were printed together in 1756 and constituted an important textbook which was translated into several languages. It incorporated the four-element theory of matter which was generally accepted at the time (phlogiston being regarded as a modification of the element fire).   (See also 818)

814. PLATINUM. Native platinum had long been known to the Indians of South America and reports of it began to reach Europe in the mid-sixteenth century. The first thorough account of the metal and its properties appeared in the *Philosophical Transactions* in 1750.

815. CRONSTEDT, A.F. (2839) In a series of papers in the periodical of the Stockholm Academy in the period 1751-64 he described many analyses of ores and minerals. He was one of the first to use the blowpipe extensively in mineral analysis. In 1751 he isolated nickel and this discovery of a new metal attracted much attention.

816. BLACK, J. (2302) "Experiments upon magnesia alba, quicklime,

and some other alcaline substances"--an article of 1756 which described a classic investigation, famous for its quantitative method, which identified "fixed air" (carbon dioxide) and its chief chemical properties, establishing that it was a new kind of "air" which could be a chemical constituent of solid substances.

1760s

817. CAVENDISH, H. (2306) "Experiments on factitious airs"--a large paper of 1766 which made a major contribution to the developing field of pneumatic chemistry. He studied Black's "fixed air", "inflammable air" (hydrogen), which he was the first to identify, and sundry "airs" evolved from the fermentation and putrefaction of organic materials. By determining a number of the physical and chemical properties of these "airs" he showed that they could be distinguished from each other and from ordinary air. This was firm evidence for the existence of different kinds of "airs"--as against the universal and hitherto unquestioned assumption of only one air.   (See also 829)

818. MACQUER, P.J. (See also 813) *Dictionnaire de chymie*. Paris, 1766. (2nd ed., rev. and enl., 1778) The first dictionary of chemistry, well written and deservedly successful.

1770s

819. SCHEELE, C.W. (2846) His greatest discovery--of what he called "fire air", i.e. oxygen--was described in his book, *Chemische Abhandlung von der Luft und dem Feuer* (Uppsala/Leipzig, 1777). Historical investigations have established that he was the first to discover oxygen--some time between 1770 and 1773--and that he told Lavoisier about it in a letter of September 1774.
     His many other discoveries (including chlorine, hydrofluoric acid, several organic acids, etc.) were published in Swedish in the periodical of the Stockholm Academy during the period 1771-86. Collections of his papers were translated into French (1785), English (1786), and German (1793).

820. PRIESTLEY, J. (2307) In 1772 he began publishing a highly important series of experiments on "different kinds of air" in the course of which he discovered nearly a dozen new gases ('nearly' because some of them he did not recognize clearly). He began with papers in the *Philosophical Transactions* but was so prolific that the Royal Society asked him to publish elsewhere and, rather unsatisfactorily, he announced his discoveries in successive volumes and new editions of a work entitled *Experiments and observations on different kinds of air*, first published in 1773.
     His most important discovery, made in 1774, was what he called "dephlogisticated air" (he never accepted Lavoisier's name 'oxygen' or the new view of chemistry in which it was so central). Associated with his discoveries of new "airs" were some experiments which constituted the first steps towards the discovery of photosynthesis, and others which led to eudiometry and gas analysis.

821. NITROGEN. Its discovery is generally attributed to D. Rutherford,

a student of Black in Edinburgh, in 1772. It was however discovered independently about the same time by Priestley, Cavendish, and Scheele.

822. LAVOISIER, A.L. (2071) The maker of the Chemical Revolution. In a series of experiments, distinguished by their logic and quantitative method, which he carried out in the period 1772-83 and published in the Paris Academy's *Mémoires*, he established: **(a)** The role of oxygen in calcination, combustion, and respiration; **(b)** The composition of atmospheric air; **(c)** The composition of sulphuric, phosphoric, and carbonic acids; and **(d)** The composition of water (interpreting and confirming Cavendish's finding--cf. 829)

      In consequence of these results he was able to identify which of the substances then known were "simple substances" (i.e. elements in the modern sense) and to overthrow the phlogiston theory. The radically different view of chemistry which ensued was incorporated in a comprehensive new system of nomenclature (cf. 836) and set out in clear and logical form in his classic *Traité élémentaire de chimie* (Paris, 1789. Trans. into English, 1790 and later; Italian, 1791; German, 1792; Dutch; and Spanish) (See also 830, 836, 837)

823. MANGANESE. Discovered in 1774 by the Swedish chemist, J.G. Gahn.

824. BERGMAN, T.O. (2844) **(a)** Published in 1775 an extensive work, based on thousands of experiments, on 'elective affinity'. It attracted much attention at the time but later chemists began to question the value of such studies. **(b)** In four treatises, published in the period 1778-80, he greatly extended and improved the methods of analysis, both qualitative and quantitative, of minerals and solutions.

1780s

825. BERTHOLLET, C.L. (2079) In a series of researches published during the 1780s he established the composition of ammonia and carried out the first thorough study of chlorine (discovered by Scheele in 1774), demonstrating its bleaching properties which he later introduced to the textile industry. He also found evidence of the existence of acids not containing oxygen --a finding which cast doubt on Lavoisier's theory of acids. (See also 833, 836, 837, 857)

826. MOLYBDENUM. Discovered in 1781 by the Swedish chemist, P.J. Hjelm, following a suggestion by Scheele.

827. KIRWAN, R. (2324) **(a)** In a series of three papers on "attractive powers" in 1781-83 he set out to measure degrees of affinity between pairs of substances (in practice, acids and bases). The combining proportions that he measured were what later came to be called equivalent weights. His findings attracted a good deal of attention and were translated into French and German.

      (b) *An essay on phlogiston and the constitution of acids.* London, 1787. A defence of the phlogiston theory against the new ideas of Lavoisier. In the following year the book was translated into French by Mme Lavoisier with addition of many critical comments by Lavoisier and his followers. The next year Kirwan brought out a new edition including an English translation

of the French comments with his remarks on them. In 1791 he
capitulated to the anti-phlogistonists.

828. FOURCROY, A.F. de. (2090) An able chemist and author of many
papers (cf. 838) but best known for his teaching and his very
influential textbook, *Elémens d'histoire naturelle et de chimie*
(Paris, 1782. Four later editions. Trans. into English, Italian,
German, and Spanish). The third edition (1788) was completely
rewritten in terms of the new chemistry of Lavoisier and did
much to disseminate it.   (See also 836, 837, 849)

829. CAVENDISH, H. (See also 817) Following an observation by Priestley,
he established that water is formed when "inflammable air"
(hydrogen) is burnt in ordinary air or in "dephlogisticated air"
(oxygen). He published his finding in the *Philosophical Transactions* in 1784 but his work had already been discussed in print
in the previous year by Priestley. Both men interpreted the
finding in terms of the phlogiston theory, as did James Watt
about the same time. Lavoisier heard of it in 1783 and immediately saw its great importance for his new theory.

830. "MÉMOIRE sur la chaleur"--a famous paper, published jointly by
Lavoisier and Laplace in 1783, which laid the foundations of
thermochemistry. As well as discussing theoretical considerations concerning heat, they described the construction and use
of an ingenious calorimeter (their term) by means of which they
were able to determine specific heats of substances, heats of
reaction, and the amounts of heat liberated in combustion and
respiration experiments.

831. TUNGSTEN. Discovered in 1783 by the Spanish mineralogist, J.J.
d'Elhuyar, shortly after he had been working on a related topic
in Sweden with Bergman and Scheele.

832. KLAPROTH, M.H. (2561) The successor of Bergman as the leading
analytical chemist in Europe. From 1785 to 1815 he published
over two hundred papers in various periodicals; they were
collected in his *Beiträge zur chemischen Kenntniss der Mineral-
körper* (6 vols, Berlin, 1795-1815. Partial English trans.,
1801-04. Partial French trans., 1807). Besides the many new
analytical procedures he introduced he discovered the elements
uranium, zirconium, titanium, and cerium in the form of their
oxides, and elucidated the composition of many substances,
including compounds of strontium, chromium, beryllium, and
tellurium.

833. "MÉMOIRE sur le fer", 1786. By Berthollet, Monge, and Vandermonde.
A classical memoir which first established that the difference
between iron and steel is due to carbon.

834. *ENCYCLOPÉDIE méthodique ... Chymie, pharmacie, et métallurgie.*
6 vols. Paris, 1786-1815. The chemical section of this vast,
topically-arranged encyclopaedia. Volume 1 was written by
Guyton de Morveau and the following volumes by Fourcroy.

835. BLAGDEN'S LAW--the lowering of the freezing point of a solution
is proportional to the concentration of the solute. Discovered
in 1788 by Charles Blagden, secretary of the Royal Society.
(Nearly a century later the discovery was developed by Raoult
--963)

836. *MÉTHODE de nomenclature chimique*. Paris, 1787. (Trans. into English, 1788; Italian, 1790; and German, 1793) The new system of nomenclature which incorporated Lavoisier's theory (and in its essentials is still used today). The book was written by Lavoisier together with his first 'converts'--Guyton de Morveau, Berthollet, and Fourcroy. The moving spirit was probably Guyton de Morveau who five years earlier had advocated a much-needed reform of the nomenclature of chemistry and had proposed a scheme which was as systematic as the state of knowledge then allowed--but could now be greatly improved.

837. *ANNALES de chimie*. The first French periodical for chemistry (and for many years the most important one). It was begun in 1789 by Lavoisier and his supporters--Guyton de Morveau, Berthollet, Fourcroy, and Monge--as a vehicle for the new anti-phlogistic chemistry.

.... LAVOISIER. *Traité élémentaire de chimie*, 1789. See 822.

1790s

838. VAUQUELIN, N.L. (2102) From 1790 until his death in 1829 he published 305 papers in his own name and 71 jointly with his friend, Fourcroy. His work was chiefly analytical, both inorganic and organic, and his *Manuel de l'essayeur* (Paris, 1799; several editions and translations) was widely used.

In 1798, in the course of his many mineral analyses, he discovered the element chromium and prepared several of its compounds. In the same year he converted the mineral beryl into an earth (beryllium oxide) which he recognized as a compound of a new element (metallic beryllium was isolated by others in 1828). In 1806 he described the first amino acid, asparagine.

839. PROUST, J.L. (2089) Published a large number of papers, mostly in the period 1790-1820. He made many contributions to quantitative analysis and is best known for his formulation of the law of definite proportions from 1794 onward. He upheld the validity of the law in a controversy during 1801-07 with Berthollet who maintained that the constituents of a compound could unite in a continuous range of proportions. The controversy was "a model for scientific polemics in its polite tone and the presentation of new experimental evidence on both sides." (J.R. Partington)

840. RICHTER, J.B. (2563) His pioneering work in stoichiometry (his term), dealing with the combining proportions of acids and bases, was published in two multi-part books over the period 1791-1802. Because of his speculative approach, combined with his obscurity and verbosity, his writings had almost no influence. The value of his work was however perceived by E.G. Fischer who added a summary of it to his German translation (of 1802) of Berthollet's small work, *Recherches sur les lois de l'affinité*. In 1803 Berthollet included Fischer's summary in his important *Essai de statique chimique* (857) and in this way Richter's work became generally known.

841. TENNANT, S. (2323) By means of quantitative combustion experiments

he established in 1796 that diamond is chemically identical
with carbon.   (See also 848)

842. SAUSSURE, N.T. de. (2902) In three papers in 1797 he developed
the work of Senebier (1338) in terms of Lavoisier's new chemistry, and showed conclusively that, in the presence of light,
green plants use carbon dioxide and water to form plant tissues,
liberating oxygen in the process.   (See also 859)

1800s

843. ELECTROCHEMISTRY. Volta's discovery (634) of the "voltaic pile"
--the first means of supplying a continuous current of electric-
ity--had a profound effect on chemistry as well as on physics.
Very soon after Volta's announcement in 1800 W. Nicholson and
A. Carlisle found that water could be decomposed by means of
the pile into hydrogen and oxygen. This highly significant
discovery opened up the new field of electrochemistry which
was rapidly explored by many chemists.

844. RITTER, J.W. (2571) One of the first experimenters in electrochemistry. In 1800, some months after the celebrated experiment of Nicholson and Carlisle (843), he also electrolysed
water and was the first to collect the hydrogen and oxygen
separately. He then found that the electric current would
deposit copper from a solution of copper sulphate and his
subsequent experiments led him to the idea of the electrochemical series of metals. Among other things he also discovered
the phenomenon of polarization of electrodes.
        Ritter held the view that the source of voltaic elec-
tricity was chemical action, in opposition to Volta and others
who maintained that the source was the actual physical contact
of dissimilar metals. The conflict between the two views took
many years to resolve.

845. DAVY, H. (2339) In the period 1800-10 he carried out some famous
investigations (described in his Bakerian Lectures to the Royal
Society, 1806-10) in electrochemistry. In particular he isolated
for the first time the elements sodium and potassium, followed
by magnesium, calcium, strontium, and barium.
(See also 861)

846. THENARD, L.J. (2119) Author of many papers, chiefly between 1800
and 1825, on numerous aspects of inorganic chemistry and of
some on organic chemistry (especially on esters). He discovered
hydrogen peroxide and examined its properties, and investigated
the catalytic properties of platinum and other metals. In a
notable collaboration with Gay-Lussac in the period 1808-11 he
devised a method for preparing potassium and sodium in substantial quantities, made an extensive investigation of their properties, discovered the element boron, and did pioneering work
in photochemistry and the combustion analysis of organic compounds.   (See also 866)

847. WOLLASTON, W.H. (2336) In the period 1800-05 he discovered a
method (now the basis of powder metallurgy) of making platinum
malleable and so enabling it to be fabricated. He derived a
fortune from his exploitation of the process and did not publish

it until just before his death in 1828. In the course of his researches on the process he discovered the elements palladium (in 1802) and rhodium (in 1804). (See also 848 and 865)

848. TENNANT, S. (See also 841) Was a partner of Wollaston (847) in the platinum business. From the residues left after the extraction of platinum from the raw material he isolated the new elements, iridium and osmium, in 1804.

849. FOURCROY, A.F. de. (See also 828) *Système des connaissances chimiques.* 11 vols. Paris, 1801-02. A masterly treatise.

850. HENRY, W. (2338) (a) *An epitome of chemistry.* London, 1801. (From the 6th ed., 1810, it was entitled *The elements of experimental chemistry.* 11th ed., 1829) In its time the most popular textbook of chemistry in English. (b) In 1802 he proposed the law dealing with the solubility of gases in liquids which came to be known as Henry's law. (c) From 1805 to about 1825 he investigated the composition of various mixtures of gases (coal gas, wood gas, fire damp, etc.) and identified many of the constituents and studied their properties.

851. DALTON, J. (2331) In three papers read to the Manchester Literary and Philosophical Society in 1801 he put forward the law of partial pressures in gaseous systems and (independently of the discovery announced by Gay-Lussac at about the same time) the law that heat expands all gases to the same extent.
     Another paper on a closely related topic which he presented to the Society in 1803 included a mention of the particle weights of different gases--the germ of his atomic theory. He went on to develop the theory and gave an extended presentation of it in his *New system of chemical philosophy* (2 parts, London, 1808-10).

852. NIOBIUM. Discovered in 1801 by the English chemist, C. Hatchett (who called it columbium).

853. THOMSON, T. (2337) *A system of chemistry.* 4 vols. Edinburgh, 1802. (6th ed., 1820. Trans. into French and German) The first comprehensive treatise in English to incorporate the new chemistry of Lavoisier. In the successive editions it kept abreast of new developments and from the third edition (of 1807) it included Dalton's atomic theory. Thereafter Thomson became Dalton's chief supporter, enthusiastically advocating the atomic theory in the later editions of his *System* and in his periodical *Annals of Philosophy.*

854. GAY-LUSSAC, J.L. (2120) (a) Established in 1802 (at about the same time as Dalton and following the unpublished and incomplete work of Charles) that all gases expand equally with the same increase in temperature. (b) In collaboration with Alexander von Humboldt (then living in Paris) he established in 1805 that exactly two volumes of hydrogen combine with one volume of oxygen in the formation of water. (c) Announced in 1809 the law of combining volumes which is named after him--that gases combine in simple proportions by volume. (He added several other examples to the paradigm case of hydrogen and oxygen) (d) Collaboration with Thenard: see 846. (e) In a large number of papers published up to about 1840 he made numerous important contributions to inorganic chemistry. (See also 864 and 876)

855. BERZELIUS, J.J. (2852) **(a)** His first researches, from 1802 onward,
     were in the new field of electrochemistry and, in conjunction
     with his later analytical work, led to his important theory of
     electrochemical dualism (see 860d). **(b)** In 1808 he published
     (in Swedish) the first volume of his textbook which developed
     through five editions up to his death in 1848. As well as
     being a masterly survey of the current state of knowledge it
     also expounded his own theories and reported his analytical
     results. From 1820 onward the successive editions were trans-
     lated into German, and some into French, and extracts from it
     were sometimes published as separate books. It was for many
     years the most authoritative treatise in the whole field of
     chemistry.   (See also 860)

856. TANTALUM. Discovered in 1802 by the Swedish professor, A.G.
     Ekeberg.

857. BERTHOLLET, C.L. (See also 825) *Essai de statique chimique*.
     Paris, 1803. An effective and influential criticism of the
     idea of affinity and the affinity tables which had been so
     favoured by eighteenth-century chemists (e.g. 805, 824a). He
     established that, contrary to previously accepted ideas, many
     reactions are reversible and that reactions generally are
     affected by the relative quantities of the reacting substances
     and by temperature, and can also be affected by solubility and
     other factors. These correct conclusions, however, led him to
     some incorrect inferences which involved him in a controversy
     with Proust from 1801 to 1807 (see 839).

858. *DICTIONNAIRE de chimie*. 4 vols. Paris, 1803. By C.L. Cadet de
     Gassicourt.

859. SAUSSURE, N.T. de. (See also 842) *Recherches chimiques sur la
     végétation*. Paris, 1804. A pioneering work on the chemical
     constituents of plants and the beginnings of plant biochemistry.
     These subjects he developed in a long series of papers up to
     about 1840.

1810s

860. BERZELIUS, J.J. (See also 855) **(a)** From 1810 his main research
     effort was devoted to the analysis of a large number of com-
     pounds, originally in order to ascertain whether the law of
     constant proportions was valid, and then--after his acceptance
     of Dalton's atomic theory--in order to determine atomic weights.
     He was a very skilful and painstaking analyst and his final
     values, published in 1826, for the atomic weights of the forty-
     nine elements then known were (apart from the uncertainties of
     current theory) remarkably accurate.
         (b) In pamphlets and periodical articles over the period
     1811-14 he developed a system of symbols for the elements which
     gradually became accepted and, in its essentials, is the system
     still used.
         (c) In 1818 he discovered selenium.
         (d) *Essai sur la théorie des proportions chimiques et
     sur l'influence chimique de l'électricité*. Paris, 1819. A
     translation of part of the first edition of his Swedish text-
     book; he arranged to have it done when he was in Paris in 1819

in order to make his work better known (there was also a German translation in 1820). In addition to giving an account of the atomic theory and of his determination of atomic weights, it was the chief exposition of his doctrine of electrochemical dualism--i.e. that chemical compounds are made up of two components, one positively and the other negatively charged. He applied this idea to the whole of chemistry. (See also 874)

861. DAVY, H. (See also 845) (a) In 1810 he overthrew Lavoisier's theory of acids by establishing--insofar as a negative could be established--that muriatic acid (now called hydrochloric acid) contained no oxygen. He went on to show that "oxymuriatic acid" contained none either and renamed it chlorine and clarified its relation to muriatic acid; he was of the opinion that it was an element analogous to oxygen.

   (b) *Elements of agricultural chemistry*. London, 1813. A pioneering work which gave an important impulse to the adoption of scientific methods in agriculture.

   (c) In 1815, in consequence of disastrous explosions in coal mines, he made some investigations of flame and the ignition of methane, and invented his famous safety lamp.

862. PELLETIER, P.J. (2128) From about 1810 until his death in 1842 he conducted a highly successful series of investigations of the chemical constituents of numerous plant products such as drugs, resins, gums, etc. From 1817 he worked in collaboration with J.B. Caventou. Their many discoveries included chlorophyll, caffeine, and several alkaloids such as strychnine, brucine, cinchonine, and quinine.

863. AVOGADRO, A. (1935) In 1811, following the work of Gay-Lussac (854b,c), he put forward his famous hypothesis that equal volumes of all gases, under the same conditions of temperature and pressure, contain the same number of molecules. In subsequent papers up to 1821, in French and Italian periodicals, he repeatedly drew attention to it and used it successfully to deduce the molecular formulae of many compounds. It was nevertheless neglected until 1858 when Cannizzaro pointed out its great importance (917).

864. IODINE. Its discovery in 1811 by the French industrial chemist, B. Courtois, was reported by his friends, N. Clément and C.B. Desormes in 1813. In the following year its chemistry was thoroughly investigated by Gay-Lussac who established that it was an element (as did Davy independently).

865. WOLLASTON, W.H. (See also 847) In an influential paper in 1813 he initiated the virtual abandonment of Dalton's atomic theory by most chemists for nearly fifty years. (From the modern standpoint Dalton's theory was radically incomplete because it lacked the concept of the molecule and was unable to deduce molecular formulae; consequently its main feature, the determination of atomic weights from analytical data, was frustrated) Instead of Dalton's atomic weights Wollaston proposed the use of 'equivalents', or combining weights, derived directly from the analytical data without--so he thought--any theoretical assumptions. In fact there were hidden assumptions and his method was no better than Dalton's. It seemed to be better

however and was widely adopted, especially because he invented
a useful logarithmic slide-rule for facilitating calculations
using equivalents.

866. THENARD, L.J. (See also 846) *Traité de chimie élémentaire, théorique et pratique.* 4 vols. Paris, 1813-16. (6th ed., 1834-36.
Trans. into German, Italian, and Spanish) An important textbook.

867. PROUT'S HYPOTHESIS. It was suggested by William Prout in 1815
that the atomic weights of all elements are whole-number mult-
iples of the atomic weight of hydrogen taken as unity. He
added the subsidiary hypothesis that hydrogen might be the
primary matter from which all other elements are derived.
Prout's hypothesis was very attractive in the early
nineteenth century when atomic weights generally were still
uncertain. Later in the century however the facts seemed to
be against it (see for example 923). Nevertheless it continued
to be suggestive and in the twentieth century it was vindicated
at a higher level of sophistication.

868. GMELIN, L. (2581) *Handbuch der theoretischen Chemie.* 3 vols.
Frankfurt, 1817-19. A systematic, concise, all-inclusive
compendium of chemical compounds, both inorganic and organic.
The fourth edition, which began publication in 1848, was entitled
Handbuch der Chemie; after Gmelin's death in 1853 it was continued by K. Kraut et al., totalling thirteen volumes by 1870.
(This edition was translated into English by H. Watts) In the
fifth edition (1871-86) the organic section was dropped and the
title changed to *Handbuch der anorganischen Chemie.* This monumental work has continued in successive editions to the present.

869. LITHIUM. Discovered in 1817 by J.A. Arfvedson, a student of
Berzelius.

870. CADMIUM. Discovered in 1817 by F. Stromeyer, professor of chemistry at Göttingen.

871. LAW OF DULONG AND PETIT. (See 647) Though the law was not
accurate enough to determine atomic weights by itself it was
nevertheless valuable in indicating their magnitude (often the
main problem) and was so used to good effect by several chemists.

1820s

872. FARADAY, M. (2358) His earliest researches were chiefly in chemistry. Among other things he discovered benzene and some other
compounds of interest, and liquefied chlorine and later other
gases.   (See also 885)

873. MITSCHERLICH, E. (2601) His discovery of isomorphism (i.e. that
compounds with the same crystal form are analogous in chemical
composition) was made in 1819 and developed in the following
years. At the time it was unexpected--because identity of
crystal form was thought to mean identity of composition--and
important because, in revealing composition, it provided a
valuable tool in the difficult problem of determining atomic
weights; Berzelius made much use of it for this purpose.
Mitscherlich later found that the phenomenon was not so general
as he had thought, and that only some elements can substitute
for others in crystals.

874. BERZELIUS, J.J. (See also 860) **(a)** *Jahres-Bericht über die Fortschritte der Chemie.* An annual review which he published from
1822 until his death in 1848. (The original was in Swedish but
it was best known in the German translation) It was a very
capable and authoritative survey which was widely read and
highly influential. **(b)** He isolated silicon in the amorphous
state in 1824. (It was obtained crystalline by Deville in 1854)
In 1829 he also discovered thorium.

875. CHEVREUL, M.E. (2127) **(a)** *Recherches chimiques sur les corps
gras d'origine animale.* Paris, 1823. A comprehensive and
thorough investigation of the chemistry of the fats, fatty acids,
and the saponification process--all of which had hitherto been
almost completely unknown. As a result of Chevreul's work the
fats were the first group of naturally-occurring organic compounds whose chemical nature was well understood. His work
also had important industrial consequences.
    (b) *Considérations générales sur l'analyse organique.*
Paris, 1824. Reflections on the methods and rationale of
research in organic chemistry, arising out of his experience
in the foregoing investigation which constituted a model study
in the new field. The book was concerned mainly with the per-
ennial problems of the organic chemist in separating mixtures,
purifying substances, and ascertaining their purity. Though
its conclusions and recommendations soon came to be taken for
granted it was of fundamental importance at the time and was
recognized as such.

876. GAY-LUSSAC, J.L. (See also 854) In several papers in the period
1824-32 he virtually created the basic methods of volumetric
analysis which have been used ever since.

877. DUMAS, J.B.A. (2136) In 1826 he devised a method (which is still
used) for determining the vapour density of any substance that
can be vaporized. The determination of vapour density appeared
to provide a promising way of arriving at atomic weights--the
leading problem in chemistry at the time. Dumas' efforts in
this direction, after some initial successes, finally ended in
frustration (due, as transpired much later, to the almost universal inability at the time to make the distinction between
atoms and molecules). The failure was a serious one and accent-
uated most chemists' lack of confidence in the atomic theory.
(See also 886)

878. BALARD, A.J. (2137) An investigation of the properties of the
recently-discovered element, iodine, and its compounds in sea-
water, led him in 1826 to the discovery of bromine. He established its elementary nature and showed that its properties
were analogous to those of chlorine and iodine.

879. WÖHLER, F. (2602) **(a)** In 1827 he was the first to isolate aluminium (at least in any quantity) and to give an account of its
properties. **(b)** His well-known synthesis of urea in 1828 did
not have the importance ascribed to it by nineteenth-century
historians (in supposedly overthrowing vitalism) though it was
of interest at the time. It was significant in another way,
however, in that urea proved to have the same composition as
ammonium cyanate but was clearly a different compound--a case

of what Berzelius in 1830 named isomerism. (It was another
case of isomerism that first brought Wöhler and Liebig together
in the late 1820s)   (See also 884)

880. "LAW OF TRIADS" Proposed in 1829 by J.W. Döbereiner (professor
     at Jena) on the basis of his observation of regular increments
     of atomic weight in elements of similar properties, notably
     calcium/strontium/barium and chlorine/bromine/iodine.

1830s

881. VANADIUM. Discovered in 1830 by the Swedish professor, N.G.
     Sefström, an associate of Berzelius.

882. LIEBIG, J. (2603) (a) During the 1830s he greatly improved and
     simplified the method of combustion analysis for the determin-
     ation of the proportions of carbon and hydrogen in organic
     compounds. This development was fundamental for the subsequent
     growth of organic chemistry.
          (b) In an important paper in 1838 on organic acids he
     clarified the confused issue of the general nature of acids,
     supporting the view that hydrogen was the significant element
     in their constitution, and establishing that there could be
     dibasic and tribasic acids as well as monobasic.
     (See also 884 and 893)

883. GRAHAM, T. (2373) (a) In a study in the early 1830s of the diff-
     usion of gases he discovered the law later named after him--
     that the rate of diffusion of a gas is inversely proportional
     to the square root of its density. He pointed out that mixtures
     of gases could be separated by virtue of their different rates
     of diffusion. He also studied the effusion and transpiration
     of gases.
          (b) In a masterly paper in 1833 he clarified the hither-
     to quite confused subject of the differences between the three
     phosphoric acids and laid the foundations of the concept (devel-
     oped by Liebig) of polybasic acids.

884. "UNTERSUCHUNGEN über das Radikal der Benzoesäure"--a famous paper,
     published jointly by Wöhler and Liebig in 1832, which opened a
     way into what Wöhler called "the dark forest of organic chem-
     istry." It established the existence of the benzoyl radical
     (containing carbon, hydrogen, and oxygen) which persisted
     through many reactions and formed part of the constitution of
     many compounds. This finding gave strong support to the nascent
     radical theory of the composition of organic compounds.

885. FARADAY, M. (See also 872) A series of investigations he made in
     1833 into the electrical conductivity of solutions and molten
     salts led to his discovery of the two laws of electrochemistry.
     The significance of these laws was not however generally real-
     ized for many years. He also developed the terminology of
     electrochemistry, in conjunction with Whewell, proposing such
     terms as 'electrode', 'anode', 'ion', etc.

886. DUMAS, J.B.A. (See also 877) (a) Put forward in 1834 the import-
     ant idea of substitution reactions (in organic chemistry)
     according to which the hydrogen in a compound could be replaced

by an equivalent amount of chlorine or some other element.

(b) From an investigation of wood spirit (methyl alcohol) in 1835 he showed that a whole series of compounds could contain one radical, such as methyl.

(c) Hitherto, like other chemists, he had assumed that Berzelius' theory of electrochemical dualism, which was eminently suitable for many inorganic compounds, also applied to organic compounds. By the late 1830s, however, the rapid accumulation of indications to the contrary led him to reject Berzelius' doctrine and espouse what was called the unitary theory.

887. LAURENT, A. (2153) His classification and system of nomenclature of organic compounds, and his general theory of their constitution (his nucleus theory), were developed in numerous papers from the mid-1830s and presented in final form in his posthumous *Méthode de chimie* (Paris, 1854. English trans., 1855)

888. DANIELL, J.F. (2347) Invented in 1836 the electrochemical cell --later well known as the Daniell cell--which was a great improvement over the ordinary voltaic cell because, by eliminating polarization, it provided an even current at a constant voltage. As well as being the prototype of later standard cells, it made a big practical contribution to research on all aspects of current electricity and also gave rise to commercial applications.

889. HESS, G.H. (2985) In the late 1830s he began research in thermochemistry along the lines initiated by Lavoisier and Laplace (830) and using an ice calorimeter like theirs. His studies of reaction heats led him in 1840 to the law (later called Hess' law) that the heat evolved in a chemical change is the same whether the change takes place directly in a single step or indirectly through several intermediate steps. (The law is actually a special case of the principle of conservation of energy which was then unknown) Hess' work did much to stimulate further investigations in thermochemistry.

890. GROVE, W.R. (2372) Introduced in 1839 a new electrochemical cell of higher voltage than the Daniell cell. The Grove cell was subsequently used widely in research.   (See also 900)

1840s

891. MOSANDER, C.G. (2856) Showed in 1839 that the earth called ceria, which in 1803 had been isolated by both Berzelius and Klaproth, was actually a mixture of true ceria (cerium oxide) and another earth which he called lanthana. Two years later he found that the latter was a mixture of true lanthana (lanthanum oxide) and a brown earth which he called didymia (in 1885 it was separated into praseodymia and neodymia). In 1843 he investigated an earth which had been discovered by Gadolin in 1794 and found that it could be separated into the oxides of yttrium, erbium, and terbium.

892. BUNSEN, R.W. (2616) From about 1840 onward he developed methods of gas analysis which he used to investigate the performance of blast furnaces (both in Germany and, in collaboration with Lyon Playfair, in England). The field of gas analysis, which he

virtually created, received its definitive treatment in his
Gasometrische Methoden (Brunswick, 1857; English trans., 1857).

893. LIEBIG, J. (See also 882) **(a)** *Die organische Chemie in ihre
   Anwendung auf Agricultur und Physiologie*. Brunswick, 1840.
   (7th ed., 1862--the last by Liebig. English trans., 1840.
   French trans., 1841) A work of great importance for plant
   biochemistry and agricultural chemistry. It had a big influence
   on agriculturalists, especially in England.
       (b) *Die Thierchemie; oder, Die organische Chemie in ihre
   Anwendung auf Physiologie und Pathologie*. Brunswick, 1842.
   (English trans., 1842) An important survey of animal metabolism,
   nutrition, and food chemistry.
       (c) *Chemische Briefe*. Heidelberg, 1844. (6th ed.,
   1878. English trans.: *Familiar letters on chemistry*, 1843 and
   later) A very successful popularization of chemistry and its
   applications in industry, agriculture, and medicine.

894. FRESCENIUS, C.R. (2636) **(a)** *Anleitung zur qualitativen chemischen
   Analyse*. Bonn, 1841. (Seventeen editions by the time of his
   death in 1897, each one an improvement. Translated into many
   languages. The English translation went to ten editions) The
   book owed its sensational success to his unique system of analysis (still extant) which answered the needs of the times,
   especially with the growth of applied chemistry.
       (b) *Anleitung zur quantitativen chemischen Analyse*.
   Brunswick, 1845. (6th ed., 1887. Translated into many languages. The English translation went to seven editions)

895. KOLBE, H. (2638) An outstanding experimenter who from 1842 until
   his death in 1884 published many papers which included discoveries of several important reactions and methods of synthesis.
   (See also 924)

896. *HANDWÖRTERBUCH der reinen und angewandten Chemie*. 9 vols & supp.
   Brunswick, 1842-64. Edited by J. Liebig, J.C. Poggendorff, and
   F. Wöhler. For a later version see 946.

897. GERHARDT. C.F. (2164) His valuable contributions to the systematization of organic chemistry--especially the concept of homologous series and the theory of types--were developed from the
   early 1840s and presented in final form (just before his early
   death) in his *Traité de chimie organique* (4 vols, Paris, 1853-56.
   German trans., 1854-58)

898. HOFMANN, A.W. (2637) From 1843 to the 1880s he published over
   350 papers describing many new organic compounds and reactions.
   He was the leading figure in the exploitation of the chemical
   riches of coal tar and from the late 1850s was prominent in the
   new field of synthetic dyes.   (See also 911)

899. RUTHENIUM. Discovered in 1844 by the Russian professor, K.K.
   Klaus.

900. GROVE, W.R. (See also 890) Presented in 1846 the first experimental evidence for thermal dissociation, showing that steam
   in contact with a very hot platinum wire is dissociated into
   hydrogen and oxygen.

901. REGNAULT, H.V. (2152) *Cours élémentaire de chimie*. 4 vols Paris,

1847-49. (4th ed., 1854) An outstanding textbook which was translated into five languages. For his work on the physical properties of gases see 675.

902. BABO'S LAW--the vapour pressure of a liquid is lowered by dissolving a substance in it, and the lowering is (in general) proportional to the concentration of the solute. Announced in 1847 by L.H.C. von Babo, professor at Freiberg. The phenomenon was later investigated further by Raoult (963b).

903. PASTEUR, L. (2172) In the course of a crystallographic investigation in 1848 he found that a salt of the recently-discovered racemic acid crystallized in two forms which were mirror images of each other. He separated the two forms by hand and found that in solution one of them rotated plane-polarized light to the right and the other to the left by an equal amount. One of the two crystal forms of racemic acid proved to be identical with the well-known tartaric acid which is metabolized by living organisms, whereas the other form of racemic acid was not metabolized.

These experiments, and the idea of molecular asymmetry which he based on them, provided the foundation for the later development of stereochemistry (cf. 948, 949). They also led to his seminal work on fermentation (1805).

904. WURTZ, C.A. (2165) Prepared (primary) aliphatic amines for the first time in 1849. The discovery was soon extended by Hofmann (911) with important theoretical consequences. (See also 914)

1850s

---

905. GRAHAM, T. (See also 883) His early work on the diffusion of gases led him to study the diffusion of liquids, especially solutions, during the 1850s. This in turn led him to the technique of dialysis and so to a pioneering investigation of the nature of colloidal solutions.

906. BUNSEN, R.W. (See also 892) During the 1850s, in collaboration with his English student, Henry Roscoe, he conducted pioneering experiments in photochemistry, studying the light-induced combination of hydrogen and chlorine. (See also 920)

907. FRANKLAND, E. (2393) **(a)** From about 1850 to about 1880 he published much substantial research, notably on organometallic compounds and methods of synthesis. **(b)** In 1852, in the context of his work on organometallic compounds, he pointed out that "no matter what the character of the uniting atoms may be, the combining power of the attracting element ... is always satisfied by the same number of these atoms." This number he called 'atomicity', a term later replaced by 'valency.' It was to prove a concept of fundamental importance in the subsequent development of chemical theory.

908. HOPPE-SEYLER, F. (2657) **(a)** In numerous papers from the 1850s to the 1880s he dealt with many aspects of the new field of physiological chemistry, notably haemoglobin and its reaction with oxygen and carbon monoxide, intracellular oxidation processes, and lecithin and its properties. **(b)** *Handbuch der*

physiologisch und pathologisch-chemischen Analyse. Berlin,
1858. (6th ed., 1893) A valuable aid to research in the new
field. The analytical methods were drawn from chemistry,
physics, and medicine.   (See also 952)

909. ODLING, W. (2414) During the 1850s he took a prominent part in
the research and discussions that were going on, especially in
London and Paris, concerning the type theory, valency, and
other issues at the forefront of chemical theory. In a notable
lecture in 1853 he introduced the idea of the 'methane type,'
an idea later taken up by Kekulé with great effect.

910. WILHELMY, L.F. (2635) A pioneer of chemical kinetics. In 1850
he studied the hydrolysis of sucrose in aqueous solution under
the catalytic action of various acids (a very suitable reaction
for the purpose since it could easily be followed with a polarimeter). He successfully derived an expression for the velocity
of the reaction and also investigated the effect of temperature
on it. His work was neglected until 1884 when Ostwald drew
attention to it.

911. HOFMANN, A.W. (See also 898) Following Wurtz' discovery of the
primary aliphatic amines (904) he showed in 1850 that primary,
secondary, and tertiary amines can be prepared from ammonia by
reaction with alkyl halides. This finding established what
came to be called the 'ammonia type' which provided a leading
example for the theory of types.

912. WILLIAMSON, A.W. (2392) **(a)** His synthesis of ethers (by reaction
of sodium alcoholates with alkyl halides), discovered in 1850,
clarified the hitherto misunderstood relation between alcohols
and ethers. It also established the 'water type' (comprehending
ethers, alcohols, and various other classes of compounds) as
one of the pillars of the theory of types.
    (b) His study, published in 1852, of the long-known
'continuous etherification process' (preparation of ether by
the action of sulphuric acid on alcohol) established for the
first time that a catalysed reaction proceeds via an intermediate product involving the catalyst, rather than by means of
some 'catalytic force.' In this and other respects his study
is now regarded as a first step towards the later investigations
of reaction mechanisms.

913. HITTORF, J.W. (2634) In a series of investigations in 1853-59
he studied the migration of ions during electrolysis by measuring the changes in concentration of the electrolyte near the
anode and cathode. He thus formulated the concept of transport
numbers (or relative capacity for carrying electricity) of ions.

914. WURTZ, C.A. (See also 904) **(a)** Discovered in 1854 the synthesis
of alkanes later named after him (i.e. the reaction of alkyl
halides with sodium). At the time the reaction had some theoretical significance in connection with the radical theory.
(b) In 1856 he prepared the first known dihydroxy alcohol
(ethylene glycol) which led him to conclude that glycerol is a
trihydroxy alcohol. These formulations had considerable theoretical significance at the time and helped to establish the
concept of valency.   (See also 941 and 972)

915. DEVILLE, H. (2166) The first to prepare aluminium in quantity.
In 1854 he devised the original industrial method for making
it (by heating aluminium chloride with metallic sodium). His
process was used for many years until supplanted by the electrolytic process.   (See also 938)

916. PERKIN, W.H. (2418) His discovery in 1856 of mauve, the first
synthetic dye, was accidental but was made within the context
of the exploration of the chemical properties of the constituents of coal tar. The synthetic dyestuffs industry which,
through Perkin's enterprise, resulted from the discovery grew
very rapidly from the late 1850s onward. Besides its economic
and technological importance, the industry was to have an
immense influence on the subsequent development of organic
chemistry, both cognitively and socially.

917. CANNIZZARO, S. (1945) "Sunto di un corso di filosofia chimica"
(published in the periodical *Nuovo Cimento* in 1858 and reprinted
as a pamphlet the next year; it was copies of this pamphlet
that were distributed at the Karlsruhe Congress in 1860--see
926). An outline of Cannizzaro's lecture course which set out,
logically and clearly, how the use of Avogadro's Hypothesis,
together with a firm distinction between atoms and molecules,
could lead to a consistent and soundly-based system of chemical
theory.

918. KEKULÉ, A. (2654) In a famous paper in 1858 he laid the foundations of the theory of chemical structure by proposing that
carbon has a constant valency of four, and that carbon atoms
can link together to form chains. His ideas were presented at
length in his influential *Lehrbuch der organischen Chemie* which
appeared in parts from 1859 onward.   (See also 937)

919. COUPER, A.S. (2415) "On a new chemical theory"--a famous paper
of 1858, effectively simultaneous with Kekulé's, in which he
also proposed that carbon has a valency of four and that carbon
atoms can link together to form chains. He was the first to
use lines between atomic symbols to represent what were later
called valency bonds (in marked contrast to Kekulé's clumsy and
misleading symbolism) and he was much clearer than Kekulé about
the philosophical difference between the new structural theory
and the theory of types which preceded it.

1860s

---

920. BUNSEN, R.W. (See also 906) In 1859-60, in collaboration with
his Heidelberg colleague, G.R. Kirchhoff (see 704), he opened
up the field of spectrum analysis. With this powerful tool he
soon discovered two new elements, caesium and rubidium, and in
subsequent years it made possible the discovery of several
other elements by various chemists.

921. BAEYER, A. von. (2656) A great experimenter who from 1860 onward
published many papers on numerous kinds of organic compounds.
Because of his early association with Kekulé he was one of the
first to exploit the new theory of chemical structure. He is
best known for his classical work in deducing the structure of
indigo. Other major investigations included his studies of

uric acid derivatives, polyacetylenes, and oxonium salts. He also discovered the class of phthalein dyes and put forward his well-known strain theory relating the stabilities of cyclic compounds to the size of the angles in their rings.

922. KÜHNE, W.F. (2670) **(a)** About 1860 he began his studies of proteins in the course of which he separated egg albumin into different fractions and investigated the properties of myosin; he also showed that myoglobin is related to haemoglobin (see also 1777). **(b)** *Lehrbuch der physiologischen Chemie*. Leipzig, 1868. One of the first textbooks of biochemistry.

923. STAS, J.S. (2816) Spent most of his career on the determination of the atomic weights of several elements (C, N, Cl, S, K, Na, Pb, Ag) with unparalleled accuracy. His results were published in two large memoirs in 1860 and 1865. Though he began the work with "an almost complete confidence in the exactness of Prout's principle" (see 867) the fact that none of his atomic weights turned out to be whole numbers led him finally to dismiss it as "a pure illusion."

924. KOLBE, H. (See also 895) Expounded his theory of the constitution of organic compounds in a major paper in 1860. Though his elaborate system contained numerous insights and had some predictive successes it proved to be a failure (partly because it was based on some outmoded ideas of Berzelius). It was completely superseded by the theory of chemical structure which Kolbe never accepted and attacked bitterly.

925. BERTHELOT, M. (2176) **(a)** *Chimie organique fondée sur la synthèse*. Paris, 1860. A monumental work which included his extensive researches in synthetic organic chemistry over the previous ten years and for the first time presented the subject in a coherent and integrated way.
       (b) In collaboration with L. Péan de Saint-Gilles he published in 1862-63 a pathbreaking investigation of physical aspects of the reaction of alcohols with acids to form the corresponding ester and water. They established that the reaction never went to completion but arrived at a state of equilibrium. The velocity of the reaction, they found, was proportional to the amounts of alcohol and acid present and they attempted to put this finding in mathematical form. Their work led to the subsequent establishment of the law of mass action (935).   (See also 956)

926. KARLSRUHE CONGRESS, September 1860. By mid-century the highly confused state of chemical theory, and especially the lack of agreement on atomic weights, was becoming intolerable. On the initiative of Kekulé, a three-day international conference, attended by about 140 chemists, was held at Karlsruhe to attempt to resolve the confusion. No agreement was reached but at the end of the conference a friend of Cannizzaro distributed copies of his "Sunto" (917); the subsequent perusal of it (or of translations) by some of the participants--notably Lothar Meyer (934a)--eventually achieved what the congress had set out to do.

927. CROOKES, W. (2416) Using the new tool of spectrum analysis, he discovered the element thallium in 1861 and subsequently isolated it and studied its properties.

928. ANDREWS, T. (2374) Formulated the concept of critical temperature
which he discovered in the case of carbon dioxide in 1861. With
considerable experimental skill he established the continuity of
the gaseous and liquid states and elucidated it by the use of
pressure-volume isotherms. He gave an account of his work in
the Royal Society's Bakerian lectures for 1869 and 1876.

929. BUTLEROV, A.M. (2990) "The chemical structure of compounds"--a
paper (in Russian and German) read at a scientific congress at
Speyer in 1861. Butlerov was one of the main founders of the
theory of chemical structure, presenting it in a more complete
form than Kekulé and Couper. His ideas were expressed fully in
his treatise on organic chemistry, published in Russian in
1864-66 and in German in 1867-68.

930. AZO COMPOUNDS. In 1862 the German-English chemist, P.J. Griess,
discovered the diazotization reaction which could be used for
a number of purposes, including the preparation of azo compounds.
It was a discovery of major importance for the new synthetic
dye industry for it brought into existence the big and important
class of azo dyes.

931. *A DICTIONARY of chemistry.* 5 vols. London, 1863-68. (2nd ed.,
1888-94) Edited by H. Watts.

932. MARKOVNIKOV, V.V. (2998) As a disciple of Butlerov his researches
during the 1860s were directed to substantiating the structural
theory, chiefly by studying cases of isomerism. This led him
in 1869 to the discovery of his well-known rule concerning the
addition reactions of unsymmetrical alkenes.

933. INDIUM. Discovered in 1863 by F. Reich, professor at the Freiberg
Mining Academy, and his assistant, H.T. Richter.

934. MEYER, L. (2655) **(a)** *Die modernen Theorien der Chemie.* Breslau,
1864. (5th ed., 1884. Trans. into Russian, English, and French)
The best and most influential presentation of the great clarif-
ication of chemical theory which took place in the early 1860s.
The book had its origin in the Karlsruhe Congress and Cannizzaro's
famous "Sunto" (see 926).
    (b) The first edition of the above work already contain-
ed a primitive version of what came to be known as the periodic
classification of the elements (made possible by the newly
clarified atomic weights). In 1868--if not earlier--Meyer drew
up a much more developed version in preparation for the second
edition of his book; it was very similar to Mendeleev's famous
table which appeared in 1869. Meyer did not publish his version
(which included his notable graph relating atomic volume to
atomic weight) until 1870 and made no claim to priority. He
must however be ranked with Mendeleev as discoverer.

935. LAW OF MASS ACTION. Discovered by C.M. Guldberg (2869) and P.
Waage who extended and refined the earlier investigation of
Bertholet and Péan de Saint-Gilles (925b). They announced
their findings initially in a Norwegian periodical in 1864
and then wrote an expanded version in French in 1867, but in
a very obscure publication. Consequently their work did not
become known for many years and the law was rediscovered by
several other people, including van't Hoff in 1878.

936. "LAW OF OCTAVES" Proposed in 1864/65 by the English industrial
     chemist, J. Newlands. He pointed out that if the elements were
     listed in the order of their atomic weights a pattern could be
     discerned in which similar properties recurred after each group
     of seven elements. His paper was rejected by the Chemical Soc-
     iety but its significance became apparent after the establish-
     ment of the periodic classification.

937. KEKULÉ, A. (See also 918) Proposed in 1865 his famous cyclic
     formula for benzene, with alternate single and double bonds.
     He suggested that this formula was characteristic of aromatic
     compounds and also introduced the terms 'nucleus' and 'side
     chain.'

938. DEVILLE, H. (See also 915) *Leçons sur la dissociation*. Paris,
     1866. An account of the work on dissociation which he had
     been doing since the late 1850s. He found that many compounds
     (water, carbon dioxide, hydrogen chloride, etc.) dissociate at
     high temperatures, the products recombining at reduced temper-
     atures. As well as explaining anomalous vapour densities (a
     point of some theoretical importance) his findings were to be
     valuable for the future field of chemical kinetics.

939. FRIEDEL, C. (2178) Began in 1866 a long series of pioneering
     investigations of silicon compounds, his primary concern being
     to ascertain to what extent the new theories being developed
     for carbon compounds were applicable to them.   (See also 954)

940. LECLANCHÉ CELL. Introduced in 1867 by the French engineer,
     G. Leclanché.

.... HELIUM--discovery of. See 460.

941. *DICTIONNAIRE de chimie pure et appliquée*. 5 vols. Paris, 1869-78.
     (2nd ed., 1892-1908) Edited by C.A. Wurtz.

942. MENDELEEV, D.I. (2991) **(a)** Like Lothar Meyer (934) he had attend-
     ed the Karlsruhe Congress where he had been impressed by Cann-
     izzaro's arguments, and a few years later, while writing a
     general treatise, he discovered the periodic classification of
     the elements. His discovery was announced to the Russian Chem-
     ical Society in 1869. Because of various anomalies and imper-
     fections his classification was not widely accepted at first,
     but from 1875 onward the discovery of new elements whose
     properties he had successfully predicted (see 950b, 957, 974)
     led to its general adoption.
           (b) *Osnovy khimii* ["Principles of chemistry"]. 2 vols.
     St. Petersburg, 1868-71. (8th ed., 1906. Trans. into German,
     English, and French) His celebrated treatise, focussed on the
     periodic classification.

1870s

943. MEYER, V. (2706) From 1870 until his death in 1897 he published
     over 300 papers containing many significant discoveries in
     organic chemistry. He is best known for his work on aliphatic
     nitro compounds, oximes, and stereochemistry, and for his
     discovery of thiophene and exploration of its properties. He
     also devised a widely-used method of measuring vapour densities
     for the determination of molecular weights.   (See also 986)

944. KOHLRAUSCH, F. (2677) From about 1870 he carried on a series of
     investigations on the electrical conductivity of solutions
     using alternating current which obviated the confusing polar-
     ization effects of direct current. Among other things he estab-
     lished (in 1876) that every ion has a characteristic mobility
     independent of whatever ion it had been associated with in the
     original salt--his "law of independent migration of ions." His
     findings did much to prepare the way for Arrhenius' theory of
     electrolytic dissociation.

945. NUCLEIN. The Swiss biochemist, J.F. Miescher, announced in 1871
     his discovery of a non-protein phosphorus-containing substance
     in the nuclei of cells which he named nuclein. It was later
     investigated by A. Kossel (959).

946. *NEUES Handwörterbuch der Chemie. Auf Grundlage des von Liebig,
     Poggendorff und Wöhler....* [see 896] 10 vols. Brunswick,
     1871-1930. Edited by H.C. von Fehling.

947. WISLICENUS, J. (2679) **(a)** In 1873, as a result of his research
     on the formerly confused topic of the naturally-occurring lactic
     acids, he established that there are two of them with different
     physical properties but the same structural formula. He conclu-
     ded that this could be explained "only on the ground that the
     difference is due to a different arrangement of the atoms in
     space." It was this finding that in the following year led
     van't Hoff to his stereochemical ideas. **(b)** Following the
     publications of van't Hoff and Le Bel in 1874, Wislicenus was
     the leader in gaining recognition for the new field of stereo-
     chemistry and subsequently in developing it. **(c)** In the late
     1870s he published several notable papers on synthetic methods,
     especially the classic methods using acetoacetic ester and
     malonic ester.

948. VAN'T HOFF, J.H. (2796) Following the work of Wislicenus on the
     lactic acids he published in 1874 his famous paper which laid
     the foundations of stereochemistry--the key ideas being **(a)** that
     the four valency bonds of a carbon atom do not lie in a plane
     but are arranged three-dimensionally as if they were directed
     towards the corners of a regular tetrahedron, and **(b)** that a
     molecule containing a carbon atom with four different groups
     attached to it will be asymmetric and so will be capable of
     existing in two forms which are mirror images of each other
     and, as Pasteur had shown (903), capable of exhibiting optical
     activity. The same ideas were put forward independently by
     Le Bel.
         In the same paper van't Hoff also defined the different
     kind of stereoisomerism known as geometrical or *cis-trans*
     isomerism.   (See also 968)

949. LE BEL, J.A. (2199) In 1874, at almost the same time as van't
     Hoff and independently of him, he put forward essentially the
     same ideas concerning the 'asymmetric carbon atom' and its
     relation to optical activity.

950. BOISBAUDRAN, F. (2188) **(a)** *Spectres lumineux*. Paris, 1874. On
     the application of spectroscopy to the study of the chemical
     elements. **(b)** In 1875 he detected spectroscopically a new

element which he named gallium. Several months later he isolated it in metallic form and investigated its physical and
chemical properties. They were found to correspond closely to
those predicted by Mendeleev for the hypothetical element 'eka-
aluminium'--a situation which did much to bring about the general
acceptance of Mendeleev's periodic classification. **(c)** From
1879 until the late 1890s he used spectroscopic methods in the
difficult field of the rare earth elements, discovering samarium
in 1880 and dysprosium in 1886.

951. CLAISEN, L. (2707) A outstanding exponent of synthetic organic
chemistry. From 1874 to 1926 he published a large number of
papers on organic reactions and synthetic methods, including
the group of reactions later known under the name of the
Claisen condensation.

.... GIBBS, J.W. Pioneering work on chemical thermodynamics--see 721.

952. HOPPE-SEYLER, F. (See also 908) *Physiologische Chemie*. 4 parts.
Berlin, 1877-81. An authoritative treatise by a leader in the
field.

953. *A TREATISE on chemistry*. 8 vols. London, 1877-92. (German
trans., 1877-91) By H.E. Roscoe and C. Schorlemmer. A well-
known standard work.

954. FRIEDEL, C. (See also 939) With his American collaborator, J.M.
Crafts, he discovered in 1877 what came to be known as the
Friedel-Crafts reaction--a versatile reaction valuable for both
pure and applied chemistry.

955. WURTZ, C.A. (See also 914) *La théorie atomique*. Paris, 1879.
(4th ed., 1886. The English translation of 1880 went to six
editions by 1892) An influential account of the developments
in chemical theory in the third quarter of the century.
(See also 972)

956. BERTHELOT, M. (See also 925) *Essai de mécanique chimique fondée
sur la thermochimie*. Paris, 1879. A comprehensive treatise
on thermochemistry incorporating his own numerous researches
done over the previous ten years. As was also the case with
his rival, Julius Thomsen (962), his theoretical principles
were thermodynamically defective but his calorimetric methods
and results constituted a major contribution. He continued to
work in the field almost to the end of the century.

957. SCANDIUM. Discovered in 1879 by the Swedish chemist, L.F. Nilson.
He found its properties to coincide almost exactly with those
predicted by Mendeleev for 'eka-boron.'

1880s

958. WALLACH, O. (2705) About 1880 he began the work on essential
oils that was to bring him the Nobel Prize thirty years later
and recognition today as the pioneer of polyisoprenoid chemistry.
A skilful experimenter, he was able to bring order into a
confused and difficult field, isolating and characterizing the
main terpenes, studying their relationships, and eventually
ascertaining their structures.

959. KOSSEL, A. (2739) During the 1880s and 1890s he conducted analytical investigations of the nuclein (nucleoprotein) that had been discovered by J.F. Miescher (945). Initial hydrolysis yielded a basic protein and an acidic non-protein product (nucleic acid); further hydrolysis of the latter gave various purine and pyrimidine bases, including adenine, thymine, cytosine, and uracil; he also found that a sugar was present but was unable to identify it. The protein component of nuclein, which he named histone, yielded on hydrolysis a mixture of amino acids, several of which he identified. After 1900 his work was chiefly on proteins.

960. BEILSTEIN, K.F. (2992) Gmelin's famous *Handbuch* (868), which had originally covered the whole of chemistry, became restricted to inorganic chemistry from the fifth edition, published from 1871 onwards. Possibly as a result, Beilstein launched his *Handbuch der organischen Chemie* (in two volumes) in 1880-82. It listed the physical and chemical properties of all known organic compounds, classified according to their molecular formula and chemical type. The first edition contained 15,000 compounds and the subsequent editions increased greatly in size. In 1900 the massive task of compilation was taken over by the Deutsche Chemische Gesellschaft which continues it to the present.

961. *ENCYCLOPÉDIE chimique*. 10 vols. Paris, 1881-1905. Edited by E. Frémy.

962. THOMSEN, J. (2865) *Thermochemische Untersuchungen*. 4 vols. Leipzig, 1882-86. The fruit of thirty years of work and over 3,500 calorimetric measurements. Together with the contemporary work of Berthelot (956) it represented a new epoch in the development of thermochemistry.

963. RAOULT, F.M. (2177) (a) Established in 1882 that the lowering of the freezing point of a solution of a particular substance is proportional to the concentration of the dissolved substance in molecular terms (cf. 835). He thereby introduced a valuable method for the determination of molecular weights which was rapidly taken up by other chemists. (Later the analogous phenomenon of the elevation of boiling point was found to provide an experimentally simpler method) (b) In 1886-87 Raoult went on to establish a similar relationship for the lowering of the vapour pressure of solutions (cf. 902). His experimental results were found to fit exactly the equation deduced in 1886 by van't Hoff from thermodynamic considerations. (See also 983)

964. FISCHER, E. (2708) In the early 1880s he began two lines of research, on the purines and the sugars, which he carried on in parallel for over twenty years. His achievements in the two fields brought him the Nobel Prize in 1902 and were of fundamental importance for the future of biochemistry. (See also 997)

965. CLARKE, F.W. (3045) One of the founders of geochemistry. His appointment as chief chemist to the U.S. Geological Survey in 1883 enabled him to launch an extensive and long-continued programme of rock analyses. Besides their significance for geology, the data so obtained enabled him in 1891 to make an

estimate of the relative abundance of the chemical elements in
the earth's crust. He continued to improve his estimate in
subsequent years.

966. LE CHÂTELIER, H.L. (2200) Announced in 1884 a major contribution
      to physical chemistry which came to be known as Le Châtelier's
      Principle. He later found that he had been partly anticipated
      by Willard Gibbs (721) but in a very abstract, mathematical way
      that had prevented Gibbs' work becoming widely known. Thereafter
      Le Châtelier did all that he could to spread a knowledge of it.

967. MOISSAN, H. (2209) In 1884 he began his researches on compounds
      of fluorine and two years later accomplished the difficult and
      dangerous task of isolating the element by means of an electro-
      lytic process. (Several earlier attempts by others had cost at
      least two lives) His work in the field was summed in his mono-
      graph, *Le fluor et ses composés* (Paris, 1900). (See also 984)

968. VAN'T HOFF, J.H. (See also 948) (a) His first researches in
      physical chemistry, begun in the late 1870s, were discussed in
      his *Etudes de dynamique chimique* (Amsterdam, 1884). In addition
      to presenting results of fundamental importance in reaction
      kinetics, the book also dealt with equilibrium states in thermo-
      dynamic terms, arriving at the concept of free energy as a
      measure of chemical affinity. (b) In the late 1880s he studied
      the work on osmotic pressure done by botanists (see 1396 and
      1398) and perceived the analogy between dilute solutions and
      gases. The exploration of this analogy led him to some brill-
      iant results.   (See also 994)

969. PERKIN, W.H., Jnr. (2461) In the mid-1880s, while working in
      Baeyer's institute in Munich, he succeeded in synthesizing
      compounds containing cyclobutane and cyclopropane rings, thus
      disproving the general view that such rings could not exist.
      (Baeyer subsequently used his results in the formulation of
      his important strain theory)

970. ARRHENIUS, S. (2878) The first version of his celebrated theory
      of electrolytic dissociation appeared in 1884. During the
      following years (up to 1890 for the most part) he improved it
      considerably, both through his own researches and through
      making use of important results obtained by others, especially
      Van't Hoff and Ostwald. For several years the theory met with
      strong and widespread opposition but "what Huxley was to Darwin,
      Ostwald became to Arrhenius."

971. OSTWALD, W. (2710) (a) *Lehrbuch der allgemeinen Chemie*. 2 vols.
      Leipzig, 1885-87. (2nd ed., 1891-1902) A famous work which
      did much to establish physical chemistry as a coherent subject
      by gathering together and integrating the scattered results of
      the previous fifty years. (b) *Grundriss der allgemeinen Chemie*.
      Leipzig, 1889. (4th ed., 1909. English trans., 1890 & later)
      An influential textbook of physical chemistry. (c) Ostwald
      was the first major supporter of Arrhenius' theory of electro-
      lytic dissociation. From the mid-1880s onward he made much use
      of conductivity measurements, arriving at his dilution law in
      1888.   (See also 977)

972. WURTZ, C.A. (See also 955) *Traité de chimie biologique*. Paris,
      1885. An important treatise.

973. BECKMANN, E.O. (2709) **(a)** Discovered in 1886 the 'Beckmann re-arrangement' of ketoximes into amides by the action of acidic catalysts. The mechanism of the rearrangement posed a problem of considerable theoretical interest for many years. **(b)** In connection with his researches on the above-mentioned problem he devised in 1888 a convenient technique for determining molecular weights, based on Raoult's results (963a). His apparatus and special thermometer were widely adopted and named after him.

974. GERMANIUM. Discovered in 1886 by C. Winkler, professor at the Freiberg Mining Academy. He found its properties to correspond closely to those predicted by Mendeleev for 'eka-silicon.'

975. ROOZEBOOM, H. (2801) In an important series of publications from 1887 until his death in 1907 he made extensive use of Willard Gibbs' phase rule (721) in the study of chemical equi-libra. His work was especially significant in its application to alloys and above all to the iron-carbon system. His findings were summed up in his *Die heterogenen Gleichgewichte vom Stand-punkte der Phasenlehre* (2 vols, Brunswick, 1901-04; further volumes were added by his students after his death)

1890s

976. NERNST, W. (2737) **(a)** Through the late 1880s and the 1890s he published numerous important papers on several aspects of physical chemistry, especially the theory of electrochemical cells, ionic mobilities, phase equilibria, and the theory of solubility products. **(b)** *Theoretische Chemie vom Standpunkte der Avogadroschen Regel und der Thermodynamik.* Göttingen, 1893. (15th ed., 1926. Several translations) The leading textbook of physical chemistry for many years.

977. OSTWALD, W. (See also 971) **(a)** During the 1890s he and his students did important wide-ranging experimental work on catalysis. **(b)** *Die wissenschaftlichen Grundlagen der analyt-ischen Chemie.* Leipzig, 1894. (English trans., 1895 & later) Through its application of the principles of physical chemistry and the theory of electrolytic dissociation it constituted a major revision of the teaching of analytical chemistry.

978. NENCKI, M. (3003) Among his numerous biochemical investigations at St. Petersburg during the 1890s was his study of the degrad-ation products of haemoglobin. A fellow Pole, L.P.T. Marchlewski, was concurrently carrying out a similar investigation in England with chlorophyll, and it became apparent that the two series of products were related. Nencki and Marchlewski collaborated at a distance and finally they both obtained the same degradation product, haemopyrrole.

979. PERKIN, W.H., Jnr. (See also 969) **(a)** From about 1890 until his death in 1929 he made numerous important contributions to the elucidation of the chemical structures of a variety of natural products. **(b)** *Organic chemistry.* London, 1894. (Many reprint-ings. 2nd ed., 1929) With F.S. Kipping. A highly successful textbook.

980. WALKER, J. (2462) **(a)** From about 1890 onward he published many

papers on aspects of the new physical chemistry of solutions,
such as ionization constants, amphoteric electrolytes, and
osmotic pressure. **(b)** *Introduction to physical chemistry.*
Edinburgh, 1899. (Many later editions) An important textbook
which did much to establish the subject in the English-speaking
world.

981. HANTZSCH, A.R. (2736) **(a)** In 1890, with his student, A. Werner
(2918), he extended stereochemical conceptions to compounds of
trivalent nitrogen. Their evidence was drawn from many previously puzzling cases of isomerism in oximes, azo compounds, etc.
(b) *Grundriss der Stereochemie.* Leipzig, 1893. (Trans. into
French and English) **(c)** Application of his stereochemical ideas
to diazo compounds in 1894 led to a long controversy with Ludwig
Bamberger in the course of which Hantzsch devised some of the
main techniques of physical organic chemistry. In particular
he made an extensive study of the relation between absorption
spectra and chemical constitution.

982. LE BLANC, M.J.L. (2738) **(a)** From 1891 onward he carried on a
series of studies of electrochemical polarization, in the
course of which he invented the hydrogen electrode in 1893.
(b) *Lehrbuch der Elektrochemie.* Leipzig, 1895. (12th ed.,
1925. Trans. into French and English)

983. RAOULT, F.M. (See also 963) In his work on the 'molecular depress-
ion of freezing point' in the 1880s (963a) he had used organic
compounds. With aqueous solutions of salts the results were
anomalous, but in the early 1890s he was able to show that the
anomalies could be accounted for in terms of Arrhenius' theory
of electrolytic dissociation. He thereby provided strong
support for Arrhenius' theory.

984. MOISSAN, H. (See also 967) The electric furnace which he invented
in 1892 made possible an extensive series of researches on high-
temperature processes which included the preparation of various
carbides, nitrides, borides, etc., and the isolation (by reduction reactions) of many of the less common metals.

985. WERNER, A. (2918) "Beitrag zur Konstitution anorganischer Verbind-
ungen"--the epochal paper of 1893 in which he put forward the
basic ideas of his celebrated co-ordination theory. In many
subsequent papers he and his students assembled a steadily
growing body of evidence for it, and his work became widely
known through his *Neuere Anschauungen auf dem Gebiete der
anorganischen Chemie* (Brunswick, 1905; 5th ed., 1923. English
trans., 1911)

986. *LEHRBUCH der organischen Chemie.* 2 vols. Leipzig, 1893-1903.
By V. Meyer and P. Jacobson. A well-written treatise which
remained a standard work for many years.

987. CLAISEN, L. (See also 951) Isolated in 1893 the two forms--*keto*
and *enol*--of a tautomeric compound, thereby greatly clarifying
the understanding of tautomerism.

988. STRUTT, J.W. (*Lord* RAYLEIGH) (2437) Noticing that the density
of nitrogen obtained from air was slightly different from that
of nitrogen obtained from its compounds, he suspected that the

discrepancy might be due to an unknown gaseous constituent of air. He set to work to extract it and in 1894 (in partial collaboration with William Ramsay) he succeeded in isolating the new element, argon.

989. RAMSAY, W. (2442) Was associated with Lord Rayleigh (988) in the discovery of argon in atmospheric air in 1894. In the following year he obtained another inert gas from the uranium mineral, cleveite; spectroscopic examination revealed that it was identical with the unknown element that Lockyer had identified in the sun in 1868 and named helium (460). In 1898, by fractional distillation of liquid air, Ramsay and his student, W.W. Travers, isolated neon, krypton, and xenon.

990. WALDEN, P. (3013) Discovered in 1896 the phenomenon later known as 'the Walden inversion'--i.e. the conversion, by means of appropriate reagents, of an optical isomer into its enantiomorph (or rather, a derivative thereof). The phenomenon was universally recognized as very significant but, despite many efforts, it remained unexplained until the 1930s.

991. BUCHNER, E. (2740) *Alkoholische Gärung ohne Hefezellen*. 1897. An account of the famous investigation in which he proved that a cell-free extract of yeast can ferment sugar. It established that living cells are not essential for fermentation (as Pasteur and others had thought) and marked the beginning of enzyme chemistry.

992. BERTRAND, G. (2213) Discovered in the late 1890s a class of oxidizing enzymes to which he gave the name of 'oxidases.' He found that they contained manganese and this fact, together with his demonstration that lack of manganese inhibits growth, led him to conclude that manganese formed an essential part of the enzymes. This in turn led to his important concept of trace elements and their metabolic significance.

993. SABATIER, P. (2210) In 1897, in collaboration with J.B. Senderens, he began his investigations of the use of heterogeneous catalysis in organic chemistry--notably the hydrogenation reaction. His researches were summed in his monograph, *La catalyse en chimie organique* (Paris, 1913).

994. VAN'T HOFF, J.H. (See also 968) *Vorlesungen über theoretische und physikalische Chemie*. 3 vols. Brunswick, 1898-1900. An edition of his lectures which was translated into several languages and had a extensive influence.

.... POLONIUM AND RADIUM. Discovered by the Curies in 1898--see 753b.

995. ACTINIUM. Discovered in 1899 by A. Debierne, working in the Curie's laboratory.

996. THIELE, J. (2753) Proposed in 1899 the concept of "partial valencies" to explain the type of addition reactions exhibited by compounds containing a conjugated system (his term) of double bonds; his proposal also helped to explain the peculiar pattern of reactions of benzene. Thiele's ideas became generally accepted and assumed greater significance many years later.

997. FISCHER. E. (See also 964) At the end of the 1890s he began his classic researches on amino acids and polypeptides.

1.07   EARTH SCIENCE IN GENERAL

Including geodesy, geomagnetism, geophysics, meteorology, climatology, oceanography, and some aspects of geography.

Mineralogy and geology are included here until 1730 but thereafter are listed separately in Sections 1.08 and 1.09.

1460s

.... PLINY. *Historia naturalis.* See 1217.

1470s

998. ARISTOTLE. (See entry 1) **(a)** *Meteorologia.* A treatise dealing with many features of the earth (not only with meteorology in the modern sense). It was highly influential in early modern times. **(b)** *De mineralibus.* A small work mistakenly ascribed to Aristotle. (It was actually a medieval translation of an Arabic treatise)

999. PTOLEMY. (ca. 100-170) His famous *Geographia* (often entitled *Cosmographia* in the early printed editions) first appeared in print in 1475 as the Latin translation made (about 1406) from the rediscovered Greek text. It was highly popular: six more editions appeared before 1500 and many subsequently.

1490s

1000. THEOPHRASTUS. (ca. 371-287 B.C.) *De lapidibus.* A treatise on mineralogy. It was included in the editions of Theophrastus' *Opera* which appeared in Greek and Latin from 1497 onward. Though only a small work (a part of a larger one now lost) it was very influential in early modern times.

1500s

1001. *BERGBÜCHLEIN.* ca. 1500. (Published anonymously without indication of place or date. Various later editions) The first important work on mining engineering and significant in the beginnings of geology. Generally ascribed to Ulrich Rülein von Calw, municipal physician in the famous mining town of Freiberg in Saxony.

1510s

1002. WERNER, J. (2478) Published in 1514 a collection of tracts by himself and others on mathematical geography and cartography.

1520s

1003. APIAN, P. (2481) *Cosmographia seu descriptio totius orbis.*
      Landshut, 1524. Included astronomical and descriptive geog-
      raphy, surveying, cartography, meteorology, etc. Very popular;
      translated into all the main European languages.

1530s

1004. GEMMA FRISIUS, R. (2805) *De principiis astronomiae et cosmo-
      graphiae.* Antwerp, 1530. (Numerous reprints. Translated
      into several languages) Chiefly astronomical geography.

1540s

1005. MAUROLICO, F. (1835) *Cosmographia.* Venice, 1543. Notable
      especially for its proposal for measuring the earth (which
      was put into effect in the 1660s by Picard—373a).

1006. AGRICOLA, G. (2488) *Opuscula.* Basel, 1546. A collection of
      small works written at various times from 1530 onward and
      containing important accounts of mining engineering, mineral-
      ogy, and geology.   (See also 1007)

..... HARTMANN, G. Investigation of terrestrial magnetism. See 570.

1550s

1007. AGRICOLA, G. (See also 1006) *De re metallica.* Basel, 1556.
      His masterpiece, incorporating his earlier writings. A famous
      work on mining and metallurgy, with numerous woodcut illustra-
      tions, and a landmark in the early history of geology, mineral-
      ogy, and chemistry.

1560s

1008. MERCATOR, G. (2806) Published in 1569 a map of the world based
      on his new projection designed for the use of navigators.

1570s

1009. *THEATRUM orbis terrarum.* Antwerp, 1570. (Many later editions)
      The first comprehensive atlas of the world, containing seventy
      maps. By Abraham Ortelius, a renowned geographer.

1580s

..... NORMAN, R. Discovery of magnetic dip. See 575.

1590s

1010. BONAVENTURA, F. (1870) **(a)** *De causa ventorun motis.* Urbino,
      1592. Dealt with the theories of Aristotle and Theophrastus
      on the causes of winds. **(b)** *Pro Theophrasto atque Alexandro
      Aphrodisensi ... apologia.* Urbino, 1592. Defended the
      ancients' meteorological theories against the moderns.
      (c) *Anemologiae pars prior.* Urbino, 1593. "Essentially a

Latin translation of Theophrastus' *De ventis* and *De signis*, with long and detailed commentaries on the two works."
"All of Bonaventura's writings on meteorology are marked by an attempt to determine the precise meaning of the ancient texts through philological techniques, with apparently little effort being made to utilize experience and observation to verify their truth." (*DSB*)

1600s

1011. BOODT, A.B. de. (2936) *Gemmarum et lapidum historia*. Hanau, 1609. An early attempt at a systematic account of minerals. It described about 600 from actual observation.

..... GILBERT, W. *De magnete*. See 576.

1620s

1012. DOMINIS, M.A. de. (1873) *Euripus; seu, De fluxu et refluxu maris*. Rome, 1624. One of the many works in the sixteenth and seventeenth centuries which attempted to explain the tides, but better than most. He held that the tides were caused by the moon and sun acting on the sea like magnets, and tried to work out this idea in detail, with some success. The book also dealt with the figure of the earth.

1630s

1013. GELLIBRAND, H. (2235) *A discourse mathematicall on the variation of the magneticall needle*. London, 1635. Announced the discovery of the secular change in the magnetic declination.

1640s

1014. LAET, J. de. (2811) *De gemmis et lapidibus*. Leiden, 1647. Evidently based on the work of Boodt (1011).

1650s

1015. VARENIUS, B. (2771) *Geographia generalis*. Amsterdam, 1650. (Many editions, translations, and summaries) Mathematical, astronomical, and regional geography. A standard text for over a century.

1016. MERCATOR, N. (2248) *Cosmographia; sive, Descriptio coeli et terrae*. Danzig, 1651. A leading textbook of physical geography.

1660s

1017. STENSEN, N. (1900) *De solido intra solidum naturaliter contento prodromus*. Florence, 1669. (English trans. by H. Oldenburg, 1671) A pioneering work in geology, full of remarkable insights. Stensen was the first to point out that the earth's crust exhibits a chronological history which can be deciphered by study of the strata. He argued against the common view that mountains grow out of the earth like trees, maintaining

that they are formed by vast changes in the earth's surface.
He suggested moreover that many rocks are formed by sedimentation and that the fossils found in them are the remains of
living organisms. And he also discovered the constancy of
interfacial angles in quartz crystals.

1670s

1018. SCILLA, A. (1899) In a book of 1670 which is now regarded as
one of the classics of geology he gave an account of his
admirable study of fossiliferous rocks and adumbrated some
key ideas on the formation of sedimentary rocks, the succession of strata, the organic origin of fossils, and their
relations to species now living.

..... PICARD, J. *Mesure de la terre.* See 373.

1019. PERRAULT, P. (2018) *De l'origine des fontaines.* Paris, 1674.
Solved the old problem of the origin of springs by proving
that rainfall is more than enough to supply the flow of springs
and rivers; this he did by measuring the annual rainfall over
the drainage area of the Seine and comparing it with the total
annual flow of the river. The book thus constitutes the
beginning of the study of the hydrologic cycle.

1680s

1020. BURNET, T. (2280) *Telluris theoria sacra.* London, 1631. The
character of the book is indicated by the title of the English
translation (1684): *The Sacred Theory of the Earth, containing
an account of the Original of the Earth and of all the General
Changes which it hath already undergone or is to undergo till
the Consummation of all Things.* This theological cosmogony
aroused much controversy, chiefly because of its allegorical
use of Scripture. It was however widely read and it helped
to set the scene for the beginnings of geology, especially by
disseminating the idea that the earth has a history.

1021. HALLEY, E. (2274) **(a)** "A theory of the variation of the magnetic
compass." In this paper of 1683 and a related one in 1692 he
proposed that the earth has a magnetic core with a slightly
different period of diurnal rotation from that of its outer
shell. **(b)** "An historical account of the trade winds and
monsoons...." A paper of 1686 which contained the first
meteorological chart of the winds and made the suggestion that
they are caused by solar heating of the atmosphere.

1022. MARIOTTE, E. (2016) *Traité du mouvement des eaux.* Paris, 1686.
A treatise in five parts, of which the first was an original
and important discussion of meteorological and hydrological
topics. (The other parts dealt with fluid mechanics)

1690s

1023. RAY, J. (2254) *Miscellaneous discourses concerning the dissol-
ution and changes of the world.* London, 1692. The revised
second edition (1693) had the title: *Three physico-theological
discourses.*

1024. WOODWARD, J. (2286) *Essay toward a natural history of the earth*.
London, 1695. (3rd ed., 1723) Elaborated a theory of the
formation of the earth's crust by sedimentation (whence strat-
ification) following a universal deluge. He held that fossils
were the remains of organisms once living--a view still not
common. The book was widely read and was translated into Latin.
(See also 1031)

1025. WHISTON, W. (2287) *A new theory of the earth, from its original
to the consummation of all things*. London, 1696. (5th ed.,
1737. Trans. into Latin and German) A cosmogony in which he
sought to explain the Biblical stories of the Creation, Flood,
and final conflagration in terms of the Newtonian world-system.
Like the work of Burnet (1020), on which it drew heavily, the
book was vigorously disputed but also widely read.

1026. *LITHOPHYLACI Britannici ichnographia*. London, 1699. By E.
Lhwyd. An illustrated catalogue of some seventeen hundred
fossils (mainly shells) in the (old) Ashmolean Museum, Oxford,
of which he was the keeper. In an appendix he discussed the
topical question of the origin of fossils.

1700s

1027. SCHEUCHZER, J.J. (2893) From about 1700 until his death in 1733
he wrote a large number of works on the mineralogy, geology,
and natural history of Switzerland, especially the Alps. His
descriptions of fossils were particularly significant and he
is regarded as one of the founders of palaeontology.

1028. HOOKE, R. (2266) *The posthumous works*. London, 1705. The
largest part of this collection consists of Hooke's lectures
and writings, spread over a period of thirty years, on geological topics. He clarified current ideas about "figured stones,"
distinguishing what we now call fossils from crystalline minerals. Fossils, he argued, were the remains of creatures once
living and not *lusus naturae* as was then often said. He suggested that the surface of the earth had undergone great changes
(due, he thought, to earthquakes), that the resulting environmental changes could produce alterations in living organisms,
and that consequently much could be learnt about the history
of the earth from the study of fossils.

1720s

1029. CASSINI, J. (2039) *Traité de la grandeur et de la figure de la
terre*. Paris, 1720. An account of the measurement of an arc
of meridian, about 9° in length, extending from Dunkirk to the
Pyrenees--a project that had been initiated by Picard, continued
by his father (G.D. Cassini), and completed by himself. On the
basis of this work he reiterated his father's view that the
earth is elongated at the poles, in contradiction to the
Newtonians who maintained that it is flattened at the poles.

1030. MARSILI, L.F. (1914) (a) *Histoire physique de la mer*. Amsterdam, 1725. The first treatise on oceanography. (b) *Danubius
... observationibus geographicis*.... Amsterdam, 1726.

(French trans., 1744) A study of the Danube and its valley,
including physical geography, hydrology, and natural history.

1031. WOODWARD, J. (See also 1024) **(a)** *Fossils of all kinds digested
into a method.* London, 1728. A classification of 'fossils'
(in the old meaning of the term). It included about two hundred minerals. **(b)** *An attempt towards a natural history of the
fossils of England.* London, 1729. A detailed catalogue of
his large collection of 'fossils', giving localities of the
specimens. His classification was influential for a long time.

From 1730 mineralogy and crystallography are listed
in Section 1.08 and geology in Section 1.09.

1730s

1032. "CONCERNING the cause of the general trade winds"--a paper of
1735 by G. Hadley which, by taking account of the rotation of
the earth, made an important addition to Halley's explanation
(1021b) of the trade winds.

1033. MAUPERTUIS, P.L.M. de. (2049) *Sur la figure de la terre déterminée par les observations de Messieurs de Maupertuis, Clair-
ault, Camus, Le Monnier et Outhier ... au cercle polaire.*
Paris, 1738. The report of the famous expedition to Lapland
to measure an arc of meridian near the North Pole (see 2049).

1740s

1034. CLAIRAUT, A.C. (2052) *Théorie de la figure de la terre.* Paris,
1743. The classical work on the subject. It contained a
detailed elaboration of the Newtonian approach, with deductions
of formulae for the variation of gravity with latitude; these
were shown to agree with the results of pendulum measurements.

1035. *LA FIGURE de la terre, déterminée par les observations de Messieurs De la Condamine et Bouguer ... envoyés par ordre du roi
au Pérou.* Paris, 1749. The report of the expedition to Peru
to measure an arc of meridian near the equator (see 2048 and
2050).

1750s

1036. TARGIONI-TOZZETTI, G. (1920) **(a)** *Relazione d'alcuni viaggi fatti
in diverse parti della Toscana.* 6 vols. Florence, 1751-54.
(b) *Prodromo della corografia e della topografia della Toscana.*
Florence, 1754. Works on the physical geography and natural
history of Tuscany. He was a pioneer in geomorphology and
the evolution of landscapes.

1037. CASSINI DE THURY, C.F. (2055) *Avertissement ou introduction à
la carte générale et particulière de la France.* Paris, 1755.
An account of the great map of France--a massive undertaking
which was not quite completed by the time of the Revolution.

1770s

1038. FRANKLIN, B. (3016) During his eight crossings of the Atlantic
he made many observations of various marine phenomena and, in

particular, collected information from sailors about the
nature and course of the Gulf Stream. In the early 1770s
(while he was postmaster-general) he had the first chart of
the Gulf Stream printed. During his later Atlantic crossings
he made series of temperature measurements at and below the
surface. He also made a contribution to meteorology with his
studies of cloud formation and the electrification of clouds.

1039. LAMBERT, J.H. (2542) "Anmerkungen und Zusätze zur Entwerfung
der Land- und Himmelscharten"--an article of 1772 which made
an important contribution to the theory of map construction
and projections.

1040. MASKELYNE, N. (2309) Made in 1774 the first significant attempt
to measure the density, and hence the mass, of the earth. By
astronomical means he measured the deviation from the vertical
of two plum-lines situated on opposite sides of a symmetrical
mountain range. From the deviations (which were due to the
mountain's gravitational attraction) and an estimate of the
mountain's density he was able to arrive at an estimate of
the earth's density. (His result was superseded in 1798 by
a famous experiment of Cavendish--see 1043)

1780s

1041. METEOROLOGICAL STATIONS. In the early 1780s the Societas Meteorologica Palatina (see 2547) organized a network of some forty
meteorological stations in many countries, including Russia
and the United States. It issued instructions for the standardization and use of the measuring instruments and published
the stations' observations until 1792. The accumulated data
were later used by Humboldt and others in the first climatol-
ogical studies.

1790s

1042. *ENCYCLOPÉDIE méthodique ... Géographie physique.* 5 vols & atlas.
Paris, 1794-1828. A section of this enormous, topically
arranged encyclopedia. Written by N. Desmarest.

1043. CAVENDISH, H. (2306) Made in 1798 the first accurate determination of the mean density, and hence the mass, of the earth
(following the earlier estimate of Maskelyne--1040). He used
a torsion balance to which was attached a suspended beam with
lead balls on either end; these two moveable balls were attracted by two stationery lead balls, and the period of oscillation
of the torsion balance gave a sensitive means of measuring the
attraction. He arrived at a figure for the earth's mean density of 5.48 times that of water--very close to the best modern
figure. The experiment was also of course an absolute measure
of gravity.

1800s

..... HUMBOLDT, A. von. See 1242.

1044. DELAMBRE, J.B.J. (2094) *Base du système métrique décimal; ou,
Mesure de l'arc du méridien compris entre les parallèles de
Dunkerque et Barcelone.* 3 vols. Paris, 1806-10. The final
report on the Paris Academy's great project, begun in 1792.

1820s

1045. DANIELL, J.F. (2347) **(a)** Invented in 1820 a new dew-point hygrometer which was generally adopted. **(b)** *Meteorological essays and observations*. London, 1823. (3rd ed., 1845) Included his observations on the behaviour of the atmosphere, and especially the trade winds. The third edition contained discussions of the meteorological effects of solar radiation and the cooling of the earth.

1046. SABINE, E. (2352) *An account of experiments to determine the figure of the earth by means of the pendulum vibrating seconds in different latitudes*. London, 1825. As a participant in various expeditions he took measurements at seventeen positions around the earth.   (See also 1051)

1047. DOVE, H.W. (2605) From the late 1820s to the late 1850s he developed the first widely influential system of meteorological concepts and theories. Though there was a reaction against his ideas (and especially his central idea of the meeting of tropical and polar atmospheric currents) in the next generation they have been largely rehabilitated in the twentieth century.

1830s

1048. REDFIELD, W. (3023) Published in 1831 his theory that a storm is a vortex, with the winds blowing anti-clockwise (in the northern hemisphere) around a centre which moves in the direction of the normal prevailing winds. He later made another important discovery concerning the paths of tropical hurricanes.

1049. HUMBOLDT, A. von. (2567) In the early 1830s, largely due to his influence, geomagnetic observatories (generally combined with meteorological observatories) were set up in several European countries and in the Russian Empire (cf. 1050a). In 1836 he wrote to the president of the Royal Society of London suggesting the establishment of a worldwide network of such observatories. As a result the British government set up observatories in several of its colonies (cf. 1051b).   (See also 1052)

1050. GAUSS, C.F. (2570) **(a)** In conjunction with Humboldt (the originator of the scheme), Gauss and his Göttingen colleague, W.E. Weber, established the Magnetischer Verein to co-ordinate the activities of the international network of geomagnetic observatories. The Verein's *Resultate* appeared in six volumes from 1836 to 1841; it included many papers by Gauss and Weber, as well as their joint *Atlas des Erdmagnetismus* (1840).
      (b) In 1839 Gauss published his *Allgemeine Theorie des Erdmagnetismus* in which, as a result of a mathematical treatment of the data, he was able to express the magnetic potential at any point on the earth's surface.
(See also 1053)

1051. SABINE, E. (See also 1046) **(a)** *Report on the variations of the magnetic intensity observed at different points of the earth's surface*. London, 1838. **(b)** Largely through his influence (added to that of Humboldt--see 1049) the British government set up magnetic observatories in several of its colonies in

the late 1830s. It was data from these observatories that
enabled him in 1852 to discover a relation between magnetic
disturbances and sunspots--see 1059 and 1060.

1840s

1052. HUMBOLDT, A. von. (See also 1049) (a) *Asie centrale. Recherches
sur les chaînes de montagnes et la climatologie comparée.*
Paris, 1843. A work arising from his expedition of 1829 and
presenting a valuable account of a relatively unknown region.
(b) His many contributions to all the earth sciences--notably
physical geography and climatology--were surveyed in his great
work, *Kosmos* (see 1249).

1053. GAUSS, C.F. (See also 1050) Published in 1844 and 1847 two
important periodical articles containing a mathematical treatment of the general problem of map projection. (This work
was an outcome of his practical experience in the triangulation
of the kingdom of Hanover in the 1820s)

1054. WEATHER MAPS. The first synoptic weather map was published in
1846 by E. Loomis, a professor in New York. The introduction
of such maps had profound effects on meteorology.

1055. *PHYSICAL geography.* London, 1848. By Mary Somerville. A work
by a brilliant expositor of science which won praise from
Humboldt and others. Among other things it supported the new
geology of Lyell and his followers.

1056. STOKES, G.G. (2387) Published in 1849 a mathematical study of
the variation of gravity over the earth's surface without
making any assumptions about its interior. Clairaut's theorem
was derived as a special case.

1850s

1057. FERREL, W. (3039) (a) Made substantial improvements in the 1850s
and 1860s to Laplace's theory of tides. In 1880, while he was
a member of the U.S. Coast Survey, he designed an analogue
machine to predict tidal maxima and minima. (b) In several
papers in the late 1850s he constructed the general mathematical theory of the motion of a body with respect to the rotating earth and applied it to atmospheric and oceanic circulation.
This work, which he continued to develop, opened a new epoch
in meteorology.   (See also 1071)

1058. TERRESTRIAL MAGNETIC CYCLE. J. von Lamont, director of the
Bogenhausen Observatory (Munich), concluded in 1850 from his
long-continued observations of terrestrial magnetism that
variations in the earth's magnetism occur in cycles of about
ten years. It was already known that there is a cycle of
similar length for sunspots (cf. 440) and a connection between
the two cycles was immediately suspected.

1059. SABINE, W. (See also 1051) Discovered in 1851 that the daily
variation in the intensity of the earth's magnetism is the
resultant of two combined variations, one stemming from within
the earth and the other from outside it.

1060. EFFECTS OF SUNSPOTS. A striking correspondence between the
      periodic variations of sunspots and various magnetic disturb-
      ances on earth was noticed in 1852 by three people independ-
      ently--A. Gautier, E. Sabine (cf. 1051), and J.R. Wolf.

1061. MAURY, M.F. (3034) *The physical geography of the sea*. New York
      and London, 1855. (Many reprintings over the following thirty
      years. Translated into six languages) A pioneering work in
      oceanography and marine meteorology. Highly popular with
      mariners and the general public but criticised by many scien-
      tists: though it was valuable as a compilation of data, many
      of its interpretations were incorrect or implausible. Never-
      theless the book gave a major stimulus to the subject.

1062. BUYS BALLOT'S LAW--in the northern hemisphere winds circulate
      anti-clockwise around low-pressure areas and clockwise around
      high-pressure areas. Formulated (in different but equivalent
      terms) by the Dutch meteorologist, C.H.D. Buys Ballot in 1857.

1063. FORBES, E. (2395) *The natural history of the European seas*.
      London, 1859. (Posthumous) A pioneering work in oceanography
      including both biological and physical aspects.

<hr>

1860s

1064. GALTON, F. (2401) *Meteorologica; or, Methods of mapping the
      weather*. London, 1863. Included the first thorough study of
      anticyclones (his term). Galton was also the first to use
      flow lines on meteorological charts.

1065. THOMSON, C.W. (2422) *The depths of the sea. An account of the
      general results of the dredging cruises of H.M.SS. "Porcupine"
      and "Lightning" during 1868-70*. London, 1873. The successive
      cruises of the two ships in the North Atlantic were authorised
      by the Admiralty at the request of the Royal Society. Thomson
      worked on board in collaboration with W.B. Carpenter and J.G.
      Jeffreys, and the importance of their discoveries, in both
      physical and biological oceanography, raised interest in the
      subject and set the scene for the much more ambitious and
      far-ranging *Challenger* Expedition (see 1068).

1066. OCEANOGRAPHIC CRUISES. *Blake* (U.S. Coast Survey ship). Cruises,
      1868-80.

1067. BUCHAN, A. (2430) **(a)** *The handy book of meteorology*. Edinburgh,
      1868. A widely-used textbook. **(b)** "Mean pressure and prevail-
      ing winds of the globe"--an important paper of 1869.
      (See also 1078)

1870s

<hr>

1068. *CHALLENGER* EXPEDITION. The famous oceanographic expedition of
      H.M.S. *Challenger*, carried out in 1872-76 through the co-oper-
      ation of the Admiralty and the Royal Society. The head of the
      scientific staff was the marine biologist, C.W. Thomson (cf.
      1065). The voyage totalled 68,890 miles (127,600 km) and, in
      addition to innumerable observations of various kinds, 492 deep
      soundings and 133 dredgings were made. The *Challenger Report*,
      presenting the scientific results, was published in fifty

volumes in 1880–95. The scope and thoroughness of the research and the importance of the results made the expedition a major landmark in the history of oceanography.

1069. HANN, J.F. (2964) *Die Erde als Ganzes, ihre Atmosphäre und Hydrosphäre.* Prague, 1872. (5th ed., 1897) The successive editions, extensively revised, included much of his own original work in all aspects of meteorology and especially atmospheric dynamics. (See also 1075)

1070. *ÉTUDES sur les mouvements de l'atmosphère.* 2 vols. Christiania, 1876–80. By H. Mohn (2870) and C.M. Guldberg (2869). An important theoretical work in dynamical meteorology. It took account of the Coriolis effect and included a treatment of friction between air currents and the earth. The Mohn–Guldberg equations became well known.

1071. FERREL, W. (See also 1057) *Meteorological researches.* 3 vols. Washington, 1877–82. Included important applications of the general theory he had established in the late 1850s.

1880s

1072. HELMERT, F.R. (2683) *Die mathematischen und physikalischen Theorien der höheren Geodäsie.* 2 vols. Leipzig, 1880–84. A monumental treatise (it was reprinted in 1962).

1073. OCEANOGRAPHIC CRUISES. **(a)** *Albert* (British ship); cruise 1881–82. **(b)** *Hirondelle* (ship belonging to Prince Albert I of Monaco); cruises 1885 onward. **(c)** *Rambler* (British ship); dredging cruise 1888.

1074. STEWART, B. (2412) In an important review article on terrestrial magnetism in 1882 he suggested that the daily and seasonal variations in the earth's magnetic field are caused by electric currents flowing in the rarefied air of the upper atmosphere (later identified as the ionosphere). His hypothesis—the first identification of external influences on the earth's magnetic field—stimulated research in atmospheric physics and was eventually verified.

1075. HANN, J.F. (See also 1069) **(a)** *Handbuch der Klimatologie.* Stuttgart, 1883. (4th ed., rev., 1932. English trans., 1903) The standard work in the subject for many years. It correlated a very large amount of data from all over the world and incorporated the valuable statistical approach he had developed. **(b)** *Atlas der Meteorologie.* Gotha, 1887.

1076. VOEYKOV, A.I. (2999) **(a)** *Climates of the earth, particularly Russia* [In Russian]. St. Petersburg, 1884. A major work which, in addition to presenting a wealth of factual description, also tried to identify the causes of climatic phenomena. **(b)** *Snow cover: Its effect on climate and weather* [In Russian]. St. Petersburg, 1885. The first important work on the subject.

1077. HENSEN, V. (2692) From the late 1880s he developed methods for estimating the quantity of plankton (his term) at various depths in the ocean. In 1889 he led a large expedition to survey the distribution of plankton in the Atlantic.

1078. BUCHAN, A. (See also 1067) "Report on atmospheric circulation, based on observations made on board H.M.S. *Challenger*"-- a major article in the *Challenger Reports* (vol. 2, 1889) which co-ordinated and interpreted the great mass of data collected.

1890s

1079. MURRAY, J. (2447) *Deep-sea deposits.* (1891) One of the volumes of the *Challenger Reports*, written in collaboration with the Belgian geologist, Alphonse Renard. Murray had collected and studied deposits from the floor of the ocean during the *Challenger* Expedition and after its return he sent specimens of them to Renard, a petrographer of established reputation, for expert examination. The resulting monograph by the two authors was a monumental work which opened up the field of marine sedimentology.

1080. FOREL, F.A. (2914. "The founder of limnology") *Le Léman: Monographie limnologique.* 3 vols. 1892-94. An intensive study of the physical and biological features of Lake Geneva (Lac Léman) and the first such study of a lake. His *Handbuch der Seenkunde: Allgemeine Limnologie* of 1901 was the first textbook of limnology.

1.08 MINERALOGY AND CRYSTALLOGRAPHY

From 1730. Earlier items are included in Section 1.07 (Earth Science in General).

1740s

1081. WALLERIUS, J.G. (2840) *Mineralogia*. Stockholm, 1747. (In Swedish. Translated into several languages) The first major treatise in the subject. It dealt with chemical properties as well as external appearances.

1750s

1082. CRONSTEDT, A.F. (2839) *Försök til mineralogie*. Stockholm, 1758. (Translated into several languages) An essay on the classification of minerals which considerably clarified the subject. It also attempted to combine the findings of chemical analysis with the study of external characteristics. The work was praised by A.G. Werner who had it newly translated into German in 1780 with his own additions.

1770s

1083. WERNER, A.G. (2554) *Von den äusserlichen Kennzeichen der Fossilien*. Leipzig, 1774. (English trans., 1805) A manual for the field worker, giving a classification of minerals based on their visible characteristics. It continued in use well into the nineteenth century.   (See also 1087)

1780s

1084. ROMÉ DE L'ISLE, J.B.L. (2072) **(a)** *Cristallographie*. 4 vols. Paris, 1783. A pioneering work which described over 450 crystal forms. It made extensive use of measurements of interfacial angles and affirmed their constancy.
(b) *Des caractères extérieurs des minéraux*. Paris, 1784. A supplement to the foregoing work. It maintained that, despite the contributions of chemistry, crystal form together with density and hardness were sufficient to characterize any mineral species.

1085. DAUBENTON, L.J.M. (2062) *Tableau méthodique des minéraux*. Paris, 1784. (8th ed., 1800) An influential treatise by a leading teacher of the subject.

1086. KIRWAN, R. (2324) *Elements of mineralogy*. London, 1784. (2nd ed., revised and much enlarged, 2 vols, 1794-96) The classification was based on chemical analysis.

1087. WERNER, A.G. (See also 1083) His mineral system was first
      published by one of his disciples in a periodical article in
      1789, and later versions of it were incorporated in some of
      Werner's own works through which it became well known. The
      final version was published posthumously as *Letztes Mineral-
      System* (Freiberg, 1817).

1790s

1088. *NEUES mineralogisches Wörterbuch*. Hof, 1798. By F.A. Reuss.

1800s

1089. HAÜY, R.J. (2092) **(a)** *Traité de minéralogie*. 4 vols & atlas.
      Paris, 1801. The first volume presented his theory of the
      crystal state and the following three applied it to the class-
      ification of minerals. (The revised and enlarged second ed-
      ition of 1822 was of Vols II-IV, his *Traité de cristallographie*,
      published at the same time, being in effect the second edition
      of Vol. I)
           (b) *Tableau comparatif des résultats de la cristall-
      ographie et de l'analyse chimique relativement à la classifi-
      cation des minéraux*. Paris, 1809.

..... *LEHRBUCH der Mineralogie*. By F.A. Reuss. See 1129.

1090. WOLLASTON, W.H. (2336) Invented the reflecting goniometer in
      1809. It made possible a much greater accuracy in the measure-
      ment of crystal angles than the contact goniometer used hither-
      to, and consequently had an important effect on the subsequent
      development of crystallography.   (See also 1092)

1810s

1091. WEISS, C.S. (2572) Developed some important new concepts in
      crystallography in a long series of papers from about 1810 to
      1820. In contrast to Haüy he emphasized the growth of crystals,
      and he conceived the idea of crystallographic axes denoting
      directions of growth and serving also as a basis for definition
      and classification. The system he elaborated made a fundamental
      contribution to crystallography and had much influence in
      Germany (as against Haüy's system).

1092. WOLLASTON, W.H. (See also 1090) Proposed in 1812 a general
      theory of crystal structure in which crystals are built up
      from spherical units packed together in different ways -- a
      theory unlike the one proposed by Haüy. Though quite specu-
      lative, and incapable of being substantiated at the time, his
      theory was found generations later to be largely correct.

1093. BREITHAUPT, J. (2583) *Über die Echtheit der Krystalle*. Freiberg,
      1815. A study of pseudomorphs. The phenomenon had been recog-
      nized earlier but this was the first extensive investigation
      of it.   (See also 1100)

1820s

1094. MITSCHERLICH, E. (2601) Discovery of isomorphism: see 873.
      He later discovered dimorphism (the occurrence of more than

one crystal form for a single substance). These discoveries refuted Haüy's principle that identity of crystal form meant identity of chemical composition. Mitscherlich also observed the different degrees of thermal expansion along dissimilar axes of a crystal.

1095. MOHS, F. (2946) **(a)** *Die Charaktere der Klassen ... der natur-historischen Mineral-Systems*. Dresden, 1820. Proposed a new system of classification in which his hardness scale (which originated in 1812) featured prominently. His scale was more successful than his system which was in the Linnean style and based on physical characteristics, largely neglecting chemical composition. **(b)** *Grund-Riss der Mineralogie*. 2 vols. Dresden, 1822-24. Made some significant contributions to crystallography.

1830s

1096. NAUMANN, K.F. (2604) *Lehrbuch der reinen und angewandten Krystallographie*. Leipzig, 1830. Included several important contributions of his own. (See also 1098)

1097. DANA, J.D. (3027) *A system of mineralogy*. New Haven, Conn., 1837. His system, combining the chemical and crystallographic aspects of the subject, was widely adopted. The book was extraordinarily successful and has continued through many revisions and expansions to the present day.

1840s

1098. NAUMANN, K.F. (See also 1096) *Elemente der Mineralogie*. Leipzig, 1846. (14th ed., 1901) A highly successful textbook which combined the crystallographic approach of Mohs with the chemical approach of Berzelius.

1099. BRAVAIS, A. (2154) Made an exhaustive study in 1848 of the geometrical properties of space lattices and arrived at the conclusion that there are only fourteen lattices (or arrangements of points in space) which are compatible with the orderly arrangements of atoms in crystals. (See also 1101)

1100. BREITHAUPT, J. (See also 1093) *Die Paragenesis der Mineralien*. Freiberg, 1849. The pioneering work on the subject.

1850s

1101. BRAVAIS, A. (See also 1099) From an analysis of the relations between ideal space lattices and the actual symmetries of crystals he deduced in 1851 that there can only be thirty-two classes of crystal symmetries.

1860s

1102. TSCHERMAK, G. (2962) Through the 1860s he carried out a comprehensive programme of studies on the crystal form, physical properties, and chemical composition of the most important rock-forming minerals. His work on the feldspars was especially notable. (See also 1105)

1870s

..... ZIRKEL, F. and ROSENBUSCH, H.   See 1193 and 1194.

1103. GROTH, P. (2711) **(a)** *Tabellarische Übersicht der einfachen Mineralien*. Brunswick, 1874. (4th ed., 1898) **(b)** *Physikalische Krystallographie*. Leipzig, 1876. (4th ed., 1905) Two textbooks of major importance. The latter especially contained much original work of his own.

1104. SOHNCKE, L. (2680) *Die Entwicklung einer Theorie der Krystallstruktur*. Leipzig, 1879. By introducing two new symmetry elements (screw axes and glide planes) he extended the lattice theory of Bravais (1099) from fourteen to sixty-five possible lattices or space groups.

1880s

1105. TSCHERMAK, G. (See also 1102) **(a)** *Die mikroskopische Beschaffenheit der Meteoriten*. Vienna, 1883. Based on the researches on meteorites that he had been carrying on since 1870. It became the standard work on the subject. **(b)** *Lehrbuch der Mineralogie*. Vienna, 1884. (5th ed., 1897)

1106. GOLDSCHMIDT, V. (2712) *Index der Kristallformen der Mineralien*. 3 vols. Berlin, 1886–91. A catalogue of the crystal forms of all known minerals.   (See also 1110)

1890s

1107. FYODOROV, E.S. (3006) Published in 1890 a work (in Russian) containing the first deduction of the 230 crystallographic space groups. In the following year the same conclusion was independently reached by Schoenflies. In subsequent articles in the *Zeitschrift für Kristallographie* Fyodorov compared his work with that of Schoenflies and went on to give a mathematical definition of thirty-two point groups for six crystallographic systems.

1108. SCHOENFLIES, A.M. (2741) *Kristallsysteme und Kristallstruktur*. Leipzig, 1891. A mathematical treatment of the arrangement of points in space which led led him to the recognition of the 230 possible space groups (cf. 1107)

1109. BARLOW, W. (2463) From 1883 he had been publishing papers attempting to explain the structure of crystals in terms of the close packing of atoms. In the late 1880s he took up the mathematical approach of Sohncke (1104) and subsequently arrived at a deduction of the 230 space groups independently of Fyodorov and Schoenflies. He did not publish his conclusions until 1894 however.

1110. GOLDSCHMIDT, V. (See also 1106) *Kristallographische Winkeltabellen*. Berlin, 1897. Tables of crystal angles measured by means of new or improved instruments. (His major work, *Atlas der Kristallformen*, appeared in nine volumes in 1913–23)

1.09 GEOLOGY

Including palaeontology.

From 1730. Earlier items are included in Section 1.07 (Earth Science in General)

1730s

1111. *TELLIAMED; ou, Entretiens ... sur la diminution de la mer, la formation de la terre.... Amsterdam, 1738.* An 'ultra-Neptunian' theory of the earth according to which the earth's surface was shaped by the action of a universal ocean and its gradual diminution. The work had an appreciable influence on later geological theorists. The author was Benoît de Maillet, a diplomat who made many geological observations in the course of his extensive travels.

1740s

1112. MORO, A.L. (1919) *Dei crostacei e degli altri corpi marini che si trovano sui monti.* Venice, 1740. Elaborated a Plutonic theory of the earth, inspired by the phenomena of earthquakes and volcanic activity in the Mediterranean region, especially the upheaval of new volcanic islands where the emerging rocks became covered with shellfish which were subsequently covered by volcanic dust--an illustration of the origin of fossils. The book attracted much attention.

1113. GUETTARD, J.E. (2061) Author of many papers (some of them of book length) in the *Mémoires* of the Paris Academy from 1746 to his death in 1786. He is best known for his ambitious plan to construct a geological map of France, an undertaking at which he laboured all his working life. (It was published in part in 1780) Among his many discoveries the most notable was his realization that the mountains of Auvergne are extinct volcanoes.

1750s

1114. LEHMANN, J.G. (2544) *Versuch einer Geschichte von Flötzgebürgen.* Berlin, 1756. An important account of stratified rocks (*Flötzgebürge*), which he recognized as being of sedimentary origin, and their distinction from what he called *Ganggebürge*--veined rocks (now called igneous rocks).

1760s

1115. ARDUINO, G. (1922) Author of numerous articles in various Italian journals from 1760 onward. His observations were chiefly concerned with stratigraphy and he was the originator

of the classification of strata into four major successive
groups--Primary, Secondary, Tertiary, and Quaternary.

1116. FÜCHSEL, G.C. (2545) "Historia terrae ac maris, ex historia
Thuringiae...." -- a long article in the *Acta* of the Erfurt
Academy for 1761. It dealt with the stratigraphy of Thuringia
but was of much wider importance for the general principles
which it contained. In particular it introduced the concept
of a formation in the sense which it still has in geology,
used the method of correlation of strata by means of index
fossils, and explicitly assumed the principle of uniformitarianism. "Füchsel's great work ... became practically the
model for the Wernerian School of geologists, and, more than
any other individual work, laid the foundation of that rapid
development of stratigraphical geology which began in Germany
in the next generation." (K. von Zittel)

1770s

1117. DESMAREST, N. (2067) In a paper of 1771 (and in later supporting papers) he presented evidence, accumulated over several
years, which established that the basalts of the Auvergne
district are of volcanic origin. This settled the 'basalt
controversy' in favour of the Plutonist position (at least in
principle--his evidence was disregarded for a long time by the
Wernerians) and was of basic importance for the subsequent
development of geological theory.

1118. PALLAS, P.S. (2977) "Observations sur la formation des montagnes
et sur les changements arrivés au globe" -- a paper in the *Acta*
of the St. Petersburg Academy for 1777. It presented a comprehensive and detailed theory of mountain formation, based on
Pallas' observations made over vast areas of Russia and Siberia.

1119. BUFFON, G.L. (2057) *Les époques de la nature*. Paris, 1778.
(Part of his *Histoire naturelle* (1234) and also published
separately. It was an elaboration of his *Théorie de la terre*
which had appeared thirty years earlier in the first volume of
his *Histoire naturelle*)
A speculative but richly suggestive scheme of the
formation and development of the earth and of living creatures
--what would today be called inorganic and organic evolution.
Especially notable are his ideas on the origin of the planetary system as a result of a collision of a comet (then thought
to be a massive body) with the sun, and his attempt to estimate
the age of the earth from experiments on the cooling of spheres.

1120. FAUJAS DE SAINT-FOND, B. de. (2080) *Recherches sur les volcans
éteints du Viverais et du Velay*. Grenoble, 1778. A thorough
study of the extinct volcanoes and associated rocks of the
region--including a mineralogical comparison of the rocks with
material ejected from active volcanoes elsewhere--which proved
the volcanic origin of basalt. His work was independent of
that of Desmarest (1117) who had reached the same conclusion
several years earlier.

1121. SAUSSURE, H.B. de. (2900) *Voyages dans les Alpes*. 4 vols.
Geneva, 1779-96. The first extensive geological investigation

of the Alps. It contained a wealth of observations and
insights concerning their structure, and the nature and origin
of mountains generally (though in his time it was not possible
to get far with these complex subjects). The work was also of
much literary merit.

1780s

1122. WERNER, A.G. (2554. "The father of historical geology") His
writings were sparse and few in number (in his later years he
became very averse to writing) but his ideas were effectively
disseminated by his many enthusiastic students. His *Kurze
Klassifikation und Beschreibung der verschiedenen Gebirgsarten*
(published in 1786 as a periodical article and later reprinted
as a pamphlet) was his only written statement of his main
theories and, despite its brevity and lack of discussion, it
had a wide influence.   (See also 1125)

1123. DOLOMIEU, D. (2093) Between 1783 and 1788 he published three
books, dealing chiefly with volcanoes and earthquakes, and
from 1784 to his death in 1801 many periodical articles on
these and various other geological and mineralogical topics.
He was a very capable geologist and took an active part in the
controversies of the time, though without initiating any major
theoretical developments.

1124. HUTTON, J. (2308) His theory of the earth, based upon the
principle of uniformitarianism, was first announced to the
Royal Society of Edinburgh in 1785 and was published in its
Transactions for 1788. His ideas, universally accepted today,
appeared at the time to be strange and alien, and in 1793 they
were vigorously attacked by the mineralogist and chemist, R.
Kirwan (1126). The attack led Hutton to write a full account
of his theory with extensive supporting evidence--his famous
Theory of the earth (2 vols, Edinburgh, 1795). It was to be
many years however before his ideas began to gain much acceptance.

1790s

1125. WERNER, A.G. (See also 1122) *Neue Theorie von den Entstehung
der Gänge*. Freiberg, 1791. A theory of the formation of ore
deposits as an extension of his general theory of the origin
of rocks.

1126. KIRWAN, R. (2324) In a paper of 1793 he made an attack on
Hutton's first outline of his uniformitarian and Vulcanist
ideas. As a result Hutton was provoked into writing his
celebrated *Theory of the earth* of 1795. Kirwan returned to
attack with another article in 1797 which he then expanded
into his *Geological essays* (London, 1799) which did him little
credit. In another book in 1800 he also attacked James Hall's
evidence for the Huttonian theory (1128).

1127. SMITH, W. (2332) By 1796 he had made the momentous discovery
that different strata could be distinguished by their characteristic fossils, even when the strata resembled each other in
other respects (or were beds of the same rock but at different
levels in the succession). By that time also his travels

around England had enabled him to recognize the general
distribution and succession of strata in much of the country
(an achievement facilitated by the character of English geology). From then onward he was gathering information for his
great undertaking--a geological map of England and Wales--but
during the long period before it was published he talked freely
of his discoveries and sketch maps, a knowledge of which was
widely disseminated by others (to some extent in print).
(See also 1141)

1128. HALL, J. (2326) In a series of papers between 1798 and 1812 he
      provided experimental evidence for Hutton's theory of the
      igneous origin of rocks. By melting samples of basalt and
      allowing them to cool very slowly he showed that a crystalline
      rock was produced--not just an amorphous glass as Hutton's
      opponents had argued. Another argument against Hutton had
      been that subterranean heating of limestone would convert it
      into lime by driving off carbon dioxide. Hall however demon-
      strated that when limestone is heated under great pressure it
      does not lose carbon dioxide but is recrystallized into a rock
      resembling natural marble.

1800s

1129. *LEHRBUCH der Mineralogie*. 8 vols. Leipzig, 1801-06. By F.A.
      Reuss. An influential textbook which, in the two volumes
      dealing with "geognosy", contained the most complete and
      authentic exposition of Werner's geological doctrines.

1130. LAMARCK, J.B. (2082) **(a)** *Hydrogéologie*. Paris, 1802. A rather
      speculative work dealing with the action of water as a geolog-
      ical agent over vast stretches of time. **(b)** *Mémoires sur les
      fossiles des environs de Paris*. Paris, 1809. A pioneering
      work in invertebrate palaeontology (mostly dating from 1802).
      The relations he discerned between fossil and living forms led
      him to think of nature as changing in time; the work thus had
      a major bearing on his theory of evolution.

1131. PLAYFAIR, J. (2325) *Illustrations of the Huttonian theory of
      the earth*. Edinburgh, 1802. A clear and well-written presen-
      tation of Hutton's great theory which was far more effective
      in bringing it to the attention of the scientific world than
      Hutton's own prolix and difficult writings. Though Playfair
      did not depart from Hutton in essentials he placed the emphases
      differently and strengthened the argument with much additional
      evidence.

1132. BUCH, L. von. (2566) *Geognostische Beobachtungen auf Reisen
      durch Deutschland und Italien*. 2 vols. Berlin, 1802-09.
      Contained a wealth of information about the geology of Germany
      and Italy which he had collected during his extensive travels.
      (See also 1136)

1133. SCHLOTHEIM, E.F. (2564) His monograph of 1804 on fossil plants
      of what later came to be called the Carboniferous formation in
      Thuringia was one of the first important works in palaeobotany.
      He concluded that though his fossils had some resemblances to
      living tree-ferns of tropical regions they nevertheless belong-
      ed to a flora that was quite extinct--a novel idea at the time.

1134. *ORGANIC remains of a former world*. 3 vols. London, 1804-11.
A comprehensive treatise on palaeontology, including much
information about British fossils. It had a wide influence.
The author was James Parkinson, a London physician of some
distinction.

1135. OMALIUS D'HALLOY, J.B.J. d'. (2815) "Essai sur la géologie du
nord de la France" -- a lengthy paper of 1808 which greatly
advanced the knowledge of the stratigraphy of the area (which
then included Belgium). The success of the work won him a
governmental commission to prepare a geological map of France
--see 1149.

1810s

1136. BUCH, L. von. (See also 1132) *Reise nach Norwegen und Lappland*.
Berlin, 1810. A travel book which established his reputation
as a writer. It included many acute observations on the geology (and climatology, etc.) of the region, especially his
inference that the Swedish coast is slowly rising.
(See also 1152)

1137. *ESSAI sur la géographie minéralogique des environs de Paris*.
Paris, 1811. By G. Cuvier and A. Brongniart. (A much enlarged
second edition appeared in 1822 under the title *Description
géologique des environs de Paris*)
Cuvier had earlier begun his investigations of the
fossils of the region (later described in his famous work of
1812--see 1138) and he needed to be able to arrange them by
period. From about 1804 he and Brongniart collaborated in a
detailed survey of the stratigraphy, but Brongniart appears
to have taken the leading part in both the field work and the
writing of the book.
The most notable features of this pioneering work in
stratigraphy were its indication that the sediments had been
laid down over a much greater period of time than had been
previously envisaged for such processes, its revelation of an
alternation of marine and freshwater conditions, and its use
of fossils for the correlation of strata (as was being done
at the same time in England by William Smith).

1138. CUVIER, G. (2106) **(a)** *Recherches sur les ossements fossiles
de quadrupèdes*. 4 vols. Paris, 1812. (4th ed., 12 vols,
1834-36) (The famous *Discours préliminaire* was translated
into English in 1813 and several times reprinted; it was also
translated into some other languages. See also (b) below)
A work of seminal importance for both palaeontology
and zoology. Cuvier's application of his principle of the
correlation of parts enabled him to reconstruct whole skeletons
of many unknown quadrupeds, thus·showing that many species had
become extinct over geological time.
(b) *Discours sur les révolutions de la surface du globe,
et sur les changements qu'elles ont produit dans le règne animal*. (Separate publication, progressively enlarged, of the
Discours préliminaire of (a); editions of it continued to
appear up to about 1830, if not later) His best known and
most discussed work. Though he assumed (on the modern view)
a relatively limited age for the earth, he was impressed by

the vast changes which, he saw, must have taken place during
its geological history--especially the extinction of animal
species. As a result he adopted the old idea of catastrophism
and gave it a powerful new impetus.   (See also 1137)

1139. *INTRODUCTION to geology.* London, 1813. (5th ed., 1838) By
R. Bakewell. The first successful English textbook of the
subject. Although it mostly followed Werner it was generally
neutral on contentious issues and showed some appreciation of
Hutton.

1140. BROCCHI, G.B. (1936) *Conchiologia fossile subappennina.* 2 vols.
Milan, 1814. A comprehensive and detailed account of the
Tertiary fossils, especially mollusca, of Italy. It also
included a review of the development of palaeontology in Italy.
The work was much used by later writers, notably Lyell.

1141. SMITH, W. (See also 1127) His great work, *Delineation of the
strata of England and Wales, with part of Scotland,* finally
appeared in 1815--the first large-scale geological map of any
country. (It compares remarkably well with modern maps of
the area)   During the next ten years it was followed by a
fine series of county maps, several charts of geological
sections across parts of the country, and two works on fossils
which set out the basis of his method: **(a)** *Strata identified
by organized fossils* (4 parts, London, 1816-19) and **(b)** *Stratigraphical system of organized fossils* (London, 1817).

1142. *TRAITÉ de géognosie.* 2 vols. Paris, 1819. (2nd ed., 1828-35)
By J.F. d'Aubuisson de Voisins. The first successful French
textbook of geology. Like Reuss' work (1129) it closely
followed Werner's lectures, though it took its examples from
the French scene.

1820s

1143. STERNBERG, K.M. von. (2947) *Versuch einer geognostisch-botan-
ischen Darstellung der Flora der Vorwelt.* 7 vols. Leipzig,
1820-33. A pioneering work in palaeobotany in which he tried
to correlate fossil plants with the botanical classifications
of existing floras.

1144. BRONGNIART, Alexandre. (2103) In some papers in the early 1820s
he followed up his important demonstration of the value of
fossils as 'markers' in the study of stratigraphy (1137). In
particular he emphasized that fossils should be used as the
primary criterion for the correlation of strata, rather than
lithology or the altitude óf the strata. Thus he showed that
the characteristic fossils of the Paris chalk were also to be
found in a hard, black limestone high up in the Alps of Savoy.
(See also 1145)

1145. *HISTOIRE naturelle des crustacés fossiles.* Paris, 1822. By
Alexandre Brongniart and the zoologist, A.G. Desmarest. The
work included the first extensive study of the trilobites, as
well as of true crustaceans, and was valuable for subsequent
investigations of Palaeozoic stratigraphy.

1146. HOFF, K.E.A. von. (2565) Published in 1822-41 a five-volume
work in which he opposed the dominant theory of catastrophism

and put forward the doctrine later termed actualism. According to this doctrine geology should study forces which are now in operation, and can be observed, and should interpret the past history of the earth in terms of them. In support of this view von Hoff assembled--chiefly from historical records extending back to antiquity--a large body of evidence of gradual geological changes of various kinds. His ideas did not gain much acceptance however, perhaps because of the literary nature and limited range of his sources.

1147. *OUTLINES of the geology of England and Wales*. London, 1822. By W.D. Conybeare (2350) and W. Phillips. A highly successful book which had a far-reaching influence. It was primarily a comprehensive and detailed survey of stratigraphy (back to the Carboniferous) based on the work of Smith (1141) but incorporating much additional information. It was effective especially in disseminating a knowledge of Smith's method of correlating strata by means of index fossils.

1148. LEONHARD, K.C. von. (2573) *Charakteristik der Felsarten*. Heidelberg, 1823. The best work on petrology yet to appear. (See also 1159)

1149. OMALIUS D'HALLOY, J.B.J. d'. (See also 1135) "Observations sur un essai de carte géologique de la France, des Pays-Bas et des contrées voisines" -- a paper of 1823 referring to his map of the area, published at the same time. The joint publications "brought the local descriptions of the geology of France into a uniform and sophisticated stratigraphic column, one that, in conjunction with the parallel efforts of Alexandre Brongniart, enjoyed wide acceptance and formed the basis for the development of Continental stratigraphy in the first half of the nineteenth century." (*DSB*)

1150. BUCKLAND, W. (2348) *Reliquiae diluvianae; or, Observations on the organic remains ... attesting the action of an universal deluge*. London, 1823. A book which identified the latest of Cuvier's 'world-catastrophes' with the biblical Flood. It was very popular with the general public and its thesis of a universal deluge was accepted by many geologists at the time. (The considerable evidence that it presented was ultimately reinterpreted in terms of the Ice Age) While continuing to uphold 'diluvial geology' Buckland later abandoned the identification with the biblical Flood.

1151. DESHAYES, P. (2139) *Description des coquilles fossiles des environs de Paris*. 3 vols. Paris, 1824-37. An investigation of the genetic relations of many species of Tertiary molluscs to present-day species. His results enabled him to subdivide the Tertiary into three periods and were very useful to Lyell as evidence for his corresponding subdivision of the Tertiary into Eocene, Miocene, and Pliocene.

1152. BUCH, L. von. (See also 1136) (a) *Physikalische Beschreibung der Canarischen Inseln*. Berlin, 1825. A study of the complex volcanic system which created the islands. (b) In 1826 he published anonymously the first geological map of Germany, comprising forty-two sheets. It went through five editions by 1843.

1153. SCROPE, G.J.P. (2361) **(a)** *Considerations on volcanoes*. London,
      1825. **(b)** *Memoir of the geology of central France, including
      the volcanic formations of Auvergne*. London, 1827.
          The two works contained the first extensive and detailed
      study of the nature and effects of volcanic activity, and the
      first notable attempt to construct a theory of volcanic action
      and to elucidate the part that volcanoes have played in the
      history of the earth. Scrope's conclusions were in opposition
      to both the Neptunism of the Wernerians and the diluvial catas-
      trophism of Cuvier and Buckland, and appear to have had a
      significant influence on the development of Lyell's uniformi-
      tarian ideas.

1154. BRONGNIART, Adolphe T. (2140) **(a)** *Prodrome d'une histoire des
      végétaux fossiles*. Paris, 1828. **(b)** *Histoire des végétaux
      fossiles*. 2 vols. Paris, 1828-37.
          The two works constituted a major contribution to the
      young science of palaeobotany. "He concluded that four dis-
      tinct periods could be defined in the history of plant life
      ... marked by increasing diversity and increasing complexity
      in the groups represented. Thus the first (in modern terms,
      Upper Palaeozoic) period was dominated by vascular cryptogams;
      in the second, with only a poor flora, there were the first
      conifers; in the third (roughly, Mesozoic), the first cycads
      had appeared and, with conifers, comprised about half the
      total flora; and finally, in the fourth (Cainozoic) period,
      dicotyledons had made their first appearance and had come to
      dominate the flora." (M.J.S. Rudwick) Brongniart went on to
      make some perceptive suggestions about possible changes in the
      earth's climate and atmosphere over geological time.

1830s

1155. ÉLIE DE BEAUMONT, L. (2138) From about 1830 he developed his
      influential doctrines on the origin of mountain ranges,
      summing them up in 1852 in his three-volume *Notice sur les
      systèmes de montagnes*. While some of his ideas were valuable
      --especially his basic concept of tectonic upheaval and his
      stratigraphical method for determining the age of mountain-
      systems--much of his work was vitiated by an incongruous
      attempt to apply mathematics (at which he excelled, being a
      Polytechnicien) to geology.   (See also 1169)

1156. LYELL, C. (2360) *Principles of geology. Being an attempt to
      explain the former changes of the earth's surface by reference
      to causes now in operation*. 3 vols. London, 1830-33. (12th
      ed., 1875) The great work which established the principle of
      uniformitarianism in place of the hitherto dominant catastro-
      phism and laid the foundations of modern dynamical geology.
      It was clearly and attractively written and, with its many
      editions, had a very wide public, thus helping to make geology
      one of the most popular sciences. .

1157. SEDGWICK, A. (2349) **(a)** In 1831 he began field work in the
      mountains of north Wales and over the next few years was
      brilliantly successful in elucidating the rock-succession and
      structure of the region. He gave the name 'Cambrian' in 1835

to the oldest of the fossiliferous strata. (b) "Remarks on the structure of large mineral masses" -- a technically import-ant paper in 1835 on the effects of diagenesis and on the distinction between the types of rocks resulting from it. (See also 1164 and 1180)

1158. BRONN, H.G. (2606) **(a)** *Italiens Tertiär-Gebilde.* Heidelberg, 1831. An investigation of Tertiary molluscs very similar to that of Deshayes (1151) and reaching much the same conclusions. **(b)** *Lethaea geognostica.* 2 vols. Stuttgart, 1835-38. (3rd ed., 1850-56. After his death the work was continued by others) An important reference book giving a chronological arrangement of the stratigraphically most significant fossils with detailed descriptions and illustrations. His *Index palaeontologicus* (2 vols, Stuttgart, 1848-49), listing all known fossils, was another valuable reference book.

1159. LEONHARD, K.C. von. (See also 1148) *Die Basaltgebilde.* Stutt-gart, 1832. A thorough study of basalt and its occurrences which supplied the final proof of its volcanic origin.

1160. AGASSIZ, L. (2910) *Recherches sur les poissons fossiles.* 5 vols. Neuchâtel, 1833-44. A highly regarded work inspired by Cuvier's example. It described over 1700 species, making use of the principles of comparative anatomy. (See also 1167)

1161. CHARPENTIER, J. de. (2905) "Notice sur la cause probable du transport des blocs erratiques de la Suisse" -- a paper of 1834 which effectively launched the (not new) idea of the geological significance of glaciers. Charpentier also spread his ideas through personal contacts and his most outstanding convert, Louis Agassiz, later acknowledged his influence. Charpentier's book on the subject, *Essai sur les glaciers et sur le terrain erratique du bassin du Rhône* (Lausanne, 1841), appeared a few months after Agassiz' well-known work (1167a).

1162. ALBERTI, F.A. von. (2584) Published in 1834 a thorough litho-logical and palaeontological study of a sequence of rocks in southwest Germany with a striking threefold division--Bunter sandstone, Muschelkalks (shell limestone), and Keuper sandstone. He proposed the name Trias for the formation and it became internationally accepted as the type of the Triassic system.

1163. MURCHISON, R.I. (2359) *The Silurian system.* 2 vols. London, 1839. A classical monograph which made a major advance in stratigraphy. From his studies of ancient rocks in south Wales Murchison established the Silurian as a stratigraphical system with a particularly characteristic fossil fauna of marine invertebrates but no vertebrates or land plants -- features which indicated that it represented a distinct period in the history of life. Geologists in other countries soon found Silurian strata in many parts of the world.
    Murchison's *Siluria* (London, 1854; 5th ed., 1872) was a revised version of his book presented in a more popular way. (See also 1164 and 1174)

1164. THE DEVONIAN SYSTEM. So named by Murchison and Sedgwick in a joint paper in 1839, "On the physical structure of Devonshire." (See also 1174)

1165. EHRENBERG, C.G. (2612) *Die Bildung der ... Kreidefelsen ... aus mikroskopischen Organismen.* Berlin, 1839. Established that chalk rocks are formed from the minute calcareous shells of foraminifera and that deposits of such shells are still accumulating on the sea floor.   (See also 1183)

1840s

1166. OWEN, R. (2365) **(a)** "Fossil Mammalia" -- Part 1 of *The zoology of the voyage of H.M.S. "Beagle"* (London, 1840). The first of his many publications in palaeontology--a description of the fossils collected by Darwin in South America. **(b)** *Odontography.* See 1482. **(c)** *A history of British fossil reptiles.* 4 vols. London, 1849-84.

1167. AGASSIZ, L. (See also 1160) **(a)** *Etudes sur les glaciers.* Neuchâtel, 1840. A famous book in which he showed that in geologically recent times Switzerland had been covered by an immense ice sheet. From the evidence of glaciation in several other parts of Europe as well he put forward the idea of the Ice Age. **(b)** *Etudes critiques sur les mollusques fossiles.* 4 vols. Neuchâtel, 1840-45.

1168. *A DICTIONARY of geology and mineralogy.* London. 1840. (3rd ed., 1860) By W. Humble.

1169. *CARTE géologique de la France.* Paris, 1841. [Accompanied by] *Explication de la carte géologique de la France.* 2 vols. Paris, 1841-48. By P.A. Dufrénoy and L. Elie de Beaumont (2138) under the supervision of A. Brochant de Villiers (professor at the Ecole des Mines).
         The Ecole des Mines had been wanting to compile a geological map of France for many years but did not get official approval until 1825. The task took the two junior authors fifteen years half-time (making their field surveys in summer). The map and its *Explication* constituted a landmark in French
                                                            geology.

1170. DARWIN, C.R. (2380) **(a)** *The structure and distribution of coral reefs.* London, 1842. (3rd ed., 1889) **(b)** *Geological observations on volcanic islands.* London, 1844. **(c)** *Geological observations on South America.* London, 1846. (These three works constituted Parts 1-3 of *The geology of the voyage of the "Beagle"....*)
         Darwin first made his name as a geologist, especially with his brilliant explanation of the formation of coral atolls.

1171. HALL, J. (3026) **(a)** *Geology of New York.* Albany, 1843. One of the classics of American geology. **(b)** *Palaeontology of New York.* 13 vols. Albany, 1847-94. A massive work by a great palaeontologist on the fossil riches of the Silurian and Devonian rocks of the state. It became the standard for much of the geological exploration of the United States.

1172. DANA, J.D. (3027) In the mid-1840s he adopted the long-standing theory of the cooling and contraction of the earth as the basic cause of the formation of mountain ranges, and for many years continued to elaborate it in detail. He based his views initially on his experience of the Pacific region while a member

of the Wilkes Expedition (1838-42) and later on his studies
of the Appalachians and other mountains.   (See also 1182)

1173. HUMBOLDT, A. von. (2567) His many contributions to geology,
especially concerning volcanoes and earthquakes, were summed
up in his great work, *Kosmos* (1249).

1174. MURCHISON, R.I. (See also 1163) *The geology of Russia in Europe
and the Ural Mountains.* 2 vols. London/Paris, 1845. With
E. de Verneuil and Λ. von Keyserling (Murchison wrote the
stratigraphical section).
      Following the proposal of the Devonian system by
Sedgwick and himself (1164) Murchison went to Russia in 1840
to examine some rocks of a similar age. He found that they
substantiated the concept of the Devonian system, being under-
lain by Silurian rocks and overlain by Carboniferous. He also
discovered near the Ural Mountains a distinctive series of
strata, of wide extent and overlying the Carboniferous, to
which he gave the name Permian.

1175. FORBES, E. (2395) In a long paper in the *Memoirs* of the Geological Survey in 1846 he maintained that most of the plants and
animals of the British Isles had migrated there over land
connections from the Continent at three separate periods --
before, during, and after the glacial epoch. He is thus
regarded as one of the pioneers of biogeography.

1176. BISCHOF, G. (2582) *Lehrbuch der chemischen und physikalischen
Geologie.* 3 vols. Bonn, 1847-54. A pioneering work in geochemistry which attracted much attention and stimulated
research in the field.

1850s

1177. SORBY, H.C. (2397) During the 1850s his interest in both microscopy and geology led him to make very thin sections of rocks
for examination with a polarizing microscope. His investigations culminated in a seminal paper on the subject in 1858.
Sorby did not follow up his initiative but in 1861 he gave an
account of it to a young German geologist, F. Zirkel, who
introduced the new approach into petrology (see 1188).

1178. MALLET, R. (2376) One of the founders of seismology (his term).
During the 1850s, by means of subterranean explosions, he
measured the rate of travel of seismic waves in various kinds
of rocks and geological formations. From his study of the
great Neapolitan earthquake of 1857 he was able to deduce its
epicentre and depth. He also compiled an extensive catalogue
of recorded earthquakes from which he constructed a seismic
map of the world.

1179. NAUMANN, K.F. (2604) *Lehrbuch der Geognosie.* 3 vols. Leipzig,
1850-54. In its time the most outstanding textbook of geology
in Germany. It was especially authoritative on petrography
and was the first book of its kind to include an extended
account of "geo-tectonics", especially earthquakes.

1180. SEDGWICK, A. (See also 1180) *A synopsis of the classification
of the British Palaeozoic rocks.* Cambridge, 1851-55. A work

relating to his elucidation of the Cambrian system in the 1830s
and provoked by a subsequent controversy.

1181. STUDER, B. (2907) *Geologie der Schweiz.* 2 vols. Bern/Zurich,
1851-53. Included the first thorough description of the
structure of the Swiss Alps together with an attempt at explaining what H.B. de Saussure had called a "hopeless jumble." The
book was written to accompany the first geological map of Switzerland (published in 1853) which was based on surveys he had
made over many years in collaboration with A. Escher.

1182. DANA, J.D. (See also 1172) *On coral reefs and islands.* New
York, 1853. A product of his extensive experience with the
subject during his four years with the Wilkes Expedition. He
dealt with both the zoological and geological aspects, confirming Darwin's subsistence theory of the formation of coral
atolls (1170a), and presenting much new information on corals,
reef-building, etc.   (See also 1187)

1183. EHRENBERG, C.G. (See also 1165) *Mikrogeologie.* Leipzig, 1854.
An account of his investigations, carried out over many years,
on the formation of sedimentary rocks from the shells of uni-
cellular organisms such as diatoms and foraminifera.

1184. DAUBRÉE, A. (2167) From the mid-1850s he developed experimental
methods for the study of various mineralogical and geological
phenomena--the synthesis of minerals and numerous other geo-
chemical processes, the permeability of different rocks to
water, the fracturing of rocks and formation of joints and
faults, metamorphism and its effects, etc. He reported his
results in a succession of publications culminating in his
Etudes synthétiques de géologie expérimentale (Paris, 1879).

1184a COTTA, B. (2617) **(a)** *Die Gesteinlehre.* Freiberg, 1855. An
important treatise on petrography. (It portrayed the degree of
development of the subject just before the introduction of the
microscopical approach) **(b)** *Die Lehre von den Erzlagerstätten.*
Freiberg, 1855. A widely used treatise on the theory of ore
deposits, a field in which he made important contributions.

1860s

1185. OWEN, R. (See also 1166) *Palaeontology; or, A systematic summary of extinct animals and their geological relations.* Edinburgh, 1860.

1186. THOMSON, W. (*Lord* KELVIN) (2389) In 1862 and 1868 he launched
an attack on the principle of uniformitarianism--and with it,
Darwinian evolution--on the grounds that calculations of the
rate of cooling of the earth showed that it was not nearly so
old as the geologists and Darwinians supposed. Furthermore,
he contended, during its early history the earth must have
been much hotter than it now is, with resulting violent phenomena and vastly different climates. His arguments seemed
incontrovertible and the geologists and biologists had to
live with them as best they could until the end of the century
when, with the discovery of radioactivity, the dilemma vanished.

..... GAUDRY, A.J. Evolutionary palaeontology--see 1527.

1187. DANA, J.D. (See also 1182) *Manual of geology*. Philadelphia, 1863. (4th ed., 1895) "The bible of American geology for four decades."

1188. ZIRKEL, F. (2681) While in England in 1860 on a geological tour he happened to meet H.C. Sorby who showed him his technique of examining thin sections of rocks with a polarizing microscope (1177). Zirkel subsequently developed the technique and in 1866 published his *Lehrbuch der Petrographie*, a textbook of the hitherto usual kind of petrology (then often called petrography) in which he included an account of the microscopical approach and its potential value. The book was effective in disseminating a knowledge of the method which was soon taken up by others, notably Rosenbusch (1194).   (See also 1193)

1870s

1189. COPE, E.D. (3050) Through the 1870s and 1880s he made major contributions to vertebrate palaeontology with his many discoveries in the newly-accessible and rich fossil fields of the western United States. In particular he established that the Age of Mammals had begun much earlier than had been thought, and--with his rival, O.C. Marsh--discovered the first complete remains of the great dinosaurs of the Cretaceous period.

1190. MARSH, O.C. (3049) In the early 1870s he led several expeditions from Yale to the fossil fields of the western United States, making a wealth of important discoveries. Thereafter for many years he employed collectors to build up a great collection of fossils at Yale's Peabody Museum. Like his rival, E.D. Cope, he made numerous major contributions to vertebrate palaeontology, one of his best-known achievements being his elucidation, in the mid-1870s, of the evolutionary history of the horse. He also obtained evidence of the link between reptiles and birds.

1191. *ELEMENTE der Geologie*. Leipzig, 1871. (Many later editions) By H.G. Credner (professor at Leipzig). The leading textbook on the subject in Germany for a generation.

1192. WILLIAMSON, W.C. (2396) Published a series of papers over the period 1871-93 describing numerous fossil plants found in coal from the Upper Carboniferous coal measures of northern England. His work constituted a major contribution to palaeobotany and opened up an important field of research.

1193. ZIRKEL, F. (See also 1188) *Die mikroskopische Beschaffenheit der Mineralien und Felsarten*. Leipzig, 1873. A treatise on microscopical petrography which, together with the similar work by Rosenbusch which appeared simultaneously, laid the foundations of the subject.

1194. ROSENBUSCH, H. (2713) *Mikroskopische Physiographie der petrographisch wichtigsten Mineralien*. Stuttgart, 1873. (Vol. 2 appeared in 1877 with the title *Mikroskopische Physiographie der massigen Gesteine* and the two books went through four successively enlarged editions)

1195. GEIKIE, J. (2429) *The great Ice Age and its relation to the
      antiquity of man*. London, 1874. (3rd ed., 1894) Included
      the contention that the Ice Age had been divided by interglacial
      periods of relatively mild climate. He considered that man had
      lived in Europe during the Ice Age.

1196. SUESS, E. (2957) *Die Entstehung der Alpen*. Vienna, 1875. Made
      a new start in the attempt to understand the structure and
      formation of mountain ranges, the Alps being taken as the
      outstanding and most-studied case. Suess rejected the hitherto
      dominant ideas of Elie de Beaumont and von Buch, and proposed
      a new theory which had considerable influence and proved to be
      a fruitful stimulus to further investigations. (See also 1205)

1197. DOKUCHAEV, V.V. (3002) About 1875 he began the field studies of
      Russian soils from which he evolved the basic principles of a
      new science--pedology, or soil science. His important class-
      ification of soils was published in a series of publications
      from 1883 onward.

1198. *HANDBUCH der Palaeontologie*. [Part 1. Palaeozoology] 4 vols.
      Munich, 1876-93. By K.A. von Zittel. [Part 2. Palaeobotany]
      1 vol. Munich, 1890. By W.P. Schimper and A. Schenk.
           A monumental presentation of the systematics of
      fossilized animals and plants.   (See also 1213)

1199. LAPWORTH, C. (2444) During the 1870s he unravelled the complic-
      ated stratigraphy and structure of the Lower Palaeozoic rocks
      of Scotland by giving close attention to a particular group
      of their fossils, the graptolites--a method that was later
      adopted in other parts of the world. In 1879, concurrently
      with this work, he proposed that a new stratigraphical system,
      which he named the Ordovician, be recognized between the
      (redefined) Cambrian and Silurian; his proposal eventually
      became generally accepted.

1200. HEIM, A. (2913) *Untersuchung über den Mechanismus der Gebirgs-
      bildung*. 3 vols. Basel, 1878. An influential and authori-
      tative treatise on the dynamics of mountain building and the
      mechanics of rock deformation. It was based on his studies
      of the Alps and derived much of its impact from its excellent
      illustrations.   (See also 1207a)

1880s

1201. MILNE, J. (2445) From about 1880 he was a leading figure in
      the establishment of modern seismology, inventing and progress-
      ively improving the seismograph and promoting the establishment
      of seismographical stations around the world. His writings
      included *Earthquakes and other earth movements* (London, 1886;
      4th ed., 1898) and *Seismology* (London, 1898; 2nd ed., 1908).

1202. PENCK, A. (2714) From the early 1880s he carried out many field
      studies in the Bavarian Alps on the effect of glaciers on the
      development and form of valleys. His results were published
      in a series of periodical articles and integrated in his classic
      Die Alpen im Eiszeitalter (3 vols, Leipzig, 1901-09). It
      provided evidence for four main periods of Pleistocene glaci-
      ation, a scheme that was widely accepted for many years.
      (See also 1212)

1203. GEIKIE, A. (2420) *Text-book of geology*. London, 1882. (4th
    ed., 1903) An outstanding treatise, the successor of Lyell's
    famous *Principles*.   (See also 1215)

1204. *HANDWÖRTERBUCH der Mineralogie, Geologie und Palaeontologie.*
    3 vols. Breslau, 1882-87. By A. Kenngott.

1205. SUESS, E. (See also 1196) *Das Antlitz der Erde*. 4 vols.
    Vienna, 1883-1909. (Trans. into French and English) A masterly
    synthesis of structural geology, surveying the history of the
    earth's crust and the formation and arrangement of ocean basins,
    continents, and mountain ranges. One of the greatest books in
    the history of geology and the source of many concepts that
    are taken for granted today. As well as being highly original
    it was based on an unequalled command of the literature of the
    science.

1206. BERTRAND, M.A. (2202) Following the work of Suess (1196 and
    1205) which he greatly admired, he suggested in 1884 that some
    mountains, particularly parts of the Alps, were formed by over-
    turned folds and overthrusts. A few years later he put forward
    an interpretation of the tectonics of the European continent
    in terms of three great epochs of mountain building.

1207. MARGERIE, E. (2211) **(a)** *Les dislocations de l'écorce terrestre*.
    Zurich, 1888. With A. Heim. A glossary, constructed on hist-
    orical and systematic principles, of many technical terms in
    geological tectonics. It sought to establish equivalents in
    French, German, and English, and so to standardize the termin-
    ology of the subject. **(b)** *Les formes du terrain*. Paris, 1888.
    With G. de La Noë. A pioneering work in geomorphology which
    related surface features to geological structures and processes,
    especially water erosion.

1208. WRIGHT, G.F. (3057) *The Ice Age in North America and its bearings
    upon the antiquity of man*. New York, 1889. During the 1880s
    Wright traced the glacial boundary (by means of morainic depos-
    its) from the Atlantic to the Mississippi and also made extens-
    ive glacial investigations in Alaska and elsewhere. He held
    the views--then disputed--that: (i) The Ice Age had ended
    relatively late--about ten thousand years ago; (ii) There had
    only been one Ice Age, with waxing and waning of glaciation;
    (iii) Man had existed in North America during the Pleistocene.

1209. DAVIS, W.M. (3058) "The rivers and valleys of Pennsylvania" --
    a widely influential article of 1889 on what is now called
    geomorphology. It introduced Davis' method of landscape
    analysis and his important concept of the cycle of erosion.

1210. ISOSTASY. The term was introduced in 1889 by the American
    geologist, C.E. Dutton. Various formulations of the concept
    had long been current.

1890s

1211. SEDERHOLM, J.J. (2880) Began in 1891 the pioneering researches
    on Precambrian rocks that he was to continue for many years.

1212. PENCK, A. (See also 1202) *Morphologie der Erdoberfläche*. Stutt-
    gart, 1894. The first comprehensive textbook of geomorphology.

1213. ZITTEL, K.A. von. (2682) *Grundzüge der Paläontologie (Paläo-
      zoologie)*. 2 vols. Munich, 1895. A treatise with an emphasis
      on systematics. It was based on his massive *Handbuch* (1198).

1214. GEER, G.J. de. (2879) In a work published (in Swedish) in 1896
      he clarified the nature of the isostatic upheaval of the
      Scandinavian land mass since the Ice Age.

1215. GEIKIE, A. (See also 1203) *The ancient volcanoes of Great
      Britain*. 2 vols. London, 1897. A comprehensive survey,
      ranging in time from the Precambrian to the Tertiary. It made
      a major contribution to the study of volcanic action in the
      past.

1216. OLDHAM, R.D. (2446) His 1899 report on a great earthquake in
      India constituted an important development in seismology.
      Among other things he was able to establish that, to a close
      approximation, the earth can be treated as a perfectly elastic
      body--a point of great importance for the mathematical approach
      to the subject.

1.10 NATURAL HISTORY

1460s

1217. PLINY. (A.D. 23-79) *Historia naturalis*. For 1500 years one of the most influential of all books. It was the first scientific book to be printed--in 1469--and there were at least forty-six editions by 1550 and innumerable commentaries. It was translated into Italian as early as 1476, into French in 1562, and into English in 1601; there were also partial German translations.

1550s

1218. ALDROVANDI, U. (1846) His extensive researches in natural history (chiefly zoology) began about 1550 and were continued for some decades. They were finally published in twelve large volumes which can be regarded as successive parts of a massive encyclopaedia. Only four volumes appeared in his lifetime, the first being the *Ornithologia* (Bologna, 1600) which included his well-known experiments on embryology. The posthumous volumes were published between 1606 and 1668.

1560s

1219. ORTA, G. d'. (2924) *Coloquios dos simples e drogas he cousas medicinais da India*. Goa, 1563. (An epitome was published in Latin in 1567 and there were Italian and French translations) An account of the materia medica of India including much natural history, especially botany.

1570s

1220. HERNÁNDEZ, F. (2926) A Spanish physician who, by order of the king, spent the years 1570-77 in Mexico studying its natural history. His massive reports were taken back to Spain and deposited in the library of the Escorial. Various summaries of them were later published, the chief one appearing (in Latin) at Rome in 1628.

1221. MONARDES, N.B. (2923) *Historia medicinal de las cosas que se traen de nuestras Indias Occidentales....* Seville, 1574. (Trans. into Italian, English, Latin, and French) Primarily an account of the drugs being imported into Spain from the New World, but relevant to natural history, especially botany.

1222. ACOSTA, C. (2925) *Tractado de las drogas y medicinas de las Indias Orientales*. Burgos, 1578. (Trans. into Italian, French, and Latin) An account of the materia medica of India which built on the similar work of d'Orta (1219).

1590s

1223. ACOSTA, J. de. (2927) *Historia natural y moral de las Indias.*
    Seville, 1590. (English trans., 1604) An account of the
    natural history, geography, and ethnography of Central and
    South America.

1630s

1224. TRADESCANT MUSEUM. John Tradescant (ca. 1572-1638), for many
    years a gardner and plant-collector for members of the
    nobility, established about 1630 a garden (including exotic
    plants) and natural history museum at South Lambeth in London.
    His establishment, the first of its kind in England, became
    well known and after his death was continued by his son (1228).

1225. LAET, J. de. (2811) *Novus orbis; sive, Descriptio Indiae
    Occidentalis.* Leiden, 1633. (French trans., 1640) An account
    of the geography and natural history of the West Indies.

1640s

1226. *HISTORIA naturalis Brasiliae.* Leiden, 1648. By G. Markgraf
    and W. Piso. Edited by J. de Laet (2811). The book was a
    product of an expedition to the Dutch settlements in Brazil.

1650s

1227. *MUSEUM Wormianum.* Leiden, 1655. A catalogue of the 'cabinet
    of curiosities' or museum of natural history and archaeology
    belonging to the Danish physician and polymath, Ole Worm.
    After his death the museum passed to the King of Denmark and
    became a public institution, one of the first of its kind.

1228. TRADESCANT MUSEUM. The second John Tradescant (1608-1662)
    continued and enlarged the garden and natural history museum
    that had been established by his father (1224) and in 1656
    published a catalogue of it--*Musaeum Tradescantium; or, A
    collection of rarities....* (London, 1656). In 1674, after
    his death, the collection came into the possession of Elias
    Ashmole (2250) who took it to Oxford and displayed it in the
    (old) Ashmolean Museum.

1660s

1229. MERRETT, C. (2252) *Pinax rerum naturalium Britannicarum.*
    London, 1666. The section on animals was the first attempt
    at compiling a fauna of Britain. The section on plants
    listed over 1400 species.

1700s

1230. VALLISNIERI, A. (1905) From 1700 almost to his death in 1730
    he published numerous works on a variety of topics in natural
    history and biology. His most notable research was to confirm
    and extend Redi's investigations of the generation of insects.
    He also studied reproduction in mammals and certain plants.
    His development of the ancient conception of 'the great chain
    of being' was important in his own time and later.

1730s

1231. *LE SPECTACLE de la nature*. 8 vols. Paris, 1732-50. One of the most successful of all popularizations of natural history. It is said to have had at least fifty-seven editions in France, seventeen in England (the English translation first appeared in 1736-37), and several in other countries. The author was N.A. Pluche, an *abbé* and a private teacher of various subjects.

1232. LINNAEUS, C. (2836) *Systema naturae*. Leiden, 1735. The sub-title says that it presents "Nature's three kingdoms, divided into classes, orders, genera, and species." The first edition, which was quite short, presented his new system of classification of animals, plants, and minerals. For the plant kingdom it used his 'sexual system' which was greeted with acclaim by the botanical world and quickly adopted. The system of classification of animals was not so successful but nevertheless became widely accepted and represented a considerable advance. The system for minerals was not of much importance.

   Linnaeus progressively enlarged the work in later editions, the last (or rather the last revised by him) being the twelfth, in four volumes (Stockholm, 1766-68).

1740s

1233. LA CONDAMINE, C.M. de. (2050) *Relation abrégée d'un voyage fait dans l'intérieur de l'Amérique méridionale*. Paris, 1745. Included the first significant account to reach Europe of the production and uses of rubber.

1234. BUFFON, G.L. (2057) *Histoire naturelle*. 44 vols. Paris, 1749-1804. (Many editions, translations, excerpts, adaptations, etc.) By the time of his death Buffon had written thirty-six volumes (with some assistance from several collaborators, notably Daubenton). The remaining eight volumes—on some animal groups—were written by his disciple, Lacépède.

   This famous work was primarily a detailed description of the animal and mineral kingdoms but it also included discussions of various aspects of what would now be termed general biology and of such wide topics as the nature and value of natural history, the extent of law in nature, the origin and development of the earth and living creatures (cf. 1119), man considered as an animal, etc. It was immensely popular, largely because of the brilliance of the writing and the richness and originality of the ideas, and immensely influential—it greatly strengthened the already growing interest in natural history, and it provided a fund of ideas that later naturalists drew upon.

1760s

1235. VALMONT DE BOMARE, J.C. (2068) *Dictionnaire raisonné universel d'histoire naturelle*. 5 vols. Paris, 1764. (5th ed., 15 vols, 1800) A very successful and influential work.

1236. EXPEDITION. *Endeavour* (British ship); voyage 1768-71. See Joseph Banks—2311.

1770s

1237. PALLAS, P.S. (2977) *Reise durch verschiedenen Provinzen des russischen Reichs ... 1768-1773.* 2 vols. St. Petersburg, 1771-76. (Trans. into Russian, French, English, and Italian) A classic account of scientific exploration, describing the findings of one of the St. Petersburg Academy's major expeditions through much of Russia and Siberia.

1238. EXPEDITIONS. **(a)** *Resolution* and *Adventure* (British ships); voyage 1772-75. **(b)** *Resolution*; voyage 1776-80.

1239. BLUMENBACH, J.F. (2557) *Handbuch der Naturgeschichte.* Göttingen, 1779. (13th ed., 1832. Several translations) A highly influential textbook which gave a big stimulus to the subject in Germany. It contained much that was new, both in information and ideas, and had what would now be called an evolutionary flavour.

1790s

1240. SPALLANZANI, L. (1925) *Viaggi alle due Sicilie....* 6 vols. Pavia, 1792-97. (Trans. into French, German, and English) An account of his travels in Sicily and southern Italy in 1788-90. It contained a wealth of information on the natural history of the region, and especially on its volcanoes.

1800s

1241. *DICTIONNAIRE des sciences naturelles.* 60 vols and 12 vols of plates. Paris, 1804-30. Edited by F. Cuvier. Prospectus by G. Cuvier. Introduction by Fourcroy.

1242. HUMBOLDT, A. von. (2567) **(a)** *Voyage aux régions équinoxiales du Nouveau Continent en 1799-1804.* 34 vols (text & atlases). Paris, 1805-34. With A. Bonpland. Contents: I. Relation historique, etc. (7 vols). II. Zoologie (2 vols). III. Essai politique (3 vols). IV. Astronomie (3 vols). V. Essai sur la géographie des plantes (1 vol.). VI. Botanique (18 vols). The *Relation historique* was also published separately (English trans.: *Personal narrative of travels....* 3 vols, 1814-19). It long remained one of the most popular travel books.
      (b) *Ansichten der Natur.* Tübingen, 1808. (3rd ed., 1849. English trans.: *Aspects of nature, in different lands and different climates,* 1849) An attractively written popularization.   (See also 1249)

1810s

1243. EXPEDITION. *Uranie* and *Physicienne* (French ships); voyage 1817-20.

1820s

1244. *DICTIONNAIRE classique d'histoire naturelle.* 18 vols. Paris, 1822-31. Edited by J.B.G.M. Bory de Saint-Vincent.

1245. EXPEDITIONS. **(a)** *Coquille* (French ship); voyage 1822-25. **(b)** *Blossom* (British ship); voyage 1825-28. **(c)** *Astrolabe* (French ship); voyage 1826-29. **(d)** *Senyavin* (Russian ship); voyage 1826-29.

1830s

1246. *DICTIONNAIRE pittoresque d'histoire naturelle et des phénomènes de la nature.* 9 vols. Paris, 1833-39. Edited by F. Guérin-Méneville.

1247. *DICTIONNAIRE universel d'histoire naturelle.* 13 vols and 3 vols of plates. Paris, 1839-49. (2nd ed., 1867-69) Edited by C. d'Orbigny.

1248. EXPEDITIONS. **(a)** *Beagle* (British ship); voyage 1831-36. See Charles Darwin--2380. **(b)** *Recherche* (French ship); voyages 1835-36 and 1838-40. **(c)** *Bonité* (French ship); voyage 1836-37. **(d)** *Vénus* (French ship); voyage 1836-39. **(e)** *Sulphur* (British ship); voyage 1836-42. **(f)** *Astrolabe* and *Zélée* (French ships); voyage 1837-40. **(g)** Wilkes Expedition (an American naval expedition), 1838-42. **(h)** *Erebus* and *Terror* (British ships); voyage 1839-43.

1840s

1249. HUMBOLDT, A. von. (See also 1242) *Kosmos. Entwurf einer physischen Weltbeschreibung.* 5 vols and atlas. Stuttgart, 1845-62. (Later reprintings. Trans. into most European languages. English trans.: *Cosmos. A general survey of the physical phenomena of the universe,* 2 vols, 1845-48) The first two volumes (1845-47) constituted the main part, the later volumes being supplementary.

"Never has any natural scientist of modern times conceived a plan on a grander scale ... Romantic natural philosophy's idea of a uniform conception of nature has received in Humboldt's *Kosmos* its most glorious memorial." (E. Nordenskiöld)

1250. EXPEDITIONS. **(a)** *Samarang* (British ship); voyage 1843-46. **(b)** *Herald* (British ship); voyage 1845-51. **(c)** *Rattlesnake* (British ship); voyage 1846-50.

1850s

1251. *ENCYCLOPÉDIE d'histoire naturelle.* 22 vols and 9 vols of plates. Paris, 1850-61. (Several re-issues) Edited by J.C. Chenu.

1252. EXPEDITION. *Novara* (Austrian ship); voyage 1857-59.

1860s

1253. EXPEDITION. *Magenta* (Italian ship); voyage 1865-68.

1870s

1254. EXPEDITIONS. **(a)** *Gazelle* (German ship); voyage 1874-76. **(b)** *Vega* (Swedish ship); voyage 1878-79. **(c)** *Willem Barents* (Dutch ship); voyages 1878, 1879, 1880-84.

1880s

1255. EXPEDITION. *Vettor Pisani* (Italian ship); voyage 1882-85.

1.11 MICROSCOPY

1620s

1256. STELLUTI, F. (1874) In 1625 he made the first microscopical
      observations to be published (in a work called *Apiarium*, put
      out by the Accademia dei Lincei of which he was a leading
      member). He probably used a microscope presented to the
      Academy by Galileo.

1650s

..... SWAMMERDAM, J. Discovery of blood corpuscles: see 1592.

1660s

1257. HOOKE, R. (2266) *Micrographia; or, Some physiological descrip-
      tions of minute bodies made by magnifying glasses*. London,
      1665. "If not the first publication of microscopical obser-
      vations, *Micrographica* was the first great work devoted to
      them; and its impact rivaled that of Galileo's *Sidereus
      nuncius* half a century before. For the first time, descrip-
      tions of microscopical observations were accompanied by
      profuse illustrations ... Above all, the book suggested what
      the microscope could do for the biological sciences." (*DSB*)

..... MALPIGHI, M. The first work on microscopic anatomy: see 1593.

..... SWAMMERDAM, J. Microscopic researches on insects: see 1425, 1441.

1670s

1258. LEEUWENHOEK, A. van. (2780) From about 1671 until almost the
      end of his life in 1723 he carried out a remarkable series of
      microscopic investigations using single lenses of very short
      focal length which he made himself. His findings were commun-
      icated in a long series of letters in Dutch to the Royal Society
      of London which published English translations or summaries of
      many of them in the *Philosophical Transactions* from 1673 to 1724.
              Leeuwenhoek applied his lenses to almost everything
      that came within his range--living and non-living--and so his
      findings were very varied as well as very numerous. His most
      important achievements were his recognition of micro-organisms
      as "little animals" (in 1674), his discovery of the spermatozoa
      of many species (with a correct suggestion of their role in
      fertilization), and the re-discovery--since he was unaware of
      Malpighi's work--of the blood capillaries (in 1683) with the
      first accurate description of the red blood cells. His invest-
      igations in plant anatomy were also significant.

..... GREW, N. Microscopical investigations of plant anatomy: see 1315.

..... MALPIGHI, M. Microscopic studies in embryology and plant anatomy: see 1657 and 1314.

1690s

1259. *OBSERVATIONES circa viventia ... cum micrographia curiosa.* Rome, 1691. Included excellent illustrations of various insects and accounts of some of the microscopes of the time. By F. Buonanni, a disciple of Kircher and his successor at the Collegio Romano.

1710s

1260. JOBLOT, L. (2032) *Descriptions et usages de plusieurs nouveaux microscopes ... avec de nouvelles observations.* Paris, 1718. The first French work on microscopy. It contained descriptions of microscopes and their construction, and accounts of various observations, especially of micro-organisms. It also described some well-planned experiments with infusions which tended to disprove the doctrine of spontaneous generation.

1740s

1261. *THE MICROSCOPE made easy.* London, 1742. (5th ed., 1769. Trans. into Dutch and French) By Henry Baker, F.R.S. His *Employment for the microscope* (1753; 2nd ed., 1764) was a less successful sequel.

1262. NEEDHAM, J.T. (2300) *An account of some new microscopical discoveries.* London, 1745. Included notable studies of plant pollen and of the milt vessels of the calamary (or squid). For his microscopical work on "generation" see 1659.

1790s

1263. ACHROMATIC OBJECTIVES. One of the first true achromatic objectives was constructed in 1791 by F. Beeldsnyder, a Dutch civil servant and amateur physicist. Achromatic lenses had been invented in 1758--see 614.

1820s

1264. AMICI, G.B. (1938) From the late 1810s, when microscopes with achromatic objectives were becoming available, he devoted his efforts to trying to reduce the large amount of spherical aberration they exhibited. In the 1820s he was largely successful in doing so by empirical means.   (See also 1267)

1830s

1265. BIOLOGY AND THE MICROSCOPE. "The period between 1830 and 1850 is very important because it was during these years that the foundations of modern cell biology, of cellular pathology and of normal histology were being laid. There was a revival of interest in the applications of the microscope to science and medicine ... It was the achromatic lens which almost certainly provided the initial impetus and once the process was actually under way the other improvements in instrumentation followed on very rapidly." (S. Bradbury)

1266. LISTER, J.J. (2362) "On the improvement of compound microscopes" -- a paper of 1830 which, on theoretical grounds, solved the problem of spherical aberration in achromatic lenses. It thus made possible the design of microscopic objectives, in place of the trial-and-error methods used hitherto, and opened the way to the construction of high-powered microscopes.

1267. AMICI, G.B. (See also 1264) His continued improvements of the microscope reached an important stage in 1837 when he introduced the use of a hemispherical lens fixed to the front of the objective. This made possible a substantial increase in resolving power and also helped to engender the concept of numerical aperture as an expression of the resolving power and light-gathering capacity of a lens system. This in turn led to his development of the oil-immersion lens and the techniques of immersion microscopy.

1268. THE POLARIZING MICROSCOPE. Invented in 1834 by W.H. Fox Talbot, using a Nicol prism (invented in 1828 by W. Nicol) as the source of polarized light. After the 1850s the polarizing microscope was to become of great importance for petrography (see 1177).

1840s

1269. TEXTBOOKS. Popularizations of microscopy constituted a genre of long standing but it was not until the 1840s that textbooks of the subject began to appear. Two notable early ones were *Mikrographie; oder, Anleitung zur Kenntniss und zum Gebrauche des Mikroskops* (Tübingen, 1846) by the botanist, H. von Mohl, and *A practical treatise on the use of the microscope* (London, 1848; 3rd ed., 1855) by J.T. Quekett, professor of histology at the Royal College of Surgeons and a leading microscopist.

1850s

1270. *THE MICROSCOPE and its revelations*. London, 1856. (8th ed., 1901) By W.B. Carpenter.

1271. *THE MICROGRAPHIC dictionary. A guide to the examination ... of microscopic objects*. London, 1856. (4th ed., 1883) By J.W. Griffith and A. Henfy.

1860s

1272. SORBY, H.C. (2397) Following his introduction of microscopic methods into the study of rocks (1177) he turned in 1863 to examining the crystalline structure of iron meteorites and then of steel. He published the results of his metallographic study of steel (with microphotographs) in 1864 but the method was not taken up by metallurgists until the 1880s.

1273. *DAS MIKROSKOP. Theorie und Andwendung desselben*. 2 vols, Leipzig, 1865-67. By K.W. Nägeli and S. Schwender. A major treatise.

1870s

1274. ABBE, E. (2684) (a) By the early 1870s his technical direction of the Zeiss optical firm had resulted in all their lenses being manufactured by a standardized process instead of being

made individually. **(b)** "Beiträge zur Theorie des Mikroskops"
-- a paper of 1873 which explained why a lens system needs a
large aperture for good resolution and emphasized the importance of the light diffracted by the object under view.
(See also 1275)

..... HIS, W. Introduction of the microtome: see 1640.

1880s

1275. ABBE, E. (See also 1274) **(a)** By the early 1880s the Zeiss
      optical firm was producing advanced oil-immersion objectives
      which he had designed. **(b)** In 1886 he introduced distortion-
      free apochromatic lenses which, before the end of the century,
      made possible the practical realization of the maximum theoret-
      ical resolving power of the light microscope.

1.12 BOTANY

Including plant physiology.

<hr>

1460s

..... PLINY. *Historia naturalis*. See 1217.

<hr>

1470s

1276. ARISTOTLE. (See entry 1) *De plantis*. Actually written by
Nicolaus of Damascus (first century B.C.) but mistakenly
attributed to Aristotle until late in the sixteenth century
(cf. 1291) and consequently influential.

1277. DIOSCORIDES (fl. A.D. 50-70) *De materia medica*. The famous
herbal of antiquity; it described over six hundred plants
and their medicinal properties (real or supposed). The first
printing, in 1478, was of one of the medieval Latin translations. The Greek text appeared in print in 1499, followed
later by new Latin translations and numerous commentaries.
Translations into the vernaculars were as follows: Italian,
from 1542; German, from 1546; Spanish, from 1555; French,
from 1559. Few of the early Latin translations were illustrated but most of the vernacular ones were.

<hr>

1480s

1278. THEOPHRASTUS. (ca. 371-287 B.C.) *Historia plantarum* and *De
causis plantarum*. Two works which constituted the beginnings
of scientific botany. They were both printed in Latin translation in 1483 and subsequently reissued in various editions.

1279. HERBALS. Appeared in print in increasing numbers from the
early 1480s onward. They were generally illustrated with
woodcuts.

<hr>

1530s

1280. BRUNFELS, O. (2484) The first of the three 'German fathers of
botany' (the others were Bock and Fuchs). (a) *Herbarum vivae
eicones*. 3 vols. Strasbourg, 1530-36. A work which is generally taken to mark the emergence of scientific botany from the
herbalism of the past. (b) *Contrafayt Kreuterbuch*. 3 vols.
Strasbourg, 1532-37. A German adaptation of (a).

1281. GHINI, L. (1829) He published nothing, but through his teaching
and correspondence he played a major part in the creation of
scientific botany out of herbal pharmacy. His pupils included
several of the outstanding botanists--not only Italian--of the
next generation.

During the 1530s his teaching at Bologna, initially on "medicina practica", became increasingly botanical, and in 1539 he was appointed "professor of simples." Among other innovations he introduced (perhaps for the first time) the use of the *hortus siccus* or herbarium.   (See also 1286)

1282. CORDUS, E. (2483) *Botanologicon*. Cologne, 1534. An early attempt at systematization in the naming of plants.

1283. *DE NATURA stirpium*. Paris, 1536. A compilation based on the work of earlier authors but notable as one of the original popularizations of botany in France. The author was Jean Ruel, a Paris physician.

1284. BOCK, J. (2489) *Neu Kreutterbuch*. Strasbourg, 1539. (The enlarged third edition, 1551, had many reprintings and was translated into Latin in 1552) A pioneering work which, in its later editions, was very popular and influential.

1540s

1285. FUCHS, L. (2490) *De historia stirpium*. Basel, 1542. (In 1543 Fuchs published a German version with the title *New Kreuterbuch*) A major work, distinguished by its organized presentation, profuse and excellent illustrations, and a glossary of botanical terms.

1286. GHINI, L. (See also 1281) In 1544 he moved from Bologna to Pisa as "professor of simples" and soon after his arrival he took a major part in establishing a botanical garden there. (It ranks with that of Padua as the first in Europe) Soon afterwards he was also involved in the creation of a botanical garden in Florence.

1287. CORDUS, V. (2492) *Annotationes in Dioscoridis de materia medica libros*. Frankfurt, 1549. He departed from the purely philological interpretation of Dioscorides, hitherto prevalent, and added numerous observations of his own.   (See also 1292)

1550s

1288. TURNER, W. (2216) *A new herball*. London, 1551. A pioneering work on British plants.

1289. MATTIOLI, P.A. (1832) *Commentarii in ... Dioscoridis ... De medica materia*. Venice, 1554. A translation of Dioscorides famous work together with a special commentary and numerous illustrations. It was enormously successful and went through many editions up to the eighteenth century. "Fundamental to the work's success is its conception and execution as a practical scientific treatise. It was intended for daily use by physicians, herbalists, and others, who could find descriptions and notes on medicinal plants and herbs, Greek and Latin names and synonyms, and the equivalents in other languages." (*DSB*)

1290. DODOENS, R. (2807) *Cruydeboek*. Antwerp, 1554. (Trans. into French and English. Progressively revised and enlarged, the final version being a Latin edition of 1583) Widely influential, especially because of its excellent illustrations.

It was intermediate between a medically-oriented herbal and a scientific treatise, and instead of using an alphabetical arrangement it attempted to classify plants according to their main characteristics.

1291. SCALIGER, J.C. (1976) *In libros duos qui inscribuntur De plantis, Aristotele autore*. Paris, 1556. Text and commentary. He established that the *De plantis* was not by Aristotle (cf. 1276).　　　　　　　　　　　　　　　　　　(See also 1294)

<hr>

1560s

1292. CORDUS, V. (See also 1287) **(a)** *Historiae stirpium libri IV*. Strasbourg, 1561. Contained descriptions of about five hundred plants with attempts at systematization. **(b)** *Stirpium descriptiones liber quintus*. Strasbourg, 1563.

1293. ANGUILLARA, L. (1837) *Semplici*. Venice, 1561. Contained detailed and accurate descriptions of some 1500 plants. It was widely used, chiefly in the annotated Latin translation by G. Bauhin.

..... ORTA, G. d'. See 1219.

1294. SCALIGER, J.C. (See also 1291) *Commentarii et animadversiones in sex libros De causis plantarum Theophrasti*. Geneva, 1566. (Posthumous)　　　　　　　　　　　　　　　(See also 1299)

<hr>

1570s

1295. L'OBEL, M. de. (1980) **(a)** *Stirpium adversaria nova*. London, 1570. (Written with Pierre Pena) Contained descriptions of over 1200 plants. **(b)** *Plantarum seu stirpium historia*. 2 vols. Antwerp, 1576. A second edition of (a) with additions.

1296. MATTIOLI, P.A. (See also 1289) *Compendium de plantis omnibus*. Venice, 1571.

..... MONARDES, N.B. See 1221.

..... ACOSTA, C. See 1222.

<hr>

1580s

1297. RAUWOLF, L. (2495) *Aigentliche Beschreibung der Raisz....* Lauingen, 1582. An account of his travels in the Middle East with descriptions of the numerous plants he discovered.

1298. CESALPINO, A. (1847) *De plantis*. Florence, 1583. Generally regarded as the first real textbook of botany. In contrast to his predecessors Cesalpino paid little attention to the medicinal uses of plants and tried to present the basic elements and principles of botany. He also attempted to construct a classification based on the characteristics of the fruit and seeds. The work had a far-reaching influence on the subsequent development of botany.

1299. SCALIGER, J.C. (See also 1294) *Animadversiones in Theophrasti historias plantarum*. Lyons, 1584. (Posthumous)

1300. DALÉCHAMPS, J. (1977) *Historia generalis plantarum*. 2 vols. Lyons, 1586-87.

1590s

1301. ALPINI, P. (1865) *De plantis Aegypti.* Venice, 1592.

1302. ZALUŽANSKÝ, A. (2939) *Methodi herbariae libri tres.* Prague, 1592. An important treatise of botany which emphasized that it was a science in its own right, distinct and separate from materia medica.

1303. *THE HERBALL or general historie of plantes.* London, 1597. By John Gerard, curator of a physic garden belonging to the Royal College of Physicians. "The best-known and most often quoted herbal in the English language." A much enlarged edition, containing 2850 plants, was published in 1633.

1600s

1304. L'ECLUSE, C. de. (1978) **(a)** *Rariorum plantarum historia.* Antwerp, 1601. Described about a hundred new species. **(b)** *Exoticorum libri decem.* 3 vols. Leiden, 1605. A collection of his earlier works with some new material.

1305. COLONNA, F. (1875) *Minus cognitarum stirpium aliquot.* Rome, 1606. One of the first botanical works to make effective use of the concept of the genus.

1610s

1306. ALPINI, P. (See also 1301) *De plantis exoticis.* Venice, 1627. (Posthumous) Chiefly on the flora of Crete.

1620s

1307. BAUHIN, G. (2889) *Pinax Theatri botanici.* Basel, 1623. An important work which distinguished between genera and species, and for each species (it included over six thousand) gave comprehensive references to earlier descriptions, and established a name. Bauhin's binomial nomenclature, with a generic name followed by a specific name, was a great advance.

1308. AROMATARI, G. degli. (1881) *Epistola de generatione plantarum ex seminibus.* Venice, 1625. A pioneering work in plant embryology.

1630s

1309. VESLING, J. (1882) *De plantis aegyptiis.* Padua, 1638. A flora of Egypt which resulted from his stay in the country.

1650s

1310. BAUHIN, J. (2496) *Historia plantarum universalis.* 3 vols. Yverdon, 1650–51. (Posthumous) Contained descriptions and synonyms of over five thousand plants, some from the Near East and some from America.

1660s

1311. RAY, J. (2254) *Catalogus plantarum circa Cantabrigiam nascentium.* Cambridge, 1660. His first book. Hitherto botany in

England had been at a rather rudimentary level but this work
set new standards based on the best Continental models.
(See also 1313)

1312. JUNGIUS, J. (2511) The following are editions of some of his
lecture notes published after his death by two of his former
pupils. **(a)** *Doxoscopiae physicae minores*. Hamburg, 1662.
A brief and fragmentary work but containing some original
botanical ideas. **(b)** *Isagoge phytoscopia*. Hamburg, 1678.
A brief but important account of the morphology of higher
plants, with a comprehensive and well-designed terminology.
The work had circulated earlier in manuscript, thus Ray had
a copy in 1660.

1670s

1313. RAY, J. (See also 1311) *Catalogus plantarum Angliae*. London,
1670. A work similar in style to his Cambridge catalogue
but wider in scope. (See also 1319)

1314. MALPIGHI, M. (1897) *Anatomes plantarum idea*. 2 vols. London,
1671-79. After his numerous microscopical investigations of
human and animal anatomy he took up the study of plant anatomy
with similar success, discovering a great number of previously
unknown structures.

1315. GREW, N. (2276) **(a)** *The anatomy of vegetables begun*. London,
1672. **(b)** *An idea of a phytological history*. London, 1673.
(c) *The comparative anatomy of trunks*. London, 1675.
In his anatomical investigations he began with the
naked eye and then went on to microscopical examination.
(See also 1318)

1316. MARIOTTE, E. (2016) *De la végétation des plantes*. Paris, 1679.
A pioneering attempt at plant physiology.

..... LEEUWENHOEK, A. Microscopic investigations of plant anatomy:
see 1258.

1680s

1317. MORISON, R. (2253) *Plantarum historiae ... pars secunda*.
Oxford, 1680. (Part I was never published. Part III appeared
posthumously in 1699) Included an attempt to apply clear and
thought-out taxonomic principles to the whole plant kingdom,
but not with much success.

1318. GREW, N. (See also 1315) *The anatomy of plants*. London, 1682.
His chief work; an expansion of his three earlier works.

1319. RAY, J. (See also 1313) **(a)** *Methodus plantarum nova*. London,
1682. (Rev. ed.: *Methodus emendata*, 1703) A collection of
essays on some general features of plants and especially on
classification. **(b)** *Historia plantarum*. 3 vols. London,
1686-1704. A massive encyclopaedia of all known European
plants, including some 18,000 species. (See also 1321)

1320. MAGNOL, P. (2019) *Prodromus historiae generalis plantarum*.
Montpellier, 1689. The first work to introduce the family as
a distinct taxonomic category.

1690s

1321. RAY, J. (See also 1319) *Synopsis methodica stirpium Britannicarum.* London, 1690. A much enlarged revision of his catalogue
of 1670 (1313).

1322. RIVINUS, A.Q. (2527) *Introductio generalis in rem herbariam.*
Leipzig, 1690. Included one of the first attempts at an
artificial classification of plants, based on the form and
number of the flower petals. It was soon eclipsed by
Tournefort's classification.

1323. TOURNEFORT, J.P. de. (2033) *Elémens de botanique.* 3 vols.
Paris, 1694. (The Latin version—*Institutiones rei herbariae,*
3 vols, 1700—can be regarded as an enlarged later edition;
it ran to two further editions) A well-organized and beauti-
fully-illustrated work that made a major contribution with a
new concept of the genus, and proposed over seven hundred
genera (most of which have been retained). The practical
merits of the new system were so evident that it was soon
widely adopted. Furthermore the use of a single Latin name
for the genus, followed by a few descriptive words for the
species, was an important development in nomenclature.
    The book also proposed an artificial system of class-
ification which, because of its simplicity and ease of application,was commonly used until the time of Linnaeus.

1324. CAMERARIUS, R.J. (2528) *Epistola ... de sexu plantarum.*
Tübingen, 1694. Demonstrated the sexuality of plants by means
of many well-planned experiments. "A splendidly lucid and
powerful piece of scientific writing, which settled in principle
one of the age-long central problems of botany and formed the
starting point of many new developments." (A.G. Morton)

1710s

1325. JUSSIEU, A. de. (2046. The first member of the remarkable family
that was to dominate French botany for three generations)
Between 1712 and about 1730 he published many articles in the
Mémoires of the Paris Academy. The most outstanding was one
in 1728 which pointed out the relation between fungi and lichens.

1326. VAILLANT, S. (2041) His flora of the Paris region was published
a few years after his death but his chief contribution was to
support and publicize Camerarius' doctrine of the sexual
function of flowers. His inaugural lecture at the Jardin du
Roi in 1717, proposing the new doctrine for the first time in
France, caused a sensation and had much to do with securing
its general acceptance. (With the help of Boerhaave the text
of the lecture, with a Latin translation, was published at
Leiden in 1718) Vaillant was also largely responsible for
developing an appropriate terminology for the parts of flowers
that corresponded to the new understanding of their function.

1720s

1327. HALES, S. (2295) *Vegetable staticks.* London, 1727. (In later
editions it was joined with his *Haemastaticks,* containing his

researches in animal physiology, under the general title
Statical essays; the joint work was widely read and was translated into the main European languages)

The foundation work of plant physiology, dealing chiefly
with transpiration and the water economy of plants. It was of
primary importance not only for the discoveries it contained
but also as an exemplar of quantitative method and systematic
experimentation in biology.

1328. MICHELI, P.A. (1915) *Nova plantarum genera*. Florence, 1729.
Described many new species of thallophytes which he classified
using Tournefort's concepts of genera and species. It also
described Micheli's microscopic investigations of their structure and gave an account of the reproduction of fungi. The
book is consequently regarded as marking the emergence of
mycology as a distinct field.

1730s

1329. LINNAEUS, C. (2836) (a) *Fundamenta botanica*. Amsterdam, 1736.
Set out his fundamental ideas on the systematization of botany.
The following works were all elaborations or applications of
the principles it laid down.

(b) *Critica botanica*. Leiden, 1737. Gave rules for
the naming of genera and species.

(c) *Genera plantarum*. Leiden, 1737. (The 5th ed.,
1754, was the last revised by Linnaeus but there were later
editions) Contained short descriptions of all known genera.
"Quickly became accepted as the basis for the definition of
existing and the naming of new genera. It provided the
stability which Linnaeus aimed to establish, and was welcomed
with something like relief by the majority of botanists."
(A.G. Morton)

(d) *Classes plantarum*. Leiden, 1738. Reviewed the
classification systems of various authors from the time of
Cesalpino (whom he regarded as the first true systematist).

1330. LOGAN, J. (3015) *Experimenta et meletemata de plantarum*.
Leiden, 1739. (English trans.: *Experiments and considerations
on the generation of plants*, London, 1747) His experiments,
using maize, provided important support for Camerarius'
doctrine of plant sexuality and pollination.

1740s

1331. DILLENIUS, J.J. (2296) *Historia muscorum*. Oxford, 1741. The
first monographic study of mosses.

1750s

1332. LINNAEUS, C. (See also 1329) (a) *Philosophia botanica*. Stock-
holm, 1751. (Many reprintings) An expanded version of his
Fundamenta botanica (1329a), discussing his theoretical ideas.
His most influential work.

(b) *Species plantarum*. 2 vols. Stockholm, 1753.
(Several later editions but not revised by Linnaeus) A monumental work giving succinct descriptions of about eight thous-

and species from many parts of the world. It incorporated
virtually all his practical achievements and led within a
decade or two to the almost universal adoption of his binomial
nomenclature. It represents the beginning of modern systematic
botany and is still accepted as the formal starting point for
botanical nomenclature.

1333. JUSSIEU, B. de. (2051) Published very little, and nothing at
all on the subject of his greatest achievement, the development of a natural system of classification: instead he expressed
it in an arrangement of living plants. In 1759 Louis XV, who
was interested in horticulture, gave him the task of making a
collection of as many species of plants as possible in the
Trianon garden at Versailles. Jussieu, who had been developing his ideas about natural classification for many years, used
the opportunity to lay out the collection as a living illustration of his system which he still regarded as tentative and
adjusted from time to time (the garden lasted until 1775).
The classification and the principles on which it was based
were well known to his nephew, A.L. de Jussieu, and it was he
who first described it in print (see 1341).

1760s

1334. KOELREUTER, G. (2551) His pioneering work on plant hybridization
was published (in four parts) in 1761-66. By means of well-
designed and controlled experiments he demonstrated the existence of hybrids and recognized that their formation is governed
by laws. The great significance of his work, however, was
not recognized for a long time.

1335. ADANSON, M. (2069) *Familles des plantes.* 2 vols. Paris, 1763-
64. "The first work to define with philosophical precision
the theory and practice of natural classification; its effect
was profound and permanent ... Adanson, more than any other
single person, effectively established natural classification
as a fundamental aim of biology, and explained the ... technique
for its realization at all taxonomic levels." (A.G. Morton)
Adanson derived much from his friend and former teacher, Bernard de Jussieu and was actually living in his house when he
wrote the book.

1770s

1336. INGENHOUSZ, J. (2316) *Experiments upon vegetables discovering
their great power of purifying the common air in the sunshine
and of injuring it in the shade and at night.* London, 1779.
A major step towards the discovery of the photosynthetic
activity of green plants. (Their "power of purifying the common
air" we now interpret as the liberation of oxygen) The book
had an immediate impact and was promptly translated into French
and German.

1780s

1337. HEDWIG, J. (2556) **(a)** *Fundamentum historiae naturalis muscorum
frondosorum.* Leipzig, 1782.
          (b) *Theoria generationis et fructificationis plantarum
cryptogamicarum.* St. Petersburg, 1784. The first significant

account of the life history and reproduction of the bryophytes (mosses and liverworts). Hitherto very little had been known about the sexual reproduction of the lower plants. (See also 1345)

1338. SENEBIER, J. (2901) *Recherches sur l'influence de la lumière solaire pour metamorphoser l'air fixe en air pure par la végétation*. Geneva, 1783. Following the work of Ingenhousz (1336) he showed that the evolution of "air pure" (oxygen) by green plants is dependent on the presence of "air fixe" (carbon dioxide). He also established that in the process the solid matter of plants is formed.

1339. THUNBERG, C.P. (2847) *Flora Japonica*. Leipzig, 1784. See 2847. (See also 1349)

1340. GAERTNER, J. (2550) *De fructibus et seminibus plantarum*. 2 vols. Stuttgart, 1788–91. A comparative study, with detailed analyses and drawings, of the fruits and seeds of over a thousand genera. It proved very useful to later systematists and aided the development of A.L. de Jussieu's natural system.

1341. JUSSIEU, A.L. de. (2081) *Genera plantarum*. Paris, 1789. A famous book which completed the development of the principles and practice of natural classification by his predecessors at the Jardin du Roi (cf. 1333 and 1335). One of the chief improvements he made to their work was the incorporation of the Linnean binomial nomenclature. The book provided such a convincing demonstration of the power and practicability of natural classification that it was soon adopted universally.

1342. *ENCYCLOPÉDIE méthodique ... Botanique*. 16 vols and 4 vols of plates. Paris, 1789–1823. The botanical section of this great, topically-arranged encyclopaedia. It was written by Lamarck and others.

1790s

1343. SPRENGEL, C.K. (2562) *Das entdeckte Geheimniss der Natur im Bau und in der Befruchtung der Blumen*. Berlin, 1793. Pollination of flowers by insects had been observed by Koelreuter (1334) but in this work Sprengel went much further and showed how the structure of flowers is adapted in minute detail to ensure effective pollination by insects. He revealed how common cross-fertilization is, and concluded that "nature appears not to have intended that any flower should be fertilized by its own pollen." His book was long neglected but in 1841 it was read by Charles Darwin who praised it highly.

1800s

1344. CANDOLLE, Augustin de. (2903) During the period 1800–29 he published a series of outstanding monographs on various plant families which first established his reputation. (See also 1353)

1345. HEDWIG, J. (See also 1337) *Species muscorum frondosorum*. Leipzig, 1801. The first systematic treatment and classification of the mosses. The work is still used today as the starting point for the nomenclature of the mosses.

1346. MIRBEL, C.F.B. de. (2104) *Traité d'anatomie et de physiologie végétales*. Paris, 1802. The first comprehensive treatment of plant anatomy since the work of Grew and Malpighi. It did not advance much beyond them however, largely because the microscope remained undeveloped. But it did revive the field and stimulated investigations by other botanists.

1347. HUMBOLDT, A. von. (2567) *Essai sur la géographie des plantes*. Paris, 1805. (Part V of *Voyage aux régions équinoxiales*.... see 1242a) A pioneering work.

1348. KNIGHT, T.A. (2333) In a paper of 1806 he described experiments on eliminating the influence of gravity on germinating seeds and the discovery of what is now called geotropism. He attached the growing plants to the rim of a vertical wheel driven by a water-wheel and rotating continuously at a speed of 150 r.p.m. --the predecessor of the klinostat. His investigation constituted the beginning of the study of tropic responses.

1349. THUNBERG, C.P. (See also 1339) *Flora Capensis*. 2 vols. Uppsala, 1807-23. See 2847.

1810s

1350. BROWN, R. (2341) From 1810 to the late 1820s he published many periodical articles on taxonomic subjects in which he made major improvements in the classification of species into genera and families. His articles contained--in a rather unsystematic way--many penetrating observations on morphology, embryology, and plant geography, as well as several important discoveries. One of his chief discoveries was that the ovule in conifers and their allies (gymnosperms) is naked, not being enclosed in an ovary at any stage; this led to the recognition of the fundamental distinction between gymnosperms and angiosperms.   (See also 1358)

1351. ACHARIUS. E. (2849) **(a)** *Lichenographia universalis*. Göttingen, 1810. **(b)** *Synopsis methodica lichenum*. Lund, 1814. Two pioneering works in the study of lichens.

1352. MOLDENHAWER, J.J.P. (2850) *Beiträge zur Anatomie der Pflanzen*. Kiel, 1812. The product of eighteen years of research and a major contribution to plant anatomy. Many of his findings were made possible by the simple but effective technique of maceration which he introduced. He showed that plants are built up entirely of cells, of different shapes and sizes, each enclosed by a continuous wall of its own, without pores --this being so even for the elongated fibres. He also analysed the structure of the vascular bundle and made the import- and distinction between vascular tissue and parenchyma (or ground tissue), a distinction which proved valuable to later anatomists.

1353. CANDOLLE, Augustin de. (See also 1344) *Théorie élémentaire de la botanique*. Paris, 1813. An important exposition of the principles of natural classification in the tradition of A.L. de Jussieu. The influence of Cuvier was also evident, both in the general approach and in particular features--thus Candolle used for plants the concept of homologous parts which

Cuvier had introduced for animals. One of Candolle's main principles was that plant anatomy, rather than physiology, should constitute the basis of classification, and one of his chief innovations was his doctrine of symmetry in floral morphology. His system of classification was accepted almost universally during the first half of the century.

1820s

1354. CANDOLLE, Augustin de. (See also 1353) **(a)** *Essai élémentaire de géographie botanique*. 1820. One of the main early works in the field, building on Humboldt's pioneering essay (1347). **(b)** *Prodromus systematis naturalis regni vegetabilis*. 21 vols. Paris, 1824–73. (He wrote most of the first seven volumes and after his death the work was continued by his son, Alphonse de Candolle) A monumental treatise originally intended to cover the whole plant kingdom but eventually restricted to the dicotyledons and gymnosperms. It included many other aspects of botany besides taxonomy.

1355. FRIES, E.M. (2854) *Systema mycologicum*. 3 vols. Lund, 1821–32. A work based on his new system for classifying fungi which, to a large extent, is still valid today.   (See also 1361)

1356. AGARDH, C.A. (2853) **(a)** *Species algarum*. 2 vols. Greifswald and Lund, 1821–28. **(b)** *Systema algarum*. Lund, 1824. Important pioneering works in the classification of the algae. Much of his system still stands.

1357. AMICI, G.B. (1938) His skill as a microscopist (cf. 1264 and 1267) made possible a series of significant observations on fertilization during the early 1820s. In a number of species of angiosperms he saw pollen tubes growing out of pollen grains, penetrating the tissue of the style, and passing into the ovary towards the ovules. His observations were disputed for some time but he finally succeeded in having them accepted. (See also 1367)

1358. BROWN, R. (See also 1350) He noticed in 1827 that particles of pollen suspended in water made continuous, small, random movements. On investigation he found that the phenomenon was exhibited by microscopic particles of any material, organic or inorganic. This 'Brownian movement' remained inexplicable for many years (see 731).   (See also 1360)

1359. BRONGNIART, Adolphe T. (2140) "Recherches sur la génération et le développement de l'embryon dans les végétaux phanérogames" -- a paper of 1827 in which he confirmed and considerably extended the pioneering work of Amici (1357). His distinction between the fertilized egg and the seed was especially important.

1830s

..... DEVELOPMENT OF THE MICROSCOPE. See 1632.

1360. BROWN, R. (See also 1358) Demonstrated in 1831 that plant cells quite generally contain what he called a nucleus. For a consequence of his discovery see 1687.

1361. FRIES, E.M. (See also 1355) *Lichenographia Europea reformata.*
      Lund, 1831. Based on his new system for classifying lichens
      which was generally accepted until the 1860s when the field
      was revolutionized by microscopic discoveries.

1362. MIRBEL, C.F.B. de. (See also 1346) Since his early treatise on
      plant anatomy he had made many contributions to the subject.
      In an important paper of 1835 (which incidentally illustrates
      the great improvement in microscopy over the intervening
      period) he maintained that all the various types of mature
      cells arise by differentiation of young cells which are init-
      ially very similar to each other. He concluded that the
      general features of cell development are the same in both
      phanerogams and cryptogams, and suspected that they are the
      same in animals as in plants.

1363. DUTROCHET, H. (2129) *Mémoires pour servir à l'histoire anatom-
      ique et physiologique des végétaux et des animaux.* Paris,
      1837. The full and final account of the researches in plant
      (and animal) physiology which he had been engaged in since the
      early 1820s. His chief achievements were: **(a)** The first
      systematic investigations of osmosis (his term) and the real-
      ization of its fundamental importance; **(b)** Pioneering researches
      (which at the time weakened the hold of vitalism in botany) on
      the response of plant organs to external stimuli such as light,
      gravity, etc.; **(c)** The realization that respiration is funda-
      mentally the same in plants and animals, and the initiation
      of research in plant respiration; **(d)** The recognition that
      only cells containing green matter decompose carbon dioxide.

1364. *A FLORA of North America.* 2 vols. New York, 1838-43. By
      J. Torrey and A. Gray (3029). A work which set new standards
      for American botany and introduced natural classification in
      place of the Linnean system.

1840s

1365. THURET, G.A. (2168) Began in 1840 a long series of masterly
      investigations of marine algae. With his assistant, E. Bornet,
      he succeeded in the difficult task of elucidating the very
      varied reproductive arrangements of the several groups. His
      work was summed up in his *Notes algologiques* (1876-80) and
      Etudes phycologiques (1878).

1366. SCHLEIDEN, J.M. (2622) *Grundzüge der wissenschaftliche Botanik.*
      2 vols. Leipzig, 1842-43. (Later editions were entitled
      Die Botanik als inductive Wissenschaft. 4th ed., 1861. English
      trans., 1849 and later)
          A textbook of major importance. As well as attacking
      Naturphilosophie and upholding an inductive methodology, it
      presented a new and original approach to the teaching of botany
      which was widely adopted. It also included an elaboration of
      the ideas about plant cells which Schleiden had first proposed
      in 1838 (see 1687).

1367. AMICI, G.B. (See also 1357) In the mid-1840s he followed up
      his earlier work on fertilization with a series of observations
      on orchids (which are very suitable for the purpose). He found

that in this family an ovum is present in the embryo-sac before pollination, that the ovum is stimulated to develop by something brought by the pollen tube (which does not enter the embryo-sac), and that this development leads to the formation of the embryo.

1368. GAERTNER, K.F. von. (2574) **(a)** *Versuche ... über die Befruchtungs-organe der vollkommeneren Gewächse*. Stuttgart, 1844.
(b) *Versuche ... über die Bastarderzeugung im Pflanzenreich*. Stuttgart, 1849.
The two works were closely related, the first dealing with the process of fertilization and the accompaning changes in the flower, and the second with hybridization. In the latter work he remarked on "regularities" over several generations of hybrids but did not study them systematically. As the first comprehensive treatment of hybridization it was closely studied by Darwin and had a major influence on Mendel.

1369. NÄGELI, K.W. von. (2640) **(a)** See 1689. **(b)** His studies of sexual reproduction in the cryptogams during the mid-1840s largely elucidated this difficult and puzzling subject.
(See also 1380)

1370. HOOKER, J.D. (2400) From 1844 until the end of the century he published numerous important books and papers on taxonomy and plant geography. (See also 1383 and 1391)

1371. GRAY, A. (3029) *A manual of the botany of the northern United States*. Boston, 1848. (Five editions during his lifetime, followed by subsequent editions to the present day)
(See also 1364 and 1379)

1372. AGARDH, J.G. (2859) *Species, genera et ordines algarum*. 6 vols. Lund, 1848-1901. A major work on the taxonomy of the algae which he based on studies of their germination, development, morphology, and environmental conditions.

1373. HOFMEISTER, W. (2642) *Die Entstehung des Embryo der Phanerogamen*. Leipzig, 1849. Corrected earlier errors and established the main features of the process of reproduction in the angiosperms.
(See also 1375)

1850s

1374. BOUSSINGAULT, J.B. (2141) In the course of a long-continued investigation of the origin and utilization of nitrogenous substances in plants he showed in the 1850s that plants obtain their nitrogen, not from the atmosphere, but from nitrates in the soil. He showed further that, provided nitrates were present, no organic or carbon-containing material was necessary in soil for plant growth to take place; thus the carbon in plants must come entirely from the carbon dioxide of the atmosphere.
He had earlier (in the late 1830s) found that legumes fix atmospheric nitrogen, and in 1859 he discovered that nitrogen could also be fixed in plant-free, fertile soil. The explanation of this finding had to await the development of microbiology.

..... MOHL, H. von. Researches in plant cytology: see 1690.

1375. HOFMEISTER, W. (See also 1373) *Vergleichende Untersuchungen....*
      Leipzig, 1851. The enlarged second edition was published only
      in English: *On the germination, development and fructification
      of the higher Cryptogamia, and on the fructification of the
      Coniferae.* London, 1862.
              A book which brought about a major transformation in
      botany by demonstrating the existence of a basic unitary
      pattern of reproduction in the whole plant kingdom, thus
      removing the barriers between lower and higher plants. Hitherto
      botanists had sought unsuccessfully to interpret the lower
      plants (ferns, mosses, etc.) in terms of the higher (flowering)
      plants. Hofmeister took the opposite approach, establishing
      first that there is an essential unity in the varied reproduct-
      ive processes of the lower plants (with an alternation of
      sexual and asexual generations). He then went on to show that
      reproduction in the conifers (gymnosperms) exhibits features
      intermediate between the lower and higher plants, and finally
      pointed out resemblances between the angiosperms and the lower
      plants.   (See also 1386)

1376. PRINGSHEIM, N. (2641) During the 1850s, concurrently with others
      (notably Thuret; cf. 1365) he established that sexual reprod-
      uction is a general phenomenon in the algae. In some cases he
      actually observed the spermatozoid penetrating into the ovum,
      thus establishing that fertilization involves a real fusion of
      the two gametes.   (See also 1387)

1377. DE BARY, A. (2659) *Untersuchungen über die Brandpilze.* Berlin,
      1853. A pioneering work on plant pathology in which he main-
      tained that the fungi associated with rust and smut diseases
      are the cause, not the effects, of the diseases. The book
      marked the beginning of his important researches on the fungi,
      a group then little understood.   (See also 1384)

1378. CANDOLLE, Alphonse de. (2908) *Géographie botanique raisonnée.*
      Paris/Geneva, 1855. A major work on plant geography, intro-
      ducing new methods and concepts.   (See also 1401)

1379. GRAY, A. (See also 1371) (a) "Statistics of the flora of the
      northern United States" -- a paper of 1856 written partly in
      response to a request from Darwin for information about the
      American flora. (b) "Relations of the Japanese flora to that
      of North America" -- a notable paper of 1858 in which he
      explained the relations he had discovered in terms of plant
      movements around the Arctic during the preglacial, glacial,
      and postglacial periods.

1380. NÄGELI, K.W. von. (See also 1369) Published in 1858 a very
      original paper on starch grains in which--from an exhaustive
      study of their structure, physical properties, and mode of
      swelling in water--he arrived at his micellar theory. Accord-
      ing to this theory starch consists of micelles, or building
      blocks (which today would be called macromolecular aggregates),
      packed together in a regular arrangement; he also suggested
      that cellulose was similarly constituted. The theory was well
      received (at least initially) and stimulated other investig-
      ations of plant ultrastructure.   (See also 1385)

1860s

1381. SACHS, J. von. (2660) **(a)** In a classical paper in 1860 he showed that plants can be grown from seed in water (without soil) if supplied with the necessary chemical nutrients (a fact that many botanists at first found hard to believe). He used the method of water culture to establish which were the mineral elements needed by plants, as well as the basic facts of nitrogen nutrition.

    (b) During the early 1860s he carried out a highly important series of investigations on photosynthesis and carbon metabolism in plants. He showed that chlorophyll is not diffused in the tissues but contained in special bodies (later called chloroplasts), that in most plants starch is the first visible product of photosynthesis, that it is formed only in association with the chlorophyll in the chloroplasts, and that its formation is absolutely dependent upon the presence of carbon dioxide and the immediate action of light. He showed further that, in appropriate conditions, starch is converted into sugar which is transported to the tissues and from which, he inferred, all the carbon-containing constituents of the plant arise.

    (c) *Handbuch der Experimentalphysiologie der Pflanzen.* Leipzig, 1865. The first textbook of plant physiology--based largely upon his own work. Its influence was enormous.

    (d) *Lehrbuch der Botanik.* Leipzig, 1868. (4th ed., 1874. English trans., 1875) A lucidly-written textbook which accorded a major place to plant physiology--along with morphology, systematics, and a discussion of the relevance to botany of Darwin's theory of evolution. The book had a deep influence on the development of botany as a comprehensive discipline. (See also 1389)

1382. DARWIN, C.R. (2380) **(a)** *On the various contrivances by which ... orchids are fertilised by insects, and on the good effects of intercrossing.* London, 1862. A detailed investigation of how the structure of flowers favours cross-pollination by insects. **(b)** *On the movements and habits of climbing plants.* London, 1865. A study of the mechanism and function of twining and climbing, and their role as adaptions. (See also 1393)

1383. *GENERA PLANTARUM.* 3 vols. London, 1862-83. By G. Bentham (2363) and J.D. Hooker (2400). A monumental work which established an authoritative system of genera and species which has become largely incorporated into modern systems of classification. It was based on Candolle's system and was the product of many years of work at Kew Gardens, using the unique resources of that great institution. It still remains a standard work.

1384. DE BARY, A. (See also 1377) *Morphologie und Physiologie der Pilze, Flechten und Myxomyceten.* Leipzig, 1866. An account of his researches on the fungi since his monograph of 1853. Around 1860 he had demonstrated sexual reproduction in the group and had also discovered the alternation of generations in the rust fungi. His elucidation of the life-cycles of many fungi and his studies of host-parasite interactions had major

implications for plant pathology. The classification of the
group which he developed is to a large extent still used and
he is regarded as the virtual founder of mycology.
(See also 1395)

1385. NÄGELI, K.W. von. (See also 1380) From a survey of stem and
root structure in a number of different kinds of plants in
1868 he arrived at several important generalizations, notably
the distinction between meristemic (or formative) tissues
and structural tissues which have largely ceased multiplying.
He also recognized phloem as a fundamental tissue, characterized by the presence of sieve-tubes and associated with xylem
to form a vascular bundle.

1386. HOFMEISTER, W. (See also 1375) *Allgemeine Morphologie der
Gewächse*. Leipzig, 1868. The first general treatise on
plant morphogenesis.

1387. PRINGSHEIM, N. (See also 1376) Observed in 1869 what he considered to be the most primitive form of sexual reproduction in
plants--the conjugation of the zoospores (or swarm spores) of
the colonial alga *Pandorina*. This finding, together with his
earlier work, substantiated his view that the basic modes of
reproduction in the lower cryptogams were the same as in the
higher cryptogams and phanerogams.

1388. SCHWENDENER, S. (2658) *Die Algentypen der Flechtengonidien*.
Basel, 1869. Announcement of his discovery that lichens
consist of an alga and a fungus in an association of mutual
benefit (the term 'symbiosis' was coined later). The idea
was indignantly opposed by the lichen systematists but was
eventually accepted.   (See also 1392)

1870s

1389. SACHS, J. von. (See also 1381) Through the 1870s Sachs and his
students at Würzburg took the leading part in the rapid expansion of research in plant physiology. The chief direction of
his research was the study of external influences, especially
light and temperature, on growth, germination, transpiration,
and other processes.   (See also 1402)

1390. BREFELD, O. (2685) At the beginning of the 1870s he initiated
a massive series of researches into the life-cycles, comparative morphology, and systematics of fungi which continued
until 1907. In the course of the work he developed improved
techniques of pure culture (some of which were adopted for
bacteriological use by Koch). As well as making a major
contribution to mycology, his work had far-reaching implications for plant pathology. His results were published from
1872 onward in successive parts of his *Botanische Untersuchungen über Schimmelpilze* which ended with Part XV in 1912.

1391. HOOKER, J.D. (See also 1370) *Flora of British India*. 7 vols.
London, 1872-97. A classic work, one of the most outstanding
of his many taxonomic treatises.

1392. SCHWENDENER, S. (See also 1388) *Das mechanische Prinzip im
anatomischen Bau der Monocotylen*. Leipzig, 1874. An exposition of the mechanical functions of cells and tissue elements

and of the mechanical principles governing the structure and strength of plants. A leading theme was the attainment of maximum rigidity from a minimum of plant material.

1393. DARWIN, C.R. (See also 1382) **(a)** *Insectivorous plants*. London, 1875. A study of some of the most extraordinary adaptations in the plant kingdom. **(b)** *The effects of cross and self fertilisation in the vegetable kingdom*. London, 1876. A work which incorporated the results of over ten years of experimental breeding and demonstrated the occurrence of hybrid vigour. **(c)** *The different forms of flowers on plants of the same species*. London, 1877. A demonstration of how polymorphism in flowers favours cross-pollination. (See also 1399)

..... STRASBURGER, E.A. Researches in plant cytology: see 1694.

1394. EICHLER, A.W. (2686) *Blüthendiagramme*. 2 vols. Leipzig, 1875–78. An outstanding work on the comparative morphology of flowers. (See also 1403)

1395. DE BARY, A. (See also 1384) **(a)** *Vergleichende Anatomie der Vegetationsorgane der Phanerogamen und Farne*. Leipzig, 1877. A comprehensive survey based on his many years of research in the field, and a landmark in plant anatomy. Much of the systematic terminology of the field is due to him. **(b)** *Die Erscheinung der Symbiose*. Strasbourg, 1879. A monograph on symbiosis (his term) and its various forms.

1396. PFEFFER, W. (2688) *Osmotische Untersuchungen*. Leipzig, 1877. An account of his fundamental researches on the osmotic relations of plant cells. A brilliant experimenter, he devised semi-permeable membranes supported on the wall of a pot of unglazed porcelain (the celebrated "pepper pots") and developed a method for the accurate measurement of osmotic pressure. (His measurements were later used for the theory of solutions by the physical chemist, van't Hoff; cf. 968b) He recognized the function of the cell's external plasma membrane in controlling the entry of solutes into the cell. (See also 1400)

1397. ENGLER, A. (2687) *Versuch einer Entwicklungsgeschichte der Pflanzenwelt*. 2 vols. Leipzig, 1879–82. "The first attempt at a genetic and historical theory of the origin of the floristic diversity of the Northern Hemisphere." (*DSB*)

1880s

..... DEVELOPMENT OF THE MICROSCOPE. See 1645.

1398. VRIES, H. de. (2797) Through the 1880s he continued the work in plant physiology, especially on osmosis, that he had begun in the 1870s while working with Sachs. He studied the phenomenon of plasmolysis (his term) and introduced the plasmolytic method for investigating the osmotic and turgor properties of cells. (His determination of the isotonic coefficients of many solutes had major implications for the work of his Amsterdam colleague, van't Hoff, on the physical chemistry of solutions; cf. 968b)

1399. DARWIN, C.R. (See also 1393) *The power of movement in plants*. London, 1880. (With his son, Francis Darwin) Described experiments in exposing shoots and root-tips to light.

1400. PFEFFER, W. (See also 1396) *Pflanzenphysiologie.* 2 vols.
      Leipzig, 1881. A comprehensive treatise which included much
      original work of his own. It became a standard work.

1401. CANDOLLE, Alphonse de. (See also 1378) *Origine des plantes
      cultivées.* Paris, 1882. A work which drew on his expertise
      in plant geography.

1402. SACHS, J. von. (See also 1389) *Vorlesungen über Pflanzen-
      physiologie.* Leipzig, 1882. A comprehensive survey of the
      development of plant physiology over the previous twenty-five
      years. "Sachs had, to the highest degree, the capacity of
      uniting the essential elements in the work of many investigators
      into a general theory ... Plant physiology came of age with
      Sach's brilliant synthesis." (A.G. Morton)

1403. EICHLER, A.W. (See also 1394) Put forward in 1886 a classifi-
      cation system in which the plant kingdom is divided into
      four principal divisions: Thallophyta, Bryophyta, Pteridophyta,
      and Spermatophyta, the last being subdivided into Angiospermae
      and Gymnospermae. His classification eventually gained world-
      wide acceptance.

1404. ENGLER, A. (See also 1397) *Die natürlichen Pflanzenfamilien.*
      Many vols. Leipzig, 1887–1915. With K. Prantl (died 1893)
      and many collaborators. (2nd ed., 1924–60) A monumental
      taxonomic survey of the plant kingdom. It continues in use
      to the present.

1890s

1405. ENGLER, A. (See also 1404) *Syllabus der Pflanzenfamilien.*
      1892. A one-volume abridgement of the great *Natürlichen
      Pflanzenfamilien.* It has continued through many editions to
      the present day as a standard reference book.

1406. *LEHRBUCH der Botanik.* Jena, 1894. (Many later editions. English
      trans., 1898) By E.A. Strasburger, F. Noll, H. Schenck, and
      A.F.W. Schimper. A famous and widely-influential textbook.

1407. DIXON, H.H. (2473) In collaboration with the physicist, John
      Joly, he put forward in 1894 the tension theory of the ascent
      of sap. In subsequent years Dixon (working alone) continued
      to improve the theory in detail.

1408. WARMING, E. (2874) *Plantesamfund.* Copenhagen, 1895. (German
      trans., 1896; 2nd ed., 1902. English trans.: *Oecology of
      plants. An introduction to the study of plant communities,*
      1909) The book which is generally regarded as initiating
      the study of plant ecology.

1409. SCHIMPER, A.F.W. (2717) *Pflanzengeographie auf physiologischer
      Grundlage.* Jena, 1898. (English trans., 1903) A study of
      the world's vegetation in which the distribution of plants
      was correlated with their physiological response to their
      environment, especially the climate. It was a major contrib-
      ution to the new science of plant ecology and widely influ-
      ential.

1.13 ZOOLOGY

Including comparative anatomy.

1460s

..... PLINY. *Historia naturalis.* See 1217.

1470s

1410. ARISTOTLE. (See entry 1) **(a)** *Historia animalium.* **(b)** *De partibus animalium.* **(c)** *De generatione animalium.* Besides these three major works (sometimes published together under the general title *De animalibus*) there were some minor works, especially *De motu animalium* and *De incessu animalium.*

1540s

1411. TURNER, W. (2216) *Avium praecipuarum ... historia.* Cologne, 1544. A list of birds mentioned by Pliny and Aristotle with many observations of his own on northern European species.

1412. LONGOLIUS, G. (2758) *Dialogus de avibus.* Cologne, 1544. (Posthumous; edited by William Turner)

1550s

..... ALDROVANDI, U. See 1218.

1413. BELON, P. (1972) **(a)** Two books on marine animals, 1551 and 1553. **(b)** *L'histoire de la nature des oyseaux.* Paris, 1555. An important work on birds. His well-known depiction of the homologies between the skeletons of a human and a bird has won him the title of the founder of comparative anatomy.

1414. GESNER, C. (2884) *Historia animalium.* 5 vols. Zurich, 1551–87. A massive encyclopaedia of some 4,500 pages, with abundant illustrations. It received universal acclaim and remained in use for many generations.

1415. WOTTON, E. (2217) *De differentiis animalium.* Paris, 1552. An encyclopaedic compilation of zoological knowledge. It was much used, especially its section on insects.

1416. RONDELET, G. (1971) *Libri de piscibus marinis.* 2 parts. Lyons, 1554–55. A pioneering work in ichthyology which was influential for over a century. It contained detailed descriptions, with illustrations, of nearly 250 kinds of marine animals—chiefly fish, but also invertebrates, seals, and whales.

1417. SALVIANI, I. (1849) *De piscibus.* 2 vols. Rome, [ca. 1555].
A description of Mediterranean fishes, illustrated with
impressive copper engravings.

1590s

1418. RUINI, C. (1871) *Anatomia del cavallo.* 2 vols. Bologna, 1598.
(Many editions and translations) An outstanding account of
the anatomy of the horse and one of the first works in non-
human anatomy.

1600s

1419. TOPSELL, E. (2231) *The historie of foure-footed beastes and
serpents.* London, 1607. Largely derived from Gesner's
Historia animalium. Though it contained nothing original,
Topsell's book had considerable influence in England because
it was the first significant work on zoology in the language.

1630s

1420. MOFFETT, T. (2225) *Insectorum, sive minimorum animalium, theatrum.*
London, 1634. (Posthumous) A systematic treatise on entomology.

1640s

1421. SEVERINO, M.A. (1877) A surgeon and anatomist who considered
that human anatomy should be studied in conjunction with non-
human. His *Zootomia Democritea* (Nuremberg, 1645) was a
pioneering work in comparative anatomy, including invertebrates
as well as vertebrates.

1650s

1422. JONSTON, J. (2773) **(a)** *De piscibus et cetis.* 1649. **(b)** *De
avibus.* 1650. **(c)** *De quadrupedibus.* 1652. **(d)** *De serpentibus
et draconibus.* 1653. (All published at Frankfurt) Though
they did not contain much that was original the four works
were useful compilations. There were many editions and
translations of them.

1660s

1423. GOEDAERT, J. (2769) *Metamorphosis naturalis; ofte, Historische
beschrijvinge....* 3 vols. Middelburg, 1662-69. (In Dutch.
English trans., 1682, and Latin trans., 1685, both by Martin
Lister. French trans., 1700) A pioneering work in entomology
which described the life-histories of a wide variety of insects.

1424. REDI, F. (1896) **(a)** *Osservazioni intorno alle vipere.* Florence,
1664. A thorough investigation of snake venom and its effects.
Regarded as the beginning of experimental toxicology.
(b) *Esperienze intorno alla generazione degli insetti.* Florence, 1668. A famous investigation disproving the accepted
idea of spontaneous generation of insects (and lower animals
generally).   (See also 1431)

1425. SWAMMERDAM, J. (2775) *Historia insectorum generalis; ofte,
Allgemeene verhandeling....* Utrecht, 1669. (In Dutch. French
trans., 1682. Latin trans., 1685) The first part of a projected

larger work which he did not live to publish but which eventually appeared in 1737-38; see 1441.

1670s

1426. PERRAULT, C. (2020) From almost the beginning of the Paris
 Academy (see 2010) its anatomist members were engaged in a
 long series of dissections of various animals. Like all the
 work of the Academy in its early decades this was a collective
 undertaking and the results were published in the name of the
 Academy as *Mémoires pour servir à l'histoire naturelle des
 animaux* (2 vols, Paris, 1671-76). Nevertheless the work is
 often to attributed to Perrault who was the leader of the
 group, at least in the first years. (Other anatomist members
 of the Academy who participated were L. Gayant, J. Pecquet,
 and J.G. Duverny) Their results constituted a major contribution to descriptive comparative anatomy.

1427. RAY, J. (2254) *Ornithologia*. London, 1676. (Published under
 Willughby's name—see 2267—but largely by Ray) Ray's pioneering attempt to classify birds according to both habitat and
 anatomy was a particularly significant feature of the work.
 (See also 1433)

1428. LISTER, M. (2277) *Historia animalium Angliae*. London, 1678.
 Included pioneering investigations in many areas of invertebrate zoology.   (See also 1432)

1680s

1429. TYSON, E. (2281) *Phocaena; or, The anatomy of a porpess ...
 With a preliminary discourse concerning anatomy and a natural
 history of animals*. London, 1680. The preliminary discourse
 was an important statement of the principles and methodology
 of comparative anatomy.   (See also 1435)

1430. BLASIUS, G. (2774) *Anatome animalium terrestrium variorum*.
 Amsterdam, 1681. The first general systematic treatise on
 comparative anatomy. Though it contained some original
 observations it was chiefly based on the writings of others.

1431. REDI, F. (See also 1424) *Osservazioni intorno agli animali
 viventi che si trovano negli animali viventi*. Florence, 1684.
 A thorough investigation of parasites and parasitism.

1432. LISTER, M. (See also 1428) *Historia ... conchyliorum*. 2 vols.
 London, 1685-92. Consisted almost entirely of illustrations
 of mollusks with very little text. It was followed in 1695-96
 by three anatomical supplements which were of more scientific
 value.

1433. RAY, J. (See also 1427) *De historia piscium*. Oxford, 1686.
 Published under Willughby's name—see 2267—but largely by Ray)

1690s

1434. RAY, J. (See also 1433) *Synopsis animalium quadrupedum et
 serpentini generis*. London, 1693. (A companion work *Synopsis
 avium et piscium* was published posthumously in 1713)
 (See also 1437)

1435. TYSON, E. (See also 1429) *Orang-outang, sive homo sylvestris;
      or, The anatomy of a pygmie compared with that of a monkey,
      an ape, and a man.* London, 1699. A pioneering work of great
      significance for comparative morphology and physical anthrop-
      ology. (The "pygmie" which he dissected was a young chimpanzee)

1700s

1436. DUVERNEY, J.G. (2026) Around 1700 he presented several important
      memoirs to the Paris Academy on the circulatory and respiratory
      systems in cold-blooded vertebrates such as frogs, snakes, etc.

1710s

1437. RAY, J. (See also 1434) *Historia insectorum.* London, 1710.
      (Posthumous)

1720s

1438. VALENTINI, M.B. (2529) *Amphitheatrum zootomicum.* Frankfurt,
      1720. An extensive work on the comparative anatomy of verteb-
      rates.

1730s

1439. KLEIN, J.T. (2535) *Naturalis dispositio Echinodermatum.* Danzig,
      1734. A pioneering account of the sea urchins, both living
      and fossil. A revised and enlarged edition was published in
      1778 by N.G. Leske.

1440. RÉAUMUR, R.A. (2042) *Mémoires pour servir à l'histoire des
      insectes.* 6 vols. Paris, 1734-42. The work of a great
      naturalist and one of the most monumental ever to be written
      on the subject. It contained a wealth of information about
      the life-histories and behaviour of a wide range of insects
      and other small invertebrates (his use of the term 'insect'
      was much wider than the modern use). His indefatigable
      researches included skilfully contrived experiments as well
      as long and patient observations. Characteristically, he
      emphasized the practical value of entomological knowledge,
      most of all in the case of bees; this aspect of the work
      contributed much to its wide appeal.

1441. SWAMMERDAM, J. (See also 1425) *Bybel der natuure ... Biblia
      naturae.* 3 vols. Leiden, 1737-38. (Parallel text in Dutch
      and Latin. Trans. into German, 1752, and English, 1758)
      His great work on entomology, edited from his manuscripts by
      Boerhaave fifty-seven years after his death, but still not
      out of date.
            Swammerdam was concerned to refute the current ideas
      that insects lack internal structure, that they originate by
      spontaneous generation, and that they develop only by abrupt
      metamorphoses without any organic growth. To do so he assembled
      a massive body of evidence, chiefly microscopic, and in the
      process greatly developed the techniques of microdissection.
      Though he was not primarily concerned with taxonomy his
      extensive studies of insect development laid the foundations
      for later classifications.

1442. ARTEDI, P. (2837) *Ichthyologia*. Leiden, 1738. (Posthumous; published by Linnaeus) Important especially for its discussion of natural classification and its clear conception of the genus. (An extract entitled *Genera piscium* was published by J.J. Wahlbaum in 1792)

1740s

1443. TREMBLEY, A. (2787) *Mémoires pour servir à l'histoire d'un genre de polypes d'eau douce*. Leiden, 1744. A study of the hydra, which he proved to be an animal (not a plant as previously thought) from its movements and feeding processes. He found that it reproduces by an asexual process of budding hitherto unknown in the animal kingdom. His most famous observation was that when the animal was cut in two, each separate part regenerated a complete individual. "It is not easy today to recapture the sense of utter amazement caused by the realization that an animal could be multiplied by cutting it in pieces." (*DSB*)

1444. BONNET, C. (2898) *Traité d'insectologie*. Paris, 1745. Most notable for its studies of insect metamorphosis and reproduction, and especially the discovery of parthenogenesis in aphids. (See also 1448)

1445. DAUBENTON, L.J.M. (2062) His important work in comparative anatomy was almost all published in the famous *Histoire naturelle* (1234) in which he was Buffon's chief collaborator. His contributions extended over the period 1749–67 (Vols III–XV) and dealt with the anatomy of 182 species of quadrupeds. Such comparative accounts of many different animals based on a uniform plan was a new development which was to be valuable for the future. Daubenton was also one of the first to use the methods of comparative anatomy with animal fossils.

1750s

1446. GEER, C. de. (2841) *Mémoires pour servir à l'histoire des insectes*. 7 vols. Stockholm, 1752–78. Used the same title as Réaumur's great work of which it was in effect the sequel.

1760s

1447. LYONET, P. (2786) *Traité anatomique de la chenille qui range le bois de saule*. The Hague, 1760. A study of the microscopic anatomy of the common goat moth caterpiller. Its thoroughness and accuracy made it something of a classic.

1448. BONNET, C. (See also 1444) (a) *Considérations sur les corps organisés*. Amsterdam, 1762. (b) *Contemplation de la nature*. Amsterdam, 1764. (c) *La palingénésie philosophique*. Geneva, 1769. Three works of biological philosophy which had considerable influence for many years. His "incapsulation" theory of preformation in embryology, arising from his discovery of parthenogenesis (1444), was basic to most of his speculations and he applied it to the whole of living nature.

1770s

1449. VICQ D'AZYR, F. (2083) In a large number of memoirs published
over the period 1772–88 he made many contributions to comparative anatomy, especially of mammals, carrying forward the
systematization of the subject which had been begun by Daubenton.

1450. FABRICIUS, J.C. (2848) One of the most outstanding entomologists
of the eighteenth century. Between 1775 and 1805 he published
ten major works on the classification of insects, describing
and naming some ten thousand species. He applied Linnaeus'
methods to entomology (the titles of his books are modelled on
those of Linnaeus) and attempted to construct a system based
on naturally defined genera. He also wrote some books of a
more theoretical nature, such as his *Philosophia entomologica*
(Hamburg/Kiel, 1778).

1780s

1451. MÜLLER, O.F. (2845) In the period 1781–86 he published three
monographs on groups of microscopic animals about which almost
nothing had hitherto been known—the Infusoria (which included
algae, bacteria, etc.), the Tardigrada, and similar groups of
minute crustaceans. He made an attempt at the difficult task
of describing them systematically and classifying them.

1452. *ENCYCLOPÉDIE méthodique*. The zoological sections of this vast,
topically-arranged encyclopaedia were written by numerous
collaborators and were as follows: **(a)** *Histoire naturelle*.
10 vols of text and 10 vols of plates, 1782–1827. **(b)** *Histoire
naturelle des vers*. 4 vols of text and 3 vols of plates, 1792–
1832. **(c)** *Système anatomique*. 4 vols of text and 1 vol. of
plates, 1792–1830.

1453. CAMPER, P. (2789) Made many contributions to the comparative
anatomy of the vertebrates, especially the primates (and also
to human anatomy and anthropology). His works were mostly
written in Dutch; those on comparative anatomy were included
in a collection translated into German: *Sämmtliche kleine
Schriften* (3 vols, Leipzig, 1784–90; translated in turn into
French in 1803).

1454. MONRO, A. (2304) *The structure and physiology of fishes explained
and compared with those of man and other animals*. Edinburgh,
1785. A notable treatise.

1455. BLOCH, M.E. (2552) *Ichthyologie; ou, Histoire naturelle des
poissons*. 12 vols. Paris, 1785–97. (Another ed., Berlin,
1795–97)

1456. LACÉPÈDE, B.G.E. (2095) His chief works were the last eight
volumes of Buffon's *Histoire naturelle* (1234), published over
the period 1788–1804. He began them at Buffon's invitation
and after the latter's death completed his plan for the great
work. They comprised volumes on oviparous quadrupeds, snakes,
fishes (five volumes), and cetaceans.

1790s

1457. LATREILLE, P.A. (2105) *Précis des caractères génériques des insectes disposés dans un ordre naturel.* Brive, 1796. An exposition of his approach to the classification of insects by means of the "natural method" (in contrast to the artificial systems of Linnaeus and others). It constituted an important step in entomology.   (See also 1463)

1458. CUVIER, G. (2106) *Tableau élémentaire de l'histoire naturelle des animaux.* Paris, 1797. His first book: a semi-popular survey based on his lectures. It introduced many of his basic ideas about classification and comparative anatomy--notably his famous principle of the correlation of parts, and the principle that the functions and habits of an animal determine its anatomical form (not the reverse, as held by his rival, Geoffroy Saint-Hilaire).

1800s

1459. CUVIER, G. (See also 1458) *Leçons d'anatomie comparée.* 5 vols. Paris, 1800-05. (With C. Duméril and G.L. Duvernoy) A major contribution to comparative anatomy, introducing new methods and approaches. His critical attitude and avoidance of speculation were in sharp contrast to most of his predecessors and contemporaries.   (See also 1467)

1460. LAMARCK, J.B. (2082) **(a)** *Système des animaux sans vertèbres.* Paris, 1801. A major revision of the classification of the invertebrates (which had been left in an unsatisfactory state by Linnaeus). **(b)** Theory of evolution: see 1522.
(See also 1465)

1461. BLUMENBACH, J.F. (2557) *Handbuch der vergleichenden Anatomie.* Göttingen, 1805. (Many later editions. English trans., 1807) The first textbook of comparative anatomy. He was the chief founder of the subject in Germany.

1462. OKEN, L. (2575) **(a)** *Die Zeugung.* Bamberg, 1805. **(b)** *Beiträge zur vergleichenden Zoologie, Anatomie und Physiologie.* 2 parts. Bamberg, 1806-07. (With D.G. Kieser) **(c)** *Lehrbuch der Naturphilosophie.* Jena, 1809. (3rd ed., 1843. English trans., arranged by Richard Owen, 1847)
        Oken's works constitute one of the most outstanding examples of the effects on biology of the *Naturphilosophie* of Schelling and others. Though suffused with speculations of an extreme--even bizarre--kind, his writings nevertheless contained many suggestive biological ideas which proved to have a valuable and constructive influence, most notably in connection with the cell theory.

1463. LATREILLE, P.A. (See also 1457) *Genera crustaceorum et insectorum secundum ordinem naturalem.* 4 vols. Paris, 1806-09. His main work: the execution of the programme he announced in his *Précis* of 1796.

1810s

1464. PALLAS, P.S. (2977) *Zoographia Rosso-Asiatica*. 3 vols. St.
      Petersburg, 1811–31. (Posthumous) A zoological geography of
      the Russian Empire, containing the fruits of many years of
      exploration and research.

1465. LAMARCK, J.B. (See also 1460) *Histoire naturelle des animaux
      sans vertèbres*. 7 vols. Paris, 1815–22. A major treatise
      based on his many years of research in the field.

1466. *MÉMOIRES sur les animaux sans vertèbres*. 2 vols. Paris, 1816.
      A brillant study of the comparative anatomy of the inverteb-
      rates which served as a model for many years. The author was
      Jules Savigny (a promising young zoologist whose career was
      terminated by illness soon after the work was published).

1467. CUVIER, G. (See also 1459) *Le règne animal*. 4 vols. Paris,
      1817. (3rd ed., 20 vols, 1839–49. German and English trans-
      lations in various editions and adaptations) A monumental
      treatise incorporating the fruits of his researches on both
      living and fossil animals. (The inclusion of fossil animals
      was a major new feature in such a work) Though his general
      scheme of classification is no longer used it then constituted
      the biggest advance since Linnaeus. He rejected the idea of
      a single continuous system covering the whole animal kingdom
      and used instead four main groups that were quite distinct.
      He thus broke away from the idea of the scale of nature which
      had been so popular in the eighteenth century.
      (See also 1473)

1468. *HISTOIRE naturelle des mammifères*. 4 vols. Paris, 1819–42.
      By E. Geoffroy Saint-Hilaire (2107) and Frédéric Cuvier.

1469. ALTERNATION OF GENERATIONS. Discovered in molluscs and tunicates
      by the French–German naturalist, A. von Chamisso, in 1819.
      (For the general phenomenon see 1483)

1820s

1470. RETZIUS, A.A. (2857) From about 1820 to 1840 he did much work
      in microscopic comparative anatomy, notably on primitive
      vertebrates such as *Amphioxus* and the slime eel, and later on
      the structure of the teeth in various higher animals.

1471. MECKEL, J.F. (2576) *System der vergleichenden Anatomie*. 6 vols.
      Halle, 1821–31. A comprehensive treatise including many
      discoveries of his own. He was the leading comparative anat-
      omist of the time in Germany.

1472. GEOFFROY SAINT-HILAIRE, E. (2107) *Essai de classification des
      monstres*. Paris, 1821. A pioneering work in scientific
      teratology. His subsequent work in the field (which included
      the beginning of experimental embryology) was carried further
      by his son; see 1478.   (See also 1468 and 1476)

1473. CUVIER, G. (See also 1467) *Histoire naturelle des poissons*.
      25 vols. Paris, 1828–49. (With A. Valenciennes) The founda-
      tion of modern ichthyology. Much of his classification is
      still retained today.

1474. *ZOOLOGICAL researches and illustrations.* 5 parts. Cork, 1828–34. By J.V. Thompson, an army surgeon. The work made several important contributions to marine zoology—in particular that metamorphosis takes place in the Crustacea (not only in the Insecta as had previously been thought), that cirripeds are Crustacea (not Mollusca), and that the group which Thompson called the Polyzoa (now Bryozoa) form a distinct group of marine invertebrates.

1830s

1475. SARS, M. (2860) From 1830 to 1860 he published many papers on various aspects of marine zoology, of which he was one of the pioneers. His researches included the elucidation of the life-histories of many marine invertebrates (which incidentally provided major evidence for Steenstrup's famous work (1483) on the alternation of generations) and the sensational discovery —made by dredging—of many deep-sea organisms whose existence had previously been unsuspected.

1476. GEOFFROY SAINT-HILAIRE, E. (See also 1472) **(a)** *Principes de philosophie zoologique discutés en mars 1830 à l'Académie.* Paris, 1830. His account of the issues in his famous dispute with Cuvier. The dispute was the culmination of the long-standing antagonism between the two men and had philosophical, religious and political ramifications. Of the scientific issues the most significant was Geoffroy's espousal of the evolutionary ideas of his friend Lamarck, and Cuvier's implacable opposition to them.
       (b) *Recherches sur les grands sauriens trouvés à l'état fossile.* Paris, 1831. A work which can be regarded as the beginning of evolutionary palaeontology. It formed part of his case against Cuvier.

1477. OWEN, R. (2365) **(a)** *Memoir on the pearly nautilus.* London, 1832. The first of his many publications on descriptive anatomy and a classic study of a very rare organism.
       (b) *Descriptive and illustrative catalogue of the physiological series of comparative anatomy.* 5 vols. London, 1833–40. A monumental catalogue of the museum of the Royal College of Surgeons (including John Hunter's museum). It is still in use. (See also 1482)

1478. GEOFFROY SAINT-HILAIRE, I. (2142) *Histoire ... des anomalies chez l'homme et les animaux ... ou Traité de tératologie.* 3 vols . Paris, 1832–37. The first treatise on teratology (his term)—a continuation and amplification of his father's work (1472).

1479. MILNE-EDWARDS, H. (2157) *Histoire naturelle des crustacés.* 3 vols. Paris, 1834–40. A classic work on the comparative anatomy and systematics of the Crustacea. It long remained definitive.   (See also 1492)

1480. AGASSIZ, L. (2910) **(a)** *Monographie d'échinodermes vivans et fossiles.* 4 vols. Neuchâtel, 1838–42. **(b)** *Histoire naturelle des poissons d'eau douce de l'Europe centrale.* 2 vols. Neuchâtel, 1839–42.   (See also 1484)

_____
1840s
_____

1481. MÜLLER, J. (2610. The famous physiologist) After 1840 he turned
      away from the physiological, histological, and pathological
      researches that had previously occupied him, and devoted his
      energies to comparative anatomy and zoology. His many achieve-
      ments in these fields, especially concerning the lower verteb-
      rates and marine zoology, were no less remarkable than his
      earlier work.

1481a LOVÉN, S. (2861) Following his earlier researches on marine
      zoology he concentrated during the 1840s on the anatomy and
      embryology of the molluscs. His work in the field became
      well known.

1482. OWEN, R. (See also 1477) *Odontography; or, A treatise on the
      comparative anatomy of the teeth. Their physiological relations,
      mode of development, and microscopic structure in the verteb-
      rate animals.* 2 vols & atlas. London, 1840–45. A major work
      which proved generally useful to zoologists and palaeontologists.
      (See also 1495)

1483. STEENSTRUP, J.S. (2866) Published in 1842 a work in both Danish
      and German which was translated into English in 1845 as *On the
      alternation of generations.* He summarized his findings thus:
      "Certain animals, notably jelly-fish and certain parasitic
      worms, habitually produce offspring which never resemble their
      parent but which, on the other hand, themselves bring forth
      progeny which return in form and nature to the grandparents
      or more distant ancestors." The work attracted much attention
      and other instances of the phenomenon were soon found in
      various groups of invertebrates. (Some instances had been
      found earlier, e.g. 1469; see also 1475)

1484. AGASSIZ, L. (See also 1480) *Nomenclator zoologicus, continens
      nomina systematica generum animalium tam viventium quam fossil-
      ium.* 12 parts & index. Solothurn, 1842–46. Written with
      several collaborators.   (See also 1496)

1485. SIEBOLD, C.T.E. von. (2619) *Lehrbuch der vergleichenden Anatomie
      der wirbellosen Thiere.* Berlin, 1848. One of the first import-
      ant textbooks on the comparative anatomy of the invertebrates.
      It incorporated major systematic reforms (and among other
      things recognized the Protozoa as single-celled organisms).
      (See also 1490)

1486. LEUCKART, R. (2644) *Über die Morphologie und die Verwandtschafts-
      verhältnisse der wirbellosen Thiere.* Brunswick, 1848. A pion-
      eering work in the application of comparative morphology to
      systematics in the field of the invertebrates. It took a
      major step in setting up five fundamental types: Coelenterata,
      Echinodermata, Annelida, Arthropoda, and Mollusca.
      (See also 1501)

_____
1850s
_____

1487. *ENCYCLOPÉDIE d'histoire naturelle.* 22 vols & 9 vols of plates.
      Paris, 1850–61. (Several reissues) By J.C. Chenu.

1488. HUXLEY, T.H. (2403) **(a)** In the period 1850–54 he published some twenty papers on the marine invertebrates he had collected and studied during his voyage on H.M.S. *Rattlesnake* (1846-50). His main contributions were to establish the Coelenterata as a major distinct group and to reorganize the classification of the Ascidiacea and the Cephalous Mollusca.

(b) After 1854 he turned to vertebrate zoology, his first major essay in the new field being his 1858 Croonian lecture to the Royal Society--"On the theory of the vertebrate skull"--which disproved the Goethe-Oken-Owen theory of the construction of the skull from vertebrae.   (See also 1497)

1489. DARWIN, C.R. (2380) *A monograph on the sub-class Cirripedia.* 2 parts. London, 1851-54. A painstaking investigation of many species of barnacles. (In the same period he also published some monographs on fossil barnacles) Among other things the investigation was relevant to the problem of the definition of a species, the variations of species, and the variations of individuals within a species.   (See also 1506)

1490. SIEBOLD,C.T.E. von. (See also 1485) Published a pioneering work on parasitism in 1854 in which he showed that different stages in the life-history of a parasite can take place in different hosts.

1491. FABRE, J.H. (2181) From 1855 he carried on a long series of entomological researches which were described in his *Souvenirs entomologiques* (10 vols, Paris, 1879-1907). His main general finding was the pervasive importance of inherited instinct in insect behaviour.

1492. MILNE-EDWARDS, H. (See also 1479) *Introduction à la zoologie générale.* Paris, 1858. "In this book Milne-Edwards set forth his principal discoveries. These concern the variations that obtain between animal groups, variations which in the final analysis display a great fundamental principle, the law of the division of labor within organisms ... In the lower animals the same tissue can adapt to different functions ... But in animals of higher zoological order, this ability tends to disappear and is progressively replaced by a specialization of the tissues...." *(DSB)*

1493. GEGENBAUER, C. (2664) *Grundzüge der vergleichenden Anatomie.* Leipzig, 1859. In the second, much revised, edition of 1870 the perspective of the book was transformed as a result of Gegenbauer's adoption (during the early 1860s) of Darwinian evolution.   (See also 1502)

1494. BRONN, H.G. (2606) *Die Klassen und Ordnungen des Thier-reichs.* Many vols. Leipzig, 1859-. An ambitious attempt to system-atize the whole animal kingdom, including fossil as well as living animals (he was chiefly a palaeontologist: see 1158). After his death in 1862 the work was continued by W. Kefer-stein et al.

1495. OWEN, R. (See also 1482) *On the classification and geographical distribution of the Mammalia.* London, 1859. (See also 1504)

1496. AGASSIZ, L. (See also 1484) *An essay on classification.* London, 1859. A brilliant work but ill-omened in that it appeared in the same year as Darwin's theory of evolution (which Agassiz never accepted).

1860s

1497. HUXLEY, T.H. (See also 1488) During the 1860s he made valuable improvements in the classification of numerous groups of vertebrates, both living and fossil, and brought order into several confused areas.

1498. BATES, H.W. (2404) Published a classic paper in 1861 on mimicry --the imitiation by some species of insects of other species which are unpalatable or dangerous to predators, or even imitation of inanimate objects like sticks. His successful explanation of the phenomenon in terms of natural selection provided powerful support for Darwin's new theory.

1499. WALLACE, A.R. (2402) Following his return to England from the Malay Archipelago in 1862 he published a detailed account of the distribution of animals in the region, including a discussion of the boundary later known as the Wallace line. (See also 1509)

1500. HAECKEL, E. (2690) From 1862 until the late 1880s he published a series of monographs on marine organisms--chiefly radiolarians, medusae, and sponges. In all, he described some four thousand new species, making a substantial contribution to the knowledge of the lower invertebrates.

1501. LEUCKART, R. (See also 1486) *Die menschlichen Parasiten*. 2 vols. Leipzig, 1863-76. (3rd ed., 1886-1901) A classical treatise on parasitology. It described the complicated life-histories of many parasites and showed that some human diseases are caused by them.

1502. GEGENBAUER, C. (See also 1493) *Untersuchungen der vergleichenden Anatomie der Wirbelthiere*. 3 parts. Leipzig, 1864-72. Comprised several major investigations in the evolutionary interpretation of comparative anatomy. (They are said to have been taken as models by a whole generation of research students) (See also 1507)

1503. MÖBIUS, K.A. (2663) *Die Fauna der Kieler Bucht*. 1865. A pioneering work on ecology. It described various sections of the Kiel estuary, systematically surveying the animal and plant life at different depths and positions, and introduced the concept of ecosystem ("Lebensgemeinschaft" or "Bioconose").

1504. OWEN, R. (See also 1495) *On the anatomy and physiology of vertebrates*. 3 vols. London, 1866-68. A monumental work, the most important of its kind since Cuvier.

1870s

1505. LANKESTER, E.R. (2449) From about 1870 until his retirement in 1907 he published several books and some two hundred papers covering almost the whole field of zoology. His studies of invertebrate morphology and embryology were especially noteworthy. He also made important contributions to comparative anatomy, parasitology, and anthropology.

1506. DARWIN, C.R. (See also 1489) *The expression of emotion in man and animals*. London, 1872. An important work in animal psychology.   (See also 1514)

1507. GEGENBAUER, C. (See also 1502) *Grundriss der vergleichenden Anatomie*. Leipzig, 1874. (2nd ed., 1878) A textbook derived from the second edition of his *Grundzüge* (1493). It became the standard textbook of evolutionary morphology. (See also 1520)

1508. DOHRN, A. (2691) *Der Ursprung der Wirbelthiere und das Princip des Functionswechsels*. Leipzig, 1875.

1509. WALLACE, A.R. (See also 1499) *The geographical distribution of animals*. London, 1876. A major treatise, synthesizing the contemporary knowledge of the subject, including his own important contributions. Its evolutionary approach established the subject in its modern form. (See also 1512)

1510. *THIERLEBEN. Allgemeine Kunde des Thierreichs*. 10 vols. Leipzig, 1876-80. (3rd ed., 1890-93) By A.E. Brehm, one-time director of the Hamburg zoological garden.

1511. MÜLLER, F. (2662) Extended the knowledge of mimicry with an explanation in 1878 of certain special cases which Bates (1498) had not been able to explain. Like Bates, he accounted for the phenomena in terms of natural selection. (The kind of phenomena he dealt with was later called Müllerian mimicry to distinguish it from the better-known Batesian mimicry)

1880s

1512. WALLACE, A.R. (See also 1509) *Island life; or, The phenomena and causes of insular faunas and floras, including a revision and attempted solution of the problem of geological climates*. London, 1880. A much-admired sequel to his *Geographical distribution of animals* (1509), resembling it in its evolutionary approach but differing in its sharper focus--geographically (and what would now be called ecologically)--and in its concentration on species rather than larger groups. The discussion of Ice Ages was especially notable.

1513. RETZIUS, M.G. (2875) *Das Gehörorgan der Wirbelthiere*. 2 vols. Stockholm, 1881-84.

1514. DARWIN, C.R. (See also 1506) *The formation of vegetable mould through the action of worms*. London, 1881. "A pioneer study in quantitative ecology." It gave measurements of the (surprisingly large) amounts of soil pulverized and brought to the surface by worms, thus establishing their important role.

1515. LUBBOCK, J. (2423) **(a)** *Ants, bees, and wasps*. London, 1882. **(b)** *On the senses, instincts, and intelligence of animals*. London, 1888. Pioneering investigations of insect behaviour.

1516. ROMANES, G.J. (2450) *Animal intelligence*. London, 1882. A survey of a large number of previously recorded observations of aspects of animal behaviour which Romanes considered were indicative of intelligence. Though the work was later criticised for its want of an experimental approach, tendencies to anthropomorphism, and general lack of rigour, it was nevertheless an advance at the time.

1517. WIEDERSHEIM, R. (2719) *Grundriss der vergleichenden Anatomie*

der Wirbelthiere. Jena, 1884. (3rd ed., 1907. English trans.,
1886) A famous textbook. It included embryological and
phylogenetic development.

1890s

1518. MORGAN, L. (2466) **(a)** *Animal life and intelligence*. London,
1890–91. **(b)** *An introduction to comparative psychology*.
London, 1895. **(c)** *Habit and instinct*. London, 1896.
Important works in animal psychology. He was one of
the first in the field to use an experimental as well as an
observational approach.

1519. *THE CAMBRIDGE Natural history*. 10 vols. London, 1895–1910.
Edited by S.F. Harmer and A.E. Shipley.

1520. GEGENBAUER, C. (See also 1507) *Vergleichende Anatomie der
Wirbelthiere*. 2 vols. Leipzig, 1898–1901. His *magnum opus*,
written at the end of his career. It summed up all his
researches and gave a full expression of his aims and ideas.

1521. WHITMAN, C.O. (3061) From 1898 until his death in 1910 he
published pioneering work in ethology.

1.14   EVOLUTION AND HEREDITY

<u>1800s</u>

1522. LAMARCK, J.B. (2082) *Philosophie zoologique*. 2 vols. Paris, 1809. The chief statement of his theory of evolution. Another important exposition of it was in the introduction to his *Histoire naturelle des animaux sans vertèbres* (1465).

<u>1830s</u>

..... GEOFFROY SAINT-HILAIRE, E. Evolutionary palaeontology: see 1476b.

<u>1840s</u>

1523. *VESTIGES of the natural history of creation*. London, 1844. Published anonymously; the author was Robert Chambers, an Edinburgh publisher, amateur geologist, and philosophically-minded journalist. It went through many editions up to 1887, the eleventh (in 1860) being the last to be revised by Chambers. There were also American editions and translations into German and Dutch.

      The book presented a sweeping view of the evolution of the earth (beginning with a discussion of the nebular hypothesis) and of living creatures. Though often incorrect in detail it made a plausible and interesting case, and became a sensational best-seller. Despite the attacks on it by scientists as well as theologians it familiarized the general public with evolutionary thinking and did much to prepare the way for the eventual acceptance of Darwin's theory.

<u>1850s</u>

1524. WALLACE, A.R. (2402) (a) "On the law which has regulated the introduction of new species", a paper on the origin of species written in 1855, before he had conceived the idea of natural selection. It attracted little overt attention but Darwin noticed its similarity to his own ideas.

      (b) "On the tendency of varieties to depart indefinitely from the original type", a paper presenting the idea of the origin of species through natural selection which he sent (in manuscript) to Darwin in mid-1858. It was read, together a paper by Darwin on the same subject, at a meeting of the Linnean Society on 1 July 1858. (The two papers were published together in the Society's *Journal* soon afterwards)
      (See also 1536)

1525. DARWIN, C.R. (2380) *On the origin of species by means of natural selection*. London, 1859. (6th ed., 1872) The celebrated work

which established natural selection of favorable inherited
variations as the mechanism of evolution.   (See also 1534)

1526. BRONN, H.G. (2606) (A well-known palaeontologist; cf. 1158)
*Untersuchungen über die Entwicklungsgeschichte der organischen
Welt*. Stuttgart, 1858. A masterly synthesis of the contemporary knowledge of the distribution of fossilized organisms
over geological time and, based on it, some significant
speculations about progressive development and the origin of
species.

1860s

..... BATES, H.W. Explanation of mimicry in terms of natural selection:
see 1498.

1527. GAUDRY, A.J. (2182) *Animaux fossiles et géologie de l'Attique*.
2 vols. Paris, 1862-67. An account of numerous fossil vertebrates of Miocene age found in Attica (Greece). Gaudry convin-
cingly demonstrated the geneological relations of five large
groups of animals and their descent from more primitive
ancestors in earlier geological epochs, thus providing detailed
palaeontological evidence for evolution. He later surveyed
the field of evolutionary palaeontology in his *Enchaînements
du monde animal dans les temps géologiques* (3 vols, Paris,
1878-90), an elegantly written semi-popular work which was
widely read.

1528. HUXLEY, T.H. (2403) His role in the great evolution controversy
is well known. Many of his writings on the subject are included
in his *Collected essays* (see 55). The most important was his
Evidence as to man's place in nature (London, 1863), extending
the theory of evolution to man (which Darwin had avoided doing
in his *Origin*).

1529. MÜLLER, F. (2662) *Für Darwin*. Leipzig, 1864. (English trans.:
Facts and arguments for Darwin, London, 1869) A small book
presenting the results of his researches on the embryology and
interrelationships of various species of Crustacea. He thought
that the complex processes of their embryonic development
provided a "historical document" which confirmed Darwin's
theory. His enthusiastic support was very welcome to Darwin
who had the book translated into English. Its chief consequence
was to stimulate Haeckel to propound his doctrine of recapit-
ulation (1531).

1530. MENDEL, G. (2958) His famous paper on experiments with plant
hybrids was read to the Naturforscher Verein in Brünn early
in 1865 and was published in its *Verhandlungen* in 1866. As
is well known, his brilliantly-designed experiments--conducted
over a period of ten years--yielded results which were eventually to establish the science of genetics but were disregarded
until 1900.

1531. HAECKEL, E. (2690) *Generelle Morphologie der Organismen*. Berlin,
1866. A speculative, quasi-philosophical elaboration of the
Darwinian theory of evolution. Its most significant feature
was the doctrine of recapitulation--that ontogeny (the embryonic
development of the individual organism) recapitulates or epito-

mises phylogeny (the evolutionary development of the species). This idea attracted much attention at the time and gave a considerable stimulus to embryological research. It was later abandoned however. (See also 1549)

1532. HYATT, A. (3051) From 1866 until the late 1880s he published many papers on fossil invertebrates, especially cephalopods, dealing with their systematics and evolutionary development. From the beginning he was an outstanding exponent of neo-Lamarckism, rejecting Darwinian natural selection and holding that the fossil record exhibited patterns of directed change that could not be the result of adaptation to the environment. New and advantageous characteristics, he contended, arose from the activities of the animals themselves and were passed on to their progeny. He also held that many species exhibit a sequence of youth, maturity, and old age leading to extinction --a view accepted by many palaeontologists.

1533. COPE, E.D. (3050) (A well-known palaeontologist; cf. 1189) From 1868 until his death in 1897 he published over fifty articles and some books on evolution. He was a leading member of the American neo-Lamarckian school whose general tenets were first formulated by A. Hyatt (1532) and were favored especially by palaeontologists.

1534. DARWIN, C.R. (See also 1525) *The variation of animals and plants under domestication.* 2 vols. London, 1868. One of his chief works--an extended treatment of a topic he had introduced in the *Origin.* It included his attempt at a theory of heredity. (See also 1537)

1535. GALTON, F. (2401) *Hereditary genius.* London, 1869. This work and his later writings on the subject (extending to about 1890) represent the first sustained scientific investigation of human heredity. As we now know, it could not get far because of the great complexity of the subject and the lack of an adequate theory of inheritance in Galton's time. However his quantitative approach to the study of variation was ultimately to prove highly fruitful (see 1542).

1870s

1536. WALLACE, A.R. (See also 1524) **(a)** *Contributions to the theory of natural selection.* London, 1870. A collection of his papers on the origin of species written from 1855 onward. **(b)** *The geographical distribution of animals*--see 1509. (See also 1540)

1537. DARWIN, C.R. (See also 1534) *The descent of man, and selection in relation to sex.* 2 vols. London, 1871. Extended the doctrine of evolution to man (a topic he had avoided in the *Origin*). The second part of the work discussed sexual selection as complementary to natural selection.

1538. WEISMANN, A. (2693) *Studien über Descendenz-Theorie.* 2 vols. Leipzig, 1875-76. (English trans., with a preface by Darwin, 1882). A detailed discussion of various themes in evolutionary biology. (See also 1541)

1539. GRAY, A. (3029) *Darwiniana*. New York, 1876. A collection of his influential essays and reviews upholding Darwinism. He was the leading botanist in the United States and Darwin's agent and chief supporter in the country.

1880s

1540. WALLACE, A.R. (See also 1536) **(a)** *Darwinism. An exposition of the theory of natural selection with some of its applications*. London, 1889. A masterly review of thirty years of research on evolution. His views on the subject were very close to those of Darwin (but with an important difference about man) **(b)** *Island life*. See 1512.

1541. WEISMANN, A. (See also 1538) *Essays upon heredity and kindred biological problems*. Oxford, 1889. A collection of his essays translated into English. From about 1880 he had been elaborating his germ-plasm theory in a series of essays. His ideas continued to develop and the full statement of his theory came in 1892--see 1548.

1542. GALTON, F. (See also 1535) *Natural inheritance*. London, 1889. The culmination of his investigations of inheritance and variation (in plants and animals as well as humans) on which he had been engaged for many years. The book contained a full statement of the statistical concepts and techniques he had developed, notably his mathematical treatment of correlation and regression. Through its influence on Weldon (1543) and Pearson (1544) it can be said to have initiated the science of biometrics and to a large extent statistics generally.

1890s

1543. WELDON, W.F.R. (2468) One of the founders of biometrics, largely due to the influence of Galton. From 1890 until his death in 1906 he published a series of papers applying statistical techniques to measurements of variation in physical character-istics, differential death-rates, etc., in animals, with the over-all aim of developing the Darwinian theory of evolution.

1544. PEARSON, K. (2467) In the early 1890s he was stimulated by the work of Galton and by his association with his zoological colleague, Weldon, to begin a line of investigation which was to establish the science of biometrics and lead to a big development in statistics generally. This he achieved in a series of about a hundred papers published in the period 1893-1906 with the main purpose of elucidating the phenomena of heredity and contributing to the Darwinian theory of evolution.

1545. VRIES, H. de. (2797) About 1890 he terminated his work in plant physiology (1398) and directed his research exclusively to the study of heredity and variation. He began with a book, *Intra-cellulare pangenesis* (1889), discussing existing knowledge in the field and proposing a theory of his own which turned out remarkably prescient. By 1896 he had discovered the phenomenon of mutation (his term) and had independently rediscovered Mendel's laws but he did not publish the work until he came across Mendel's paper early in 1900.

1546. BATESON, W. (2469) From the late 1880s he gradually gave up
the conventional morphological research he had been doing and
turned towards a study of heredity and variation. The evidence
he collected in the field was presented in his book of 1894--
Materials for the study of variation. He then began a programme
of experiments on hybridization and in May 1900, during a search
of the literature, discovered Mendel's paper of 1866.
Bateson subsequently became one of the leaders of the
new science of genetics (his term).

1547. *THE COLOURS of animals, their meaning and use*. London, 1890.
By E.B. Poulton (professor of zoology at Oxford). A thorough
study of the significance of protective coloring and mimicry.
The conclusions were firmly Darwinian.

1548. WEISMANN, A. (See also 1541) *Das Keimplasma. Eine Theorie der
Vererbung*. Jena, 1892. (English trans.: *The germ-plasm. A
theory of heredity*, 1893) The germ-plasm was the name he
gave to a hypothetical substance which he believed was responsible for heredity, being transmitted from parent to offspring.
His theory was formulated in terms of the current knowledge of
cytology and he located the germ-plasm in the chromosomes.
The theory led him to predict the phenomenon of meiosis, a
prediction which was soon afterwards shown to be correct. It
also led him to take a firm stand against the Lamarckian idea
of the inheritance of acquired characteristics.
Weismann's last major work, *Vorträge über Descendenz-
theorie* (Jena, 1902. English trans.: *The evolution theory*,
1904) was a discussion of evolution in relation to his germ-
plasm theory.

1549. HAECKEL, E. (See also 1531) *Systematische Phylogenie*. 3 vols.
Berlin, 1894-96. A sequel to his *Generelle Morphologie* of
1866 and a survey of the developments in the study of phylogeny
over the intervening period. As in the earlier work, he sought
evidence for phylogeny more from embryology and morphology than
from palaeontology.

1550. CORRENS, C. (2755) Began a study of the xenia effect in hybridization in 1894. After early experiments with maize he changed
to garden peas which gave much more straightforward results,
and in October 1899 he independently discovered Mendel's laws.
A few weeks later he found Mendel's paper of 1866. Correns
subsequently became an outstanding geneticist.

1551. TSCHERMAK, E. (2971) As an agriculturist he was interested in
the breeding of vegetables and in 1898 began experiments on
the hybridization of garden peas. In the following year he
independently discovered Mendel's laws and soon afterwards
found Mendel's paper. He later became an important geneticist
and plant-breeder.

1.15 HUMAN ANATOMY

Including histology.

Comparative anatomy is included under Zoology.

1470s

1552. GALEN. (A.D. 129/30-199/200) The printing of his numerous individual works began in 1473 and by 1600 the editions of them totalled about seven hundred. The *Opera* were printed in Latin in 1490 and in Greek in 1525. His anatomical doctrines, disseminated indirectly through the writings of others as well as directly through his own works, were dominant until the time of Vesalius (and later in many quarters).

1553. IBN SĪNĀ (called AVICENNA). (980-1037) *Canon medicinae.* Perhaps the most famous textbook in the history of medicine; it was first printed in 1473. Its anatomical sections, which owed much to Galen, were the most widely studied anatomical texts through the sixteenth century.

1490s

1554. *FASICULUS medicinae.* Venice, 1491. By Joannes de Ketham (fl. 1460). The first anatomical book to have printed illustrations.

1500s

1555. LEONARDO DA VINCI. (1824) His extensive researches in anatomy were mostly begun in the 1500s and continued until his death in 1519. However in this field, as in others, his work was never brought to the stage of publication.

1556. BENEDETTI, A. (1825) *Historia corporis humani.* Venice, 1502. A widely-used work.

1510s

1557. BERENGARIO DA CARPI, G. (1827) *Anothomia Mundini noviter impressa ac per Carpum castigata.* Bologna, 1514. An edition of the dissection manual of Mondino de' Luzzi, written in 1316.

1520s

1558. BERENGARIO DA CARPI, G. (See also 1557) **(a)** *Commentaria cum amplissimis additionibus super anatomia Mundini.* Bologna, 1521. "The first work since the time of Galen to display any considerable amount of anatomical information based upon personal investigation and observation ... the most important forerunner of Vesalius' *Fabrica.*" (*DSB*) **(b)** *Isagogae breves ...*

in anatomia humani.... Bologna, 1522. A condensation of (a),
intended as a manual for his students.

1530s

1559. MASSA, N. (1830) *Liber introductorius anatomiae*. Venice, 1536.
Though not illustrated, it was the best short textbook of
anatomy at the time.

1560. DRYANDER, J. (2491) One of the first to make illustrations from
his own dissections. His *Anatomiae, hoc est, corporis humani
dissectionis, pars prior* (Marburg, 1537) included numerous
woodcut illustrations. Though never completed, it was one of
the most important pre-Vesalian atlases.

1540s

1561. CANANO, G.B. (1841) *Musculorum humani corporis picturata
dissectio*. Ferrara, 1541 (or 1543). A work of considerable
originality. It was one of the first anatomies to be illustrated with copperplate engravings.

1562. ESTIENNE, C. (1973) *De dissectione partium corporis humani*.
Paris, 1545. Completed before the appearance of Vesalius'
Fabrica. It included many original observations.

1563. VESALIUS, A. (1833) *De humani corporis fabrica*. Basel, 1543.
The celebrated book which revolutionized anatomy.

1564. *COMPENDIOSA totius anatomie delineatio*. London, 1545. The
first of many plagiarisms of Vesalius' *Fabrica*, its main
feature being the imposing engravings by the author, Thomas
Geminus, a London engraver and printer. It had considerable
influence and its illustrations were plagiarized in turn by
several later anatomical works.

1550s

1565. VALVERDE, J. de. (1842) *Historia de la composición del cuerpo
humano*. Rome, 1556. Largely based on Vesalius but with many
corrections and additions. Its translations into Italian and
Latin were widely read.

1566. COLOMBO, R. (1840) *De re anatomica*. Venice, 1559. (Numerous
reprintings) A well-written (though not illustrated) treatise
by a capable anatomist. It was inevitably strongly influenced
by Vesalius' *Fabrica* but it made a number of improvements and
added some new discoveries, notably the lesser circulation of
the blood.

1560s

1567. GUIDI, G. (1839) *De anatome corporis humani*—an important
treatise written about 1560 but published posthumously in his
Ars medicinalis of 1611.

1568. FALLOPPIO, G. (1850) *Observationes anatomicae*. Venice, 1561.
A commentary on Vesalius' *Fabrica*, correcting some errors and
adding many new discoveries, especially concerning the skull
and the reproductive organs.

1569. EUSTACHI, B. (1838) *Opuscula anatomica.* Venice, 1564. His only published work. The major treatise on which he had been engaged was never published and most of his work was lost, though the magnificent engravings that he had had prepared were rediscovered in the early eighteenth century. "Had these plates appeared in 1552, when completed, his name would have stood by the side of Vesalius as one of the founders of modern anatomy." (C. Singer)

1570s

1570. COITER, V. (2497) *Externarum et internarum principalium humani corporis partium tabulae.* Nuremberg, 1572. This and another work (of complicated bibliography) published in 1575 included his important researches in embryology and comparative anatomy as well as human anatomy.

1571. VAROLIO, C. (1858) *De nervis opticis.* Padua, 1573. Described a new approach to the dissection of the brain and some discoveries resulting from it.

1572. *THE HISTORIE of man, sucked from the sappe of the most approved anathomistes.* London, 1578. Contained nothing original but of some significance as one of the earliest anatomical textbooks in English. The author was John Banister, the anatomical lecturer of the Barber-Surgeons' Company.

1573. ARANZIO, G.C. (1851) *Observationes anatomicae.* Basel, 1579. Contained significant findings in physiology as well as anatomy.

1580s

1574. ALBERTI, S. (2499) **(a)** *De lachrimis.* Wittenberg, 1581. A study of the lachrimal apparatus. **(b)** *Historia plerumque partium humani corporis.* Wittenberg, 1583. A textbook containing some original observations of his own. **(c)** *Tres orationes.* Nuremberg, 1585. Included an account of his important work on the venous valves.

1575. PLATTER, F. (2887) *De corporis humani structura.* Basel, 1583. Important especially for his work on the anatomy and physiology of the eye.

..... PICCOLOMINI, A. See 1710.

1600s

1576. FABRICI, G. (1857) He published his researches in a number of monographs which appeared from 1600 until his death in 1619; they were intended to be parts of a massive treatise which he did not live to complete. They included several works of the first rank in human and comparative anatomy, embryology, and physiology. He is perhaps best known for his work in embryology (1654).

1577. DU LAURENS, A. (1984) *Historia anatomica humani corporis.* Paris, 1600. (Many later editions, both of the original and of the French translation) A widely-used textbook which generally adhered to the traditional Galenic position on disputed issues.

1578. CASSERI, G. (1866) **(a)** *De vocis auditusque organis historia anatomica.* Ferrara, 1600-01. **(b)** *De quinque sensibus liber.* Venice, 1609. The two works constituted a major contribution to the anatomy and physiology of the sense organs.

1579. BAUHIN, G. (2889) *Theatrum anatomicum.* Frankfurt, 1605. (Many reprintings; new ed., 1621) The best anatomical textbook of its time.

1610s

1580. BARTHOLIN, C. (2826) *Anatomicae institutiones corporis humani.* Wittenberg, 1611. A useful manual which from 1641 became more widely known when his son, Thomas Bartholin, brought out a succession of revised and enlarged editions (see 1584).

1620s

1581. SPIEGEL, A. van den. (1878) *De humani corporis fabrica.* Venice, 1627. An important treatise.

1582. ASELLI, G. (1879) *De lactibus.* Milan, 1627. Discovery of the lacteal (or chylous) vessels and their function.

1640s

1583. VESLING, J. (1882) *Syntagma anatomicum.* Padua, 1641. (Many editions. Trans. into Dutch, German, and English) A successful textbook which included some original findings of his own.

1584. BARTHOLIN, T. (2827) *Institutiones anatomicae.* Leiden, 1641. His first revised edition of his father's textbook (1580). (3rd ed., 1651)   (See also 1588)

1585. FOLIUS, C. (1890) *Nova auris internae delineatio.* Venice, 1645. A thorough account of the middle ear.

1650s

1586. HIGHMORE, N. (2245) *Corporis humani disquisitio anatomica.* The Hague, 1651. The first anatomical textbook to accept Harvey's theory of the circulation of the blood. It also included some significant observations of his own.

1587. PECQUET, J. (2009) *Experimenta nova anatomica.* Paris, 1651. Announced his discovery of the thoracic duct in dogs by vivisection--a development that considerably enhanced the knowledge of the newly-discovered lymphatic system.

1588. BARTHOLIN, T. (See also 1584) *De lacteis thoracis.* Copenhagen, 1652. Discovery of the thoracic duct in man, following Pecquet's discovery of it in dogs.

1589. RUDBECK, O. (2829) *Nova exercitatio anatomica.* Uppsala, 1653. A short work describing his discovery of the thoracic duct (which he made earlier than Bartholin but published later).

1590. GLISSON, F. (2239) *Anatomia hepatis.* London, 1654. A monograph on the anatomy of the liver, with a theory (in Aristotelian terms) of its operation.   (See also 1602)

1591. WHARTON, T. (2255) *Adenographia; sive, Glandularum totius corporis descriptio*. London, 1656. The first thorough description of the glands of the whole body.

1592. SWAMMERDAM, J. (2775) Observed and described red blood corpuscles in 1658 (his first important discovery, made before he even began his medical course). (See also 1600)

1660s

1593. MALPIGHI, M. (1897) *De pulmonibus observationes anatomicae*. Bologna, 1661. The first work in microscopic anatomy. An investigation of the fine structure of the lungs, in the course of which he observed the capillary network connecting the small arteries with the small veins, thus providing a major confirmation of the theory of the circulation of the blood.

In two small works published in 1665-66 he continued his microscopic investigations, describing among other things the fine structure of the brain and the fibres of the central nervous system—work which sketched the outlines of neuro-anatomy. In another investigation he discovered (independently of Swammerdam) the red blood corpuscles. In further researches in the late 1660s he investigated the microscopic anatomy of organs such as the liver, spleen, and kidneys, and of bone and skin. (This and much of his later work was described in a succession of short reports sent to the Royal Society of London)

1594. BELLINI, L. (1901) *Exercitatio anatomica de structura et usu renum*. Florence, 1662. Led by his mechanistic ideas he discovered the tubular structure of the kidney which indicated its function.

1595. STENSEN, N. (2777) **(a)** *Observationes anatomicae*. Leiden, 1662. **(b)** *De musculis et glandulis*. Copenhagen, 1664. **(c)** *Elementorum myologiae specimen*. Florence, 1666/7. Three works containing the results of two major series of researches, the first on the glandular and lymphatic system, and the second on the muscular system; in both he made many important discoveries. **(d)** *Discours ... sur l'anatomie du cerveau*. Paris, 1669.

1596. WILLIS, T. (2256) *Cerebri anatome; cui accessit Nervorum descriptio et usus*. London, 1664. "The foundation document of the anatomy of the central and autonomic nervous systems. It greatly surpassed, in the detail and precision of its descriptions, the fragmentary treatments of the brain that had preceded it. As a text it continued to be used until the late eighteenth century." (*DSB*)

1597. RUYSCH, F. (2776) *Dilucidatio valvularum in vasis lymphaticis et lacteis*. The Hague, 1665. On the existence of valves in the lymphatic vessels. He continued to publish works on anatomy, containing numerous discoveries, up to 1724 and was renowned for his anatomical preparations and for his refinement of the technique of injection.

1598. GRAAF, R. de. (2778) *De virorum organis generationi inservientibus*. Leiden, 1668. A detailed and accurate account of the male reproductive system. (See also 1601)

1670s

1599. KERCKRING, T. (2521) *Spicilegium anatomicum*. Amsterdam, 1670. Contained some important findings, especially on the development of the foetal bones.

1600. SWAMMERDAM, J. (See also 1592) *Miraculum naturae; sive, Uteri muliebris fabrica*. Leiden, 1672. A study of the female genital organs, describing among other things the ovarian follicles. It also introduced the technique of wax injection.

1601. GRAAF, R. de. (See also 1598) *De mulierum organis generationi inservientibus*. Leiden, 1672. A work of fundamental importance on the anatomy and physiology of the female reproductive system.

1602. GLISSON, F. (See also 1590) *Tractatus de ventriculo et intestinis*. London/Amsterdam, 1677. A treatise on the anatomy and physiology of the abdomen and intestines. Among other things it was notable for its use of the concept of irritability.

..... LEEUWENHOEK, A. Investigations in microscopic anatomy: see 1258.

1680s

1603. GLASER, J.H. (2890) *Tractatus de cerebro*. Basel, 1680. A treatise, containing numerous original observations, on the anatomy and physiology of the brain and central nervous system.

1604. DUVERNEY, J.G. (2026) *Traité de l'organe de l'ouie*. Paris, 1683. A thorough study of the ear, with a theory of hearing.

1605. VIEUSSENS, R. (2025) *Nervographia universalis*. Lyons, 1684. A detailed and accurate description of the nervous system.

1606. BIDLOO, G. (2781) *Anatomia humani corporis*. Amsterdam, 1685. The first large-scale anatomical atlas since Vesalius. It consisted of 105 fine copperplate engravings with brief descriptions.

1690s

1607. HAVERS, C. (2282) *Osteologia nova; or, Some new observations of the bones*. London, 1691. A systematic and detailed treatise, notable especially for its description of the microscopic structure of the bones and its discussion of their physiology and growth.

1700s

1608. BAGLIVI, G. (1907) *De fibra motrice*. Perugia, 1700. An account of his microscopic research on muscle fibres. He was the first to distinguish between smooth and striped muscle, and between the main kinds of fibres. The book also included a report on his physiological investigations of saliva, bile, and blood.

1609. VALSALVA, A.M. (1906) *De aure humana tractatus*. Bologna, 1704. A major treatise on the anatomy of the ear. He introduced the terms 'external', 'middle', and 'internal', and as well as an exhaustive anatomical treatment he also gave an account of the physiology of hearing.

1610. MORGAGNI, G.B. (1910) *Adversaria anatomica prima*. Bologna, 1705. (It was followed by five later *Adversaria* ["Notes"] over the period 1717-19) A highly regarded work which contained many discoveries in microscopic anatomy. (See also 1619)

1710s

1611. WINSLOW, J. (2036) Between 1711 and 1743 he published about thirty memoirs, some of considerable importance, on various aspects of anatomy and physiology. (See also 1614)

1720s

1612. SANTORINI, G.D. (1909) *Observationes anatomicae*. Venice, 1724. A work by a master dissector. It contained many new discoveries of anatomical details as well as corrections of the mistakes of earlier anatomists.

1613. POURFOUR DU PETIT, F. (2043) Since the time of Willis' *Cerebri anatome* (of 1664), and perhaps earlier, it had been thought that the lateral sympathetic nerve chain originated from the cranial nerves. By means of experiments on dogs, Pourfour showed definitely in 1727 that, wherever the sympathetic chain originated, it was not in the cranium.

1730s

1614. WINSLOW, J. (See also 1611) *Exposition anatomique de la structure du corps humain*. Paris, 1732. (Many later editions and several translations) A well-known work which condensed and systematized the anatomical knowledge of the time. It remained in use for a century or more.

1740s

1615. HALLER, A. von. (2896) *Icones anatomicae*. 4 parts. Göttingen, 1743-54. Very fine engravings with brief descriptions,

1616. ALBINUS, B.S. (2785) **(a)** *Tabulae sceleti et musculorum corporis humani*. Leiden, 1747. **(b)** *Tabulae VII uteri mulieris gravidae*. Leiden, 1748. **(c)** *Tabulae ossium humanorum*. Leiden, 1753.
    Three sets of copperplate illustrations, engraved by one of the best Dutch artists of the time, which remain unsurpassed for their beauty and accuracy. They are said to have cost Albinus a fortune.

1750s

1617. BORDEU, T. de. (2063) *Recherches anatomiques sur les différentes positions des glandes et sur leur action*. Paris, 1752. A pioneering work on the glands and their secretions. It recognized that the secretions could influence other parts of the body. (See also 1620)

1618. MONRO, A. (2304) Established in 1757 the distinction between the lymphatic and circulatory systems. (See also 1622)

1760s

1619. MORGAGNI, G.B. (See also 1610) *De sedibus et causis morborum per anatomen indagatis.* Venice, 1761. (English trans.: *The seats and causes of diseases as investigated by anatomy,* 1769) A classic work which laid the foundations of pathological anatomy.

1620. BORDEU, T. de. (See also 1617) *Recherches sur le tissu muqueux.* Paris, 1767. A well-known treatise on connective tissue (which he called mucous tissue).

1770s

1621. HUNTER, J. (2312) *Treatise on the natural history of the human teeth.* 2 parts. London, 1771–78. A systematic investigation into the origin and growth of the teeth which advanced the knowledge of the subject considerably.   (See also 1624)

1780s

1622. MONRO, A. (See also 1618) **(a)** *Observations on the structure and functions of the nervous system.* Edinburgh, 1783. Contained a number of original discoveries as well as many good descriptions. **(b)** *A description of the bursae mucosae of the human body.* London, 1788. A highly regarded work.

1623. SÖMMERRING, S.T. (2558) From 1784 to 1809 he published many significant investigations, chiefly on neuroanatomy and the brain, and also on the sense organs. His publications were notable for their excellent illustrations. Some of them were also characterized by extravagant speculations in the philosophical style of the time.   (See also 1627)

1624. HUNTER, J. (See also 1621) *Observations on certain parts of the animal oeconomy.* London, 1786. A collection of researches on various topics in human and comparative anatomy.

1625. MASCAGNI, P. (1931) *Vasorum lymphaticorum corporis humani historia et iconographia.* Siena, 1787. A comprehensive treatise, containing many original observations, on the lymphatic vessels. It was highly regarded, especially because of its many fine illustrations.

1626. SCARPA, A. (1930) *Anatomicae disquisitiones de auditu et olfactu.* Pavia, 1789. Important researches on the auditory and olfactory apparatus in fishes, birds, reptiles, and man. (See also 1628)

1790s

1627. SÖMMERRING, S.T. (See also 1623) *Vom Baue des menschlichen Körpers.* 5 vols. Frankfurt, 1791–96. (2nd ed., 1796–1801. Another ed., 1839–45. Latin trans., 1794–1801) A major encyclopaedia of anatomy.

1628. SCARPA, A. (See also 1626) *Tabulae neurologicae.* Pavia, 1794. An outstanding account of much of the nervous system, especially important for its first adequate delineation of the nerves of the heart.

1800s

1629. BICHAT, X. (2108) **(a)** *Traité des membranes en général*. Paris, 1800. **(b)** *Anatomie générale*. 4 vols. Paris, 1801. **(c)** *Traité d'anatomie descriptive*. 5 vols. Paris, 1801-03
Bichat is sometimes called the father of histology. Though he did not use the microscope he conceived of the organs of the body as being formed from simple functional elements which he called tissues. He distinguished twenty-one of them, characterized by their texture and various other properties. He visualized a new anatomy and a new pathology based on knowledge of the tissues and their particular susceptibilities to disease.

1810s

1630. BELL, C. (2343) *Idea of a new anatomy of the brain*. London, 1811. (A pamphlet of which only a hundred copies were distributed. It was expanded in 1830 into his *Nervous system of the human body*) Bell was the first to suggest a fundamental division between sensory and motor functions in the nervous system. By experiments on animals he established that the anterior roots of the spinal nerves are motor in function, while the posterior roots are sensory. His work was overlapped by that of Magendie, which led to a bitter priority dispute. What came to be called the Bell-Magendie law was formulated by the latter in 1822.

1631. BURDACH, K.F. (2585) *Vom Baue und Leben des Gehirns*. 3 vols. Leipzig, 1819-26. The first two volumes dealt with the neuroanatomy of the brain and spinal cord, making numerous significant contributions to the subject, while the third was an attempt at elaborating the physiology of the nervous system.

1830s

1632. DEVELOPMENT OF THE MICROSCOPE. Until the early nineteenth century simple microscopes (consisting of a single lens in a tube, commonly with a focusing arrangement and various accessories) were generally used. Compound microscopes remained of little value because of the phenomena of chromatic and spherical aberration. By the 1830s however these defects had been overcome through the efforts of many people, especially Amici (1264, 1267) and Lister (1266), and the way was open for the development of the compound microscope into a high-powered instrument --a development of fundamental importance for anatomy and related sciences.

1633. MÜLLER, J. (2610) Published in 1830 the results of an investigation of the microscopic anatomy of the glands which he had carried out with one of the new microscopes. His monograph had a considerable effect on the emerging subject of histology but his chief influence on the subject was through the students he inspired--including Henle, Schwann, and Kölliker.

1634. PURKYNĚ, J.E. (2609) Acquired a new achromatic microscope in 1832, one of the best available at the time, and began a series of investigations in histology which continued to the

mid-1840s. As one of the pioneers of the subject he did much
to develop its basic techniques as well as making a multitude
of discoveries, especially in the histology of the nervous
system. His work in the field was also important in setting
the scene for Schwann's formulation of the cell theory.

1840s

1635. BOWMAN, W. (2405) **(a)** "On the minute structure and movement of
voluntary muscle" -- a paper of 1840 which, though a pioneering
investigation, became a classic, definitive study in muscle
histology. **(b)** "The Malpighian corpuscles of the kidney" --
a paper of 1842 in which he elucidated the minute structure of
the kidney and argued that its structure revealed its function
--the filtration of the blood and secretion of urine. This
persuasive derivation of function from structure had an important influence at the time.

1636. HENLE, J. (2620) *Allgemeine Anatomie*. Leipzig, 1841. The first
systematic treatise of histology. It included many of his own
discoveries, especially in connection with the epithelial
system which he had studied intensively. The book accepted
and made use of the Schleiden-Schwann cell theory.

1637. *THE MICROSCOPIC anatomy of the human body*. 2 vols. London,
1846-49. The first English textbook on the subject. By
A.H. Hassall.

1850s

1638. KOELLIKER, R.A. (2645) **(a)** *Mikroskopische Anatomie; oder,
Gewebelehre des Menschen*. 3 parts. Leipzig, 1850-54. A
treatise on histology based on his own researches. His aware-
ness of the structure of cells and his emphasis on the significance of the nucleus helped to set the scene for the emergence
of the new science of cytology. **(b)** *Handbuch der Gewebelehre
des Menschen*. Leipzig, 1852. (6th ed., 1899-1902. Trans.
into French, English, and Italian) An important textbook.

1860s

1639. BROCA, P.P. (2183) Proved in 1861 the localization of speech
function at a site in the left cerebral hemisphere. His later
papers on cerebral localization and the comparative anatomy
of the brain were collected after his death in his *Mémoires
sur le cerveau de l'homme et des primates* (Paris, 1888).

1870s

1640. HIS, W. (2665) *Beschreibung eines Mikrotoms*. 1870. A description of the microtome which he had invented in 1866. Though
similar devices had been made earlier His was the main agent
in the effective introduction of the microtome.

1641. STAINING AGENTS. The new synthetic dyes were introduced as
histological stains by the pathologist, C. Weigert, in 1871.
Their use soon spread to bacteriology (cf. 1811b).

1642. GOLGI, C. (1951) In 1873 he devised a valuable technique for the controlled staining of nerve cells and fibres using silver nitrate. The technique made possible a whole series of researches during the next ten years in which he elucidated the fine structure of the nervous system, hitherto largely unknown. On the basis of his anatomical findings he was able to propose a general theory of the functioning of the nervous system.

1643. RETZIUS, M.G. (2875) *Studien in der Anatomie des Nervensystems und des Bindegewebes.* 2 vols. Stockholm, 1875-76. With A. Key.   (See also 1648)

1644. *TRAITÉ technique d'histologie.* Paris, 1875. (2nd ed., 1889. German trans., 1888) The leading textbook in the subject for many years. By L.A. Ranvier (professor at the Collège de France).

1880s

1645. DEVELOPMENT OF THE MICROSCOPE. Its continued development (cf. 1632) reached a new stage in the 1880s with the introduction --largely due to E. Abbe (1275)--of advanced oil-immersion lenses and distortion-free apochromatic lenses. As a result the maximum possible resolving power of the light microscope was attained before the end of the century.

1646. RAMÓN Y CAJAL, S. (2931) From about 1886 to 1904 he pursued fundamental investigations into the fine structure of the nervous system, using initially Golgi's staining technique (1642) and later other techniques. He disproved the generally-held network theory of nerve interconnection, establishing that the nervous system is made up of discrete, individual units. This work made possible the recognition of the neuron's fundamental role in the functioning of the nervous system.

1890s

1647. HIS, W. (See also 1640) *Die anatomische Nomenklatur.* Leipzig, 1895. His was largely responsible for the Basle 'Nomina Anatomica.'

1648. RETZIUS, M.G. (See also 1643) *Das Menschenhirn. Studien in der makroskopischen Morphologie.* 2 vols. Stockholm, 1896. An outstanding work on the gross anatomy of the brain.

1649. *HANDBUCH der Anatomie des Menschen.* 32 parts. Jena, 1896-1934. A major collective work. Edited initially by K. von Bardeleben.

1.16 EMBRYOLOGY

Human and comparative.

Including "generation"

1470s

..... ARISTOTLE. *De generatione animalium.* See 1410.

1550s

1650. RUEFF, J. (2883) *De conceptu et generatione hominis.* Zurich, 1554.

1560s

1651. ALDROVANDI, U. (1846) Apparently the first in modern times to use Aristotle's method of studying the development of the chick in the egg, day by day. His observations were included in his *Ornithologia* (1218).

1652. ARANZIO, G.C. (1851) *De humano foetu opusculum.* Rome, 1564.

1570s

1653. COITER, V. (2497) Following Aldrovandi's lead he made a thorough study of the development of the chick. His results were published in 1572 in his collection of anatomical researches (see 1570).

1600s

1654. FABRICI, G. (1857) **(a)** *De formatione ovi et pulli.* (Written before (b) but published posthumously in 1621) A study of the development of the chick which surpassed the work of his predecessors. **(b)** *De formatu foetu.* 1604. An extensive study of the comparative morphology of the foetus in man and several other mammals, birds, and sharks. The first work of its kind.

1650s

1655. HARVEY, W. (2232) *Exercitationes de generatione animalium.* London, 1651. A major advance in the understanding of generation and development.

1656. HIGHMORE, N. (2245) *The history of generation.* London, 1651. An account of his researches carried on partly in conjunction with Harvey. His work was significant especially because of the use he made of the microscope.

1670s

1657. MALPIGHI, M. (1897) After his explorations in microscopic
      anatomy (1593) he turned his attention to embryology, his
      chief work in the field being *De formatione pulli in ovo*
      (London, 1673), the first microscopic study of the development
      of the chick. It revealed a multitude of previously unknown
      structures and phenomena.

1700s

..... VALLISNIERI, A. See 1230.

1740s

1658. MAUPERTUIS, P.L.M. de. (2049) *Vénus physique*. n.p., 1745. An
      argument   against the generally accepted idea of preformation.
      "Maupertuis argued convincingly that the embryo could not be
      preformed, either in the egg or in the animalcule (spermatozoon)
      since hereditary characteristics could be passed down equally
      through the male or the female parent ... A strict mechanist,
      although a believer in the epigenetic view of the origin of
      the embryo, he looked for some corporeal contribution from
      each parent as a basis of heredity." (*DSB*)

1659. NEEDHAM, J.T. (2300) Described in 1748 an experiment which
      attracted much interest and was widely discussed for many
      years. He boiled some mutton broth in a flask that was then
      corked to exclude any airborne sources of life. When the
      flask was opened a few days later the liquid was swarming with
      "animalcules." He did similar experiments with infusions of
      other animal and vegetable materials and obtained the same
      result. To Needham, as to his ally, Buffon, and to many others,
      the experiments seemed to constitute a proof of spontaneous
      generation (a conclusion that was however opposed by Spallanzani
      --see 1661a). Needham went on, in a work of 1750, to elaborate
      a speculative theory combining spontaneous generation, vitalism,
      and epigenesis.

1750s

1660. WOLFF, C.F. (2546) *Theoria generationis*. Halle, 1759. A work
      that was eventually responsible for establishing the concept
      of epigenesis. During the eighteenth century however the
      opposed concept of preformation was so strongly entrenched
      that the book received hardly any notice. It was not until
      1812, when Wolff's ideas were taken up by J.F. Meckel (2576),
      that they began to make an impression.
           Wolff based his argument on evidence obtained from
      both plants and animals. He showed that the organs of plants
      (leaves, stipules, etc.) develop out of undifferentiated tissue
      at the tip of the growing shoot, and that likewise in the chick
      embryo the organs arise gradually from tissue that appears
      quite homogeneous. It was his detailed work on the latter
      theme during the 1780s that proved most convincing.

1760s

1661. SPALLANZANI, L. (1925) (a) In his *Saggio di osservazioni micros-copiche* of 1765 he opposed the claim of Needham and Buffon (1659) to have proved the occurrence of spontaneous generation. He repeated Needham's experiments, varying the conditions, and found that if a glass flask containing an infusion was completely sealed by means of a blowpipe and heated in boiling water for over half an hour, the liquid remained free of "animalcules" indefinitely, provided the flask was not opened. (The matter was not completely settled however for Needham and others argued that the long heating destroyed some property of the air that sustained life)

         (b) In a work of 1768, which formed part of his programme of research on "generation", he reported the results of his experiments on the regeneration of cut-off parts of animals, a phenomenon discovered by Trembley (1443). He worked with a range of species and found that lower animals have greater regenerative powers than higher, and young individuals greater than adults (of the same species). The lowest organisms, he found, can exhibit extraordinary regenerative capacities. He also studied the related topic of transplantation and was even able to transplant the head of one snail onto the body of another.   (See also 1662)

..... BONNET, C. See 1448.

1780s

1662. SPALLANZANI, L. (See also 1661) *Dissertazioni di fisica animale e vegetabile.* 2 vols. Modena, 1780. (For a brief account of Vol. 1 see 1736) Vol. 2 contained dissertations on "generation" including studies of the reproductive process in amphibians, of artificial insemination in lower animals and in a dog, and of the nature of spermatozoa--but here his strong preformationist views prevented him understanding their function.

1810s

1663. PANDER, C.H. (2587) In his doctoral dissertation of 1817 on the early development of the chick embryo he distinguished three layers in the blastoderm (his term) from which the organs of the embryo arise. Following Wolff, he named the layers *Blätter* (leaves)--a consequence of Wolff's comparison of the development of plants and animals (1660). Pander never followed up this pioneering work but it provided the starting point for the famous researches of his friend, K.E. von Baer (1667).

1820s

..... GEOFFROY SAINT-HILAIRE, E. Pioneering work in teratology: see 1472.

1664. PREVOST, J.L. (2909) (a) "Sur les animalcules spermatiques de divers animaux" -- a paper of 1821 written jointly with J.B. Dumas (later a famous chemist). It demonstrated for the first time that spermatozoa originate in tissues of the male sex

glands. (b) "Sur la génération" -- a series of papers in 1824
also written jointly with Dumas. They contained the first
detailed account of the segmentation of the frog's egg and
discussed the general phenomena in the development of the
fertilized egg.

1665. RATHKE, M.H. (2607) From the early 1820s he made many significant discoveries in comparative embryology, the best known
being that of gill slits and gill arches in the embryos of
birds and mammals in 1825. These correspond to a fish-like
stage--an illustration of von Baer's principle of corresponding
stages (1667c).

1666. PURKYNĚ, J.E. (2609) In the period 1825-32 he made some close
studies of the early development of the avian egg. His discovery in 1825 of the germinal vesicle (later recognized as the
nucleus of the unripe ovum) established a significant link
between the large eggs of birds and the small ova of other
animals, and led on to the famous discovery of von Baer in 1827.

1667. BAER, K.E. von. (2586) His researches were done in the period
1819 to 1834 (after which he gave up embryology) and were
discussed in his treatise, *Über die Entwickelungsgeschichte
der Thiere* (2 vols, Königsberg, 1828-37), which became the
classic work in the subject. Besides many particular discoveries his chief contributions were as follows.
        (a) The discovery in 1827 of the mammalian egg in the
ovary (hitherto undiscovered, despite many attempts extending
back to Harvey, because of its minute size). This solved a
long-standing puzzle and brought the reproductive processes of
mammals into line with those of other animals.
        (b) His germ-layer theory, which was a development of
the findings of Pander (1663) and was in turn further developed
by Remak (1671).
        (c) His idea of corresponding stages--that embryos of
one species can resemble embryos (but not adults) of another,
and that the younger the embryos the greater the resemblance.
This was in line with his epigenetic principle that development
proceeds from general to special, and from homogenous to
heterogenous.

1830s

..... DEVELOPMENT OF THE MICROSCOPE. See 1632.

..... GEOFFROY SAINT-HILAIRE, I. Treatise on teratology: see 1478.

1668. WAGNER, R. (2611) Discovered the nucleolus in the ova of several
species of mammals in 1835. He was one of the first to use the
achromatic microscope to investigate spermatozoa and in 1837
published some very accurate illustrations of them; this work
was especially significant because at the time the nature and
function of spermatozoa were still in doubt.

1840s

1669. KOELLIKER, R.A. (2645) (a) Showed in 1840 (while still a student)
that spermatozoa are true sexual entities, formed in special
cells, and not parasites in the seminal fluid as had often been

thought. He also demonstrated their cellular nature.
(b) *Entwicklungsgeschichte der Cephalopoden*. Zurich, 1844.
A monograph containing a detailed and accurate account of
cell-division and embryonic development in the cephalopods.
(See also 1672)

1670. BISCHOFF, T.L.W. (2621) *Entwicklungsgeschichte der Säugetiere
und des Menschen*. Leipzig, 1843. A treatise on the embryology
of mammals, including man. Between 1842 and 1854 he also
published a series of monographs on the development of the
fertilized ovum in various mammals.

1850s

1671. REMAK, R. (2646) *Untersuchungen über die Entwicklung der Wirbelthiere*. 3 parts. Berlin, 1850-55. Included an important
study of the development of frog's eggs in which he showed
that the egg is a cell which divides into new cells--the
division beginning from the nucleus--and the new cells in turn
divide. In the embryonic development of vertebrates generally
he distinguished three germinal layers in the early embryo
(developing the ideas of von Baer)--the *ectoderm*, giving rise
to the skin and nervous system; the *mesoderm*, giving rise to
muscular, skeletal, and excretory systems; and the *endoderm*,
giving rise to the notochord and digestive system.

1860s

1672. KOELLIKER, R.A. (See also 1669) *Entwicklungsgeschichte des
Menschen und der höheren Thiere*. Leipzig, 1861. (2nd ed.,
1879; extensively revised) A comprehensive treatise based on
his lectures. The first treatise on comparative embryology.

1673. SIEBOLD, C.T.E. von. (2619) *Über Parthenogenesis*. Munich, 1862,
A pioneering work on parthenogenesis in insects.

1674. KOVALEVSKY, A.O. (3000) From 1865 until his death in 1901 he
published over a hundred substantial articles on comparative
embryology. "The importance of Kovalevsky's studies was
quickly recognized by Baer, who nevertheless criticized their
evolutionary tone; by Haeckel, who was greatly excited and
generalized them well beyond Kovalevsky's conclusions into
his own theory of the gastrula; and by Darwin, who saw them
as providing embryological proofs for his theory of descent."
(*DSB*)

1675. HIS, W. (2665) *Untersuchungen über die erste Anlage des Wirbel-
tierleibes*. Leipzig, 1868. Contained the first attempt to
provide a causal explanation of embryonic development--relying
entirely on observation.

1870s

1676. HIS, W. (See also 1675) *Unsere Körperform und das physiologische
Problem ihrer Entstehung*. Leipzig, 1874. A collection of
essays dealing with current controversies in embryology,
especially concerning Haeckel's 'biogenetic law' (1531)--which
His rejected--and the new mechanistic-physiological approach.
(See also 1681)

1677. *THE ELEMENTS of embryology*. London, 1874. (2nd ed., 1883.
      German trans., 1876) By M. Foster and F.M. Balfour. An
      important textbook.

1678. HERTWIG, O. (2720) In 1875, from his experiments on the fertil-
      ization of sea-urchin's eggs, he perceived that the essential
      factor was the fusion of the nuclei of the ovum and spermato-
      zoon. He also found that the ovum was fertilized by only one
      spermatozoon.   (See also 1680 and 1683)

1679. FOL, H. (2916) In the late 1870s, using sea-urchins, he showed
      that the nucleus of an egg is not a separate and autonomous
      cell within the egg, as had been suggested, but rather an
      important functional part of the egg. He was the first to
      see a spermatozoon penetrate an egg, and established that
      fertilization is effected by a single male cell. He suggested
      (as had Hertwig) that the nuclei of all the cells in the
      growing embryo are descended from the original nucleus formed
      by the fusion of the egg and sperm nuclei.

..... BÜTSCHLI, O. See 1695.

1680. HERTWIG, O. (See also 1678) and HERTWIG, R. (2721) From 1878
      into the early 1880s they jointly published a series of papers
      on the germ-layer theory in which they questioned the existence
      of the mesoderm and supported Haeckel's gastrea theory. Their
      well-known coelom theory, dealing with the development of the
      two-layered embryo, won general acceptance at the time but was
      later found to be too schematic.

<hr>

1880s

..... DEVELOPMENT OF THE MICROSCOPE. See 1645.

..... BENEDEN, E. van. See 1696.

1681. HIS, W. (See also 1676) *Anatomie menschlicher Embryonen*. 3 vols.
      Leipzig, 1880-85. An exhaustive study of the development of
      the human embryo.

1682. BALFOUR, F.M. (2451) *A treatise on comparative embryology*.
      2 vols. London, 1880-81. (German trans., 1881) A masterly
      treatise which was highly regarded and had a wide influence.

1683. HERTWIG, O. (See also 1678) **(a)** In 1885, influenced by the ideas
      of the botanist, Nägeli, he suggested that egg and sperm nuclei
      were the bearers of heredity as well as the essential agents
      in the process of fertilization. **(b)** *Lehrbuch der Entwicklungs-
      geschichte des Menschen und Wirbeltiere*. 2 vols. Jena, 1886-
      88. (10th ed., 1915. Trans. into French, English, and Italian)
      A highly regarded treatise.

1684. ROUX, W. (2722) In a famous experiment in 1888 he initiated a
      long series of researches into what he called *Entwicklungs-
      mechanik*. In a developing frog's egg, which had just segmented
      into two cells, he destroyed one of the cells with a hot needle:
      the other cell then developed as a half-embryo. The fact that
      a different result was obtained later by Driesch from a similar
      experiment (1686) served to underline the significance of the
      experimental approach.

Due to the efforts of Roux and others who followed his lead the field of experimental embryology began to grow rapidly in the 1890s, especially as it promised to uncover cause-effect relations in embryonic development--something that the traditional method of pure observation could hardly do.

1890s

1685. WILSON, E.B. (3069) Published in 1890-92 some researches on the early development of invertebrates in which he pioneered the method of cell lineage: the cell-by-cell tracing of different kinds of tissues from individual precursor cells.

1686. DRIESCH, H. (2756) Performed in 1891 a famous experiment inspired by that of Roux (1684). By violently shaking sea-urchin eggs which had developed to the two-cell stage he was able to separate the two cells: each of them then grew into a whole (but half-sized) larva. This result, together with corresponding results from later experiments, converted the philosophically-minded Driesch from a mechanist to a vitalist position. His experimental results and philosophical conclusions were presented in 1908 in his *Science and philosophy of the organism* (the Gifford lectures delivered at the University of Aberdeen).

1.17 CELL THEORY AND CYTOLOGY

1830s

1687. SCHLEIDEN, J.M. (2622) "Beiträge zur Phytogenesis" -- an article
of 1838 which aroused wide interest. It took as its starting
point Robert Brown's discovery of the cell nucleus (1360), the
importance of which Schleiden perceived. His article constituted a major step in the emergence of the cell theory--most of
all in his conception of the plant as a community of cells--
even though his ideas about the origin of cells were later
found to be erroneous.

1688. SCHWANN, T. (2623) *Mikroscopische Untersuchungen....* Berlin,
1839. (English trans.: *Microscopical researches into the
accordance in the structure and growth of animals and plants*,
1847) A famous work which extended the cell theory from the
plant to the animal kingdom. Schwann accepted the conclusions
reached by Schleiden concerning plants and successfully undertook the more difficult task of demonstrating the cellular
constitution of animal tissue generally.

1840s

1689. NÄGELI, K.W. von. (2640) Having worked with Schleiden in the
early 1840s he accepted his idea of free-cell formation.
Observing the formation of cells by division, however, he for
a time thought that both processes of cell formation existed.
But in 1846 he published a paper rejecting Schleiden's idea
and concluding that cells are formed only by division.
(See also 1699)

1850s

..... KOELLIKER, R.A. See 1638a.

1690. MOHL, H. von. (2608) *Grundzüge der Anatomie und Physiologie der
vegetabilischen Zelle.* Brunswick, 1851. An important survey
of the state of knowledge of plant cytology, including his own
substantial contributions. He had been the first to propose
that new cells are formed by cell division, a process that he
had observed in lower plants, and he was also one of the first
to see intracellular movements. In 1846 he had introduced the
term 'protoplasm' for the colloidal material surrounding the
nucleus and making up the bulk of the cell. Being a skilled
microscopist he was able to show that all parts of plants are
cellular in structure, a point that had been much debated.

1691. *DIE CELLULARPATHOLOGIE.* Berlin, 1858. A famous work by the
great pathologist, Rudolf Virchow (whose career belongs to
the history of medicine rather than the history of science).

Though it dealt with pathology, the book--and Virchow's advocacy before and after its publication--was a major influence in
gaining acceptance for the cell theory among biologists in
general. Virchow upheld the version of the theory which maintained (against earlier versions) that all cells originate
from pre-existing cells.

1860s

1692. SCHULTZE, M. (2666) In two papers in 1861 and 1863 he initiated
      a new stage in the development of the cell theory with a
      revision of the idea of the cell. The older view had conceived
      of the cell largely in terms of its membrane or wall (thus the
      word 'cell' primarily means a space enclosed by walls) but
      Schultze redefined it as a lump of protoplasm with a nucleus.
      He introduced the term 'protoplasm'--hitherto used only by
      botanists--as a universal term for the fundamental substance
      of all cells, animal as well as plant. This view of protoplasm
      as "the physical basis of life" was widely accepted and some-
      times misused in unwarranted speculations. (It later became
      apparent that protoplasm was not a single substance and that
      the way forward in cytology lay in investigation of the nucleus)

1870s

1693. CELL DIVISION. First described in detail in 1873 by the zoologist,
      F.A. Schneider, in the course of his research on the life
      history of flatworms. He gave a description of the successive
      stages of the process, with drawings of the nucleus and chromo-
      somal strands.

..... HERTWIG, O. See 1678.

..... FOL, H. See 1679.

1694. STRASBURGER, E.A. (2695) *Zellbildung und Zelltheilung*. Jena,
      1875. (3rd ed., 1880) An account of his extensive researches
      in plant cytology, the successive editions being substantially
      revised. It described in detail the phenomena of cell division
      and pointed out how similar the process is in plants and animals
      (the process in animals was then being studied by several
      investigators). The third edition enunciated the principle
      that nuclei can arise only from the division of other nuclei.
      (See also 1698)

1695. BÜTSCHLI, O. (2723) Published in 1876 a comprehensive account
      of the process of cell division in animals. He also studied
      fertilization, establishing that only one spermatozoon enters
      the ovum, and that the nuclei of the male and female germ
      cells fuse together.

1880s

..... DEVELOPMENT OF THE MICROSCOPE. See 1645.

1696. BENEDEN, E. van. (2821) In several papers in the 1880s he
      showed that: **(a)** fertilization consists in the fusion of two
      half-nuclei, one from the ovum and one from the sperm; **(b)** the
      two nuclear membranes do not break down, with consequent fusion,

until chromosomes (to use the later term) have appeared; and
(c) each half-nucleus contributes half of the total number of
chromosomes found in the fertilized ovum. He also showed
that the number of chromosomes is the same for each body cell
in any particular organism (the number in the ova and spermatozoa being half the number in the body cells), and thought that
the number is characteristic of the species.

1697. FLEMMING, W. (2694) *Zellsubstanz, Kern und Zelltheilung*.
Leipzig, 1882. A major work on cell division in animals,
describing in detail the process of mitosis (his term). He
made use of the new synthetic dyes for selective staining,
and introduced the term 'chromatin' for the part of the
nuclear substance that takes up certain stains.

1698. STRASBURGER, E.A. (See also 1694) Concluded from his studies
of fertilization in plants in 1884 that the filaments (i.e.
chromosomes) in the nucleus are the bearers of heredity.
(The same conclusion was reached at about the same time by
workers in animal cytology, such as Hertwig and Koelliker)

1699. NÄGELI, K.W. von. (See also 1689) *Mechanisch-physiologische
Theorie der Abstammungslehre*. Munich, 1884. A famous work.
While his ideas on evolution were unfruitful, his theory of
a cytological mechanism of heredity included ideas which,
though inexact on a later view, were suggestive and stimulating
(cf. 1683a), and turned out many years later to be essentially
correct.

1700. RABL, C. (2970) Published in 1885 a detailed description of the
process of cell division, expressing the view that the chromosomes (to use the later term) retain their individuality and
persist even when they seem to disappear during the resting
stage. This idea was soon found to be fundamental to an
understanding of the process.

1701. BOVERI, T. (2748) Inspired by the work of van Beneden (1696)
he began in 1885 a series of researches on the chromosomes.
Three years later he succeeded in demonstrating their individuality (i.e. that they continue to exist as components of the
nucleus even when they become invisible).   (See also 1703)

1702. CHROMOSOMES. The word was introduced in 1888 by the anatomist,
H.W. Waldeyer-Hartz.

1890s

1703. BOVERI, T. (See also 1701) In 1890 he confirmed and extended
van Beneden's conclusion that at fertilization the ovum and
sperm contribute equal numbers of chromosomes to the new cell.
This research, together with his earlier work, directed
attention to the chromosomes as the bearers of heredity, rather
than the nucleus as a whole. (His important demonstration of
the differential value of the chromosomes in inheritance was
made in 1902)

1704. WILSON, E.B. (3069) (a) In 1892 he began studies of chromosome
movements in cell division, especially spindle formation and
the origin of centrosomes. (b) *The cell in development and
heredity*. New York, 1896. (2nd ed., 1900. 3rd ed., 1925;

extensively revised) One of the most outstanding biological
treatises of the late nineteenth century; it had a profound
influence on the subsequent development of cytology and experimental biology generally. In particular it concentrated
attention on the cell nucleus, and especially the chromosomes,
as the bearers of heredity, thereby aiding the general acceptance of Mendelian theory after 1900.

1705. HERTWIG, O. (2720) *Die Zelle und die Gewebe*. 2 vols. Jena,
      1893-98. (The subsequent editions were entitled *Allgemeine
      Biologie*. 7th ed., 1923) An important textbook.

1706. GOLGI, C. (1951) As a consequence of his work on the detailed
      anatomy of the nervous system (1642) he discovered in 1898 a
      small, complex structure within the cytoplasm of nerve cells.
      This structure, known as the Golgi apparatus or Golgi complex,
      has since been found in all cells (except bacteria) and is
      now known to play an important part in the operations of the
      cell.

1.18 PHYSIOLOGY

Human and comparative.

Plant physiology is included under Botany.

1470s

1707. GALEN. (See 1552) His physiological system continued to be
     dominant until the time of Harvey.

1540s

1708. FERNEL, J. (1968) *De naturali parte medicinae*. Paris, 1542.
     (A revised version was included in his *Medicina* of 1554 and
     a later version in his posthumous *Universa medicina* of 1567)
     An important work in Galenic physiology which was widely
     influential.

1550s

1709. SERVETUS, M. (1974) A theologian and medical practitioner whose
     chief work, *Christianismi restitutio* (1553), contained the
     discovery of the lesser (or pulmonary) circulation of the
     blood presented in a theological context; he also came very
     close to discovering the complete circulation. Most copies
     of his book were burned because of its heretical character
     and it was not until 1694 that his discovery became generally
     known. (When Harvey published his discovery of the general
     circulation in 1628 he did not know of Servetus' work)

1580s

..... PLATTER, F. See 1575.

1710. PICCOLOMINI, A. (1852) *Anatomicae praelectiones*. Rome, 1586.
     Publication of his lecture course. It included much physiology,
     chiefly derived from Galen and Aristotle but with some original
     ideas. The book had an influence on Harvey.

1600s

..... CASSERI, G. See 1578.

1610s

1711. SANTORIO, S. (1867) *De medicina statica*. Venice, 1615. (About
     forty reprintings as well as several translations) A famous
     book which introduced quantitative experimentation into human
     biology. Santorio was a friend of Galileo and seems to have
     been influenced by his ideas.

1620s

1712. HARVEY, W. (2232) *Exercitatio anatomica de motu cordis et
      sanguinis in animalibus*. Frankfurt, 1628. A small book
      containing the revolutionary discovery of the circulation
      of the blood. (His work of 1649 with a similar title can be
      regarded as a second edition. Both works were translated
      into English in 1653)

1630s

1713. DESCARTES, R. (1991) In the early 1630s, if not before, he
      formed his physiological ideas--in close accord with his
      philosophical ideas--and elaborated them in a work entitled
      Traité de l'homme. Though it was not published until after
      his death (the French original appeared in 1664, preceded in
      1662 by the Latin translation) many of the ideas in it were
      reflected in his philosophical works, beginning with the
      Discours de la méthode in 1637 and continuing through his
      career. Thus the powerful conception of a completely mechan-
      istic physiology began gradually to make its impact from the
      late 1630s onward, though its full statement did not appear
      until the 1660s. However speculative and ill-founded it was
      in many of its details, the ultimate impact of the new view
      was enormous: the souls, faculties, principles, etc., of
      traditional physiology were swept away, to be replaced by
      pipes and pumps.

1640s

1714. ENT, G. (2240) *Apologia pro circulatione sanguinis*. London,
      1641. One of the first works to combat the antagonism to
      Harvey's doctrine with a detailed defence of it. He had a
      distinctive approach of his own which was especially signifi-
      cant in his treatment of respiration and body heat.

1650s

1715. BARTHOLIN, T. (2827) **(a)** *Vasa lymphatica ... in animalibus*.
      Copenhagen, 1653. **(b)** *Vasa lymphatica in homine*. Ibid.,
      1654. In the two works he established that the lymphatics
      (which had already been described anatomically) form a separate
      physiological system, previously unknown.

1716. CHARLETON, W. (2246) *Natural history of nutrition, life and
      voluntary motion*. London, 1659. The first textbook of
      physiology in English.

1660s

1717. SYLVIUS, F. (2770) In his very influential teaching at Leiden
      from 1658 until his death in 1672, as in his books, he was
      the outstanding representative of the iatrochemical school.
      According to the iatrochemical doctrines all physiological
      activities were due to such chemical processes as effervescence,
      fermentation, and especially the reaction between acids and
      alkalies. Though highly speculative these ideas inspired some
      solid physiological research by Sylvius' pupils and others.

1718. CROONE, W. (2269) *De ratione motus musculorum*. London, 1664. Described experiments which indicated that muscular contraction was brought about by something passing from the brain along a nerve to the muscle where it evidently had some sort of chemical effect.

1719. LOWER, R. (2268) *Tractatus de corde*. London, 1669. (Many editions up to 1749) Contained important observations and experiments on the heart and its movements, the circulation of the blood, and the effect of respiration on the blood flowing through the lungs. The book had a major effect on physiological thought.

1670s

1720. GRAAF, R. de. (2778) *Tractatus anatomico-medicus de succi pancreatici natura et usu*. Leiden, 1671. A pioneering investigation of the pancreatic juice. (He devised a means of collecting it from dogs)

1721. MAYOW, J. (2270) *Tractatus quinque medico-physici*. Oxford, 1674. Contained some very perceptive investigations of the "nitro-aerial spirit" (which today can be regarded as corresponding to oxygen) and its role in respiration and the production of animal heat.

1680s

1722. BORELLI, G.A. (1885) *De motu animalium*. 2 vols. Rome, 1680-81. The first major application of Cartesian principles to physiology, and the pioneering work of the school of iatromechanics (or iatrophysics). It was a detailed account of the muscles and their mechanical action in walking, swimming, etc., and also of the heart and vascular system as a single hydraulic system, with rates of blood-flow, etc. The work concluded with some less successful speculations on such topics as respiration, digestion, nervous action, etc.

1690s

1723. BELLINI, L. (1901) *Opuscula aliquot*. Leiden, 1695. A very influential statement of the iatromechanical conception of physiology.

1700s

1724. STAHL, G.E. (2525) *Theoria medica vera*. Halle, 1708. In marked contrast to most of his contemporaries Stahl rejected mechanistic physiology and developed a thoroughly vitalistic system which was in line with his important chemical ideas.

1725. BOERHAAVE, H. (2782) *Institutiones medici*. Leiden, 1708. (Many editions and several translations) A highly influential work because of his great fame as a teacher. The physiology section was a comprehensive and well-organized treatise, closer to modern ideas than the works of Stahl and Hoffmann. Though Boerhaave's approach was mechanistic he kept theory in the background as much as possible and had a strong sense of the limitations of contemporary knowledge.

1710s

1726. POURFOUR DU PETIT, F. (2043) In 1710 he concluded from many
observations on human patients and dissections of dogs that
the movements of a limb are brought about by "animal spirits"
coming from the side of the brain opposite the limb.

1727. HOFFMANN, F. (2526) *Medicina rationalis systematica*. 2 vols.
Halle, 1718-20. Presented an influential system of mechanistic
physiology.

1730s

1728. HALES, S. (2295) *Haemastaticks* [Included in his *Statical essays*
--see 1327]. London, 1733. Hales brought to animal physiology
the same quantitative approach and skill in the design of
experiments that he had used so effectively in plant physiology.
His work in this field dealt chiefly with blood-pressure, the
speed of the blood-stream, and factors affecting it. Though
not so fundamental as his work in plant physiology it was
nevertheless a substantial contribution to the physiology of
the blood.

1740s

1729. HALLER, A. von. (2896) *Primae lineae physiologiae*. Göttingen,
1747. (Several later editions. Trans. into French & English)
A concise textbook which proved very popular and was still in
use by 1800.   (See also 1731)

1750s

1730. WHYTT, R. (2301) *The vital and other involuntary motions of
animals*. Edinburgh, 1751. A work of major importance in the
development of neurophysiology. He gave the first clear
description of what later came to be called reflex action and
discussed various kinds of reflexes--including the one now
named after him which he was able to localize. He also showed
that the spinal cord is involved in reflex action.

1731. HALLER, A. von. (See also 1729) **(a)** Though the concept of
irritability had been mentioned by some earlier writers it was
Haller who demonstrated it experimentally in 1753 and effectively introduced it. He made the distinction between sensibility (nerve impulse) and irritability (muscular contraction).
(b) *Elementa physiologiae corporis humani*. 8 vols. Lausanne,
1757-66. A synthesis of the whole of the physiological knowledge of the time. His greatest work.

1760s

1732. FONTANA, F. (1926) *De irritabilitatis legibus*. Lucca, 1767.
An extensive investigation of the phenomena which had been
designated by Haller (1731a) as sensibility and irritability.
It made a substantial contribution.

1770s

1733. SPALLANZANI, L. (1925) *De' fenomeni della circolazione*. Modena,
1773. A treatise on the dynamics of the circulation. It
included accounts of over 300 experiments.   (See also 1736)

1734. "ON THE ELECTRIC property of the torpedo" -- a paper of 1773
which contained the first accurate description of the torpedo
fish and its electric organs--a topic of considerable significance for the future of physiology. The author was John Walsh,
a naturalist and fellow of the Royal Society.

1780s

1735. GALVANI, L. (1927) Began in 1780 a series of investigations in
the little-known field of the electric stimulation of nerves
and muscles. Using parts of dissected frogs, especially the
legs, he soon discovered a number of quite unexpected phenomena.
(On a modern interpretation he revealed three different kinds
of electric effects, chiefly because the frogs' legs acted as
very sensitive electroscopes, indicating by their twitch the
presence of even a minute current) One of his discoveries
occurred when he hung some partly dissected frogs on an iron
railing by means of brass hooks--the legs twitched. This and
other discoveries he explained in terms of "animal electricity,"
a subtle fluid similar to ordinary electricity but present in
nerves and muscles. In 1791 he published his discoveries and
conclusions in book form.

    The most important reaction to his book was that of
Volta who in 1793 developed his theory of contact electricity
(see 634) and denied the existence of "animal electricity."
A controversy ensued in which--as it turned out much later--
both sides were largely in the right. Apart from the highly
important lead that it gave to Volta, Galvani's work was
significant in opening the way to the nineteenth-century
developments in electrophysiology.

1736. SPALLANZANI, L. (See also 1733) *Dissertazioni di fisica animale
e vegetabile*. 2 vols. Modena, 1780. Vol. 1 was a treatise
on digestion, containing accounts of a large number of experiments (including experiments on himself) which added greatly
to the knowledge of the subject. He concluded that the fundamental factor is the dissolvent action of the "gastric juice"
(his term). (For a brief account of Vol. 2 of the work see
1662)   (See also 1740)

1737. PROCHÁSKA, G. (2944) *Commentatio de functionibus systematis
nervosi*. Prague, 1784. An account of the operations of the
nervous system. His concept of reflex action involved a
sensorium commune which co-ordinated and passed on to the
motor nerves the sensations received by the brain.

1790s

1738. LAVOISIER, A.L. (2071) "Premier mémoire sur la respiration des
animaux" -- an account of an investigation conducted jointly
with A. Seguin and reported to the Paris Academy in 1790
(published in 1793). In his earlier work (822, 830) Lavoisier
had elucidated the chemistry of respiration and solved the
ancient puzzle of animal heat. In the present memoir (which,
because of political events, had no sequel) he initiated the
modern study of metabolism. With Seguin he measured the
amount of oxygen consumed by a human subject (Seguin himself)

showing that it increases with exercise and during periods of
digestion, and also with increase in environmental temperature.

1739. YOUNG, T. (2330) In his researches on physiological optics,
published mostly between 1791 and 1801, he established the
mechanism of the accommodation of the eye, and found the cause
of astigmatism. He also studied colour perception, concluding
that the retina responds to colours in general as blends of
three principal colours—blue, green, and red. This three-
colour theory of vision was later developed by Maxwell (1764).

1800s

1740. SPALLANZANI, L. (See also 1736) *Mémoires sur la respiration.*
Geneva, 1803. (Posthumous; ed. and trans. by J. Senebier)
His experiments established that the conversion of oxygen into
carbon dioxide takes place in the tissues (not in the lungs
as Lavoisier had thought) and that the gases are transported
in solution in the blood.

1810s

1741. LEGALLOIS, J.J.C. (2121) *Expériences sur le principe de la vie.*
Paris, 1812. Included an investigation of the action of the
vagus nerve on respiration. He located the respiratory centre
in the medulla.

1742. MAGENDIE, F. (2130) **(a)** From 1813 until about 1830 he made a
multitude of discoveries covering virtually the whole extent
of physiology as it then existed. He was a skilful and
original experimenter and his contribution to the suject was
as much in the experimental techniques he created as in his
factual discoveries. He is best known for his part in establishing the Bell-Magendie law (1630), for pioneering work in
pharmacology, and for his investigations of the functions of
the cerebellum and the cerebrospinal fluid.

    (b) *Précis élémentaire de physiologie.* 2 vols. Paris,
1816-17. (4th ed., 1836. Several translations) A highly
influential textbook. It opposed the excessive speculation
characteristic of the time and put a strong emphasis on experimental findings.

1820s

1743. FLOURENS, P. (2144) During the 1820s he carried out a brilliant
series of investigations on the brain and nervous system of
vertebrates which made a major contribution. His chief
conclusions were that the higher psychic and intellectual
faculties reside in the cerebral hemispheres, the co-ordination
and regulation of movements in the cerebellum, and the control
of vital functions (especially respiration) in the medulla.

1744. PURKYNĚ, J.E. (2609) **(a)** *Beobachtungen und Versuche zu die
Physiologie der Sinne.* Berlin, 1825. An important work on
subjective sensory phenomena. **(b)** In the late 1820s he made
some pioneering studies in the field of physiological pharmacology, investigating the effects produced by such substances
as opium, belladonna, digitalis, etc.

1745. WEBER, E.H. (2588) **(a)** *Wellenlehre*. Leipzig, 1825. Written
with his brother, the physicist W.E. Weber. A thoroughgoing
application of hydrodynamics to the circulation of the blood.
(For his later work in the field see 1765)
    (b) In the mid-1820s he began a long series of studies
of the sense of touch. His important concept of the threshold,
or "just-noticeable difference" (the minimum difference perceiv-
able between similar stimuli) enabled him to take a quantitative approach which yielded many valuable conclusions. His
achievements in the field (which he summarized in 1846--see
1762b) make him one of the founders of experimental psychology.

1746. MÜLLER, J. (2610) *Zur vergleichenden Physiologie des Gesichts-
sinnes des Menschen und der Tiere*. Leipzig, 1826. A brilliant
work which made a wide impression and immediately established
his reputation. It contained much new material on animal and
human vision, including investigations of the compound eyes of
insects and crustaceans, and some remarkable analyses of human
sight. The most notable feature of the book was what Müller
called his "law of specific nerve energies", namely that the
kind of sensation induced by the stimulation of a sensory
nerve depends, not on the mode of stimulation, but on the
nature of the sense organ with which the nerve is connected.
Thus, stimulation of the optic nerve by any means produces
only luminous sensations, stimulation of the nerve of hearing
produces only auditory sensations, and so on. This was a
physiological conclusion with epistemological implications:
we do not perceive the external world "as it really is" but
only the effects it has on our senses. (See also 1748)

1747. BURDACH, K.F. (2585) **(a)** See 1631. **(b)** *Die Physiologie als
Erfahrungswissenschaft*. 6 vols. Leipzig/Königsberg, 1826-40.
A comprehensive treatise which included generation and embryology as well as physiology. Sections of the work were written
by several collaborators (including von Baer, Rathke, Johannes
Müller, and R. Wagner) under Burdach's direction. Though
never completed (ten volumes had been planned) it was widely
read.

1830s

1748. MÜLLER, J. (See also 1746) **(a)** Took up research in 1831 on the
Bell-Magendie law (1630), a topic that had become confused
because of priority disputes and other reasons. By using frogs,
which proved to be much more suitable than other experimental
animals, he greatly clarified the situation. This work led
him on to other investigations in neurology, including an
elucidation of reflex action.
    (b) *Handbuch der Physiologie des Menschen*. 2 vols (in
several parts). Coblenz, 1833-40. (Some later eds. Trans.
into English and French) A critical survey of the state of
knowledge of the subject which included many findings of his
own. It was widely influential and provided a big stimulus,
both to physiological research and to its application in
clinical practice.

1749. HALL, M. (2381) From 1832 until about 1850 he carried out a
series of researches on many animal species into the phenomena

of reflex action (his term). He concluded that **(a)** the spinal
cord consists of a chain of units, each of which acts as an
independent reflex arc; **(b)** the functioning of each arc depends
on the actions of the sensory and motor nerves and the section
of the spinal cord from which the nerves arise; and **(c)** the
arcs interact with each other and with the brain to produce
co-ordinated movement.

1750. *EXPERIMENTS and observations on the gastric juice and the physiology of digestion.* Plattsburg (U.S.A.), 1833. (Another ed.,
Edinburgh, 1838. German trans., 1834) By William Beaumont,
a surgeon in the U.S. Army. One of his patients had suffered
a gunshot wound which, after it healed, left a permanent large
fistula giving access to his stomach. Beaumont took advantage
of this situation to conduct a long series of experiments,
reported in his book, which added greatly to knowledge of the
digestive process. The book was well received by European
physiologists and made a considerable impression.

1751. SCHWANN, T. (2623) During the period 1834-39, when he was
working in Müller's laboratory in Berlin, he **(a)** devised a
means of measuring muscular contraction which provided an
important example of a quantitative approach in physiology;
(b) discovered the digestive enzyme ("ferment"), pepsin, the
first enzyme preparation from animal sources; **(c)** observed
the multiplication of yeast cells in a fermenting solution of
sugar and concluded that fermentation is the result of life
processes (a conclusion vigorously attacked by Liebig and
other chemists); and **(d)** made his famous contribution to the
cell theory (see 1688).

1752. WEBER, W.E. (2600) *Mechanik der menschlichen Gehwerkzeuge.*
Göttingen, 1836. Written with his brother, the anatomist
E.F. Weber. An important work on the physiology of human
locomotion.

1753. MATTEUCCI, C. (1941) His pioneering researches in electrophysiology, carried out between 1836 and 1844, were surveyed in his
Traité des phénomènes électriques des animaux (Paris, 1844).

1754. WAGNER, R. (2611) *Lehrbuch der Physiologie.* Erlangen, 1838.
(7th ed., 3 vols, 1885-87. English trans., 1841) (cf. 1755)

1840s

1755. *HANDWÖRTERBUCH der Physiologie.* 5 vols. Brunswick, 1842-53.
Edited by R. Wagner (cf. 1754). A compendium of the physiological knowledge of the time. It included many extensive
review articles written by various authors, including some
of the leading figures in the field.

1756. DU BOIS-REYMOND, E. (2648) From the beginning of his research
career in 1842 until its end in the late 1860s he devoted
himself to electrophysiology ("animal electricity"), a subject
which extended back to Galvani but hitherto had not developed
much. His intimate knowledge of physics was a major factor
in enabling him to lay the foundations, both instrumentally
and conceptually, of all subsequent work in the field. His

research results were collected in his *Untersuchungen über thierische Elektrizität* (2 vols in several parts, Berlin, 1848–84). In his famous preface to Volume 1 he launched a massive attack on the prevalent theories of vitalism and offered his own researches as an example of the new physico-chemical reductionist approach.

1757. BOWMAN, W. (2405) *The physiological anatomy and physiology of man.* 2 vols in 4 parts. London, 1843–56. With R.B. Todd (professor of physiology at King's College, London). The first work on physiology in which histology played a prominent part.

1758. BERNARD, C. (2170) In the period from 1843 to the late 1850s he made numerous important discoveries, the chief of which related to (a) the role of the pancreas in digestion, (b) the formation of glycogen by the liver (*inter alia* the first demonstration of the body's ability to synthesize complex compounds), and (c) the action of the sympathetic nerve system in regulating the blood-supply to different parts of the body (now called the vaso-motor mechanism).

A profound thinker as well as a skilful and successful experimenter, he also enriched physiology by his creation of numerous new concepts, the most famous being that of the *milieu intérieur*; it encapsulates his great insight that the complex functions of the various organs are closely inter-related and are all directed to maintaining the constancy of internal conditions despite external change. (See also 1767)

1759. LUDWIG, C. (2647) From 1844 until the 1860s he published much work in his own name but later let his numerous students publish the ideas he suggested to them. He is best known for his researches on the cardiovascular system but he made major contributions to many areas of physiology. A special feature of his work was his mechanical skill which, together with a good knowledge of physics, enabled him to invent various aids and measuring devices for physiological research, notably the kymograph or rotating drum (now used in many automatic record-ers) which he used to record variations in arterial blood pressure. He was also the first to keep animal organs alive *in vitro.* (See also 1766)

1760. BRÜCKE, E.W. (2955) One of the most versatile physiologists of the time, with an unusually wide range of interests. From the mid-1840s until his death in 1892 he published many papers and several books on numerous aspects of physiology and related subjects, including plant physiology, microscopic anatomy, and physics. His *Vorlesungen über Physiologie* (2 vols, Vienna, 1873–75; 3rd ed., 1881–84) included much of his own work.

1761. MAYER, J.R. (2630) In a highly original and wide-ranging essay, printed as a pamphlet in 1845 (because it was rejected by editors of scientific journals), he applied his concept of the conservation of "force" (i.e. energy; see 680) to the living world. He pointed out that the sun is the ultimate source of force, its heat and light being converted by plants into latent chemical force; animals then consume plant material as food, assimilating the chemical force by digestion and expending it in muscular action and body heat.

1762. WEBER, E.H. (See also 1745) **(a)** In 1845 he and his brother,
the anatomist E.F. Weber, made the unexpected discovery that
stimulation of the vagus nerve had an inhibitory effect on
heartbeat. "It was the first instance of nerve action causing
inhibition of an autonomic activity, rather than exciting it.
It became an important milestone in the evolution of physiology
not only for its significance to the circulation but also
because its discovery brought to light a hitherto unknown but
essential kind of nerve action." (*DSB*)
    (b) "Tastsinn und Gemeinfühl" -- an article in Wagner's
Handwörterbuch der Physiologie (see 1755), and later reprinted
separately, which drew general attention to his pioneering
work on the sense of touch (1745b).   (See also 1765)

1763. BROWN-SÉQUARD, C.E. (2171) His chief work was done in the 1840s
and the 1850s. He is best known for his researches in neurophysiology, especially of the spinal cord, and for his discovery
of the importance of the adrenal glands. The latter discovery,
and his subsequent realization that "internal secretion" pro-
vides a control mechanism independent of the nervous system,
gives him a claim to be one of the founders of endocrinology.

1850s

1764. MAXWELL, J.C. (2413. The great physicist) During the 1850s he
experimented on colour vision and, reviving the three-colour
hypothesis of Thomas Young (1739), created the first quantitative theory of the subject. He also made some other contributions to physiological optics.

1765. WEBER, E.H. (See also 1762) *Über die Anwendung der Wellenlehre
auf die Lehre vom Kreislauf des Blutes*. Leipzig, 1850. A
survey of his later work in the application of hydrodynamics
to the circulation of the blood (cf. 1745). His most notable
finding was that the pulse is a wave in the arteries that is
propagated much more rapidly than the flow of the blood.

1766. LUDWIG, C. (See also 1759) *Lehrbuch der Physiologie des Menschen*.
2 vols. Heidelberg, 1852-56. (2nd ed., 1861) An influential
textbook which rejected the (often vitalistic) philosophical
approach characteristic of physiological textbooks hitherto
and sought to explain biological phenomena in terms of physics
and chemistry.

1767. BERNARD, C. (See also 1758) His famous lectures at the Collège
de France were published in ten volumes of *Leçons* over the
period 1855 to 1879 (the last posthumously). They had a far-
reaching influence.   (See also 1776)

1768. HELMHOLTZ, H. von. (2631) *Handbuch der physiologischen Optik*.
3 vols. Leipzig, 1856-67. An exhaustive treatise resulting
from his researches in the field (he invented the ophthalmo-
scope in 1851 and the ophthalmometer in 1855). It covered
all previous work on the subject and was characterized by his
familiarity with mathematics and physics and his keen philosophical insight.   (See also 1774)

1769. FICK, A. (2667) His monograph, *Die medizinische Physik* (Brunswick, 1856), published almost at the beginning of his career,

announced his physical approach to physiology and adumbrated the areas of research in which he was to work until the 1890s --chiefly the mechanics of muscles, various aspects of the generation of body heat, diffusion phenomena, physiological optics, haemodynamics, and aspects of electrophysiology.

1770. PFLÜGER, E. (2668) In his researches on electrophysiology in association with Du Bois-Reymond in the late 1850s he established what came to be called "Pflüger's law of convulsion", relating to the effect of a current applied to a nerve. (See also 1775)

1860s

1771. MAREY, E.J. (2190) During the 1860s he developed various applications of the technique of graphical recording in physiology. His invention of the sphygmograph in 1860 proved especially valuable for clinical medicine.   (See also 1789)

1772. HERING, E. (2696) (a) *Beiträge zur Physiologie. Zur Lehre vom Ortsinne der Netzhaut.* 5 parts. Leipzig, 1861-64. An investigation of visual space-perception in which he contended (against Helmholtz and others) that it is due to the capabilities of the retina. (b) *Die Lehre vom binokularen Sehen.* Leipzig, 1868. (c) In collaboration with J. Breuer he discovered in 1868 the reflex actions in the lungs, mediated by the vagus nerve, which cause the alternation of inspiration and expiration during breathing.   (See also 1778)

1773. VOIT, C. von. (2669) Between 1862 and 1873 he and his colleague, Max Pettenkofer, carried out a classical series of investigations on metabolism. They used a special respiration chamber large enough to accomodate a man and designed for experiments on measurement of material balance (input and output) in both nutrition and respiration concurrently. They were thereby able to achieve the first accurate measurements of gross metabolism in man and other animals, and so laid the foundation for subsequent research in both metabolism and nutrition.

1774. HELMHOLTZ, H. von. (See also 1768) *Die Lehre von den Tonempfindungen als physiologische Grundlage für die Theorie der Musik.* Brunswick, 1863. During the previous ten years he had made a number of important discoveries in physiological acoustics (especially the resonance theory of hearing) and in this masterly work he applied them to the theory of music, explaining many well-known phenomena. The book became famous and went through many editions in several languages.

1775. PFLÜGER, E. (See also 1770) From the mid-1860s he studied the experimentally-difficult subject of gas exchange in the blood and in the cells. He was able to establish that the fundamental oxidation process takes place in the cells (not in the blood or lungs as had previously been thought). This and his other findings in the field were presented in a major article, "Über die physiologische Verbrennung in den lebenden Organismen", in 1875.

1776. BERNARD, C. (See also 1767) *Introduction à l'étude de la médecine expérimentale.* Paris, 1865. A celebrated work which is largely

a meditation, in lucid and attractive style, on the philosophy
of biology and extending to the philosophy of science generally.

1870s

1777. KÜHNE, W.F. (2670) **(a)** From about 1870 he carried on a series
of researches on digestion, especially the breaking down of
protein molecules. In the course of the work he isolated
trypsin and established its role in pancreatic secretion. He
proposed the term 'enzyme' in 1878. **(b)** In a brilliant investigation in 1878 he elucidated the chemical and physiological
functioning of visual purple (the pigment of the retina).

1778. HERING, E. (See also 1772) From the 1870s until after 1900 he
continued his investigations on the physiology of the senses,
especially vision (challenging the Young-Helmholtz theory of
colour vision).

1779. "ÜBER die elektrische Erregbarkheit des Grosshirns" -- a classic
paper published in 1870 by the psychiatrist, E. Hitzig, and
the anatomist, G.T. Fritsch. In an investigation of the
electrical excitability of the cerebral cortex of the dog they
established conclusively the localization of function in the
cortex. Their paper opened up the field of the electrophysiology of the cerebral cortex.

1780. BERT, P. (2191) **(a)** *Leçons sur la physiologie comparée de la
respiration*. Paris, 1870. **(b)** *La pression barométrique:
Recherches de physiologie expérimentale*. Paris, 1878. The
latter was a classical work on the physiological effects of
high and low atmospheric pressure. Its practical value for
the exploration of the upper atmosphere and the ocean depths
continued well into the twentieth century.

1781. MACH, E. (2960) **(a)** *Optisch-akustische Versuche*. Prague, 1873.
(b) *Grundlinien der Lehre von den Bewegungsempfindungen*.
Leipzig, 1875. Researches on visual, aural, and kinesthetic
sensation--conducted in the borderlands of physics, physiology,
and psychology.

1782. ENGELMANN, T.W. (2697) **(a)** From 1873 to 1895 he carried on a
notable series of researches on muscle contraction, and proposed
a hypothesis in which the contraction of striped muscle was
attributed to a flow of fluid from the isotropic bands to the
anisotropic bands. **(b)** In 1875 he devised a famous experiment
which proved conclusively that the heartbeat is myogenic (i.e.
originates in the heart muscle, not from an external nerve
impulse).

1783. LANGLEY, J.N. (2456) From 1875 to 1890 he made systematic
studies of glandular secretion and the secretory glands,
especially the salivary glands.   (See also 1794)

1784. FERRIER, D. (2453) *The function of the brain*. London, 1876.
An account of his classic researches, begun in 1873, on the
detailed mapping of localized areas of function in the cerebral
cortex. His results were obtained from systematic experiments
on many species of vertebrates, chiefly primates.

1785. FOSTER, M. (2431) *A text-book of physiology*. London, 1877. (6th ed., 1900. Trans. into German, Italian, and Russian) A standard work, highly regarded for its style and judgment.

1786. RICHET, C.R. (2204) From the late 1870s onward he made substantial contributions to the knowledge of digestion, respiration, the regulation of body heat, and a number of other physiological and medical topics. (See also 1798)

1787. *HANDBUCH der Physiologie*. 12 vols. Leipzig, 1879–83. Edited by L. Hermann, H. Aubert, and C. Eckhard.

1880s

1788. ARSONVAL, A. d'. (2205) In the course of his physiological researches during the 1880s and 1890s he constructed many notable kinds of apparatus and instruments (winning the Prix Montyon of the Paris Academy in 1882 for the ingenuity of his devices). Of his numerous electrical instruments the best known was the highly sensitive galvanometer which he developed in 1882. His contributions to biophysics were summed up in his *Traité de physique biologique* (Paris, 1903).

1789. MAREY, E.J. (See also 1771) As part of a long-term programme of research on animal locomotion (begun in 1868) he invented and developed the ciné camera in the early 1880s. Until his death in 1904 he made extensive and imaginative use of it to study motion generally in the animal world.

1790. RUBNER, M. (2725) From 1883 onward he made accurate measurements of the calorific value of various foodstuffs and *inter alia* established that fats and carbohydrates are interchangeable as energy sources. He developed original ideas on metabolism from the standpoint of energy and demonstrated the validity of the principle of conservation of energy in living organisms. In connection with studies on heat regulation he found that the heat production of mammals is proportional to their surface area, and investigated heat loss from body surfaces.

1791. GASKELL, W.H. (2454) **(a)** In a well-known paper of 1883 he put forward powerful evidence that the heart's action is essentially myogenic rather than neurogenic. **(b)** "On the structure, distribution and function of the nerves which innervate the visceral and vascular systems" -- a classic paper of 1886 which established the main features of the involuntary nervous system.

1792. SHERRINGTON, C.S. (2471) The first phase of his famous researches on the nervous system began in the late 1880s and was summed up in his monograph of 1906, *The integrative action of the nervous system*.

1793. LOEB, J. (3071) His first major line of research, on animal tropisms, was initiated in 1888 (inspired by the botanists' work on plant tropisms). His second main theme, the famous experiments on artificial parthenogenesis, began in 1899.

1890s

1794. LANGLEY, J.N. (See also 1783) From 1890 until his death in 1925
he carried on a long series of investigations on the involuntary
nervous system.

1795. VASSALE, G. (1959) Began in 1890 the researches in endocrinology
which he continued almost until his death in 1913. He was
concerned especially with the secretions of the thyroid, para-
thyroid, and adrenal glands.

1796. ENGELMANN, T.W. (See also 1782) In the early 1890s he resumed
research on the heart, greatly advancing the knowledge of its
detailed operation and function.

1797. (SHARPEY-) SCHÄFER, E.A. (2455) **(a)** In 1894, in collaboration
with G. Oliver (a practising physician), he discovered the
existence and properties of adrenaline, and went on subsequently
to further important work in endocrinology. **(b)** *Text-book of
physiology*. 2 vols. Edinburgh, 1898–1900. An important
collective work edited by Schäfer.

1798. *DICTIONNAIRE de physiologie*. 10 vols. Paris, 1895–1928.
Edited by C. Richet et al.

1799. VERWORN, M. (2757) Though his experimental work, begun in 1889,
was extensive and wide-ranging, he was basically concerned
with a rather speculative theoretical scheme which he called
"cellular physiology." His *Allgemeine Physiologie. Ein Grundriss der Lehre vom Leben* (Jena, 1895) reflected his theoretical
concerns. It was widely read and gained some support for his
ideas.

1800. *ELEKTROPHYSIOLOGIE*. Jena, 1895. (English trans., 2 vols, 1896–
98) By W. Biedermann. The first comprehensive treatise on
the subject.

1801. PAVLOV, I.P. (3008) [*Lectures on the function of the main food-
digesting glands*. In Russian]. St. Petersburg, 1897. (English trans., ca. 1900; 2nd ed., 1910) An account of his work
on the physiology of digestion, carried out over the previous
twelve years. (His famous work on conditioned reflexes was
done after 1900)

1.19 MICROBIOLOGY

Pathology, immunology, and medicine generally are mostly excluded.

1740s–1780s

..... NEEDHAM, J.T. See 1659.

..... SPALLANZANI, L. See 1661a.

..... MÜLLER, O.F. See 1451.

1830s

1802. BASSI, A.M. (1937) Published in 1835 the results of his investigations into silkworm disease, showing that the disease is
contagious, being caused by a parasitic fungus. He later
suggested that some infectious diseases of humans, as well as
of animals and plants, are caused by living parasites. The
significance of his discovery was generally recognised.
"He is justly regarded as the real founder of the doctrine of
pathogenic micro-organisms of vegetable origin." (W. Bulloch)

1803. EHRENBERG, C.G. (2612) Published in 1838 an impressive, well-
illustrated monograph on the Infusoria (a grouping which then
included micro-organisms of widely different kinds). Despite
his mistaken belief—partly due to the inadequacies of his
microscopes—that the Infusoria contained organs similar to
those of higher animals, his book made a considerable addition
to the knowledge of the group.

1840s

1804. DUJARDIN, F. (2158) *Histoire naturelle des zoophytes. Infusoires.*
Paris, 1841. A work which considerably advanced the study of
the Infusoria, largely because of Dujardin's skill as a microscopist. He rejected the view of Ehrenberg that they contain
organs like those of higher animals—a view that had become
generally accepted—and the resulting controversy helped to
stimulate interest in the subject. In contrast to the prevailing conception he put forward the view that the Infusoria
consist of a homogeneous, living jelly which he called "sarcode"
—a concept which eventually became accepted though his term
was later replaced by 'protoplasm.'

1850s

1805. PASTEUR, L. (2172) Began in the mid-1850s a long series of
investigations into various kinds of fermentation—lactic

(1857), alcoholic (1858-60), butyric (1861), and acetic (1861-64). His germ theory of fermentation was first announced in 1857 and received further support from his subsequent work. It aroused wide interest and, in some quarters, fierce opposition because fermentation had hitherto been generally thought to be a purely chemical process, proceeding independently of any living organisms present.

This work had many consequences, the most immediate being his discovery of anaerobic life and his development of the method of pasteurization. A long-range consequence was that it strongly suggested the germ theory of disease. (See also 1807)

1806. POUCHET, F.A. (2145) *Hétérogénie; ou, Traité de la génération spontanée, basé sur de nouvelles expériences*. Paris, 1859. A large and systematic treatise which discussed earlier work in the subject and presented an account of his numerous experiments concerning what he believed to be the requisite factors for spontaneous generation. The book aroused much interest among French scientists and prompted Pasteur to undertake his famous experiments. In 1863-64 Pouchet, with others, made vigorous attempts to refute Pasteur's conclusions.

1860s

1807. PASTEUR, L. (See also 1805) **(a)** His establishment of the germ theory of fermentation led him to oppose the idea of spontaneous generation which had been brought to the fore by Pouchet's book. In a series of celebrated papers in 1860-61 he presented accounts of skilfully designed experiments which constituted a massive challenge to the idea. There followed in 1863-64 a dispute between Pasteur and Pouchet from which the latter finally withdrew.

(b) In 1865 Pasteur was asked by the French government to investigate diseases of the silkworm which were causing havoc in the silk industry. He worked at the problem for several years and though he did not succeed in isolating the causative organism he was able to recommend effective countermeasures. Apart from its major economic importance the investigation gave support to the germ theory of disease and set the scene for his subsequent work on animal diseases.

1808. DAVAINE, C.J. (2159) In 1863, stimulated by Pasteur's work on fermentation, he began a series of investigations of the "filiform bodies" (bacilli) which he and others had previously observed in the blood of animals suffering from anthrax. Over the next ten years he accumulated much solid evidence in support of the view that the disease was caused by the microorganisms, but he was unable to answer all the objections raised (because he did not know that anthrax bacteria can form highly resistant spores). Nevertheless he is to be regarded as one of the main founders of the germ theory of disease. Pasteur acknowledged the stimulus he had received from his work.

1870s

1809. COHN, F.J. (2649) Between 1872 and 1876 he published four papers under the general title, "Untersuchungen über Bacterien", which can be regarded as the beginning of bacteriology as a distinct field (not yet of medical importance). He established that bacteria can be classified into genera and species, thus bringing order into a hitherto chaotic field, and investigated their nutrition and other characteristics. One of his most significant discoveries was that some of them can form spores that are highly resistant to heat.

1810. TYNDALL, J. (2411) As an offshoot of his physical experiments on the scattering of light by fine dust (see 719) he constructed a special cabinet in which the air could be made free of dust, and showed in 1876 that infusions exposed to the dust-free air remained unaffected even for months. In another brilliant investigation in the following year he proved that bacteria can have phases, one in which they are easily destroyed by heat and another in which they are remarkably resistant to it (due to spore formation). He presented an illuminating account of his work in the field in his influential book *Essays on the floating matter of the air in relation to putrefaction and infection* (London, 1881). "Tyndall's researches gave the final blow to the doctrine of spontaneous generation as much if not more than those of Pasteur." (W. Bulloch)

1811. KOCH, R. (2726) **(a)** Following the researches of Davaine, and Cohn's discovery of the ability of some bacteria to form resistant spores, he succeeded in 1876 in elucidating the complete life-history of the anthrax bacillus. His work provided the first clear proof of a particular disease being caused by a particular micro-organism. **(b)** In his second paper, in 1877, he greatly improved the methods of staining bacteria (using the new synthetic dyes), devised various microscopic techniques, and introduced microphotography into bacteriology. **(c)** In a monograph of 1878 he described his experiments on the aetiology of wound infection in animals. (See also 1813)

1812. PASTEUR, L. (See also 1807) Initiated a series of researches on animal diseases in 1877 which was to occupy him until his death in 1895. He began with a study of the cause of anthrax but presumably as a result of the work of Koch he turned towards investigations of how infective disease is produced in animals, a direction which led to his famous discoveries in immunology during the 1880s.

1880s

1813. KOCH, R. (See also 1811) Following his appointment in 1880 to the Imperial Department of Health in Berlin, he and his associates published highly important papers on bacteriological techniques (especially methods of obtaining pure cultures using solid media such as gelatine, etc.). In 1882 he made the great discovery of the elusive tubercle bacillus and later

succeeded in the difficult task of growing it in pure culture.
Sent to Egypt and then to India in 1883 as a member of the
German Cholera Commission, he discovered the cholera vibrio
and elucidated its modes of transmission. He then resumed
his studies of tuberculosis but his attempts to produce a
curative agent were unsuccessful.   (See also 1818)

1814. GRAM STAINING METHOD. Discovered in 1884 by the Danish bacteri-
      ologist, H.C.J. Gram. It proved to be of great value in the
      classification and identification of bacteria.

1815. DE BARY, A. (2659) The methods and concepts developed in his
      important work on mycology (1384) had considerable relevance
      to bacteriology. He surveyed the field (then still regarded
      as part of botany) in his *Vorlesungen über Bacterien* of 1885.

1816. VINOGRADSKY, S.N. (3009) **(a)** From 1885 to 1889 he studied the
      morphology and physiology of sulphur bacteria, showing that
      they obtain energy by oxidizing hydrogen sulphide to sulphur
      --a process he compared with respiration. **(b)** During 1889-90
      he discovered bacteria capable of oxidizing ammonium salts to
      nitrites, and others capable of oxidizing nitrites to nitrates.
      A few years later he discovered the anaerobe *Clostridium* that
      is able to fix nitrogen from the atmosphere (cf. 1817).

1817. BEIJERINCK, M.W. (2799) Isolated in 1888 the *Rhizobium* bacillus
      which exists in nodules on the roots of leguminaceous plants
      and fixes atmospheric nitrogen. In the course of his studies
      on soil bacteria he developed (simultaneously with Vinogradsky)
      the method of enrichment culture whereby he was able to isolate
      many new kinds of bacteria--denitrifying, sulphate-reducing,
      urea-decomposing, etc. He also discovered another group of
      nitrogen-fixing bacteria, *Azotobacter*.   (See also 1819)

1890s

1818. KOCH, R. (See also 1813) Through the 1890s he and his student-
      associates continued to investigate many diseases of men and
      animals. He also led various expeditions to study major
      tropical diseases. His many outstanding disciples included
      G. Gaffky, F. Loeffler, S. Kitasato, R. Pfeiffer, E. von
      Behring, and P. Ehrlich.

1819. BEIJERINCK, M.W. (See also 1817) During the 1890s, concurrently
      with the Russian botanist, D.I. Ivanovsky, he studied the
      causative agent of mosaic disease in tobacco plants. Like
      Ivanovsky, he found that the sap of infected plants retained
      its infectious activity even after all microscopically-visible
      organisms had been filtered out. From a study of its proper-
      ties he concluded that the causative agent was of a quite new
      type (later called a virus).

SYNCHRONIC SUMMARY OF PART 1

The first column of figures gives the number of entries for each of
the sciences in the decade. The second column gives the entry numbers
(the first for the decade when there are more than one).

1450s

Astronomy	2	317+	Physics	1	567

Total entries: 3

1460s

Mathematics	1	68	Natural History	1	1217
Astronomy	1	319			

Total entries: 3

1470s

Science in General	1	1	Earth Sci. in General	2	998+
Mathematics	1	69	Botany	2	1276+
Astronomy	1	320	Zoology	1	1410
Mechanics	1	490	Human Anatomy	2	1552+
Chemistry	1	766	Physiology	1	1707

Total entries: 13

1480s

Mathematics	2	70+	Physics	1	568
Mechanics	1	491	Botany	2	1278+

Total entries: 6

1490s

Mathematics	1	72	Earth Sci. in General	1	1000
Astronomy	1	321	Human Anatomy	1	1554

Total entries: 4

1500s

Science in General	1	2	Chemistry	1	767
Mathematics	3	73+	Earth Sci. in General	1	1001
Mechanics	1	492	Human Anatomy	2	1555+

Total entries: 9

1510s

Mathematics	4	76+	Earth Sci. in General	1	1002
Astronomy	2	322+	Human Anatomy	1	1557

Total entries: 8

1520s

Mathematics	4	80+	Earth Sci. in General	1	1003
Chemistry	2	768+	Human Anatomy	1	1558

Total entries : 8

1530s

Mathematics	4	84+	Earth Sci. in General	1	1004
Astronomy	3	324+	Botany	5	1280+
Mechanics	1	493	Human Anatomy	2	1559+
Physics	1	569			

Total entries: 17

1540s

Mathematics	7	88+	Botany	3	1285+
Astronomy	4	327+	Zoology	2	1411+
Physics	1	570	Human Anatomy	4	1561+
Chemistry	2	770+	Physiology	1	1708
Earth Sci. in General	2	1005+			

Total entries: 26

1550s

Science in General	2	3+	Natural History	1	1218
Mathematics	4	95+	Botany	4	1288+
Astronomy	2	331+	Zoology	5	1413+
Mechanics	1	494	Human Anatomy	2	1565+
Chemistry	1	772	Embryology	1	1650
Earth Sci. in General	1	1007	Physiology	1	1709

Total entries: 25

1560s

Science in General	1	5	Natural History	1	1219
Astronomy	1	333	Botany	3	1292+
Chemistry	2	773+	Human Anatomy	3	1567+
Earth Sci. in General	1	1008	Embryology	2	1651+

Total entries: 14

1570s

Science in General	2	6+	Earth Sci. in General	1	1009
Mathematics	6	99+	Natural History	3	1220+
Astronomy	3	334+	Botany	2	1295+
Mechanics	3	495+	Human Anatomy	4	1570+
Physics	4	571+	Embryology	1	1653
Chemistry	3	775+			

Total entries: 32

1580s

Science in General	1	8	Physics	1	575
Mathematics	3	105+	Botany	4	1297+
Astronomy	5	337+	Human Anatomy	2	1574+
Mechanics	2	498+	Physiology	1	1710

Total entries: 19

1590s

Science in General	1	9	Earth Sci. in General	1	1010
Mathematics	4	108+	Natural History	1	1223
Astronomy	3	342+	Botany	3	1301+
Mechanics	1	500	Zoology	1	1418
Chemistry	1	778			

Total entries: 16

1600s

Mathematics	5	112+	Botany	2	1304+
Astronomy	3	345+	Zoology	1	1419
Physics	3	576+	Human Anatomy	4	1576+
Chemistry	3	779+	Embryology	1	1654
Earth Sci. in General	1	1011			

Total entries: 23

1610s

Science in General	1	10	Chemistry	3	782+
Mathematics	4	117+	Botany	1	1306
Astronomy	6	348+	Human Anatomy	1	1580
Mechanics	1	501	Physiology	1	1711
Physics	2	579+			

Total entries: 20

1620s

Science in General	5	11+	Earth Sci. in General	1	1012
Mathematics	4	121+	Microscopy	1	1256
Astronomy	1	354	Botany	2	1307+
Mechanics	3	502+	Human Anatomy	2	1581+
Physics	1	581	Physiology	1	1712
Chemistry	1	785			

Total entries: 22

1630s

Science in General	2	16+	Earth Sci. in General	1	1013
Mathematics	11	125+	Natural History	2	1224+
Astronomy	5	355+	Botany	1	1309
Mechanics	3	505+	Zoology	1	1420
Physics	2	582+	Physiology	1	1713
Chemistry	2	786+			

Total entries: 31

1640s

Science in General	2	18+	Earth Sci. in General	1	1014
Mathematics	4	136+	Natural History	1	1226
Astronomy	2	360+	Zoology	1	1421
Mechanics	1	508	Human Anatomy	3	1583+
Physics	3	584+	Physiology	1	1714
Chemistry	3	788+			

Total entries: 22

1650s

Science in General	1	20	Natural History		2	1227+
Mathematics	8	140+	Botany		1	1310
Astronomy	3	362+	Zoology		1	1422
Mechanics	1	509	Human Anatomy		7	1586+
Physics	1	587	Embryology		2	1655+
Chemistry	1	791	Physiology		2	1715+
Earth Sci. in General	2	1015+				

Total entries: 32

1660s

Science in General	1	21	Natural History		1	1229
Mathematics	7	148+	Microscopy		1	1257
Astronomy	8	365+	Botany		2	1311+
Mechanics	4	510+	Zoology		3	1423+
Physics	5	588+	Human Anatomy		6	1593+
Chemistry	6	792+	Physiology		3	1717+
Earth Sci. in General	1	1017				

Total entries: 48

1670s

Science in General	1	22	Natural History		1	1230
Mathematics	6	155+	Microscopy		1	1258
Astronomy	8	373+	Botany		4	1313+
Mechanics	3	514+	Zoology		3	1426+
Physics	3	593+	Human Anatomy		4	1599+
Chemistry	3	798+	Embryology		1	1657
Earth Sci. in General	2	1018+	Physiology		2	1720+

Total entries: 42

1680s

Science in General	1	23	Earth Sci. in General		3	1020+
Mathematics	6	161+	Botany		4	1317+
Astronomy	2	381+	Zoology		5	1429+
Mechanics	3	517+	Human Anatomy		4	1603+
Physics	2	596+	Physiology		1	1722

Total entries: 31

1690s

Science in General	1	24	Earth Sci. in General		4	1023+
Mathematics	3	167+	Microscopy		1	1259
Astronomy	1	383	Botany		4	1321+
Mechanics	4	520+	Zoology		2	1434+
Physics	2	598+	Human Anatomy		1	1607
Chemistry	2	801+	Physiology		1	1723

Total entries: 26

1700s

Science in General	1	25	Chemistry		2	803+
Mathematics	2	170+	Earth Sci. in General		2	1027+

Astronomy	3	384+	Zoology	1	1436
Mechanics	2	524+	Human Anatomy	3	1608+
Physics	3	600+	Physiology	2	1724+

Total entries: 21

1710s

Science in General	1	26	Microscopy	1	1260
Mathematics	5	172+	Botany	2	1325+
Astronomy	2	387+	Zoology	1	1437
Mechanics	1	526	Human Anatomy	1	1611
Physics	1	603	Physiology	2	1716+
Chemistry	1	805			

Total entries: 18

1720s

Mathematics	1	177	Earth Sci. in General	3	1029+
Astronomy	4	389+	Botany	2	1327+
Mechanics	4	527+	Zoology	1	1438
Physics	2	604+	Human Anatomy	2	1612+
Chemistry	2	806+			

Total entries: 21

1730s

Mathematics	2	178+	Natural History	2	1231+
Mechanics	2	531+	Botany	2	1329+
Physics	3	606+	Zoology	4	1439+
Chemistry	4	808+	Human Anatomy	1	1614
Earth Sci. in General	2	1032+	Physiology	1	1728
Geology	1	1111			

Total entries: 24

1740s

Science in General	1	27	Geology	2	1112+
Mathematics	3	180+	Natural History	2	1233+
Astronomy	3	393+	Microscopy	2	1261+
Mechanics	3	533+	Botany	1	1331
Physics	4	609+	Zoology	3	1443+
Chemistry	1	812	Human Anatomy	2	1615+
Earth Sci. in General	2	1034+	Embryology	2	1658+
Mineral. & Crystal.	1	1081	Physiology	1	1729

Total entries: 33

1750s

Science in General	2	28+	Mineral. & Crystal.	1	1082
Mathematics	3	183+	Geology	1	1114
Astronomy	7	396+	Botany	2	1332+
Mechanics	2	536+	Zoology	1	1446
Physics	4	613+	Human Anatomy	2	1617+
Chemistry	4	813+	Embryology	1	1660
Earth Sci. in General	2	1036+	Physiology	2	1730+

Total entries: 34

1760s

Science in General	1	30	Natural History	2	1235+
Mathematics	5	186+	Botany	2	1334+
Astronomy	6	403+	Zoology	2	1447+
Mechanics	3	538+	Human Anatomy	2	1619+
Physics	5	617+	Embryology	1	1661
Chemistry	2	817+	Physiology	1	1732
Geology	2	1115+			

Total entries: 34

1770s

Science in General	1	31	Mineral. & Crystal.	1	1083
Mathematics	2	191+	Geology	5	1117+
Astronomy	4	409+	Natural History	3	1237+
Mechanics	1	541	Botany	1	1336
Physics	4	622+	Zoology	2	1449+
Chemistry	6	819+	Human Anatomy	1	1621
Earth Sci. in General	3	1038+	Physiology	2	1733+

Total entries: 36

1780s

Science in General	1	32	Mineral. & Crystal.	4	1084+
Mathematics	3	193+	Geology	3	1122+
Astronomy	4	413+	Botany	6	1337+
Mechanics	3	542+	Zoology	6	1451+
Physics	5	626+	Human Anatomy	5	1622+
Chemistry	13	825+	Embryology	1	1662
Earth Sci. in General	1	1041	Physiology	3	1735+

Total entries: 58

1790s

Mathematics	6	196+	Geology	4	1125+
Astronomy	2	417+	Natural History	1	1240
Mechanics	1	545	Microscopy	1	1263
Physics	4	631+	Botany	1	1343
Chemistry	5	838+	Zoology	2	1457+
Earth Sci. in General	2	1042+	Human Anatomy	2	1627+
Mineral. & Crystal.	1	1088	Physiology	2	1738+

Total entries: 34

1800s

Science in General	1	33	Geology	7	1129+
Mathematics	5	202+	Natural History	2	1241+
Astronomy	4	419+	Botany	6	1344+
Mechanics	1	546	Zoology	5	1459+
Physics	6	635+	Evolution	1	1522
Chemistry	17	843+	Human Anatomy	1	1629
Earth Sci. in General	1	1044	Physiology	1	1740
Mineral. & Crystal.	2	1089+			

Total entries: 60

1810s

Science in General	1	34	Geology	7	1136+
Mathematics	5	207+	Natural History	1	1243
Astronomy	4	423+	Botany	4	1350+
Mechanics	1	547	Zoology	6	1464+
Physics	8	641+	Human Anatomy	2	1630+
Chemistry	12	860+	Embryology	1	1663
Mineral. & Crystal.	3	1091+	Physiology	2	1741+

Total entries: 57

1820s

Science in General	1	35	Geology	12	1143+
Mathematics	12	212+	Natural History	2	1244+
Astronomy	4	427+	Microscopy	1	1264
Mechanics	4	548+	Botany	6	1354+
Physics	15	649+	Zoology	5	1470+
Chemistry	9	872+	Embryology	4	1664+
Earth Sci. in General	3	1045+	Physiology	5	1743+
Mineral. & Crystal.	2	1094+			

Total entries: 85

1830s

Science in General	5	36+	Natural History	3	1246+
Mathematics	12	224+	Microscopy	4	1265+
Astronomy	8	431+	Botany	5	1360+
Mechanics	3	552+	Zoology	6	1475+
Physics	11	664+	Human Anatomy	3	1632+
Chemistry	10	881+	Embryology	1	1668
Earth Sci. in General	4	1048+	Cell Theory	2	1687+
Mineral. & Crystal.	2	1096+	Physiology	7	1748+
Geology	11	1155+	Microbiology	2	1802+

Total entries: 99

1840s

Science in General	3	41+	Microscopy	1	1269
Mathematics	15	236+	Botany	9	1365+
Astronomy	7	439+	Zoology	7	1481+
Mechanics	2	555+	Evolution	1	1523
Physics	16	675+	Human Anatomy	3	1635+
Chemistry	14	891+	Embryology	2	1669+
Earth Sci. in General	5	1052+	Cell Theory	1	1689
Mineral. & Crystal.	3	1098+	Physiology	9	1755+
Geology	11	1166+	Microbiology	1	1804
Natural History	2	1249+			

Total entries: 112

1850s

Science in General	4	44+	Microscopy	2	1270+
Mathematics	14	251+	Botany	7	1374+
Astronomy	5	446+	Zoology	10	1487+

1850s (cont'd)

Mechanics	2	557+	Evolution		3	1524+
Physics	14	691+	Human Anatomy		1	1638
Chemistry	15	905+	Embryology		1	1671
Earth Sci. in General	7	1057+	Cell Theory		2	1690+
Mineral. & Crystal.	1	1101	Physiology		7	1764+
Geology	9	1177+	Microbiology		2	1805+
Natural History	2	1251+				

Total entries: 108

1860s

Science in General	7	48+	Microscopy		2	1272+
Mathematics	11	265+	Botany		8	1381+
Astronomy	10	451+	Zoology		8	1497+
Mechanics	1	559	Evolution & Heredity		9	1527+
Physics	10	705+	Human Anatomy		1	1639
Chemistry	23	920+	Embryology		4	1672+
Earth Sci. in General	4	1064+	Cytology		1	1692
Mineral. & Crystal.	1	1102	Physiology		6	1771+
Geology	4	1185+	Microbiology		2	1807+
Natural History	1	1253				

Total entries: 113

1870s

Science in General	6	55+	Microscopy		1	1274
Mathematics	15	276+	Botany		9	1389+
Astronomy	9	461+	Zoology		7	1505+
Mechanics	2	560+	Evolution & Heredity		4	1536+
Physics	18	715+	Human Anatomy		5	1640+
Chemistry	15	943+	Embryology		5	1676+
Earth Sci. in General	4	1068+	Cytology		3	1693+
Mineral. & Crystal.	2	1103+	Physiology		11	1777+
Geology	12	1189+	Microbiology		4	1809+
Natural History	1	1254				

Total entries: 133

1880s

Science in General	4	61+	Microscopy		1	1275
Mathematics	13	291+	Botany		7	1398+
Astronomy	14	470+	Zoology		6	1512+
Mechanics	3	562+	Evolution & Heredity		3	1540+
Physics	15	733+	Human Anatomy		2	1645+
Chemistry	18	958+	Embryology		4	1681+
Earth Sci. in General	7	1072+	Cytology		7	1696+
Mineral. & Crystal.	2	1105+	Physiology		6	1788+
Geology	10	1201+	Microbiology		5	1813+
Natural History	1	1255				

Total entries: 128

1890s

Science in General	3	65+	Botany		5	1405+
Mathematics	13	304+	Zoology		4	1518+

Astronomy	6	484+	Evolution & Heredity	9 1543+
Mechanics	2	565+	Human Anatomy	3 1647+
Physics	18	748+	Embryology	2 1685+
Chemistry	22	976+	Cytology	4 1703+
Earth Sci. in General	2	1079+	Physiology	8 1794+
Mineral. & Crystal.	4	1107+	Microbiology	2 1818+
Geology	6	1211+		

Total entries: 113

PART 2

THE SOCIAL DIMENSION

Arrangement by Countries or Regions

Persons are placed in the chronological arrangement according to the
decade in which each began his career. Many of the career sketches
are limited by lack of biographical information, especially in the
earlier periods. To avoid excessive repetition, the term 'university'
is used only where necessary; generally the names of cities are used
to designate the universities in those cities.

2.01 ITALY

1450s

1820. BIANCHINI, Giovanni. fl. 1450-64. (Astronomy 317) The leading
    Italian astronomer of the period. He taught at Ferrara during
    the 1450s and 1460s. When Peuerbach was in Italy in the early
    1450s he visited Bianchini, as did Regiomontanus in the 1460s.

1821. MARLIANI, Giovanni. died 1483. (Physics 567) M.D., Pavia, ca.
    1440. Professor of medicine at Pavia for many years and
    finally court physician to the Duke of Milan.

1460s

1822. LEONICENO, Nicolò. 1428-1524. (Not in Part 1) M.D., Padua,
    1453. From 1464 he taught various subjects, finally medicine,
    at Ferrara. A leading Greek scholar, he was one of the chief
    pioneers in the movement to establish accurate texts of the
    ancient Greek physicians, freed from Arabic and medieval
    corruptions, and to provide faithful Latin translations. His
    work was done in the late fifteenth century and his editions
    of Galen and Hippocrates were published during the 1510s, near
    the end of his long life.

1470s

1823. PACIOLI, Luca. ca. 1445-1517. (Mathematics 72, 75) Franciscan.
    Teacher of mathematics in various cities, at several univers-
    ities for short periods, and at the ducal courts of Urbino and
    Milan.

1480s

1824. LEONARDO DA VINCI. 1452-1519. (Mechanics 491. Anatomy 1555)
    The celebrated painter, sculptor, engineer, and architect.
    He received only an elementary education (no Latin) and an
    apprentice's training; otherwise he was self-educated.

1825. BENEDETTI, Alessandro. ca. 1450-1512. (Anatomy 1556) Studied
    and taught at Padua and later practised medicine at Venice.

1490s

1826. FERRO (or FERREO), Scipione. 1465-1526. (Mathematics 73)
    Lecturer in arithmetic and geometry at Bologna from 1496
    until his death.

1500s

1827. BERENGARIO DA CARPI, Giacomo. ca. 1460-1530. (Anatomy 1557,
    1558) Graduated in medicine at Bologna in 1489, then practised

as a surgeon. From 1502 he was lecturer in surgery and anatomy at Bologna. He was very successful both as a teacher and as a surgeon.

1520s

1828. TARTAGLIA, Niccolò. 1500–1557. (Mathematics 92, 98. Mechanics 493) Brought up in poverty and almost entirely self-educated. From about 1520 he was a private teacher of mathematics in various cities, chiefly Venice.

1829. GHINI, Luca. ca. 1490–1556. (Botany 1281, 1286) Studied medicine at Bologna and in 1527 was appointed a lecturer there in practical medicine. For his subsequent career see 1281 and 1286.

1830. MASSA, Niccolò. 1485–1569. (Anatomy 1559) Graduated in medicine at Padua and practised as a physician in Venice.

1530s

1831. CARDANO, Girolamo. 1501–1576. (Mathematics 87, 90. Mechanics 495. Encyclopaedism 3) Studied at Pavia and Padua, taking his M.D. at the latter in 1526, and eventually became an outstanding physician with a European-wide reputation. He was appointed professor of medicine at Pavia in 1543 and then, after some political difficulties, at Bologna in 1562.

1832. MATTIOLI, Pietro Andrea. 1501–1577. (Botany 1289, 1296) Graduated in medicine at Padua in 1523. He was a physician in several Italian cities and, from 1554 to 1570, at the court of the Emperor in Prague.

1833. VESALIUS, Andreas. 1514–1564. (Anatomy 1563) A native of Brussels. After studying at Louvain and Paris he gained the M.D. at Padua in 1537 and was thereupon appointed lecturer in surgery and anatomy. In 1543, after the publication of his great book, he became physician to the Emperor Charles V and later to Philip II of Spain.

1540s

1834. FERRARI, Ludovico. 1522–1565. (Mathematics 91) As a youth in humble circumstances he was employed by Cardano as his amanuensis. He owed much of his education, especially in mathematics, to Cardano. From 1540 he was a private teacher of mathematics in Milan but he continued to collaborate with Cardano.

1835. MAUROLICO, Francesco. 1494–1575. (Mathematics 103. Physics 574. Cosmography 1005) A Benedictine monk in Sicily. From the 1550s, besides his original work, he published editions and translations of several Greek mathematical and astronomical works as well as various commentaries. In 1569 he was appointed professor of mathematics at the University of Messina.

1836. BOTANICAL GARDENS, PISA AND PADUA. The first botanical gardens. They were established at almost the same time in the mid-1540s. For the early directors see 1286, 1837, 1847, 1848.

1837. ANGUILLARA, Luigi. ca. 1512–1570. (Botany 1293) From 1539 to 1544 he was a disciple of Ghini, first at Bologna and then at

Pisa. In 1546 he was appointed the first director of the botanical garden at Padua and held the post with distinction until 1561 when he gave it up because of a quarrel. He then became herbalist to the Duke of Ferrara.

1838. EUSTACHI, Bartolomeo. ca. 1505-1574. (Anatomy 1569) Apparently he studied medicine at the Archiginnasio della Sapienza (the Roman university of the Renaissance period). He began to practise about 1540 and some ten years later became professor at the Sapienza.

1839. GUIDI, Guido. 1508-1569. (Anatomy 1567) After practising medicine in Rome and Florence he went to Paris in 1542 and became physician to François I and the first professor of medicine at the new Collège Royal. In 1547, following the death of François, he returned to Italy and became professor of philosophy and medicine at Pisa.

1840. COLOMBO, Realdo. ca. 1510-1559. (Anatomy 1566) After serving an apprenticeship to a surgeon for seven years he studied medicine at Padua (anatomy under Vesalius), graduating in 1541. He became professor of anatomy and surgery at Padua in 1544, then at Pisa from 1545, and finally at the Archiginnasio della Sapienza in Rome from 1548.

1841. CANANO, Giovan Battista. 1515-1579. (Anatomy 1561) Graduated in medicine at Ferrara in 1543 and was lecturer there until 1552. He then became physician to the Pope and subsequently returned to Ferrara as chief physician at the ducal court.

1842. VALVERDE, Juan de. ca. 1520-1588. (Anatomy 1565) Born and educated in Spain, then studied at Padua under Vesalius and Colombo. From 1545 he was Colombo's assistant at Pisa. About 1548 he settled in Rome where he became a prominent physician.

1550s

1843. COMMANDINO, Federico. 1509-1575. (No main entry in Part 1) The most outstanding editor and translator of the classics of Greek mathematics and astronomy. He studied at Padua and Ferrara, taking a medical degree at the latter. After practising medicine for some years he was enabled to devote himself fully to scholarly activities by the patronage of a Roman cardinal and later the Duke of Urbino. At Urbino he also taught mathematics and had some important pupils (cf. 1860, 1864). His chief work was done from the late 1550s up to his death and included valuable editions and translations of works by Appolonius, Archimedes, Euclid, Ptolemy, and others.

1844. BOMBELLI, Rafael. 1526-1572. (Mathematics 101) An engineer-architect. He learnt his profession by apprenticeship.

1845. BENEDETTI, Giovanni Battista. 1530-1590. (Mechanics 494) Came from a family of high social status; educated informally but well. From 1558 until his death he was court mathematician, first to the Duke of Parma and later to the Duke of Savoy. His duties included some teaching of mathematics and science to members of the court or their families, construction of sundials, etc., and especially the provision of engineering and astrological advice.

1846. ALDROVANDI, Ulisse. 1522-1605. (Natural history 1218. Embryology 1651) After various studies, in the course of which he worked briefly with Rondelet in Rome and attended Ghini's lectures at Pisa, he took his medical degree at Bologna in 1553. He was appointed to lecture on "simples" (i.e. medicinal herbs) and extended his lecture course--which was very successful--to cover animals and minerals as well as plants. He thus gave an impetus to the systematic study of natural history generally. He was also the founder and first director of the botanical garden at Bologna. He had the support of the Pope who liberally financed the publication of his works.

1847. CESALPINO, Andrea. 1519-1603. (Botany 1298. Natural philosophy 6) M.D., Pisa, 1551. He was a pupil of Ghini who had a big influence on him, and in 1555 he succeeded Ghini as professor of medicine and director of the botanic garden at Pisa. In 1592 he went to Rome as papal physician and professor at the Sapienza.

1848. WIELAND (or GUILANDINO), Melchior. ca. 1520-1589. (Not in Part 1. His only scientific writings were in the form of correspondence) Born and educated in Germany. Studied at the University of Königsberg and then in Rome. Travelled widely and acquired a considerable reputation as a scholar. In 1561 he was appointed director of the botanic garden at Padua (in succession to Anguillara); his duties also included lecturing in medical botany. During the 1560s and 1570s he was outstanding as director of the garden and introduced many new plants and other improvements.

1849. SALVIANI, Ippolito. 1514-1572. (Zoology 1417) Professor of medicine at the Sapienza in Rome from 1551 and personal physician to several popes.

1850. FALLOPPIO, Gabriele. 1523(?)-1562. (Anatomy 1568) Studied medicine at Modena, Ferrara, and perhaps Padua. Became professor of anatomy at Pisa from 1549 and then at Padua from 1551. His teaching was very successful and attracted a number of outstanding students. He taught botany as well as anatomy and was in charge of the botanic garden at Padua.

1851. ARANZIO, Giulio Cesare. ca. 1530-1589. (Anatomy 1573. Embryology 1652) Studied medicine at Padua and Bologna, receiving the M.D. at the latter in 1556. He became lecturer in surgery at Bologna soon afterwards, and later professor of anatomy and surgery.

1852. PICCOLOMINI, Arcangelo. 1525-1586. (Physiology 1710) Graduated in medicine at Ferrara, probably in the late 1540s. He then went to Rome where he became physician to several popes and professor at the Sapienza.

1560s

1853. ZABARELLA, Jacopo. 1533-1589. (Natural philosophy 9) Graduated in philosophy at Padua in 1553. Became professor of logic and then of natural philosophy at Padua from 1564 until his death.

1854. CLAVIUS, Christoph. 1537-1612. (Mathematics 102, 105, 114. Astronomy 339) A native of Germany. Entered the Jesuit order in Rome in 1555. Professor of mathematics at the Collegio Romano from 1565 to his death. A life-long friend of Galileo despite his anti-Copernicanism. The Gregorian calendar, introduced by Pope Gregory XIII in 1582, was based on his recommendations.

1855. MOLETI, Giuseppe. 1531-1588. (Astronomy 337) Professor of mathematics at Padua from 1564 to 1584.

1856. BOTANICAL GARDEN, BOLOGNA. Established at the University by Aldrovandi in 1568.

1857. FABRICI, Girolamo (or FABRICIUS AB AQUAPENDENTE, Hieronymus). ca. 1533-1619. (Anatomy 1576. Embryology 1654) Graduated in medicine at Padua, ca. 1559. After practising surgery and giving private anatomy lessons for a few years he was appointed professor of surgery and anatomy at Padua in 1565. He held the post for fifty years and became an outstanding member of the University; one of his main contributions to it was the construction of the famous anatomical theatre in 1594. He is regarded as one of the greatest teachers of anatomy; Harvey was one of his students and gained much from him.

1858. VAROLIO, Costanzo. 1543-1575. (Anatomy 1571) M.D., Bologna, 1567. From 1569 to 1572 he was *professor extraordinarius* in surgery at Bologna. He then went to Rome, apparently in the service of the Pope.

1570s

1859. CATALDI, Pietro Antonio. 1552-1626. (Mathematics 117) Professor of mathematics at Perugia from 1572 and then, from 1584, at Bologna. Author of many textbooks and tracts on various aspects of mathematics.

1860. MONTE, Guidobaldo, *Marchese* del. 1545-1607. (Mechanics 496) A noble of Urbino. Studied at the University of Pisa in the 1560s. Subsequently, at Urbino, he was a pupil and friend of Commandino who was attached to the ducal court there until his death in 1575. In 1588 del Monte became impressed by the ability of the young Galileo whom he befriended and helped to obtain the chair at Pisa in the following year. Not only was he a valuable friend of Galileo until his death but he is now thought to have been a major influence on Galileo's ideas on mechanics.

1861. PORTA, Giambattista della. 1535-1615. (Natural magic 8) A younger son of a noble family in Naples; educated informally. Founder of one of the earliest scientific academies, the short-lived Accademia dei Segreti (or Academia Secretorum Naturae). It began some time before 1580 and was modelled on the numerous literary academies which were a prominent feature of Renaissance humanist culture. The academy consisted of Porta and several of his friends who gathered periodically at his house in Naples to study and discuss "natural curiosities." It was closed after a few years, apparently by order of the Inquisition.

In 1610 Porta became a member of the more solid Accademia dei Lincei (1872) which seems to have been inspired, at least to some extent, by Porta's earlier academy.

1580s

1862. VALERIO, Luca. 1552–1618. (Mathematics 115) Educated at the
    Jesuits' Collegio Romano where Clavius was his teacher in
    mathematics. For most of his life he taught various subjects
    in Rome, including mathematics at the Sapienza. He was a
    member of the Accademia dei Lincei and a friend of Galileo.

1863. GALILEI, Galileo. 1564–1642. (Mechanics 501, 506, 507. Astron-
    omy 348, 356. Philosophy of science 12, 16) Began the study
    of medicine at Pisa in 1581 but soon gave it up for mathematics
    which he studied under a private teacher. Subsequently, for
    a few years Galileo was himself a private teacher of mathematics
    in Florence, but in 1589 he became professor of the subject at
    Pisa. In 1592 he gained the much better chair at Padua where
    he remained until 1610. In that year he became mathematician
    and philosopher to the Grand Duke of Tuscany, a position he
    retained until his condemnation in 1633 after which he was
    confined to his home.

    Besides his unparalleled importance for the cognitive
    history of science, Galileo had a considerable influence on
    its social history, at least in Italy. Generations of Italian
    scientists were inspired by his personal example and, more
    directly, the continued development of science in Florence was
    a consequence of his activities at the court of the Medici.
    The Grand Duke, Ferdinand II (acceded 1628), and his younger
    brother, Prince Leopold, had both in their youth acquired a
    love of science from Galileo, and Florence possessed the best
    collection of physical instruments in existence, as well as
    gardens of rare plants and a remarkable menagerie. The estab-
    lishment of the Medici's academy—the Accademia del Cimento
    (1891)—in 1657 was only the formalization of the group of
    Galileo's disciples who had been led originally by Torricelli
    and, after his death in 1647, by Viviani. And the Accademia
    del Cimento (which appears to have served as a model for the
    later Paris Academy) was inspired in all its activities by
    the ideas and example of Galileo.

1864. BALDI, Bernardino. 1553–1617. (Mechanics 502) In company with
    Guidobaldo del Monte (1860) he studied mathematics under
    Commandino at Urbino in the early 1570s. Commandino's inspir-
    ation led him to make some translations of Greek scientific
    texts. From 1580 he served the Duke of Mantua and subsequently
    the Duke of Urbino as historian and biographer.

1865. ALPINI, Prospero. 1553–1616. (Botany 1301, 1306) In 1578 he
    graduated in medicine at Padua where he studied under Wieland
    (or Guilandino), the director of the botanical garden. After
    a three-year botanical expedition to Egypt he was appointed
    lecturer in "simples" at Padua in 1594. Nine years later he
    became director of the botanical garden, and after his death
    was succeeded in the post by his son. He wrote many books on
    botany and medicine.

1866. CASSERI, Giulio. ca. 1552–1616. (Anatomy 1578) After graduating
    in medicine at Padua in 1580 he became a surgeon in the town
    and also taught at the University. He declined offers of chairs

at Parma and Turin because he hoped to succeed his former teacher, Fabrizio, at Padua but he died before he could do so.

1867. SANTORIO, Santorio (or SANCTORIUS). 1561-1636. (Physiology 1711) M.D., Padua, 1582. Physician in his native Croatia and then in Venice. In 1611 he was appointed professor of theoretical medicine at Padua, a post he held with distinction until his retirement in 1624.

1590s

1868. CAMPANELLA, Tommaso. 1568-1639. (Natural philosophy, etc. 13) Dominican. Around 1590 he was active in Porta's academy (1861) in Naples and in 1593 he was briefly in Padua where he met Galileo. Because of his theological and political ideas he spent much of his life in prison.

1869. GHETALDI, Marino. 1566-1626. (Mathematics 113) Scion of a patrician family. Travelled widely. In Rome he was associated with Clavius and in Paris with Viète. He declined an offer of a chair at Louvain.

1870. BONAVENTURA, Federigo. 1555-1602. (Meteorology 1010) Educated informally. Studied Greek mathematics and natural philosophy at the court of the Duke of Urbino and later served as the Duke's ambassador at several European courts.

1871. RUINI, Carlo. ca. 1530-1598. (Zoology 1418) A member of the Bolognese aristocracy and a prominent lawyer and senator.

1600s

1872. ACCADEMIA DEI LINCEI. The first formally-organized scientific society. Founded in Rome in 1603 by Federico Cesi, a young noble. Initially it consisted of only four members but the number increased eventually to thirty-two; Galileo became a member in 1611. The Academy was financed and run by Cesi and for nearly thirty years served as a valuable support for the embryo scientific movement in Rome and more widely in Italy; Cesi's extensive correspondence and his influence in high quarters in Rome were especially helpful. Following his death in 1630 and Galileo's condemnation three years later the Academy faded out. (It was revived in 1847, after some earlier unsuccessful attempts, and in 1875 became the national academy of Italy)

1873. DOMINIS, Marco Antonio de. 1560-1626. (Physics 579. Earth science 1012) Studied at Padua and then taught mathematics, logic, and philosophy in several Italian cities. He later became a bishop.

1874. STELLUTI, Francesco. 1577-1652. (Microscopy 1256) Came from a patrician family; studied law in Rome in the 1590s. He was a member of the Accademia dei Lincei throughout its existence and the one most concerned with its publishing activities and business affairs.

1875. COLONNA (or COLUMNA), Fabio. 1567-1650. (Botany 1305) A member of a noble family and an important legal official in Zagarola. He was a member of the Accademia dei Lincei.

1610s

1876. CASTELLI, Benedetto. 1578–1643. (Mechanics 503. Physics 582)
Benedictine. Studied under Galileo at Pisa, ca. 1600. From
1611 he was closely associated with Galileo and was one of his
most loyal supporters. He was professor of mathematics at
Pisa from 1613 and then, from ca. 1626, at Rome where he was
also a consultant on hydraulics for the papal government.
At Pisa he was the teacher of Cavalieri and at Rome of Borelli
and Torricelli.

1877. SEVERINO, Marco Aurelio. 1580–1656. (Zoology 1421) Studied
medicine, chiefly at Naples, and graduated at Salerno in 1606.
After several years as a practitioner he became professor of
anatomy and surgery at Naples in 1615. He became famous as a
surgeon and published numerous books on surgery and anatomy
as well as maintaining a correspondence with many of the
leading figures in medicine and science.

1878. SPIEGEL, Adriaan van den. 1578–1625. (Anatomy 1581) A native
of Brussels. Studied at Louvain and Leiden, and then at Padua
where he graduated in medicine about 1603. After practising
in several countries he was appointed professor of anatomy and
surgery at Padua in 1616.

1879. ASELLI, Gaspare. 1581–1625. (Anatomy 1582) Studied medicine
at Pavia and became a distinguished surgeon in Milan.

1620s

1880. CAVALIERI, Bonaventura. 1598(?)–1647. (Mathematics 129) A
member of the Gesuati order. He studied mathematics under
Castelli at Pisa and became a disciple of Galileo through
whose influence he was appointed professor of mathematics at
Bologna in 1629.

1881. AROMATARI, Giuseppe degli. 1587–1660. (Botany 1308) M.D.,
Padua, 1605. Settled in Venice where he became famous both
as a physician and a man of letters.

1882. VESLING, Johann. 1598–1649. (Botany 1309. Anatomy 1583)
A native of Germany. Studied medicine and botany at Leiden
and Bologna. During the period 1628–32 he served as physician
to the Venetian representative in Egypt and took the opportunity
to investigate the botany of the country. In 1633 he was
appointed professor of anatomy and surgery at Padua and from
1638 he taught botany (instead of surgery). The botanical
garden at Padua was renovated under his direction.

1630s

1883. KIRCHER, Athanasius. 1602–1680. (Not in Part 1) Born and
educated in Germany and became a Jesuit. After teaching at
Würzburg and Avignon he was appointed professor of mathematics
at the Collegio Romano about 1638. He spent most of his life
in Rome, a centre where he was able to gather (especially from
his fellow Jesuits) much of the information he disseminated in
his numerous writings. "Some forty-four books and more than

2000 extant letters and manuscripts attest to the extraordinary variety of his interests and to his intellectual endowments. His studies covered practically all fields both in the human- ities and the sciences." (*DSB*) His books were published from about 1630 up to the time of his death. Many of them were very popular and though there was little in them that was original they served a useful purpose, in the sciences as in other fields, in disseminating recent discoveries and knowledge generally.

1884. TORRICELLI, Evangelista. 1608-1647. (Mathematics 137. Physics 584) Of humble origin. Educated first at a Jesuit college and then studied mathematics under Castelli in Rome. Through Castelli, who befriended him, he became acquainted with Galileo and worked as one of his assistants (with Viviani) during the last few months of his life. In 1642 he succeeded Galileo as mathematician to the Grand Duke of Tuscany and held the post until his untimely death.

1885. BORELLI, Giovanni Alfonso. 1608-1679. (Astronomy 369. Mechanics 510. Physiology 1722) Born in Naples of humble origin; in his youth he was apparently associated with Campanella. He went to Rome where, together with Torricelli, he was a student of Castelli; about 1635 Castelli helped to get him appointed to the public lectureship in mathematics at Messina. In 1656 he became professor of mathematics at Pisa and was a member of the Accademia del Cimento throughout its existence. In 1667 he returned to his previous position at Messina.

1640s

1886. MENGOLI, Pietro. 1625-1686. (Mathematics 140) Studied under Cavalieri at Bologna and succeeded him in the chair of math- ematics in 1648.

1887. RICCIOLI, Giambattista. 1598-1671. (Astronomy 362) Jesuit. Teacher of various subjects in Parma and Bologna. A skilful experimenter and a prolific author of works on astronomy and geography. He was the most learned and best known of the opponents of Galileo and the Copernican system.

1888. RENALDINI, Carlo. 1615-1698. (Physics 588) Originally a military engineer. In 1648 he became professor of philosophy at Pisa, and from 1667 until his death professor of mathematics as well. He was a member of the Accademia del Cimento.

1889. GRIMALDI, Francesco Maria. 1618-1663. (Physics 591) Jesuit. From 1648 he was teacher of mathematics at the Jesuit college in Bologna where he collaborated in a series of astronomical and physical investigations with the teacher of astronomy, G. Riccioli.

1890. FOLIUS (or FOLLI), Caecilius. 1615-1650. (Anatomy 1585) Graduated in medicine at Padua and became a teacher of anatomy in Venice.

1650s

1891. ACCADEMIA DEL CIMENTO. Its establishment at the court of the Medici in Florence in 1657 represented the formal organization

of a group which had already been in existence for many years
as the continuation of Galileo's band of disciples. The Academy
functioned under the aegis of Prince Leopold (younger brother
of the Grand Duke) and comprised ten members, the most outstanding being Viviani, Borelli, Redi, and Steno. It was the first
academy consisting of professional scientists and operating in
the manner of a research institute. Its research was collective,
the members working together on common tasks, and the account
of its experiments that it published in 1657 as *Saggi di naturali esperienze fatte nell'Accademia del Cimento* (translated
into English in 1684, Latin in 1730, and French in 1755) described the work of the institution as a whole. The Academy
came to an end in 1657 for reasons that are not clear--possibly
because there were not enough scientists available to continue
its activities.

1892. VIVIANI, Vincenzo. 1622-1703. (Mathematics 149) Studied mathematics with a private teacher. He was an assistant to Galileo
during the last few years of his life, and about 1650 he was
appointed mathematician to the Grand Duke of Tuscany, naturally
becoming a member of the Accademia del Cimento. In the late
1660s he was invited by Louis XIV of France to be a member of
the new Académie Royale des Sciences but declined; he was
later made one of the few foreign members of the Académie.
From 1696 he was also a foreign member of the Royal Society
of London.

1893. ANGELI, Stefano degli. 1623-1697. (Mathematics 148) A member
of the order of the Gesuati. Studied mathematics at Bologna
under Cavalieri in the mid-1640s. After holding various
administrative positions in his order he became professor of
mathematics at Padua in 1663.

1894. CASSINI, Gian Domenico. 1625-1712. (Astronomy 365) Educated
at the Jesuit college in Genoa. From 1648 he was employed in
the private observatory of a rich senator of Bologna. He
probably learnt much from the two Jesuit astronomers, Riccioli
and Grimaldi, who were then teaching in Bologna. In 1650, on
the recommendation of his patron, he was appointed professor
of astronomy at Bologna University. Though he also became
involved, as an official expert, in various engineering undertakings, he was able to accomplish much important work in
astronomy. As a result he was invited by Louis XIV--on the
recommendation of Colbert--to become a member of the new
Académie Royale des Sciences in Paris (and offered a larger
salary than the other distinguished foreign member, Huygens).
For his subsequent career in France see 2014.

1895. TACHENIUS, Otto. fl. 1640-70. (Chemistry 796) Born in Germany
and trained there as an apothecary. In 1644 (perhaps because
of the Thirty Years War) he migrated to Italy and obtained an
M.D. at Padua in 1652. He settled in Venice where he manufactured and sold drugs.

1896. REDI, Francesco. 1626-1697/8. (Zoology 1424, 1431) Graduated
in medicine at Pisa in 1647. He became head physician to the
Medici court in Florence and was on close terms with two
successive Grand Dukes. He was of course a member of the

Medici's Accademia del Cimento. A man of wide learning, he
was distinguished as a writer, poet, and linguist.

1897. MALPIGHI, Marcello. 1628-1694. (Anatomy 1593. Embryology 1657.
Botany 1314) Graduated in medicine and philosophy at Bologna
in 1653. From 1656 to 1659 he was professor of medicine at
Pisa where he came under the influence of Borelli and the
Galilean tradition. After brief periods in some other universities he became professor of medicine at Bologna in 1666.
In 1667 he was invited by the Royal Society of London to be
one of its correspondents and was elected a foreign member
two years later. Thereafter all his works were published in
London under the auspices of the Society. He spent his last
years (from 1691) in Rome as chief physician to the Pope.

1660s

1898. MONTANARI, Geminiano. 1633-1687. (Astronomy 376) After studying in Florence he graduated in law at Salzburg. He practised
law in Vienna and then in Florence, at the same time acquiring
(partly through the help of a friend) a good knowledge of mathematics and science. He became professor of mathematics at
Bologna in 1664 and then, from 1679, professor of astronomy
and meteorology at Padua.

1899. SCILLA, Agostino. 1629-1700. (Geology 1018) Well known in his
day as a painter. He was a cultured man of many interests,
including mathematics and science.

1900. STENSEN (or STENO), Niels. 1638-1686. (Anatomy 1595. Geology
1017) For his early career in Denmark, Holland, and France
see 2777. He left Paris in late 1665 and went via Montpellier
to Pisa and Florence where for some years (interrupted by two
trips home to Copenhagen) he was at the court of the Grand
Duke of Tuscany, partly as physician and partly as tutor to
the crown prince and others; in consequence he was a member
of the Accademia del Cimento. In 1675 he gave up scientific
activities and became a Catholic priest and finally a bishop.

1901. BELLINI, Lorenzo. 1643-1704. (Anatomy 1594. Physiology 1723)
Studied at Pisa, where he was influenced by Borelli, and graduated there in medicine. He was appointed professor of medicine
at Pisa in 1663, later becoming professor of anatomy.

1670s

1902. CEVA, Giovanni. 1647-1734. (Mathematics 160) Studied at the
University of Pisa. Became mathematician and engineer to the
Duke of Mantua.

1680s

1903. GUGLIELMINI, Domenico. 1655-1710. (Mechanics 523) Studied
mathematics and medicine at Bologna, graduating in the latter
in 1678. His early activities were chiefly in astronomy and
in 1686 he was appointed professor of mathematics at Bologna.
His ability in hydraulic engineering led the Senate of Bologna
to establish a special chair of hydrometry for him in 1694.

Four years later he accepted the chair of mathematics at Padua
while retaining his title (and salary) at Bologna. In 1702 he
changed to the (doubtless highly paid) chair of theoretical
medicine at Padua, declining attractive offers from the Grand
Duke of Tuscany and the Pope. In his last years he published
several medical works.

Guglielmini was one of the early members of the Accad-
emia degli Inquieti and was elected a foreign member of the
Royal Society of London and of the Academies of Paris and
Berlin.

1690s

1904. ACCADEMIA DEGLI INQUIETI. Established in Bologna in the early
1690s as the formalization of the gatherings held at the house
of E. Manfredi (1908). Morgagni joined it in 1699 and became
its head in 1704. At the same time it was reorganized and
began to hold its meetings at the mansion of L.F. Marsili
(1914). In 1714 it was incorporated in the Bologna Academy
(1911) founded in that year by Marsili.

1905. VALLISNIERI, Antonio. 1661-1730. (Natural history 1230)
Studied medicine under Malpighi at Bologna and took his M.D.
at Reggio in 1684. After further study at Bologna and elsewhere he practised medicine at Reggio, at the same time developing his interests in natural history and biology. In 1700 he
was appointed to a chair at Padua, initially of experimental
philosophy, then of practical medicine, and finally of theoretical medicine. He became widely known and was a member of
numerous academies, including the Royal Society of London.

1906. VALSALVA, Anton Maria. 1666-1723. (Anatomy 1609) Studied
medicine at Bologna, where he was Malpighi's favorite pupil,
and graduated in 1687. As a young physician in Bologna he
attended the meetings that led to the founding of the Accademia
degli Inquieti and became a member of the Academy. In 1705,
after the publication of his important book (1609), he was
appointed lecturer and demonstrator in anatomy at the University.
Morgagni was one of his pupils.

1907. BAGLIVI, Georgius. 1668-1707. (Anatomy 1608) Studied medicine
at Naples and received his M.D. (probably at Salerno) in 1688.
From 1691 to 1694 he was Malpighi's assistant in Rome. In 1696
he was appointed professor of anatomy (later of theoretical
medicine) at the Sapienza in Rome. He became very famous both
as a physician and as a teacher of medicine and was elected
to several academies, including the Leopoldina and the Royal
Society of London.

1700s

1908. MANFREDI, Eustachio. 1674-1739. (Astronomy 388, 391) Graduated
in law in 1692 but never practised. He studied mathematics
and hydraulics with Guglielmini and was self-taught in astronomy. From the early 1690s he held scientific meetings in his
house in Bologna which led to the formation of the Accademia
degli Inquieti (1904). In 1699 he became professor of mathematics at Bologna and from 1704 he was also superintendant of
hydrological affairs in the Bologna region. When the Bologna

Institute was established in 1714 he became its astronomer.
Manfredi was elected a foreign member of the Paris
Academy in 1726 and of the Royal Society of London in 1729.
In addition to his scientific achievements he was also a well-
known poet and had the distinction (especially significant
for a non-Tuscan) of being made a member of the Accademia della
Crusca of Florence.

1909. SANTORINI, Giovanni Domenico, 1681-1737. (Anatomy 1612)
Studied medicine at Bologna, Padua, and Pisa, graduating in
1701. He taught anatomy in Venice for many years and later
became a leading physician there. He was a popular teacher
and was widely regarded as a distinguished anatomist.

1910. MORGAGNI, Giovanni Battista. 1682-1771. (Anatomy 1610, 1619)
Studied medicine at Bologna in the tradition of Malpighi,
graduating in 1701. He then practised medicine in the city
and became a prominent member of the Accademia degli Inquieti.
In 1711 he was appointed professor of medicine at Padua where
he remained for the rest of his life.

1710s

1911. BOLOGNA ACADEMY. Scientiarum et Artium Institutum Bononiense
atque Academia (known in the nineteenth century as Accademia
delle Scienze dell'Istituto di Bologna). The Academy was
closely associated with the Institute, both bodies being
founded as parts of a single plan in 1714 by the Senate of
Bologna on the initiative of L.F. Marsili (1914). The Institute
was a teaching body charged with the task of disseminating a
knowledge of the experimental and observational sciences and
their applications in such fields as navigation and military
engineering (the chief reason for its creation being the
unwillingness of the University of Bologna to give such teaching). The Academy was the enlarged successor of the Accademia
degli Inquieti (1904) and was intended to complement the
Institute and to stimulate its activities and connect it with
the international world of learning. The Academy's *Commentarii*
began in 1731 and became one of Italy's chief scientific periodicals.

1912. RICCATI, Jacopo Francesco. 1676-1754. (Mathematics 172) A
Venetian noble. Graduated in law at Padua in 1696 but did
not practise and pursued his interests in mathematics. From
about 1710 he published extensively and corresponded with
mathematicians all over Europe. He became famous as a mathematician and received offers of high positions in academies,
universities, and courts, all of which he declined.

1913. FAGNANO (DEI TOSCHI), Giulio Carlo. 1682-1766. (Mathematics
174) A noble of Sinigaglia; self-taught in mathematics. His
publications made him well known and he was elected a foreign
member of the Berlin and Paris Academies and the Royal Society.

1914. MARSILI, Luigi Ferdinando. 1658-1730. (Earth science 1030)
Came from a noble family and served as a general in the army
of the Emperor until 1704. A man of wide learning and many
travels, some of which were undertaken for purposes of research.

He became an important figure in the Accademia degli Inquieti
(1904) and was the chief mover in the establishment of the
Bologna Institute and Academy (1911) in 1714.

1915. MICHELI, Pier Antonio. 1679-1737. (Botany 1328) His father
was a labourer and he had only an elementary education. He
was fortunate however in gaining the patronage of his sovereign,
the Grand Duke of Tuscany, and thereby obtaining positions in
the botanical gardens of Pisa and Florence. Though he never
had a post commensurate with his talents he won a considerable
reputation among the botanists of his time.

1740s

1916. BOSCOVICH, Ruggiero Giuseppe. 1711-1787. (Natural philosophy
28) Jesuit. Educated at the Collegio Romano and appointed
professor of mathematics there in 1740; in this capacity he
carried out several engineering and architectural projects
for the papal government. In 1759-63 he travelled widely in
Europe, spending several months in Paris (he had long been a
corresponding member of the Paris Academy) and in London where
he was elected a fellow of the Royal Society. On his return
to Italy he was appointed professor of mathematics at Pavia
and was largely instrumental in establishing the Brera Observatory (1924). In 1773, when the Jesuit order was suppressed,
he went to Paris where a post was created for him as director
of optics (in relation to astronomy) for the navy. He returned
to Italy in 1782 to supervise the publication of his five-
volume *Opera pertinenta ad opticam et astronomiam* and died
soon afterwards.
    The last great polymath in the history of science,
Boscovich published many substantial works on numerous aspects
of the mathematical and physical sciences (astronomy, mechanics,
optics, geophysics, geodesy, meteorology) but is best known
for his suggestive speculations in natural philosophy.

1917. RICCATI, Vincenzo. 1707-1775. (Mathematics 189) Son of J.F.
Riccati (1912). Became a Jesuit in 1726. From 1739 almost
until his death he taught mathematics at the Jesuit college
in Bologna. He was one of the first members of the Società
dei Quaranta (1932).

1918. BECCARIA, Giambatista. 1716-1781. (Physics 613) A member of
the teaching order of Piarists. From 1737 he taught in various
schools of his order and in 1748 was appointed professor of
physics in the University of Turin. His students there included Lagrange and he appears to have had some indirect influence
on Volta.

1919. MORO, Antonio Lazzaro. 1687-1764. (Geology 1112) A priest
who held various ecclesiastical positions.

1920. TARGIONI-TOZZETTI, Giovanni. 1712-1783. (Earth science 1036)
M.D., Pisa, 1734. Practised medicine in Florence and was also
director of the Magliabechi Library and the botanical garden.
Member of the Leopoldina Academy. His works included a history
of science in Tuscany.

1750s

1921. LAGRANGE, Joseph Louis. 1736-1813. (Mathematics 186, 197. Astronomy 407. Mechanics 544) Born in Turin of a wealthy family, partly of French origin. He initially studied law but soon gave it up for mathematics and from 1755 was teaching the subject at the artillery school in Turin. From that year he was also in correspondence with Euler (then at the Berlin Academy) about his mathematical discoveries. Euler was very helpful and was even able to have him elected, at an unprecedentedly early age, as a foreign member of the Berlin Academy. Until the mid-1760s he published his researches in the *Mélanges de Turin*, a periodical issued by a small scientific society (the predecessor of the Turin Academy) which he and others had established in 1757.

     In 1764 he was awarded the *grand prix* of the Paris Academy for an essay in celestial mechanics, and he won the prize again in 1766. By this time he was recognized as one of the leading mathematicians in Europe and in 1766 he was invited by Frederick the Great, on the recommendation of Euler and d'Alembert, to join the Berlin Academy. For his career in Berlin see 2548.

1922. ARDUINO, Giovanni. 1714-1795. (Geology 1115) Studied at Verona but did not take a degree. As a result of his practical experience he became recognized as a mining expert, in which capacity he served several Italian administrations. He finally became professor of mineralogy at Padua and a member of the Società dei Quaranta.

1760s

1923. SIENA ACADEMY. Accademia delle Scienze di Siena detta de' Fisiocritici. Its origins went back to 1691 but it did not begin its *Atti* until 1761.

1924. BRERA OBSERVATORY. The Jesuits' astronomical observatory near Milan. Founded in 1764 largely through the efforts of Boscovich who was its first director.

1925. SPALLANZANI, Lazzaro. 1729-1799. (Physiology 1733, 1736, 1740. Embryology 1661, 1662. Natural history 1240) Graduated in philosophy at Bologna in 1753/4 and after theological studies was ordained a priest in 1757. From 1763 he was professor of philosophy at Modena, and then from 1769 professor of natural history at Pavia, a post he held until his death. (At that time the University of Pavia, "the Oxford of Italy", was being rehabilitated by the Austrian government of Lombardy) His duties at Pavia included the direction of the public Museum of Natural History which he made into one of the best in Europe.

     One of the greatest figures of the time in both natural history and experimental biology, Spallanzani's interests extended into many other fields--he made, for example, some pioneering observations in vulcanology. In his late years he received many honours and was a member of numerous academies.

1926. FONTANA, Felice. 1730-1805. (Physiology 1732) Studied medicine at Padua and graduated at Bologna in 1757. After spending

some years in Rome he was appointed professor of "physics" at
Pisa in 1766. At the same time he was given the task of re-
organizing the Medici's collection of scientific apparatus,
specimens, models, etc., in Florence. As a result of his
efforts the Florence Museum of Physics and Natural History
was opened in 1775. He continued to direct the Museum and
extend its collections.

Though Fontana's most important achievement was in
physiology he also did notable work in chemistry.

1927. GALVANI, Luigi. 1737-1798. (Physiology 1735) M.D., Bologna,
1762. In 1766 he was appointed curator of Bologna University's
anatomical museum and in 1782 he became professor of obstetrics
at the Istituto delle Scienze. He was president of the Bologna
Academy from 1772. In his last years he was deprived of his
posts because of his refusal to swear allegiance to the new
Cisalpine Republic established by Napoleon.

1770s

1928. PADUA ACADEMY. Its origins went back to the sixteenth century
but in 1779 it was refounded as the Cesareo-Regia Accademia
(later Reale Accademia) di Scienze, Lettere ed Arti di Padova.
Its *Saggi* began in 1786.

1929. VOLTA, Alessandro. 1745-1827. (Physics 634) Came from a noble
Lombard family. He was largely self-taught in science and
began experimenting with electricity at an early age. In 1775
he invented the electrophore--a device, then quite significant,
for generating and storing static electricity. This and other
discoveries led to his appointment in 1777 as professor of
experimental physics at the University of Pavia, a post he
held for the rest of his career.

In 1800, following his great invention of the electric
pile, he made a triumphant visit to Paris where he demonstrated
the properties of the pile at special sessions of the Institut
National attended by Napoleon who was very impressed by it.
Napoleon made him a count and he was elected to all the leading
academies.

1930. SCARPA, Antonio. 1752-1832. (Anatomy 1626, 1628) Graduated
in medicine at Padua in 1770 and was assistant to Morgagni
until his death. He became professor of anatomy at Modena
from 1772 and then at Pavia from 1783; at both universities
he had anatomical amphitheatres constructed. As well as being
an outstanding anatomist and surgeon he was an important
teacher who had a wide influence, especially during his long
period at Pavia. In later life he received many honours and
was a foreign member of the Paris Academy, the Royal Society
of London, and the Leopoldina Academy.

1931. MASCAGNI, Paolo. 1755-1815. (Anatomy 1625) Graduated in
medicine in 1775. He was appointed professor of anatomy at
Pisa in 1779 and later taught the subject in Florence.

1780s

1932. SOCIETÀ ITALIANA. Established in Verona in 1782 as the Società

(Italiana) dei Quaranta. It was founded by the mathematician and physicist, A.M. Lorgna, and its *Memorie di matematica e di fisica* (covering the full range of the natural sciences) became the leading Italian scientific periodical of the time. From 1804 the Society was named Società Italiana delle Scienze (residente in Modena). It was, for Italy, a new type of society in that it was national in scope; moreover, being a private society with (eventually) a large membership, it was different in character from the traditional academies.

1933. TURIN ACADEMY. Founded in 1783 as the Académie Royale des Sciences de Turin. It originated from a private society formed in 1757 by J.L. Lagrange and others. The Academy was a strong, well-endowed institution and acquired a considerable reputation internationally. From 1815 it was named Reale Accademia delle Scienze di Torino.

1934. RUFFINI, Paolo. 1765-1822. (Mathematics 210) Graduated at Modena in 1788 in philosophy, medicine, and mathematics. Later in the same year he was appointed professor of mathematics at Modena and from 1799 he also taught medicine. He later became rector of the University and president of the Società Italiana dei Quaranta.

1800s

1935. AVOGADRO, Amedeo. 1776-1856. (Chemistry 863) Came from a distinguished legal family in Turin. Graduated in law in 1796 and thereafter practised, but developed an interest in mathematics and physics which he studied privately. After some early publications he became professor of natural philosophy at the College of Vercelli in 1809. In 1819 he was elected to the Turin Academy and in the following year was appointed to the chair of mathematical physics at Turin.

1936. BROCCHI, Giovanni Battista. 1772-1826. (Geology 1140) Came from a wealthy and cultured family. Studied botany at Padua and later taught natural history at a gymnasium. In 1808 he was appointed an inspector of mines in Lombardy.

1937. BASSI, Agostino Maria. 1773-1856. (Microbiology 1802) Graduated in law at Pavia in 1798 (he also studied some science under Volta, Spallanzani, etc.). He became a public servant but maintained a strong interest in scientific and agricultural topics.

1810s

1938. AMICI, Giovan Battista. 1786-1868. (Microscopy 1264, 1267. Botany 1357, 1367) Graduated as an engineer-architect at Bologna in 1807 and became a teacher of mathematics at a secondary school. In 1831, as a result of his successful work in improving optical instruments, he was appointed head of the astronomical observatory in Florence and later became director of microscopic research at the Florence Museum.

1820s

1939. MELLONI, Macedonio. 1798-1854. (Physics 667) Professor of

physics at Parma from 1824. Because of his involvement in the
unsuccessful revolution of 1830 he was forced to flee to Paris
where he lived for eight years. His important research on
radiant heat had been begun at Parma but was mostly done during
his stay in Paris where he was in close touch with the French
physicists, especially Biot.

In 1839 he was able to return to Italy and was appointed director of the Conservatory of Arts and Crafts in Naples.
Until 1848 he was also in charge of the meteorological observatory on Mount Vesuvius. He was a correspondent of the Paris
Academy and a member of the Società Italiana delle Scienze.

1830s

1940. RIUNIONE DEGLI SCIENZIATI ITALIANI. (Sometimes called Congresso
degli....) A national society which held meetings in different
Italian cities, beginning in 1839. For political reasons it
was suppressed in 1848, though in the 1860s and 1870s some
meetings were held. (The Società Italiana per il Progresso
delle Scienze, formed in 1907, regards itself as a continuation
of it)

1941. MATTEUCCI, Carlo. 1811-1868. (Physiology 1753. Though his
most important work was in electrophysiology the majority of
his many papers dealt with topics in physics) Graduated in
physics at Bologna in 1828 and then studied in Paris for some
months. He became professor of physics at Bologna from 1832,
at Ravenna from 1838, and at Pisa from 1840 until his death.
In 1844 he began a periodical with the name *Il Cimento* ("Experiment"); it was short-lived, possibly because of his involvement in the political upheaval of 1848, but in 1855 he revived
it under the title of *Il Nuovo Cimento* and it subsequently
became the leading Italian journal for physics.

After the establishment of the Kingdom of Italy in
1861 Matteucci was made a senator for life. He became general
director of telegraphs and in 1862 minister of education.

1840s

1942. BRIOSCHI, Francesco. 1824-1897. (Not in Part 1. Though he was
a capable mathematician with many publications to his credit
his chief significance was in his educational and institutional
role) Graduated at Pavia in 1845 and carried on further studies
there; from 1852 to 1861 he was professor of applied mathematics.
In 1861-62, following the establishment of the Kingdom of Italy,
he was the general secretary of the Ministry of Education, and
in the following year took a leading part in establishing the
Istituto Tecnico Superiore in Milan; he continued as its director and professor of mathematics until his death. He was made
a senator in 1865 and from 1870 to 1882 was a member of the
executive council of the Ministry of Education. He was a
member of the leading Italian academies and of some foreign
ones, and from 1884 was president of the Accademia Nazionale
dei Lincei.

Like his contemporary, Betti (1943), Brioschi did much
to revivify Italian mathematics and to improve the teaching of

the subject in high schools and universities. His own students included most of the leading Italian mathematicians of the next generation (notably Casorati, Cremona, and Beltrami).

1943. BETTI, Enrico. 1823-1892. (Mathematics 257, 277) Graduated in science and mathematics at Pisa in 1846 and two years later fought in the first war for Italian independence. From 1849 he taught mathematics in secondary schools and in 1857 was appointed professor of mathematics at Pisa where he remained for the rest of his life. From 1865 he was also director of the teachers' college in Pisa. He was a member of parliament in 1862 and subsequently the secretary of the Ministry of Education and a member of the Council for Public Instruction. He was made a senator in 1884.

Betti was an excellent teacher and both through his own teaching and his influence in the education system he made an important contribution to the revival of mathematics in Italy after the Risorgimento.

1944. SECCHI, Angelo. 1818-1878. (Astronomy 457, 469) Jesuit. His studies at the Collegio Romano included mathematics and physics which he later taught at one of the colleges of his order. In 1849 he was appointed director of the observatory of the Collegio Romano and he was able to have it rebuilt and equipped with new instruments, especially for astrophysical research. He was the author of many publications and continued working at the observatory almost until his death.

As well as being a member of the Roman Academy (the revived Accademia dei Lincei) and other Italian societies, Secchi was a foreign member of the Royal Society of London and was awarded a prize by the Paris Academy. With P. Tacchini he founded the Società degli Spettroscopisti Italiani (1953) in 1871.

1945. CANNIZZARO, Stanislao. 1826-1910. (Chemistry 917) Began the study of medicine at Palermo in 1841 and soon afterwards took part in some physiological research which led him to chemistry. In 1845 he became assistant to Piria, professor of chemistry at Pisa, and began chemical research. In 1847/48 however he joined in a revolution in his native Sicily and when it failed in early 1849 he was compelled to go into exile. After a period in Paris, where he worked with Chevreul, he was able to return to Italy in 1851.

For a few years he was a schoolteacher in Alessandria but in 1855 he became professor of chemistry at Genoa. In 1861, in consequence of Garibaldi's successful revolt, he returned to Palermo and became professor of chemistry there. His final move, to the chair of chemistry at the new University of Rome in 1871, also followed a political development--the taking of Rome and its conversion into the national capital. In the same year he was made a senator.

1850s

1946. CREMONA, (A.) Luigi (C.G.). 1830-1903. (Mathematics 270, 281) Graduated in civil engineering at Pavia in 1853. Having fought against Austrian rule in the war of 1848 he was debarred from

official positions but was eventually able to become a teacher
in a secondary school. In 1860, under the government of newly-
liberated Lombardy, he was appointed professor of higher geometry at Bologna. From 1867 he was professor at the new Technical Institute in Milan, from 1873 director of the new Polytechnic School of Engineering in Rome, and from 1877 until
his death he held the chair of higher mathematics at the
University of Rome.

One of the most outstanding Italian scientists of his
time, Cremona was a member of many academies, Italian and
foreign. He was appointed a senator in 1879 and served as a
member of the council of the Ministry of Education.

1947. SCHIAPARELLI, Giovanni Virginio. 1835-1910. (Astronomy 454,
471) Graduated in civil engineering at Turin in 1854 and
subsequently studied astronomy there. In 1857 he went at
government expense to the Berlin Observatory to work under
Encke and two years later spent some time at Pulkovo. On his
return to Italy in 1860 he was appointed to a post at the
Brera Observatory in Milan, becoming its director two years
later. After his retirement in 1900 he did important work on
the history of ancient astronomy.

Schiaparelli was a member of the Roman Academy (the
revived Accademia dei Lincei) and various other Italian bodies
and a foreign member of several academies. He twice won the
Lalande Prize of the Paris Academy and received numerous other
honours. He was made a senator of the Kingdom of Italy in 1889.

1860s

1948. BELTRAMI, Eugenio. 1835-1899. (Mathematics 265. Mechanics
560) Studied mathematics at Pavia under Brioschi in 1853-56
but had to discontinue his course in order to earn a living.
He continued to study the subject however and published his
first paper in 1862. From that year onward (following the
establishment of the Kingdom of Italy in 1861) he held a
succession of mathematical chairs: Bologna, 1862-64; Pisa,
1864-66; Bologna, 1866-73; Rome, 1873-76; Pavia, 1876-91;
Rome, 1891 until his death. He became president of the
Accademia Nazionale dei Lincei and was appointed a senator
just before his death.

1949. GEOLOGICAL SURVEY. Reale Comitato Geologico d'Italia. Established in 1868.

1950. SOCIETÀ ENTOMOLOGICA ITALIANA. Its periodical began in 1869.

1951. GOLGI, Camillo. 1843-1926. (Anatomy 1642. Cytology 1706)
Graduated in medicine at Pavia in 1865. He then did histological research in the University's laboratory of experimental
pathology until 1872 when financial difficulties obliged him
to go into medical practice. His published results however
gained him an appointment as lecturer in histology at Pavia,
and in 1880 he acceded to the chair in the subject. He built
up a vigorous research school and in 1906 was awarded the
Nobel Prize (shared with his Spanish rival, Ramón y Cajal)
for his achievements in neuroanatomy.

1870s

1952. RICCI-CURBASTRO, Gregorio. 1853-1925. (Mathematics 302) Studied at Rome, Bologna, and Pisa, taking his doctorate at the last in 1875. Soon afterwards he won a travelling scholarship which enabled him to spend a year at Munich with Klein. In 1880, after a period as an assistant at Pisa, he was appointed professor of mathematical physics (and mathematics) at Padua, a post he held until his death.

1953. SOCIETÀ DEGLI SPETTROSCOPISTI ITALIANI. Founded at Palermo in 1871 by P. Tacchini, of the Palermo Observatory, and A. Secchi (1944). Its *Memorie*, begun in 1872, was the first periodical for astrophysics.

1880s

1954. CIRCOLO MATEMATICO DI PALERMO. Founded in 1884 by G.B. Guccia, professor of mathematics at Palermo. Its *Rendiconti* began in the following year and soon won international esteem. In 1888 the Circolo introduced the category of foreign corresponding members, with the result that a large number of foreign mathematicians joined it and the *Rendiconti* became an international journal.

1955. PEANO, Giuseppe. 1858-1932. (Mathematics 299, 310) Graduated at Turin in 1880. After ten years in a junior teaching position there he became an *extraordinarius* in 1890 and full professor in 1895. From 1886 he was also professor at the military academy and he was a prominent figure in the Turin Academy. His important work in mathematical logic inspired the formation of a small school in the subject at Turin and in 1891 he began the *Rivista di Matematica* to publish the work of his school.

1956. VOLTERRA, Vito. 1860-1940. (Mathematics 292) A mathematical prodigy (one of his leading ideas was conceived when he was thirteen). Graduated in mathematical physics in 1882 at Pisa, where he was much influenced by Betti. In the following year he became professor of rational mechanics at Pisa, in 1892 professor of mechanics at Turin, and finally in 1900 professor of mathematical physics at Rome.

In recognition of his achievements Volterra was made a senator in 1905, and in 1921 he received an honorary knighthood from the King of England. He was president for a time of the Accademia Nazionale dei Lincei and received many scholarly honours. Nevertheless in 1931 he was deprived of his chair because of his anti-Fascist stand.

1957. SEGRE, Corrado. 1863-1924. (Mathematics 297) D.Phil., Turin, 1883. He continued research and teaching at Turin, becoming professor of higher geometry there in 1888. He was an important teacher, attracting many students and strengthening the revival of Italian mathematics.

1958. SOCIETÀ GEOLOGICA ITALIANA. Founded 1881.

1959. VASSALE, Giulio. 1862-1913. (Physiology 1795) Studied medicine at Modena and Turin, graduating at the latter in 1887. After holding junior teaching positions in pathology at Modena he became professor of the subject there in 1898.

1890s

1960. SOCIETÀ SISMOLOGICA ITALIANA. Its periodical began in 1895.

1961. SOCIETÀ BOTANICA ITALIANA. Its periodical began in 1892.

1962. UNIONE ZOOLOGICA ITALIANA. Its periodical began in 1890.

2.02 FRANCE

1480s

1963. CHUQUET, Nicolas. fl. late 15th century. (Mathematics 70)
Graduated in medicine at Lyons in 1484. (Very little else
is known of his life)

1500s

1964. BOUVELLES, Charles. ca. 1470-1553. (Mathematics 74, 76) A
canon and teacher of theology at Noyon (an important ecclesiastical centre in northern France).

1965. THOMAZ, Alvaro (or ALVARUS THOMAS). fl. 1509-13. (Mechanics
492) Born in Lisbon. Around 1510 he was regent of the
Collège de Coqueret (University of Paris).

1520s

1966. FINE, Oronce. 1494-1555. (Not in Part 1) Professor of mathematics at the newly established Collège Royal from 1531 until
his death. He also edited mathematical and astronomical books
for printers in Paris and elsewhere. His own books, written
from 1526 onward, were numerous but of an encyclopaedic or
elementary character and his significance was as a popularizer
of the mathematical sciences.

1530s

1967. COLLÈGE ROYAL. Founded by François I in 1530 as a means of
disseminating the new learning of the Renaissance. Since
mathematics had a place, albeit a small one, in the new humanist
pattern of education it was included in the College's teaching.
From its beginning therefore the College had an educational
significance for the embryo scientific movement in France and
from the late seventeenth century this significance greatly
increased (see 2037).
  The unique character of the College (which has persisted
with remarkably little change to the present day) was determined
by the circumstances of its formation, which in essentials
remained until the Revolution. It was founded in face of the
opposition of the entrenched and powerful University of Paris,
and it could not be allowed to compete with the University as
a qualifying institution. Because it could grant no degree
or other qualification there was no need for examinations or
entrance qualifications. Consequently its courses were open
to all. Furthermore there was no fixed curriculum and its
professors could teach whatever they were interested in. It
was probably the first institution in Europe to embody the
principles of freedom of teaching and freedom of learning.

Unlike the universities the College was not weighed
down by the past and it did not have the function of training
for the established professions. Its function was simply to
disseminate the kind of knowledge the times called for, initially Renaissance humanism and later a more developed humanism
together with an increasing amount of science--and always
mathematics.

1968. FERNEL, Jean. 1497(?)-1558. (Physiology 1708) Graduated in
medicine at Paris in 1530. Established a high reputation as
a physician and was appointed professor of medicine at Paris
in 1534.

1540s

1969. RAMUS, Petrus (or LA RAMÉE, Pierre de). 1515-1572. (Not in
Part 1) M.A., Paris, 1536. Teacher of philosophy, logic,
and rhetoric at the University of Paris and later at the
Collège Royal. He wrote many books and was best known for
his new ideas in pedagogy and logic. He also emphasized the
importance of mathematics in education and wrote several textbooks of arithmetic and geometry, chiefly in the 1560s.

1970. PELETIER, Jacques. 1517-1582. (Mathematics 93, 97) A teacher
of mathematics and various other subjects in several cities.

1971. RONDELET, Guillaume. 1507-1566. (Zoology 1416) M.D., Montpellier, 1537; eight years later he became regius professor of
medicine there. He was also personal physician to a French
cardinal with whom he did a good deal of travelling which he
used to good effect for his studies in natural history. He
was a successful teacher and several of his students became
important naturalists.

1972. BELON, Pierre. 1517-1564. (Zoology 1413) Of humble origin.
He became an apothecary and developed an interest in natural
history. Through the patronage of his local bishop he was
able to go to the University of Wittenberg in 1540 to study
botany under Valerius Cordus (he apparently did not take a
degree). In 1542 he settled in Paris as an apothecary.

1973. ESTIENNE, Charles. ca. 1505-1564. (Anatomy 1562) A member of
the Parisian family of printers and publishers. From 1530 to
1534 he was in Italy and studied classical philology at the
University of Padua; he also developed an interest in botany.
After returning to Paris he studied anatomy and medicine at
the University, graduating in 1540. Thereafter he practised
as a physician and for a few years lectured on anatomy at the
University. After 1550 however he was manager of the family
business.

1974. SERVETUS, Michael. 1511(?)-1553. (Physiology 1709) Born and
educated in Spain. Studied medicine at Paris in the late
1530s and then practised in Lyons and Vienne. His chief
concern was theology.

1550s

1975. BUTEO, Johannes. ca. 1492-1570. (Mathematics 96) A monk of
the abbey of St. Antoine de Vienne. In the 1520s he studied
in Paris under Oronce Fine.

1976. SCALIGER, Julius Caesar. 1484-1558. (Botany 1291, 1294, 1299. Encyclopaedism 4) A native of Italy. In 1524 he migrated to France and settled in Agen where he spent the rest of his life as a successful physician. He was famous for his literary works and classical scholarship. (His son, J.J. Scaliger, was one of the most celebrated classical scholars) His scientific writings date from the last few years of his life.

1977. DALÉCHAMPS, Jacques. 1513-1588. (Botany 1300) Studied medicine at Montpellier (botany under Rondelet) and gained his M.D. in 1547. From 1552 until his death he lived in Lyons. He carried on a correspondence with several leading naturalists and was active in the editing and translating of ancient scientific and medical works.

1978. L'ÉCLUSE, Charles de (or CLUSIUS). 1526-1609. (Botany 1304) Came from a wealthy family in northern France. Graduated in law at Louvain in 1548 and then worked with Rondelet at Montpellier on botanical and zoological investigations. He travelled widely, collecting botanical specimens and writing regional floras. In 1593 he was appointed professor of botany at Leiden and was instrumental in making it the chief centre for botany in Europe.

1560s

1979. VIÈTE, François. 1540-1603. (Mathematics 104, 108, 118, 138) Graduated in law at the University of Poitiers in 1560. He took up a political career, becoming councillor to the *parlement* of Brittany and later that of Tours, and in 1589 a royal privy councillor. His mathematical works were written during interludes in his political activities.

1980. L'OBEL, Mathias de. 1538-1616. (Botany 1295) Studied medicine at Montpellier in the 1560s (botany under Rondelet) and there collected his botanical material and wrote his book. In 1571 he went to the Netherlands where he practised medicine until 1584 when civil war drove him to England. He spent the rest of his life in England, finally becoming botanist to James I.

1570s

1981. RISNER, Friedrich. died ca. 1580. (Physics 578) A native of Germany. Spent many years in Paris as a pupil and collaborator of Peter Ramus.

1982. DUCHESNE (or QUERCETANUS), Joseph. ca. 1544-1609. (Chemistry 776, 780) Studied medicine, first at Montpellier and then at Basel where he graduated in 1573. He practised at Kassel, Geneva, and finally Paris where he became physician to Henry IV.

1580s

1983. GUIBERT, Nicolas. ca. 1547-1620. (Chemistry 779) A native of Lorraine. Graduated in medicine at the University of Perugia. He became well known as an alchemist and was employed as such by the Grand Duke of Tuscany, the Viceroy of Naples, the Archbishop of Augsburg, and others. However he finally lost faith in alchemy and attacked it.

1984. DU LAURENS, André. 1558-1609. (Anatomy 1577) On gaining his
      M.D. at Montpellier in 1583 he was appointed professor of
      medicine there. He later became physician to the king and
      queen.

1590s

1985. BOTANICAL GARDEN, MONTPELLIER. The first botanical garden in
      France. Established by the University in 1593. The director
      was P.R. de Belleval, the occupant of the chair of anatomy
      and botany which was created at the same time.

1600s

1986. BEGUIN, Jean. ca. 1550-1620. (Chemistry 782) Had a laboratory
      together with a school in Paris where he gave instruction on
      the new chemical remedies. His lectures were very popular, as
      was the book that resulted from them.

1610s

1987. PEIRESC, Nicolas Claude Fabri de. 1580-1637. (Not in Part 1)
      A member of the landed gentry of Provence; graduated in law
      at Montpellier and became *conseiller* in the *parlement* of
      Provence. He was an important virtuoso who did much to stimu-
      late the scientific movement in France in the 1620s and 1630s.
      Though a competent astronomer (in 1610 he was one of the first
      in France to use a telescope), his role was as a patron and
      amateur of science and learning generally. He kept in close
      touch with developments in Italy, and from his home in Provence
      he carried on an enormous correspondence within and beyond
      France. His support of Gassendi, who lived in his house while
      writing his philosophy, was especially notable.

1988. RIOLAN, Jean. 1580-1657. (Not in Part 1) Graduated in medicine
      at the University of Paris in 1604 and was soon afterwards
      appointed professor of anatomy and botany there, and also
      professor of medicine at the Collège Royal. He became a leading
      member of the Paris Faculty of Medicine and, like the Faculty
      generally, was highly conservative--if not reactionary--in his
      attitude to new developments in medicine. He was however an
      able anatomist and by his very successful teaching and widely-
      used textbooks did much to make Paris an important centre for
      anatomy.

1620s

1989. MERSENNE, Marin. 1588-1648. (Natural philosophy and the math-
      ematical sciences 14, 506) A friar of the order of Minims.
      He was educated at the Jesuit college of La Flèche and studied
      theology at the Sorbonne. From 1619 until his death he was
      stationed in Paris by his order.
           From the early 1620s his living quarters at the Minim
      convent in Paris became a recognized meeting-place for *savants*
      interested in the new science--not only Frenchmen but also
      travellers from other countries. At the same time he built up
      an extensive correspondence which continued until his death
      and through which he made contact with almost all the chief

figures in the scientific movement at the time. In an age before the existence of periodicals he seems to have been very conscious of the communal nature of science and of the waste of time and talent, and general loss of opportunity, that resulted from inadequate communication. Through his personal contacts and his correspondence he worked hard to rectify this state of affairs and he can be regarded as the first outstanding organizer of the scientific movement. The importance of his initiative and efforts was recognized both by his contemporaries and by later historians: he was dubbed "the secretary-general of learned Europe" and it was said that "To inform Mersenne of a discovery meant to publish it throughout the whole of Europe."

1990. GASSENDI, Pierre. 1592–1655. (Natural philosophy, etc. 15) After graduating in theology at Avignon in 1614 he was ordained and became a teacher of philosophy. In the 1620s he spent some years in Paris where he was a close friend of Mersenne. For several years in the 1630s he lived in the house of his patron, Peiresc (1987), in Provence, working on his philosophy. He then returned to Paris and in 1645 became professor of mathematics (actually of astronomy) at the Collège Royal, though only for a short time because of ill health.

1991. DESCARTES, René. 1596–1650. (Philosophy 17. Mathematics 135. Physics 583. Physiology 1713) A member of the *noblesse de robe* and financially independent. Educated at the Jesuit college of La Flèche (which he later called "one of the most celebrated schools of Europe") and then at the University of Poitiers where he graduated in law in 1616. After various travels he lived in Paris for a few years, in close touch with Mersenne and his circle. In 1628 he retired to Holland where he lived in seclusion for nearly all the rest of his life, but remaining in contact with Mersenne and others by means of correspondence and occasional visits.

1992. MYDORGE, Claude. 1585–1647. (Mathematics 128) As a member of a rich and famous family he held high positions in the legal profession. From about 1625 he was a firm friend of Descartes.

1993. DESARGUES, Girard. 1591–1661. (Mathematics 133) Educated apparently in his native Lyons. Became an architect-engineer and from the mid-1620s (if not earlier) was living in Paris. From about 1630 he was a member of Mersenne's circle.

1994. FERMAT, Pierre de. 1601–1665. (Mathematics 134) A member of the *noblesse de robe*. He graduated in law at the University of Orléans in 1631 and later became a leading figure in the *parlement* of Toulouse. His aristocratic style of life may have influenced his attitude to his mathematical achievements. "He never wrote for publication. Indeed, adamantly refusing to edit his work or to publish it under his own name, Fermat thwarted several efforts by others to make his results available in print. Showing little interest in completed work, he freely sent papers to friends without keeping copies for himself. Many results he merely entered in the margins of his books...." (*DSB*)

1630s

1995. ROBERVAL, Gilles Personne de. 1602-1675. (Mathematics 131. Mechanics 505) Apparently of humble origin and self-educated in mathematics. Came to Paris in 1628 and joined Mersenne's circle. Six years later he became professor of mathematics at the Collège Royal, a post he held until his death. When the Académie Royale des Sciences was established in 1666 he was selected as one of the foundation members.

1996. FRENICLE DE BESSY, Bernard. ca. 1605-1675. (Mathematics 144) *Conseiller* at the Cour des Monnaies. From the early 1630s he was a member of Mersenne's circle. When the Académie Royale des Sciences was established in 1666 he was selected as one of the foundation members.

1997. BOULLIAU, Ismael. 1605-1694. (Astronomy 360) A priest; educated in the humanities and law but with an early interest in astronomy. Settled in Paris in 1633 and joined the scientific circle. For many years he carried on astronomical observations, and was the author of several books on astronomy and mathematics which were mostly of no great importance. More significant was his role in the 1650s and 1660s as an "intelligencer" and industrious correspondent. He served as a librarian, finally at the Bibliothèque du Roi where he was associated with the incipient Académie Royale des Sciences (though not as a member). He had contacts in several countries—including Italy, England, and Holland—and in 1663 he was one of the first foreign members elected by the Royal Society of London.

1998. JARDIN DU ROI. Founded in 1635 as a garden of medicinal plants and an institution for the teaching of pharmacy. Its character as a teaching institution was shaped in the same way as that of the Collège Royal (1967): it could not be allowed to compete with the University of Paris as a qualifying institution and so its courses were open to all and had a liberal and flexible character like those of the Collège Royal. Especially significant aspects of its liberality were that its teaching was in French, rather than Latin, and included the new chemical or Paracelsian pharmacy (cf. 1999) in addition to the traditional herbal pharmacy. Thus it took within its scope the incipient sciences of botany and chemistry. It also gave instruction in anatomy for surgeons, an activity which eventually brought in comparative anatomy and zoology (cf. 2026). Despite the continued hostility of the University's Faculty of Medicine the new institution was very successful.

    Around the end of the seventeenth century the Jardin du Roi, while retaining its original function, became increasingly oriented towards science, its staff including some of the leading French botanists and chemists (cf. 2033, 2040, 2041, 2046). This trend increased greatly in the course of the eighteenth century: see 2056.

1999. DAVISON, William. 1593-1669. (Chemistry 787) A native of Scotland. M.A., Aberdeen, 1617. About 1620 he migrated to France and apparently gained the M.D., possibly at Montpellier. He later practised medicine in Paris and gave private lessons

in medical chemistry. In 1648 he was appointed *intendant* of
the Jardin du Roi and took the important step of introducing
public lectures in chemistry as part of that institution's
teaching function.

1640s

2000. CARCAVI, Pierre de. ca. 1600–1684. (Not in Part 1) Son of a
banker and bought his way into the *noblesse de robe*. He was
a person of considerable learning and cultural attainments
and after the death of Mersenne he served as an "intelligencer"
to the French scientific community, carrying on a large correspondence with such figures as Fermat, Descartes, Pascal, and
Huygens. In the early 1660s he became a protégé of Colbert
and one of his most trusted advisers on scientific matters.
In 1663 Colbert made him custodian of the Bibliothèque du Roi
and in 1666 a foundation member of the Académie Royale des
Sciences.

2001. BOURDELOT, Pierre Michon. 1610–1685. (Not in Part 1) Physician
to the Condé family in Paris. In the 1640s he established a
biweekly discussion group attended by some scientists but
mostly by prominent virtuosi, philosophers, and literary men.
The activities of his "academy" were suspended for many years
as a result of political events and the exile of his patrons,
but in 1664 he was able to make a new start. The meetings
were now more concerned with scientific matters and they served
as a useful forum for the scientific community in Paris until
they came to an end in 1684. (During that period the Académie
Royale des Sciences was in existence but it was then a small
group that was closed to outsiders)

2002. ROHAULT, Jacques. 1620–1675. (Natural philosophy 22) Educated
in his native Amiens and then in Paris where he taught himself
mathematics, eventually becoming a private tutor. He embraced
Cartesianism, apparently in the late 1640s, and became the
leading advocate of Cartesian natural philosophy. He was an
active participant in the Montmor Academy (2006) and similar
gatherings, and the lecture course on "physics" that he gave
from the mid-1650s onward was highly successful.

2003. PASCAL, Blaise. 1623–1662. (Mathematics 136, 147, 151. Physics
586, 590) Famous in the history of literature and religion as
well as in the history of mathematics and science. Educated
by his father (himself a talented mathematician). He was a
child prodigy in mathematics and began to participate in the
activities of Mersenne's circle in 1639, before he had turned
sixteen. In 1646 he adopted the religous views of the Jansen-
ists and by 1654 he had largely abandoned his scientific work
in order to devote himself to their cause.

2004. PETIT, Pierre. 1594(or 1598)–1677. (Astronomy 368. Physics
585) A senior governmental official in Paris. He was a
member of Mersenne's circle and in touch with many important
scientific figures. Later (ca. 1660) he was a member of the
Montmor Academy and an advocate of the establishment of a
royal academy of science and a royal observatory. (He himself
owned one of the best collections of telescopes and instruments

in Paris) He was also a correspondent of Oldenburg and played
an important part in developing scientific relations between
Paris and London. In 1667 he was one of the first foreign
members elected by the Royal Society.

2005. PICARD, Jean. 1620-1682. (Astronomy 370, 373) Very little is
definitely known of his early life except that in 1645 he was
associated with Gassendi and helped him in making astronomical
observations. In 1666 he was appointed a foundation member of
the Académie Royale des Sciences.

1650s

2006. MONTMOR, Henri Louis Habart de. ca. 1600-1679. (Not in Part 1)
A member of the *noblesse de robe*. A good scholar with a rich
library; patron of Gassendi in his last years. From the early
1650s he held regular meetings at his mansion in Paris to
discuss scientific and philosophical matters; many of the
savants who attended had previously been members of Mersenne's
circle (Mersenne had died in 1648). In the late 1650s the
meetings of what was now called the Académie Montmor were
formalized: a constitution was drawn up and a secretary appointed, Montmor himself being president. Most of the leading
French scientists were members, together with numerous prominent virtuosi. Foreign visitors to Paris, such as Huygens and
Oldenburg, also attended.
      Though for several years the academy provided a valu-
able focus for the scientific community in Paris, it was
plagued with dissension, partly because of a rift between the
scientists and the virtuosi. As a result it finally collapsed
in 1664. Both its earlier success and its final failure under-
lined the need for a strong academy supported by the government.

2007. MYLON, Claude. ca. 1618-1660. (Not in Part 1) A Parisian
lawyer who acted as an "intelligencer" in the 1650s, exchanging
information between mathematicians in France and Holland
(notably Huygens and van Schooten).

2008. LE FEBVRE, Nicaise. ca. 1610-1669. (Chemistry 793) Trained
as an apothecary by apprenticeship. In 1651, after some years
in the craft, he was appointed demonstrator in chemistry at
the Jardin du Roi. Among those who frequented his course were
some English royalist exiles, and after the restoration of the
monarchy in 1660 he was invited to London as "chymist" (or
apothecary) to Charles II. He became a member of the Royal
Society in 1661.

2009. PECQUET, Jean. 1622-1674. (Anatomy 1587) Studied at the Paris
Faculty of Medicine from about 1646 and made his chief discovery
(of the thoracic duct) while still a student. He took the
M.D. at Montpellier in 1652 and became a physician in Paris.
He was acquainted with a number of scientists and was appointed
a member of the Académie Royale des Sciences on its foundation.

1660s

2010. PARIS ACADEMY. Established in 1666 as the Académie Royale des
Sciences. Its creation was due to Louis XIV's great minister,

Colbert, who in the 1660s, as one of his measures for the
aggrandizement of the French monarchy, was establishing several
royal academies--for the fine arts, historical studies, etc.
The Académie des Sciences appears to have been modelled on the
Florentine Accademia del Cimento (1891)--a small, select body
of professional scientists working together as a team on pre-
determined problems. Originally it comprised only fifteen
members, including two eminent foreigners, Huygens and Cassini,
whom Colbert had induced to settle in Paris. The Academy's
researches covered the whole range of the sciences as they
then existed but its most important work was done in astronomy
and geodesy, the Paris Observatory (2022) being constructed
for its use.

    The Academy was reconstituted in 1699 (see 2034) and
until then it did not publish a periodical, some of its
collective researches being published in book form (cf. 1426).
The policy of collective research soon broke down and until
1699 most of the academicians published their individual
researches in the *Journal des Sçavans* and other periodicals.

2011. *JOURNAL DES SÇAVANS.* The first periodical. Began 5 January 1665.
It covered the whole range of learning and included a substantial amount of mathematics and science until the beginning of
the eighteenth century; the amount then decreased as a result
of the inception of the Paris Academy's *Mémoires.*

2012. THÉVENOT, Melchisédech. ca. 1620-1692. (Not in Part 1) A man
of letters with considerable cultural attainments; widely
travelled and wealthy, with a rich library. His scientific
contributions were quite minor and his importance was as a
patron and "intelligencer." He was a prominent member of the
Montmor Academy for several years and when it collapsed in
the early 1660s he held similar meetings at his own residence
as well as carrying on an extensive correspondence. He was
one of the chief movers in the lobbying that took place in
the mid-1660s for a royal academy. In 1685, soon after he
had been appointed custodian of the Bibliothèque du Roi (in
succession to Carcavi), he was made a member of the Academy.

2013. DU HAMEL, Jean-Baptiste. 1623-1706. (Not in Part 1) A scholarly
priest who was the author of many books, including several on
scientific topics. At the foundation of the Academy in 1666
he was made its secretary, a position he held until 1697. In
1698 he published (in elegant Latin) the first history of the
Academy.

2014. CASSINI, Jean Dominique (formerly Gian Domenico). 1625-1712.
(Astronomy 374) For his earlier career in Italy see 1894.
He arrived in Paris in 1669 and took up his duties in the
Academy. When the Paris Observatory commenced operations
in 1672 he became its (unofficial) director and continued
working in the Observatory until his declining years.

2015. RICHER, Jean. 1630-1696. (Astronomy 380) Nothing is known of
his early life. When the Academy was established in 1666 he
was appointed to its ranks as an *élève astronome.*

2016. MARIOTTE, Edme. died 1684. (Mechanics 514, 517. Physics 595,
596. Earth science 1022. Botany 1316) Nothing definite is

known of his early life. He was appointed to the Academy soon
after its foundation and devoted his life to it, his career of
research exemplifying the intentions of its founders.

2017. GLASER, Christophe. ca. 1615-1672(?). (Chemistry 795) A native
of Basel where he was trained as an apothecary. He settled in
Paris and established a successful business. In 1662 he was
appointed demonstrator in chemistry at the Jardin du Roi.

2018. PERRAULT, Pierre. 1611-1680. (Hydrology 1019) A lawyer and
government official in Paris. Brother of Claude Perrault (2020).

2019. MAGNOL, Pierre. 1638-1715. (Botany 1320) Graduated in medicine
at Montpellier in 1659 and took up the study of botany, soon
acquiring a considerable reputation in the subject. From 1687
he was demonstrator at the botanical garden at Montpellier and
from 1697 its director (as well as being professor of medicine).
He became a member of the Paris Academy in 1709.

2020. PERRAULT, Claude. 1613-1688. (Zoology 1426) Graduated in
medicine at Paris in 1639. He then practised medicine until
1666 when he was appointed a foundation member of the Académie
Royale des Sciences. He continued as an active member until
his death, contributing many memoirs on a wide variety of
topics. He is best known as an architect (notably of the
colonnade of the Louvre and the Paris Observatory) and was
also an accomplished inventor of machinery.

1670s

2021. LA HIRE, Philippe de. 1640-1718. (Mathematics 159, 166. Mechanics 522) His father, an artist, was one of the first disciples of Desargues and the son was influenced by Desargues' ideas
from an early age. After four years (1660-64) in Venice studying art and mathematics he returned to France and was occupied
chiefly as an artist, developing his geometrical interests in
relation to problems in architecture and stonemasonry. He was
appointed to the Academy in 1678 and about that time was becoming interested in astronomy. From 1682 he was based at the
Paris Observatory and took an active part in the programmes of
observation and the Academy's geodesic and cartographic projects.
He was also from 1682 professor of mathematics at the Collège
Royal, and in 1687 he was appointed a professor at the Académie
Royale d'Architecture.
    La Hire was author of many books and papers on a
variety of subjects and an outstanding figure in the French
scientific community of the time.

2022. PARIS OBSERVATORY. Colbert's plans for the Académie Royale des
Sciences had from the beginning included the construction of
an astronomical observatory as an important adjunct to the
Academy's work. A piece of land to the south of Paris was
bought in 1667 and the building of the observatory, on the
plan of Claude Perrault, was completed in 1672. It was well
equipped at governmental expense and its staff began on an
extensive and long-term programme of work, in geodesy and
cartography as well as astronomy.

2023. RÖMER, Ole. 1644-1710. (Astronomy 378) In the 1660s he studied
at the University of Copenhagen (mathematics and astronomy under
Erasmus Bartholin). When Picard came to Denmark to determine
the co-ordinates of Tycho Brahe's old observatory (see 373b)
Römer assisted him and in 1671 returned with him to Paris where
he was appointed to the Academy and the Observatory. He made
his famous discovery of the velocity of light there in 1675.
Six years later he returned to Denmark to become professor of
mathematics and astronomer royal.

2024. LEMERY, Nicolas. 1645-1715. (Chemistry 799) Trained as an
apothecary by apprenticeship and later studied pharmacy at
Montpellier. In 1672 he settled in Paris where he built up
a successful business. He also became well known for his
private lectures in chemistry which became extraordinarily
popular in fashionable circles. He was appointed to the
Academy in 1699, presumably because of the great success of
lectures and his book.

2025. VIEUSSENS, Raymond. ca. 1635-1715. (Anatomy 1605) M.D., Montpellier, 1670. Became chief physician at the main hospital
in Montpellier. Elected correspondent (1699) and then member
(1708) of the Paris Academy.

2026. DUVERNEY, Joseph Guichard. 1648-1730. (Anatomy 1604. Zoology
1436) Graduated in medicine at Avignon in 1667. Settled in
Paris and took part in the meetings of Bourdelot's circle.
Became a member of the Academy in 1674 and professor of anatomy
at the Jardin du Roi in 1679. His anatomical lectures became
very popular.

1680s

2027. FONTENELLE, Bernard de. 1657-1757. (Popularization of science
23) Educated at a Jesuit college. Began his career as a
littérateur with some scientific leanings and, after the
success of his *Entretiens* (23) and some other works, was
appointed secretary of the Académie Royale des Sciences in
1697. His duties included the editing of the Academy's
Mémoires and the writing of its annual *Histoire* (prefixed to
the *Mémoires*). The latter function he used very effectively
as a means of explaining and justifying science to the educated
public, especially through the *éloges* of deceased academicians
which he made into a distinct literary genre of remarkable
popular appeal. He became a member of the Académie Française,
the Académie Royale des Inscriptions et Belles-Lettres, and
some provincial academies, and was elected a foreign member
of the Royal Society of London and the Berlin Academy.

2028. ROLLE, Michel. 1652-1719. (Mathematics 167) Received only an
elementary education and taught himself mathematics. His skill
in algebra, which he displayed in articles in the *Journal des
Sçavans*, won him recognition and in 1685 he was appointed to
the Academy.

2029. VARIGNON, Pierre. 1654-1722. (Mechanics 521) M.A., Caen, 1682.
Was first a parish priest in Caen but soon afterwards settled
in Paris. He became professor of mathematics at the Collège
Mazarin in 1688 and was appointed to the Academy in the same

year. In 1704 he also became a professor at the Collège
Royal. He was important not only for his original work but
also as a teacher of higher mathematics in Paris for over
thirty years.

2030. SAUVEUR, Joseph. 1653-1716. (Physics 600) Educated at the
famous Jesuit college of La Flèche, then went to Paris where
he studied mathematics and attended the lectures of Rohault.
He became well established as a teacher of practical mathematics
(for a time he was tutor at the court of Louis XIV) and in 1686
was appointed professor of mathematics at the Collège Royal.
He was elected to the Academy in 1696 and became examiner for
the Corps du Génie in 1703.

2031. HOMBERG, Wilhelm. 1652-1715. (Chemistry 802) Of German-Dutch
origin. Studied law at Jena and Leipzig but after graduating
went to Padua to study medicine and science. He travelled in
France (where he was associated with Lemery), England (where
he worked with Boyle), Holland, and took his M.D. at Wittenberg.
After further travels in Hungary and Sweden, during which he
gained experience in mining and metallurgical chemistry, he
went to Paris where he became a protégé of Colbert and worked
with French chemists. After Colbert's death in 1683 he had to
leave Paris and practised medicine in Rome for a few years,
but in 1691 he was appointed to the Academy. The rest of his
life was spent in Paris as an academician, absorbed in chemical
research.

2032. JOBLOT, Louis. 1645-1723. (Microscopy 1260) From 1680 to 1721
he taught geometry and perspective at the Ecole des Beaux-Arts.
He associated with several leading members of the scientific
community in Paris and besides his work in microscopy was
interested in physical topics, especially magnetism.

2033. TOURNEFORT, Joseph Pitton de. 1656-1708. (Botany 1323) Studied
medicine (and botany under Magnol) at Montpellier and became a
physician in Paris. In 1683 he was appointed to the staff of
the Jardin du Roi, his duties including not only teaching but
also botanical expeditions to several parts of Europe and in
1700-02 to the Levant. He was elected a member of the Academy
in 1691. He made an important contribution to botany with his
teaching as well as with his writings.

1690s

2034. PARIS ACADEMY. (For its early years see 2010) In 1699 it was
reconstituted and considerably enlarged and strengthened. The
policy of collective research was now officially abandoned and,
rather belatedly, the Academy instituted its own periodical,
its *Histoire et Mémoires*. The *Mémoires* immediately became the
most prestigious vehicle of scientific publication while in
the *Histoire* the Academy addressed itself to the educated
public, an activity brilliantly carried on by its new secretary, Fontenelle (2027).

From its renewal and enlargement in 1699 until its
temporary abolition in 1793, during the Revolution, the Academy
was the leading scientific institution in Europe.

2035. L'HOSPITAL, Guillaume François Antoine de. 1661-1704. (Mathematics 169) A noble and a one-time cavalry officer. He had been a child prodigy in mathematics. From 1693 he was an honorary member of the Academy (honorary because he was a noble).

2036. WINSLOW, Jacob (or Jacques Bénigne). 1669-1760. (Anatomy 1611, 1614) Born and educated in Denmark. About 1696 he became demonstrator in anatomy at the University of Copenhagen. In 1698 he went to Paris where he studied anatomy and surgery under J.G. Duverney. He became Duverney's assistant at the Jardin du Roi and, having become a Catholic, was able to settle in Paris and practise medicine, which he did with much success. He was elected to the Academy in 1708, succeeded Duverney at the Jardin du Roi in 1721, and was made *docteur-régent* at the Paris Faculty of Medicine in 1728. He was largely responsible for making Paris an important centre for the study of anatomy and by mid-century he was regarded as the most outstanding anatomist in Europe.

1700s

2037. COLLÈGE ROYAL. (For its early years see 1967) By 1700 it had twenty chairs, of which eleven were for languages and other humanistic subjects, four were for medicine (including herbal pharmacy or botany and of course anatomy), three were for mathematics, pure and applied, and two were for philosophy—which from 1700, if not earlier, meant natural philosophy and eventually "general physics." Later in the century the content of science in the College's courses became greater and more diversified, with inclusion of such subjects as experimental physics, chemistry, and natural history.

Throughout the eighteenth century (and beyond) the College's teaching was of fundamental importance for the scientific movement in France. For its fortunes during the Revolution see 2096.

2038. MONTPELLIER ACADEMY. Established in 1706 as the Société Royale des Sciences. The most important of the many French provincial academies in the eighteenth century, at least for science—essentially because of the presence of the University and its important medical faculty. It was modelled on the Paris Académie des Sciences and had close links with it. Its members' best work was published in the *Mémoires* of the Paris Academy and it produced only two volumes of its own *Mémoires* during the century. Nevertheless it constituted an important forum on the local scene.

With all the other academies it was abolished in 1793. It was re-established in republican form in 1795 as the Société Libre des Sciences et Belles-Lettres, abolished again in 1815, and restored in 1846 as the Académie des Sciences et Lettres.

2039. CASSINI, Jacques (CASSINI II). 1677-1756. (Astronomy 384. Geodesy 1029) Son of G.D. Cassini (2014) and brought up in the family's quarters in the Paris Observatory; educated at the Collège Mazarin. Through his father's influence he was admitted to the Academy as an *élève* in 1694 (*associé* in 1699;

pensionnaire in 1712). In the late 1690s he travelled in Italy,
the Low Countries, and England (where he met Newton, Halley,
and Flamsteed, and became a member of the Royal Society).
Between 1700 and 1710 he gradually took over from his father
as head of the Paris Observatory and from 1706 he also held a
number of official legal positions. After 1740 he gradually
gave up his scientific work, relinquishing the management of
the Observatory to his son, C.F. Cassini (2055).

2040. GEOFFROY, Etienne François. 1672-1731. (Chemistry 805) Qualified as a pharmacist in 1694 by apprenticeship and then took
up the study of medicine, gaining the M.D. at Paris in 1704.
He was admitted to the Academy as the *élève* of Homberg in 1699
and became an *associé* later in the same year (*pensionnaire* in
1715). He built up a successful medical practice in Paris and
became professor of medicine at the Collège Royal in 1709 and
professor of chemistry at the Jardin du Roi in 1712; he held
both chairs almost up to the time of his death. From 1726 to
1719 he was also dean of the Paris Faculty of Medicine.

2041. VAILLANT, Sébastien. 1669-1722. (Botany 1326) Of humble origin.
He learnt some medicine at a provincial hospital and became an
army surgeon. In 1691 he settled in Paris as a surgeon. He
attended Tournefort's lectures on botany at the Jardin du Roi
and became secretary to Fagon, the director of the Jardin.
Fargon appreciated his ability and made him demonstrator of
plants. From 1717 he substituted for the titular professor
and his inaugural lecture made him famous in botanical circles
(see 1326).

2042. RÉAUMUR, René Antoine. 1683-1757. (Zoology 1440) Came from a
wealthy family of the lesser nobility. Studied law in Bourges,
then mathematics in Paris and became friendly with Varignon
who in 1708 nominated him as his *élève* in the Academy (he
became a *pensionnaire* three years later). Having inherited a
fortune he was able to devote himself entirely to his research
which he carried on in the ambit of the Academy.

   In 1710 he was commissioned by the Academy to compile
technological descriptions of the industries of France (as part
of a long-term task of the Academy that had originally been
given to it by Colbert). He continued these technological
investigations through most of his career with some important
results. At the same time he was pursuing various zoological
researches which he had initiated in 1709; most of these investigations became incorporated in his great work on insects (1440)
but they also encompassed several aspects of embryology and
physiology.

   As author of a vast number of article on a wide variety
of topics in the Academy's *Mémoires*, and of several important
books, Réaumur was one of the most outstanding scientists of
the first half of the eighteenth century and surely one of the
most versatile of any period. He was a leading figure in the
Academy and a foreign member of the Royal Society of London
and all the chief Continental academies.

2043. POURFOUR DU PETIT, François. 1664-1741. (Anatomy 1613. Physiology 1726) He gained the M.D. at Montpellier in 1690 and

then completed his surgical training in Paris. He served as an army surgeon until 1713 and then settled in Paris as an eye specialist. He was elected to the Academy in 1722.

1710s

2044. BORDEAUX ACADEMY. Founded in 1712 as the Académie Royale des Sciences, Belles-Lettres et Arts de Bordeaux. A well-endowed and active academy which was noted for its series of scientific prize competitions, the first such series in France and one of the biggest in Europe.

2045. DELISLE, Joseph Nicolas. 1688-1768. (Not in Part 1. A capable astronomer but important chiefly for his organizational role) Educated at the Collège Mazarin. After studying astronomy under Lieutaud he acquired some instruments and an observatory of his own. The observations he made led to his admission to the Academy in 1714. In 1718 he was appointed a professor of mathematics at the Collège Royal. Three years later, as a result of his growing reputation, he received an invitation from the Tsar of Russia to join his new academy in St. Petersburg and to set up its observatory. For his subsequent career in Russia see 2974.

     On his return to Paris he resumed both his place in the Academy and his chair at the Collège Royal, and continued his astronomical activities. He played an important organizational role in preparing for the transit of Venus in 1761.

2046. JUSSIEU, Antoine de. 1686-1758. (Botany 1325) The first member of a notable family of botanists. He studied medicine (and botany under Magnol) at Montpellier and obtained the M.D. in 1707. After further study of botany in Paris at the Jardin du Roi he was appointed professor there in 1710 and held the post until his death. He also became a member of the Academy. In addition he carried on a lucrative medical practice and the fortune he amassed helped other members of the family to devote themselves to botany, especially his two younger brothers--who were also among his students--Bernard (2051) and Joseph who became a botanical explorer of some renown.

1720s

2047. DUFAY, Charles François de Cisternai. 1698-1739. (Physics 607) Came from the nobility and was originally an army officer. In 1723, through family influence, he was elected to the Academy though he had not yet written anything; he soon justified his election however. After several memoirs on a variety of subjects, he published in 1733-37 his fundamental researches on electricity. In 1732 he was appointed director of the Jardin du Roi, the first non-medical occupant of the position. The Jardin had become run down under the previous director but he was able to restore it to a flourishing condition--see 2056.

2048. BOUGER, Pierre. 1698-1758. (Physics 605, 618. Geodesy 1035) The son of the royal professor of hydrography and a child prodigy. By 1731 he had won three prizes of the Paris Academy for essays on nautical subjects and in that year he was appoint-

ed to the Academy. In 1735 he and two other academicians
(Godin and La Condamine) were sent to Peru by the Academy to
measure an arc of meridian near the equator (cf. 2049). The
expedition was successful and in addition he carried out many
other scientific investigations (he was the first to attempt
to measure the horizontal gravitational attraction of mountains), not returning to Paris until 1744. His subsequent
writings on naval architecture and navigation were esteemed
at the time but he is now remembered for his work in photometry.

2049. MAUPERTUIS, Pierre Louis Moreau de. 1698-1759. (Geodesy 1033.
Mechanics 534. Embryology 1658) Came from a wealthy family;
educated privately. He took up mathematics and in 1723 was
admitted to the Academy as an *élève*, becoming a full member
in 1731. In 1728 he made a trip to London and was converted
from the Cartesian to the Newtonian view of the world; thereafter he was the leading advocate of Newtonianism in France.
In the controversy in the Academy about the figure of the earth
he was the chief protagonist of the Newtonian view that the
earth is slightly flattened at the poles. To settle the question the Academy in 1735-36 sent off two expeditions—one to
measure the length of a degree along the meridian in Peru, near
the equator, and the other to do the same in Lapland, near the
North Pole. Maupertuis was the leader of the latter. The
results of the two expeditions resolved the issue in favour of
the Newtonian position.
        Maupertuis' success gained him the favour of Frederick
the Great who in 1741 invited him to Berlin and made him a
(non-resident) member of the Berlin Academy. In 1745 he took
up residence in Berlin as president of the Academy but a few
years later became embroiled in some bitter controversies and
was savagely attacked by his former friend, Voltaire. As a
result his health broke down in 1753 and he died a few years
later.

2050. LA CONDAMINE, Charles Marie de. 1701-1774. (Geodesy 1035.
Natural history 1233) Came from a wealthy family; educated
at a Jesuit college in Paris. After a short military career
he established himself in the scientific community in Paris
and was appointed to the Academy in 1730. Five years later
he and two other academicians (Godin and Bouguer) were sent
to Peru by the Academy to measure an arc of meridian near the
equator (cf. 2049). After the successful completion of the
mission he made his own way back to France, travelling through
the Amazon valley and making many scientific observations of
various kinds. He arrived back in Paris in 1745 after an
absence of ten years.

2051. JUSSIEU, Bernard de. 1699-1777. (Botany 1333) Trained in
botany by his elder brother Antoine (2046) whom he accompanied
on collecting expeditions. Studied medicine and graduated at
Montpellier, taking a second M.D. at Paris in 1726. In 1722
he was appointed to a post at the Jardin du Roi where he spent
the rest of his life—teaching, conducting field courses in
the countryside around Paris, and supervising and improving
the gardens. He became well known in the scientific community
for his teaching and field excursions, and several of his

students--including his nephew, A.L. de Jussieu (2081)--later became important botanists. For this and other reasons (cf. 1333) he had a far-reaching influence on botany in France.

1730s

2052. CLAIRAUT, Alexis Claude. 1713-1765. (Mathematics 179, 180. Geodesy 1034. Astronomy 397, 402) Educated by his father who was a mathematics teacher of some distinction. His first original contribution to mathematics was made at the age of twelve and he was elected to the Academy at the unprecedented age of sixteen. He became a member of the small group of Newtonians in the Academy and a friend of Maupertuis and Voltaire. In 1736-37 he was a member of the expedition, headed by Maupertuis, which went to Lapland to measure an arc of meridian inside the Arctic Circle. He later assisted the Marquise du Châtelet in her translation of Newton's *Principia* and made important contributions to the accompaning explanations. His impressive work on the figure of the earth (1034) was one of the chief factors leading to an acceptance of Newtonianism in French scientific circles. He became a foreign member of the Royal Society of London and of the leading Continental academies.

2053. LACAILLE, Nicolas Louis de. 1713-1762. (Astronomy 401, 405) An *abbé*. Studied philosophy and theology at the Collège de Navarre but became more interested in astronomy which he taught himself. He was introduced to Jacques Cassini, head of the Paris Observatory, who appointed him as an assistant in 1737; during the next four years he took a prominent part in the Observatory's programme of geodetic measurements. His demonstrated abilities led to his appointment to the chair of mathematics at the Collège Mazarin in 1739 and his election to the Academy two years later. A small observatory was available to him at the Collège Mazarin and there he made a very large number of observations.

In 1750, at his suggestion, the Academy sent him to the Cape of Good Hope to observe the southern skies. He remained there for two years and completed a systematic survey in the course of which he determined the positions of nearly ten thousand stars as well as carrying out various other important observations. This work made him famous and on his return he was elected a member of several of the leading academies.

2054. NOLLET, Jean Antoine. 1700-1770. (Physics 609) Son of an illiterate peasant; his village *curé* set him on the course for a career in the Church. He went to Paris in 1718, graduated in theology and became an *abbé*, but decided to take up science. He made the acquaintance of Dufay and Réaumur and during the early 1730s served as assistant to each of them. With Dufay he visited England and Holland, meeting numerous scientists including Desaguliers, 'sGravesande, and Musschenbroek. On returning to Paris in 1735 he resolved to emulate the teaching methods (though not the theories) of the Newtonians. His lecture course proved to be an outstanding success and he continued to give it for many years (on occasion at the royal court at Versailles). He was elected to the Academy in 1739

and when a chair of physics was created at the Collège de
Navarre in 1759 he was appointed to it. He also had annual
lectureships at the military colleges of La Fère and Mézières
(Coulomb was one of his pupils at the latter).

2055. CASSINI (DE THURY), César François (CASSINI III). 1714–1784.
(Cartography 1037) Son of Jacques Cassini (2039); educated at
the family home in the Paris Observatory. From 1733 he assisted
his father in the Observatory's programme of geodetic measurements. In 1735 he was appointed to the Academy as a supernum-
erary assistant and subsequently progressed through the usual
grades of membership. He gradually assumed the running of the
Observatory as his father withdrew from the position during
the 1740s. After having been obliged to concede victory to
the Newtonians in the dispute about the shape of the earth,
he devoted himself for the rest of his career to the preparation of an accurate large-scale map of France--a very big
undertaking but largely accomplished by the time of his death.

2056. JARDIN DU ROI. (For its early years see 1998) Its first non-
medical director, C.F. Dufay, a prominent member of the Academy,
was appointed in 1732. Though his own researches were in the
physical sciences he was a successful administrator of the
Jardin, completing its evolution into a botanical garden and
initiating a policy of general expansion. He died in mid-
career and his last service to the institution was his greatest:
his nomination of Buffon as his successor.

From 1739 until his death in 1788 Buffon reigned (so
it was said) at the Jardin du Roi. His fame as a writer gave
him influence in the highest quarters and brought forth the
funds for further expansion. As well as enlarging and enrich-
ing the garden, he and his protégé, Daubenton, built up the
institution's natural history collection from almost nothing
to the finest museum of its kind in Europe. In addition, the
Jardin du Roi continued its development as a major centre for
both teaching and research in botany and chemistry. For its
fortunes during the Revolution see 2096.

2057. BUFFON, Georges Louis Leclerc, *Comte* de. 1707–1788. (Natural
history 1234. Geology 1119) Came from a wealthy and cultured
family. Studied law at Dijon but gave it up and went to Angers
where he studied mathematics, and probably also medicine and
botany. After some travelling he settled in Paris in 1732 but
with many visits to his estate in Burgundy; by good management
of his inheritance he became very wealthy.

His scientific interests were originally in mathematics
(he was one of the first followers of Newton in France) and it
was a work on probability theory that led to his election to
the Academy in 1734. His interests shifted to botany however
and in 1739 he was appointed director of the Jardin du Roi
(see 2056). By virtue of his position there he was also the
custodian of the Cabinet du Roi, a natural history museum,
and his famous work, the *Histoire Naturelle*, developed out of
the catalogue he made for the museum.

Buffon was elected to the Académie Française (the literary academy) in 1753 and the inaugural *Discours sur le Style*
which he presented to it is still renowned. In his later years
he was famous internationally, both as a writer and a naturalist.

1740s

2058. ALEMBERT, Jean Le Rond d'. 1717–1783. (Mathematics 187. Astronomy 395, 400. Mechanics 533, 536) Educated at the Collège Mazarin. After trying law and medicine he decided to devote himself to mathematics in which he was almost entirely self-taught. Some early work led to his election to the Academy in 1741, and in 1747 he won a prize of the Berlin Academy to which he was also elected.

During the 1740s he became well known in the salons as a conversationalist and *philosophe*. From about 1746 he was associated with Diderot in the projected *Encyclopédie*. As well as being the editor of the mathematical and scientific articles he took a large part in the enterprise generally and wrote the famous *Discours Préliminaire* that introduced the first volume in 1751. This and his other philosophical and literary works led to his election to the Académie Française in 1754. He later became its permanent secretary and most influential member.

From 1752 Frederick the Great repeatedly tried to persuade him to become president of the Berlin Academy but he declined, though for many years he continued to advise Frederick on the running of the Academy and the selection of new members. He also declined another financially-attractive offer of a position at the court of the Empress Catherine of Russia.

Famous as a mathematician, *philosophe*, and writer, and a luminary of the leading academies of Europe, d'Alembert was one of the most significant figures of the Enlightenment.

2059. ROUELLE, Guillaume François. 1703–1770. (Not in Part 1) Attended the University of Caen. Went to Paris about 1730 and became a pharmacist by apprenticeship. In 1742 he was appointed to a teaching post in chemistry at the Jardin du Roi which he held almost to the time of his death. His teaching was extraordinarily popular and was of major significance for the development of chemistry in France. Most of the men who were to be the leading chemists later in the century were his pupils, including Macquer and Lavoisier. Many non-scientists also attended his lectures, including several of the *philosophes*, notably Diderot and Rousseau. He published only a few papers but many manuscript versions of his lectures, written by his students, are still extant. He was elected to the Academy in 1744 and became a foreign member of the academies of Stockholm and Erfurt.

2060. MACQUER, Pierre Joseph. 1718–1784. (Chemistry 813, 818) M.D., Paris, 1742; he practised medicine for only a few years. After studying under Rouelle his early researches led to his election to the Academy in 1745. From 1757 to 1773 he gave a public lecture course in chemistry (jointly with A. Baumé) and then became lecturer in the subject at the Jardin du Roi. He was a foundation member of the Société de Médecine in 1776 and his numerous activities included membership of the editorial board of the *Journal des Sçavans* and various technical, consulting, and advisory posts.

2061. GUETTARD, Jean Etienne. 1715–1786. (Geology 1113) M.D., Paris, 1743. While still a student he became Réaumur's assistant and curator of his natural history collection. In 1743 he was elected to the Academy and a few years later became physician

to the Duke of Orleans. After the Duke's death his successor
continued to act as patron, granting Guettard sufficient income
to enable him to give all his time to his field trips and
research. He did a great deal of travelling in order to collect
information for his geological map of France.

2062. DAUBENTON, Louis Jean Marie. 1716-1800. (Zoology 1445. Miner-
      alogy 1085) Studied medicine at Paris and attended the anatomy
      and botany lectures at the Jardin du Roi; took his M.D. at
      Rheims in 1741. Buffon, who knew him (having been born in the
      same town), procured his election to the Academy in 1744 on
      the strength of a memoir on the classification of shellfish,
      and appointed him in 1745 to a position at the Jardin du Roi
      in charge of the natural history collection. For the next
      twenty years he was largely occupied with the investigations
      in comparative anatomy that formed part of Buffon's *Histoire
      Naturelle* (cf. 1445). He later did much research in veterinary
      science, especially sheep-breeding, and became a member of the
      Société d'Agriculture and the Société de Médecine as well as
      being elected to several foreign academies. From 1778 he held
      the chair of natural history at the Collège Royal in addition
      to his post at the Jardin du Roi.
            During the Revolution Daubenton played an important
      part in the reorganization of the Jardin du Roi, and when the
      Muséum National d'Histoire Naturelle was established in 1793
      he was elected its director by his colleagues; he was also
      appointed to its chair of mineralogy, a subject he had long
      taught at the Jardin du Roi.

2063. BORDEU, Théophile de. 1722-1776. (Anatomy 1617, 1620) M.D.,
      Montpellier, 1743. His early researches in anatomy led to his
      election as a correspondent of the Paris Academy in 1747.
      After practising in the provinces he went to Paris in 1754.
      He became a very successful physician and, in addition to his
      research in anatomy, made important contributions to the theory
      and practice of medicine.

1750s

2064. BEZOUT, Etienne. 1739-1783. (Mathematics 192) His mathematical
      ability led to his election to the Academy in 1758. From 1763
      he was examiner in mathematics for the Gardes de la Marine
      (trainee naval officers) and a few years later for the Corps
      d'Artillerie as well.

2065. BORDA, Jean Charles. 1733-1799. (Mechanics 538) A younger son
      of a noble family. Graduated at the Ecole du Génie in 1759.
      Three years later a memoir on ballistics led to his election
      to the Academy. After serving as a military engineer he joined
      the navy in 1767 and later rose to high rank. He made a number
      of improvements to nautical astronomical instruments and was
      made a member of the Bureau des Longitudes on its foundation
      in 1795. In the following years he played a leading part in
      the establishment of the metric system of weights and measures
      (the word *mètre* is said to be due to him).

2066. LALANDE, Joseph Jérôme. 1732-1807. (Astronomy 406, 421) Educa-
      ted at the Jesuit college in Lyons, then studied law in Paris

but gave it up for astronomy. He attended the lectures given
at the Collège Royal by Delisle and Le Monnier and in 1751 was
commissioned by the latter to go to Berlin and make lunar observations in concert with those scheduled to be made at the Cape
of Good Hope by Lacaille. (Berlin was chosen because it is on
the same meridian as the Cape) His success in the task, and
the resulting determination of the moon's distance, led to his
election to the Academy in 1753. In 1760 he became editor of
the *Connoissance des Temps*, which he considerably improved,
and in the same year was appointed professor of astronomy at
the Collège Royal. He held the post until his death and was
an outstanding teacher, with many distinguished pupils. During
the 1760s he played a major organizational role in the preparations for the 1769 transit of Venus.

   An indefatigable writer on many subjects, including
technology and the history of science, Lalande had a penchant
for publicity and was often in the public eye. He was also
well known for his popularizations of astronomy (including
several editions of his *Astronomie des Dames*). In the later
stage of the Revolution--after Thermidor--he worked to re-
establish French scientific life and its institutions, and
remained a prominent member of the scientific community until
his death.

2067. DESMAREST, Nicolas. 1725-1815. (Geology 1117) Educated at an
   Oratorian college in Troyes. Went to Paris in 1747 and after
   a precarious existence as a private tutor and literary hack
   obtained a post in 1757 in the governmental bureaucracy which
   supervised and controlled French industry. His ability won
   him promotion, finally in 1788 to the exalted position of
   inspector-general of manufactures in the whole of France.
   Concurrently he was developing his scientific interests and
   during his travels around France on official business he
   accumulated much geological information. His publications,
   from 1753 onward, led to his election to the Academy in 1771.

      During the Revolution his former official position
   (which had made him many enemies) led to his imprisonment and
   he narrowly escaped execution. He was later appointed to a
   governmental position however, and accorded membership in the
   Institut National.

2068. VALMONT DE BOMARE, Jacques Christophe. 1731-1897. (Natural
   history 1235) After some study of scientific subjects he
   obtained a commission from the Ministry of War to travel in
   other countries as a naturalist (and presumably also in a less
   innocent role). On his journeys through much of Europe over
   a period of several years he acquired an extensive knowledge
   of natural history and in 1756 he instituted a course of public
   lectures in the subject at the Jardin du Roi. His lectures
   proved highly successful (as did the book that was doubtless
   based on them) and he continued to give the course for over
   thirty years.

2069. ADANSON, Michel. 1727-1806. (Botany 1335) Educated at the
   Collège de Plessis-Sorbonne in Paris and attended courses at
   the Collège Royal and the Jardin du Roi. At the latter his
   teacher was Bernard de Jussieu with whom he became friendly.

He went to Senegal in 1749 as an employee of a trading company
and when he returned in 1754 he brought back thousands of
botanical and zoological specimens. He published an account
of his scientific explorations in Africa in 1757 and a few
years later was elected to the Academy. In the 1760s part of
his natural history collection was acquired by the Cabinet du
Roi (then under the direction of Buffon) for which he was
granted a life annuity.

1760s

2070. CONDORCET, M.J.A.N. Caritat, *Marquis* de. 1743-1794. (Mathematics
      194) Educated at the Collège de Navarre and concentrated on
      mathematics; became acquainted with d'Alembert who introduced
      him to the salons. His mathematical publications, beginning
      in 1765, led to his election to the Academy in 1769. In 1776
      he became the Academy's permanent secretary and in that capacity
      composed many notable *éloges* of deceased academicians. These
      were doubtless a reason for his election to the Académie Franç-
      aise in 1782. He was also a foreign member of several academies
      in other countries.

           From the mid-1770s Condorcet became deeply involved in
      the growing movement for social, economic, and political reform,
      publishing various pamphlets and influential books such as his
      Vie de Turgot (1786) and *Vie de Voltaire* (1789). In the early
      stages of the Revolution he took a very prominent part in
      politics but with the rise of Robespierre he was obliged to go
      into hiding. It was then that he wrote his best known work,
      Esquisse d'un tableau historique des progrès de l'esprit humain.
      Soon afterwards he was arrested and died in mysterious circum-
      stances.

2071. LAVOISIER, Antoine Laurent. 1743-1794. (Chemistry 822, 830,
      836, 837. Physiology 1738) Came from a wealthy family of
      Paris lawyers; was educated at the Collège Mazarin and gradu-
      ated in law at Paris in 1764. From the early 1760s however
      his interests were turning to science, especially after he
      attended Rouelle's lectures at the Jardin du Roi. In the mid-
      1760s he accompanied the geologist, Guettard (a family friend),
      on several extensive geological expeditions. In 1768, after
      some early researches, he was elected to the Academy.

           In the late 1760s he became a member of the Ferme
      Générale, a group of financiers who collected taxes for the
      government, and for the rest of his life much of his time was
      taken up with financial affairs. In 1775 he was appointed
      Régisseur des Poudres--supervisor of the manufacture of gun-
      powder for the army and navy (and especially of the saltpetre
      used in it), a task which took up much of his time over many
      years. He was able to greatly improve the quality of French
      gunpowder--doubtless a factor in France's future military
      successes. Other activities included the supervision of his
      model farm near Blois in which he demonstrated the potential-
      ities of scientific agriculture, and the proposal of various
      schemes for social and economic improvements. In these and
      other ways he was a true man of the Enlightenment.

           While carrying on his numerous public activities
      Lavoisier was at the same time conducting a long and arduous

series of experiments through which he revolutionized chemistry
and set it on its modern foundations.   At the time of his
death he had embarked on a promising new career in physiology.

2072. ROMÉ DE L'ISLE, Jean Baptiste Louis. 1736-1790. (Mineralogy
1084) After serving in the army he took up the study of science
about 1765 and through the indulgence of a wealthy patron was
able to devote himself to research. In 1785, after the publication of his important book, he was granted a government pension.
He became a member of several of the main European academies
but not of the Paris Academy--because of his opposition to
Buffon, so it was said.

1770s

2073. ROZIER'S JOURNAL. Began in 1771 under the editorship of F. Rozier
as *Observations sur la physique, sur l'histoire naturelle et
sur les arts* and continued from 1794 as *Journal de physique, de
chimie, et d'histoire naturelle*. It marked a new stage in the
history of scientific periodicals, being the first non-instit
utional journal devoted entirely to original work in science
which was able to survive and even to flourish. For a time it
became the biggest of all scientific periodicals and was
followed in other countries by journals of a similar kind.
(During the first half of the nineteenth century however such
general science journals were, with a few exceptions, superseded by specialized journals for the individual sciences)

2074. MONGE, Gaspard. 1746-1818. (Mathematics 201, 206. Chemistry
833, 837) Of humble origin. His ability as a draughtsman came
to the notice of an official and in 1765 he was appointed to a
lowly position at the prestigious Ecole Royale du Génie at
Mézières. He soon won promotion by his original methods for
solving mathematical problems concerned with fortifications,
and by 1770 he was teaching mathematics at the school. His
teaching was highly successful and contributed substantially
to forming the scientific tradition of the élite Corps du Génie.
At the same time he was developing his original ideas in
descriptive geometry and initiating researches in other branches
of mathematics.

In 1772 he became a correspondent of the Academy--not
a full member because he did not live in Paris. In 1780 however
he was made a full member and had to divide his time between
Paris and Mézières (in the far north of France). An appointment in 1783 to the important post of examiner of naval cadets
finally obliged him to give up his position at Mézières.

During the Revolution Monge was one of the leaders of
the important war effort of the scientific and technical community. Later he was also prominent in the educational planning
for the new Ecole Polytechnique in which his experience at the
Ecole du Génie and the naval schools was decisive. From its
foundation he was a dominant figure at the Ecole Polytechnique
and his lectures there, as well as at the short-lived Ecole
Normale de l'An III, were the first public disclosure of his
methods of descriptive geometry (hitherto a military secret).
He was also a foundation member of the Institut National.

From 1796 Monge's duties led him into political activities and contact with Napoleon with whom he became friendly.

He and Berthollet were Napoleon's closest associates and they
both accompanied him, as scientific attachés, on his expedition
to Egypt in 1798-99. During the Napoleonic régime Monge was
loaded with honours--all of which were stripped from him on
the restoration of the monarchy.

2075. LAPLACE, Pierre Simon. 1749-1827. (Mathematics 191, 209.
Astronomy 412, 418. Physics 639, 830) Son of a prosperous
farmer in Normandy. Studied for two years at the University
of Caen, decided to devote himself to mathematics, and left
for Paris in 1768. There he made himself known to d'Alembert
who, impressed by his ability, arranged for his appointment as
professor of mathematics at the Ecole Militaire. His mathematical papers began appearing in 1771 and led to his election to
the Academy in 1773. In 1784 he was appointed examiner of
artillery cadets, an important position not far from the seats
of power (one of his young examinees in 1785 was Napoleon
Bonaparte).

Surviving the Revolution unharmed (though threatened
during the Terror), Laplace took a leading place in the new
institutions set up in 1794-95--foundation member of the
Institut National and the Bureau of Longitudes (soon becoming
president of both) and professor at the short-lived Ecole
Normale de l'An III and at the Ecole Polytechnique where he
was the predominant figure during the late 1790s.

His highly successful *Système du Monde* appeared in 1796,
followed three years later by the first volumes of his masterly
Mécanique Céleste. These works brought him great fame and in
the eyes of many he was the leading scientist of the world.
During the Napoleonic period he was at the height of his influence in the Institut National and in French scientific life
generally. In 1807, in conjunction with his friend, Berthollet,
he established the Société d'Arcueil, a small and select group
which for a few years played an important part in the scientific
community (see 2110).

Laplace stood high in Napoleon's favour and was even
appointed Minister of the Interior for a short period in 1799;
soon afterwards Napoleon made him a member of the Senate and
later its Chancellor. These offices brought him considerable
wealth. At Napoleon's fall in 1815 he successfully transferred
his allegiance to the restored monarchy and was made a marquis
and a member of the Chamber of Peers. In 1816 he was elected
to the Académie Française, a distinction presumably due to the
literary excellence and great success of his *Système du Monde*.

2076. LEGENDRE, Adrien Marie. 1752-1833. (Mathematics 195, 196, 208,
218) Came from a wealthy family. Educated at the Collège
Mazarin where he was well taught in mathematics. He was professor of mathematics at the Ecole Militaire for a few years
(1775-80) but his private income enabled him to devote all his
time to research. In 1782 he won a prize of the Berlin Academy
and in the following year was elected to the Paris Academy.

During the Revolution he lost his fortune and thereafter
held a succession of minor governmental positions, never being
able to get a post commensurate with his talents--due, it was
said, to the jealousy of Laplace. He did however become a
member of the Institute in 1795 and of the Bureau des Longitudes in 1813.

2077. COULOMB, Charles Augustin. 1736-1806. (Mechanics 541. Physics 627) Came from a wealthy bourgeois family. Graduated in 1761 from the Ecole Royale du Génie and from 1764 to 1772 was stationed on the island of Martinique in charge of the construction of a fort. After his return to France he was responsible for engineering projects in several provincial towns, in the course of which he was able to carry out some extensive experiments on friction. This work won a prize of the Academy in 1781 and led to his election in the same year. Thereafter, as a member of the Academy, he was permanently stationed in Paris and was able to devote much of his time to research in physics. He retired from the Corps du Génie in 1791 but continued to be an active member of the Academy and, from 1795, of the Institute.

2078. GUYTON DE MORVEAU, Louis Bernard. 1737-1816. (Chemistry 836, 837) An outstanding lawyer in Dijon. From the mid-1760s, after becoming a member of the Dijon Academy, he became interested in chemistry which he taught himself from textbooks and experiments in a laboratory of his own. From 1776 he gave a public course in the subject on behalf of the Dijon Academy, wrote a textbook and many of the chemical articles in Diderot's *Encyclopédie*, and did translations. During an extended visit to Paris in 1787 his discussions with Lavoisier convinced him of the validity of the latter's new theory and he became one of its most active protagonists.

       Guyton took a prominent part in the Revolution. After local activities in Burgundy he was elected to the National Assembly in 1791 and to the National Convention in 1792, becoming president of the Committee of Public Safety until the emergence of Robespierre. During 1794 he took a leading part in the remarkable war effort of the French scientific community. He was one of the foundation members of the Institut National and taught at the Ecole Polytechnique from its beginning in 1794 until 1811, also acting as its director for some years.

2079. BERTHOLLET, Claude Louis. 1748-1822. (Chemistry 825, 833, 836, 837, 857) Born in Savoy and graduated in medicine at the University of Turin in 1768. Settled in Paris in 1772 as a physician and took another M.D. there in 1778. At the same time he was doing much research in chemistry and was elected to the Academy in 1780. In 1784 he was appointed inspector of dye works and director of the Gobelins factory; he did much to put the traditional craft of dyeing on a scientific basis and was in general a keen proponent of the application of chemistry to industry. He was the first "convert" to Lavoisier's new view of chemistry (in 1785), though a critical one. After Lavoisier's death he was the leading chemist in France.

       During the wars of the Revolution he took an important part in the war effort of the scientific community and was later a member of the committee set up to establish the Ecole Polytechnique where he subsequently taught for some years (he was not a success as a teacher however). He was a foundation member of the Institut National and became a favorite of Napoleon who bestowed many honours on him. In 1807 he and his friend, Laplace, established the Société d'Arcueil, a small and select group which for a time had important ramifications in the scientific community (see 2110).

2080. FAUJAS DE SAINT-FOND, Barthélemy. 1741-1819. (Geology 1120)
      Originally a lawyer but in the early 1770s he took up natural
      history, partly through the influence of Buffon who in 1778
      appointed him to a junior post at the Jardin du Roi. In 1785
      he became royal commissioner of mines and in 1793 professor of
      geology at the Muséum d'Histoire Naturelle.

2081. JUSSIEU, Antoine Laurent de. 1748-1836. (Botany 1341) Came to
      Paris from his native Lyons in 1765 and was thereafter in close
      touch with his uncle, Bernard de Jussieu (2051). In 1770 he
      graduated in medicine and in the same year Bernard procured
      an appointment for him at the Jardin du Roi as deputy to the
      professor of botany; he eventually succeeded to the chair.
      He was elected to the Academy in 1774.
           During the Revolution he took a small part in politics
      and was also involved in the reorganization which converted
      the Jardin du Roi into the Muséum d'Histoire Naturelle. He
      became director of the Museum (in succession to Daubenton) in
      1800. On the establishment of the Université de France in 1808
      he was made a member of its council.

2082. LAMARCK, Jean Baptiste. 1744-1829. (Zoology 1460, 1465. Evolu-
      tion 1522. Geology 1130) Left school at the age of fifteen;
      after several years in the army he studied medicine but a few
      years later gave it up for botany which he studied at the
      Jardin du Roi under Bernard de Jussieu. In 1779 he published
      a flora of France which incorporated an original method of
      identification and was well received by the public and the
      scientific community--and particularly by Buffon who thereafter
      helped him in his career, arranging his election to the Academy
      and his appointment to an assistant's post at the Jardin du Roi.
           When the Muséum d'Histoire Naturelle was established
      in 1793 Lamarck was put in charge of the section for inverteb-
      rates, his research interests having changed in this direction.
      He remained at the Museum until he was overtaken by ill-health
      and blindness after 1810.

2083. VICQ D'AZYR, Félix. 1748-1794. (Zoology 1449) M.D., Paris,
      1774. In the same year he was elected to the Academy, having
      already given proof of his talents in anatomy and physiology.
      He gave private courses in anatomy, which were very popular,
      and also did some lecturing at the Jardin du Roi. After
      successfully halting an epidemic in the Midi he was appointed
      permanent secretary of the Société Royale de Médecine in 1778.
      The *éloges* that he delivered in that capacity on past disting-
      uished physicians won him election to the Académie Française
      in 1788 (as Buffon's successor) and in the same year he was
      granted the additional distinction of being appointed personal
      physician to the Queen. During the Revolution he continued to
      serve the Queen and the strain of his situation apparently
      contributed to his death from fever.
           In his relatively short life Vicq d'Azyr made substant-
      ial contributions to both human and veterinary medicine as well
      as to comparative anatomy.

1780s

2084. SOCIÉTÉ PHILOMATIQUE DE PARIS. Began in 1788 as a small private society. It expanded during the revolutionary period (especially as a result of the suppression of the Academy) and thereafter became established as a non-official scientific society with a substantial membership. It published a bulletin featuring scientific news, etc.

2085. LAGRANGE, Joseph Louis. 1736-1813. (For entries in Part 1 see 1921. For his career in Berlin see 2548) On his arrival in Paris in 1787 to take up his position in the Academy he was received by the royal family and lodged in the Louvre. Despite his closeness to the monarchy (he was a favorite of the Queen) he passed through the Revolution unharmed. From 1790 he was chairman of the committee set up by the Constituent Assembly to standardize weights and measures (leading eventually to the adoption of the metric system) and he took a prominent part in the new institutions set up in 1794-95: foundation member of the Institut National and the Bureau des Longitudes, and professor of mathematics at the short-lived Ecole Normale de l'An III and (until 1799) at the Ecole Polytechnique. The greatest mathematician of the age became an inspiring teacher of mathematics.

Under Napoleon Lagrange was loaded with honours. His last mathematical effort was the revision and extension of his masterly *Mécanique Analitique* for a second edition, during which task he died.

2086. LACROIX, Sylvestre François. 1765-1843. (Mathematics 200, 203) Studied at the Collège des Quatre Nations and in the early 1780s became a protégé of Monge. After a number of teaching posts, mainly at naval and military colleges, he became mathematical examiner for the artillery corps in 1793 (in succession to Laplace). He was elected to the Institute in 1799 and in the same year succeeded Lagrange at the Ecole Polytechnique. He was also taught mathematics at the Ecole Centrale des Quatre Nations and later (until 1815) at the Lycée Bonaparte. When the Paris Faculté des Sciences was established in 1808 he was appointed its dean and professor of calculus. In 1815 he also became professor of mathematics at the Collège de France.

Though he made few original contributions of any importance to mathematics, Lacroix was highly influential as a teacher and writer of textbooks, most of which went through many editions.

2087. CARNOT, Lazare. 1753-1823. (Mathematics 199, 204. Mechanics 542) Graduated in 1773 at the Ecole Royale du Génie where he had been a pupil of Monge. Until the Revolution he served as a garrison officer in various provincial towns, taking part in whatever literary activities existed there. He wrote a book on the theory of machines in 1783 which attracted hardly any attention, and a more successful biography of Vauban (the great seventeenth-century military engineer) in 1784.

Early in the Revolution he took a minor part in national politics but as the military crisis deepened he became more prominent. From 1793 he was a member of the Committee of Public

Safety and earned a permanent place in French history as "the organizer of victory." He continued to hold office into the period of the Directory and was not displaced until 1797. Apart from a short period of power when the Napoleonic régime was collapsing, he devoted the last part of his life to the scientific and technical interests of his early years. From 1796 he was a member of the Institute and served on many of its committees. At the restoration of the monarchy he was obliged to flee the country and he died in exile.

Carnot is unique in the history of science in spending the last period of his life in important research after a hectic career at the heights of political power.

2088. PRONY, Gaspard Clair. 1755-1839. (Mechanics 545) Graduated at the Ecole des Ponts et Chaussées in 1780. From 1783 he worked with its famous director, J.R. Perronet, on the Neuilly bridge, and in 1790 was appointed inspector of studies at the school. When the Ecole Polytechnique was established in 1794 he was appointed professor of analysis, a position he retained until 1815, and he was a foundation member of the Institute.

From 1798 Prony was director of the Ecole des Ponts et Chaussées, and from 1805 he was a member of the official body controlling the Corps des Ponts et Chaussées. In these positions, which he held until his death, he exerted a powerful influence on the development of civil engineering in France. He received honours from Napoleon, Charles X, and Louis Philippe.

2089. PROUST, Joseph Louis. 1754-1826. (Chemistry 839) The son of an apothecary. Qualified as an apothecary by apprenticeship about 1776 and studied chemistry under Rouelle. After holding some minor posts as a teacher of chemistry he migrated to Spain in 1786 and taught at Madrid and Segovia. Most of his research was done in Spain but was published in French journals. He returned to France in 1806 and ten years later was appointed to the Institute.

2090. FOURCROY, Antoine François de. 1755-1809. (Chemistry 828, 836, 837, 849) M.D., Paris, 1780. He did not practise medicine but gave private courses on chemistry and pharmacy. In 1784 he was appointed professor of chemistry at the Jardin du Roi where his teaching was very successful. He was elected to the Société de Médecine in 1780 and to the Academy in 1785. In 1786 he became one of the earliest "converts" to Lavoisier's new view of chemistry and thereafter his teaching and textbooks were very effective in disseminating it.

During the Revolution Fourcroy was involved in politics, becoming a member of the Jacobin Club and of the Convention and its Committee of Public Instruction. Later he was a member of the planning bodies which set up the new scientific and educational institutions and became a foundation member of the Institute. He subsequently taught at the Ecole Polytechnique and the Paris Ecole de Médecine as well as retaining his chair at what was now the Muséum d'Histoire Naturelle. In 1799 Napoleon appointed him to the Council of State where he played a large part in planning and setting up the new national system of education which began in 1808.

2091. ÉCOLE DES MINES. Founded in 1783 on the model of the successful engineering school, the Ecole des Ponts et Chaussées. Because of the undeveloped state of the mining industry in France it was a small institution and around 1790 it ceased to exist in consequence of the national financial crisis. It was re-established by the Convention in 1794 but was abolished in 1802 and replaced by small, practice-oriented schools in the mining districts. In 1814 however it was re-established in Paris and thereafter became a solidly-based institution which played a major part in the development of geology and mineralogy, as well as of mining engineering, in France.

2092. HAÜY, René Just. 1743-1822. (Mineralogy 1089) Educated at the Collège de Navarre. In 1770 he became a priest and later a college teacher. He took up mineralogy as a result of attending Daubenton's lectures on the subject at the Jardin du Roi, and after some early researches was elected to the Academy in 1783. In 1795 he became a member of the Institute and in the same year was appointed to the Ecole des Mines where he taught mineralogy and physics; he later wrote a notable textbook of physics. He became professor of mineralogy at the Muséum d'Histoire Naturelle in 1802 and also at the Faculty of Sciences in 1809.

2093. DOLOMIEU, Dieudonné (or Déodat). 1750-1801. (Geology 1123) Came from the lesser nobility. In 1780 he retired from a military career to devote himself to geological investigations. For much of the 1780s he was travelling in southern Europe, especially in the Alps, and his articles and books appeared from 1783 onward. In 1795 he was appointed to the Institute and in the same year became an *ingénieur* of the Corps des Mines; soon afterwards he began teaching physical geography (i.e. geology) at the Ecole des Mines. Shortly before his death he was appointed professor at the Muséum d'Histoire Naturelle.

2094. DELAMBRE, Jean Baptiste Joseph. 1749-1822. (Geodesy 1044) A private tutor in various subjects from about 1770; largely self-taught in mathematics and astronomy. In the early 1780s he became assistant to Lalande and his subsequent astronomical achievements led to his election to the Academy in 1792. In the same year he was appointed a member of the geodetic survey which the Academy had set up to measure the arc of meridian from Dunkirk to Barcelona--a big undertaking which was not completed until 1810.

     Delambre was a foundation member of the Bureau des Longitudes and in 1803 became one of the permanent secretaries of the Institute. He was appointed professor of astronomy at the Collège de France in 1807 and until his retirement in 1815 held a number of official posts in the national education system. The last years of his life were devoted to writing his six-volume *Histoire de l'Astronomie*--"the greatest full-scale technical history of any branch of science ever written by a single individual." (*DSB*)

2095. LACÉPÈDE, B.G.E., *Comte* de. 1756-1825. (Zoology 1456) Came from the lesser nobility; educated privately. Some early works which he wrote impressed Buffon who in 1785 invited him to

collaborate on the famous *Histoire Naturelle* and appointed him
to a position at the Jardin du Roi. Lacépède wrote two volumes
and after Buffon's death in 1788 completed the project by writing six more. Buffon was always his model, in literary style
and in other ways, and he became recognized as Buffon's successor.

In the early stages of the Revolution Lacépède was
prominent in politics but had to go into exile during the
Terror. In 1794, when it was over, he was appointed to a chair
of zoology at the newly-constituted Muséum d'Histoire Naturelle
and soon afterwards became a foundation member of the Institute.
When the *Histoire Naturelle* was completed in 1804 he gave up
his scientific career and devoted his energies to public affairs
under Napoleon.

1790s

2096. INSTITUTIONS AND THE REVOLUTION. In 1793, during the Terror,
all the royal academies were abolished, as were all the univer-
sities--decayed remnants of the Middle Ages. The Collège Royal
and the Jardin du Roi, however, were spared because of their
liberal character and their long-standing service to the public.
The former was renamed the Collège de France and the latter,
in high favour with the revolutionaries, was enlarged and
reconstituted as the Muséum d'Histoire Naturelle (2097).

The end of the Terror, in July 1794, was followed by
a remarkable effort in the construction of scientific and
technical institutions. They included the Ecole Polytechnique
(2098), the Institut National (2099), the Bureau des Longitudes
(2100), and the Conservatoire des Arts et Métiers (a major
collection of machines, tools, models, and technical documents
with the purpose of stimulating industry and agriculture).
In addition the defunct Ecole des Mines (2091) was restored.

Attempts were also made to set up a system of secondary
education (with an emphasis on mathematics and science) but it
was to be a long time before such a massive social undertaking
could be brought to fruition.

2097. MUSÉUM D'HISTOIRE NATURELLE. Formed in 1793 by enlargement and
reorganization of the Jardin du Roi--which institution already
included, besides the botanical garden and chemical laboratories,
the rich museum of natural history that had been built up by
Buffon and Daubenton (see 2056). The tradition of teaching
and research inherited from the past was continued and augmented:
around 1800 the teaching staff consisted of thirteen professors
--three of zoology, three of anatomy, three of botany, two of
chemistry, one of mineralogy, and one of geology. The collections grew rapidly and the Museum continued for generations to
be the leading institution of its kind in the world. It was,
said Cuvier, "the most magnificent establishment that science
has ever possessed."

2098. ÉCOLE POLYTECHNIQUE. The famous school, founded in December
1794, which inaugurated a new era in higher technical education
and was to be of fundamental importance for the future of
mathematics and physics in France. Its predecessors were the
two engineering schools of the *ancien régime*, the Ecole du

Génie and the Ecole des Ponts et Chaussées, but it incorporated the new idea of a basic training in mathematics and science for all types of engineers. The course took three years (later two) and the graduates then went on to *écoles d'application*-- specialized schools for the different branches of engineering (military, civil, mining) as well as the school of artillery and various others. There was a formidable entrance examination and the course was demanding--mainly mathematics, pure and applied, with physics and chemistry.

The extraordinary success of the school was due to many factors, above all the torrent of social energy that the Revolution had released--the school was itself an expression of the principle of 'careers open to talent'. Most of the leading mathematicians and physicists of nineteenth-century France were graduates of it, and many of its graduates who were practising engineers made significant contributions to science.

2099. INSTITUT NATIONAL. Established in 1795 as a single body replacing the former academies. It was divided into three sections or "classes": the First Class for mathematics and science, the Second for "moral and political sciences", and the Third for literature and the fine arts. The First Class was, in effect, the continuation in republican form of the old Académie des Sciences. Because of the great prestige of science in the Republic the First Class was the largest of the three as well as having primacy of place in the Institute.

In 1816, in consequence of the restoration of the monarchy, the royal academies of the *ancien régime* were restored (and the seniority of the literary academy, the Académie Franç-aise, re-established) but with the Institute retained to form a kind of federal structure--an arrangement that has continued to the present day. (For the revived Académie des Sciences see 2122)

2100. BUREAU DES LONGITUDES. Established in 1795. A committee of astronomers, mathematicians, and naval officers which controlled the state observatories and was responsible for carrying on the practical astronomical and geodetic work formerly done under the direction of the Académie des Sciences; it also dealt with various measures for the improvement of navigation. Its formation as a central organization for astronomy initiated a new stage in the development of the science in France.

2101. FOURIER, (J.B.) Joseph. 1768-1830. (Physics 655) Educated at the military school in Auxerre and in 1789 became teacher of mathematics at the school. In 1794 he went as a student to the short-lived Ecole Normale de l'An III and when the Ecole Polytechnique was established several months later his obvious ability led to his appointment as an assistant lecturer.

In 1798 he was selected by Monge to join the party of *savants* accompanying Napoleon on his Egyptian campaign. Fourier became secretary of the Institut d'Egypte and carried out intensive research on Egyptian antiquities as well as taking part in diplomatic activities. After his return to France in 1801 he assisted in the publication of the enormous amount of Egyptian material that had been amassed and was deputed to write a lengthy historical preface to the resulting *Description de l'Egypte* (21 vols, 1808-25)

When he returned to France Fourier had wanted to resume teaching at the Ecole Polytechnique but Napoleon, aware of his administrative ability, appointed him prefect of the *département* of Isère, with his headquarters at Grenoble. This post he held with distinction from 1802 until Napoleon's fall in 1815. (Much of the work for his later classic on the mathematical physics of heat was done during this period) He then managed to get a position as director of the Bureau of Statistics in Paris. In 1817 he was appointed to the Academy and five years later became its permanent secretary. For his contribution to Egyptology he was elected to the Académie Française in 1827.

2102. VAUQUELIN, Nicolas Louis. 1763-1829. (Chemistry 838) Initially a pharmacist's apprentice. He became Fourcroy's laboratory assistant about 1784 and eventually his friend and collaborator, their first joint paper being published in 1790. During the Revolution he worked as a pharmacist and also took part in the urgent wartime production of saltpetre (for gunpowder). He then became assistant professor of chemistry at the new Ecole Polytechnique (1794-97), professor of assaying at the Ecole des Mines (1795-1801), professor of chemistry at the Collège de France (1801-04), and finally professor of applied chemistry at the Muséum d'Histoire Naturelle from 1804 onward.

Vauquelin also became director of the Ecole de Pharmacie on its establishment in 1803, and from 1811 to 1822 was professor of chemistry at the Faculté de Médecine (in succession to Fourcroy). He had been elected to the Académie des Sciences in 1793, a few days before its abolition, and in 1795 he was a foundation member of the Institut National.

2103. BRONGNIART, Alexandre. 1770-1847. (Geology 1137, 1144, 1145) Studied at the Ecole des Mines but during the Revolution became a pharmacist in the army. In 1794, after his war service, he was appointed *ingénieur des mines* and in 1800 became director of the porcelain factory at Sèvres, a post he held with distinction for the rest of his life. In 1815, after his classical work on the geology of the Paris basin, he was elected to the Academy. In his capacity of mining engineer he was promoted to *ingénieur en chef des mines* in 1818 and four years later was appointed professor of mineralogy at the Muséum d'Histoire Naturelle (in succession to Haüy).

2104. MIRBEL, Charles François Brisseau de. 1776-1854. (Plant anatomy 1346, 1362) His education and early career were disrupted by the Revolution. After a succession of short-term posts (including one at the Muséum d'Histoire Naturelle and another as head gardener at Malmaison) he became supplementary professor of botany at the new Faculté des Sciences in Paris in 1808, and in the same year was elected to the Institute. In 1829 he was appointed professor-administrator at the Jardin des Plantes.

2105. LATREILLE, Pierre André. 1762-1833. (Zoology 1457, 1463) M.A., Paris, 1780. Ordained a priest in 1786. By the mid-1790s he was well known in Paris as a naturalist and in 1798 was made organizer of the entomological collection of the Muséum d'Histoire Naturelle. He became a member of the Institute in 1814 and a few years later took over Lamarck's course on invertebrate zoology at the Museum. In 1829 he was appointed to the Museum's new chair of entomology.

2106. CUVIER, Georges. 1769-1832. (Zoology 1458, 1459, 1467, 1473; see also 1476. Palaeontology 1137, 1138) Born in a town in eastern France which then belonged to the Duchy of Württemberg; received a good education (including natural history) at a leading college in Stuttgart, graduating in 1788. For the next six years he was a private tutor to a noble family in Normandy, a post which allowed him ample leisure to pursue his studies in zoology. In 1794 he wrote to E. Geoffroy Saint-Hilaire (recently appointed professor at the Muséum d'Histoire Naturelle) who invited him to come to the Museum and collaborate in research, which he did early in 1795.

Later in 1795 Cuvier was appointed assistant professor at the Museum and in the following year was elected to the Institute. He became full professor at the Museum in 1802 and secretary of the First Class of the Institute the next year. (In the latter capacity he later wrote some important reports and many attractive *éloges*) In 1800 he succeeded Daubenton as professor of natural history at the Collège de France (in addition to his post at the Museum).

From 1800 Cuvier became heavily involved in the post-revolutionary reorganization of higher education, initially in establishing provincial *lycées*. He came into favour with Napoleon and when the Université de France was set up in 1808 he was appointed one of its councillors and had much to do with the reshaping of the Sorbonne and the establishment of provincial faculties. He received the title of *chevalier* from Napoleon and when the monarchy was restored he was not only allowed to retain his appointments but was even created a Councillor of State. He became a member of the Académie Française in 1818, a baron in 1819, and after the revolution of 1830 a peer of France.

2107. GEOFFROY SAINT-HILAIRE, Etienne. 1772-1844. (Zoology 1468, 1472, 1476) His education, at some of the chief colleges in Paris, was disrupted by the Revolution. Through the influence of Daubenton he became demonstrator in zoology at the Jardin des Plantes--as the Jardin du Roi was then known--early in 1793 (in place of Lacépède who had been forced to flee). When the Jardin became the Muséum d'Histoire Naturelle a few months later he was appointed one of the professors of zoology (the other being Lamarck).

Early in 1795 Cuvier made his début in Paris and he and Geoffroy worked together for a year, jointly publishing several papers. From 1798 to 1801 Geoffroy was one of the group of *savants* who accompanied Napoleon to Egypt, distinguishing himself on the expedition. During his absence however Cuvier had become recognized in Paris as the rising star of zoology; the tension between the two men progressively increased thereafter.

Geoffroy was elected to the Institute in 1807 and became professor of zoology at the newly reorganized Sorbonne in 1809 (in addition to his post at the Museum).

2108. BICHAT, (M.F.) Xavier. 1771-1802. (Anatomy 1629) Studied
anatomy and surgery initially in a hospital in Lyons and then,
from 1794, at the Hôtel-Dieu in Paris where he became assistant
to the chief surgeon. From 1796 he gave private lessons in
anatomy and in that year founded the Société Médicale d'Emula-
tion which included Cabinis, Corvisart, and Pinel.

1800s

2109. FACULTIES OF SCIENCE. The new system of education set up under
Napoleon in 1808 had its main strength in the secondary schools,
the *lycées*. The universities, which had all been abolished in
1793, were not restored: instead individual faculties were
created. In Paris and several provincial cities Faculties of
Letters, Science, Medicine, Law, and Theology were established.
There were no links between the different faculties: there
were only vertical links to the ministry of education which,
strange as it now seems, was named the Université de France.
The traditional conception of a university as a local, corpor-
ate body concerned with the whole range of knowledge (a conception soon to flourish in Germany) was absent from France for
almost the whole of the nineteenth century.

  The separate faculties were created as professional
training schools, the chief role of the Faculties of Science
(as of Letters) being to train teachers for the *lycées*.
Research was not part of their function.

2110. SOCIÉTÉ D'ARCUEIL. A unique society centered on the two luminaries, Berthollet and Laplace, who both lived at Arcueil (then
a village near Paris), and consisting of them and about a dozen
of their friends and protégés. It existed informally from 1801
to 1807 and formally from 1807 to 1813, and seems to have been
essentially a private research group in which bright young men
worked on the ideas of the two masters (cf. 639). Most of them
soon became outstanding in the scientific community and in
various subtle ways the Société d'Arcueil seems to have had
considerable influence.

2111. POISSON, Siméon Denis. 1781-1840. (Mathematics 207. Mechanics
547. Physics 643, 666) Graduated at the Ecole Polytechnique
in 1800 and through the influence of Laplace was appointed to
a succession of teaching posts there, becoming titular professor in 1806. He was a member of the Société d'Arcueil and,
doubtless through Laplace's patronage, was appointed to the
Bureau des Longitudes and in 1809 to the chair of mechanics
at the Paris Faculté des Sciences. In 1812 he was elected to
the Institute. From 1820 until his death he was a member of
the governing body of the national education system where he
opposed the anti-science tendencies of the royalist régime.

2112. POINSOT, Louis. 1777-1859. (Mechanics 546, 552) Graduated
from the Ecole Polytechnique in 1797 and from the Ecole des
Ponts et Chaussées three years later. He gave up engineering
in order to devote himself to mathematics, and after some
junior teaching positions was appointed to a post at the Ecole
Polytechnique in 1809; in the same year he also became an
inspector of the newly-established Université de France. In

the 1840s he was appointed to the Bureau des Longitudes and
the Conseil de l'Instruction Publique and also played a small
part in politics.

2113. NAVIER, (C.L.M.) Henri. 1785-1836. (Mechanics 550) Graduated
in 1804 at the Ecole Polytechnique and two years later at the
Ecole des Ponts et Chaussées. After working as an engineer
and an editor of engineering publications he taught applied
mechanics at the Ecole des Ponts et Chaussées from 1819,
becoming titular professor in 1830. In the following year
he was also appointed to a chair at the Ecole Polytechnique.
He was elected to the Academy in 1824. After the revolution
of 1830 he became an important consultant to the government
on engineering matters.

2114. BIOT, Jean Baptiste. 1774-1862. (Physics 642, 651) Graduated
at the Ecole Polytechnique in 1797 and was soon afterwards
appointed entrance examiner for the Ecole, a position he retained when he became professor of mathematical physics at the
Collège de France in 1800. About 1801 he became associated
with Laplace and Berthollet, and was one of the first members
of the Société d'Arcueil. In 1803 he was elected to the
Institute, and in 1808--doubtless through the influence of
Laplace--was appointed professor of astronomy at the Paris
Faculté des Sciences (though he mostly taught physics). He
was the author of several textbooks (those on astronomy and
physics being especially important) and over three hundred
papers on many aspects of physics.
     On account of his researches on ancient astronomy
(Egyptian, Babylonian, and Chinese) Biot was elected to the
Académie des Inscriptions in 1841. Presumably because of these
and other historical works (which included a history of the
sciences during the Revolution and a notable biography of
Newton) and because he was a writer of some distinction he was
elected to the Académie Française in 1856.

2115. AMPÈRE, André Marie. 1775-1836. (Physics 650. Philosophy of
science 39) A child prodigy, especially in mathematics. After
some minor teaching posts he became a *répétiteur* (assistant
lecturer) in mathematics at the Ecole Polytechnique in 1803.
In 1808 he was appointed inspector-general of the new education
system, a post he held for most of his life. He became a
member of the Institute in 1814 and was appointed professor
of experimental physics at the Collège de France in 1824.
     Ampère's researches were originally in pure mathematics
and then for a few years in chemistry. It was not until 1820
that he began the work in electrodynamics for which he became
famous. Throughout his career he was much concerned with
philosophy.

2116. MALUS, Etienne Louis. 1775-1812. (Physics 640) Son of a high
official; educated privately. Entered the Ecole Polytechnique
in 1794 and eventually graduated as an officer in the Corps
du Génie. He took part in Napoleon's campaign in Egypt and
Syria (1798-1801) and after later garrison service in several
parts of France he was able to stay in Paris, becoming an
examiner for the Ecole Polytechnique in 1805. Soon afterwards
he became a member of the Société d'Arcueil and began publish-

ing on optics; his discovery of the polarization of light was
announced in 1808. In 1810 he won a prize of the Institute
and some months later was elected to it. The Royal Society
of London awarded him its Rumford medal in 1811.

2117. ARAGO, (D.) François (J.). 1786-1853. (Physics 645) Graduated
at the Ecole Polytechnique in 1805 and, through the influence
of Laplace, was appointed secretary of the Bureau des Longitudes.
Soon afterwards he became a member of the Société d'Arcueil.
In 1809 he was elected to the Institute and in the same year
became professor of mathematics at the Ecole Polytechnique.
This post he held until 1830 when he became permanent secretary
of the Academy. He did however continue his work at the Paris
Observatory which had been put under his direction by the
Bureau des Longitudes.

As secretary of the Academy Arago was at the centre
of the French scientific community for twenty years and, by
virtue of his office, wrote many *éloges* which were widely read
(a collection of them was translated into English). Though he
made no major contributions to astronomy he was an assiduous
observer and, as director of the Paris Observatory, had a
beneficial influence on the careers of several young astronomers, notably Leverrier. He was also a successful popularizer of astronomy and--in association with his friend, A. von
Humboldt--of the earth sciences.

Following the revolution of 1830 Arago went into
politics and became an influential member of the Chamber of
Deputies, and was for a time the president of the Paris Municipal Council. At the revolution of 1848 he was a member of
the provisional government and briefly a minister.

2118. ÉCOLE DE PHARMACIE. Established in 1803. The first director
was Vauquelin. Later professors included Pelletier, Berthelot,
and Moissan.

2119. THENARD, Louis Jacques. 1777-1857. (Chemistry 846, 866) Son
of a peasant. Went to Paris in 1794, attended the public
lectures at the Collège de France and other institutions, and
became Vauquelin's laboratory assistant. He was helped by
Vauquelin and Fourcroy and in 1798 became a demonstrator at
the Ecole Polytechnique. In 1804 he succeeded Vauquelin as
professor of chemistry at the Collège de France and in 1808
was also appointed professor at the Paris Faculté des Sciences.
He was a member of the Société d'Arcueil and was elected to
the Institute in 1810.

Thenard became dean of the Paris Faculty in 1822 and
was subsequently prominent in the upper administrative levels
of the Université de France, becoming its chancellor in 1845-52.
Because of his interests in applied chemistry he was a leading
member of the Société d'Encouragement pour l'Industrie Nationale
and a member of the governing body of the Conservatoire des Arts
et Métiers. He was an influential adviser to the government
on scientific and technical matters, and received many honours,
becoming a baron in 1825 and a peer in 1832. He was elected
to the Chamber of Deputies in 1827 and 1830.

2120. GAY-LUSSAC, Joseph Louis. 1778-1850. (Chemistry 854, 864, 876)
Graduated at the Ecole Polytechnique in 1800 and in the next

year was invited by Berthollet to be his assistant in his private laboratory at his mansion in Arcueil. Gay-Lussac thus became one of the first members of the Société d'Arcueil. As well as training him in chemical research Berthollet also facilitated his future career.

After holding junior positions at the Ecole Polytechnique from 1803 Gay-Lussac succeeded Fourcroy as professor of chemistry there in 1810. In 1808 he had become professor of physics in the Paris Faculté des Sciences and had already--in 1806--been elected to the Institute. He was put in charge of the national gunpowder factory in 1818, became chief assayer at the Mint in 1829, and acted at various times as a governmental adviser. In 1832 he gave up his physics chair at the Faculty and took the chair of general chemistry at the Muséum d'Histoire Naturelle. During the 1830s he was elected several times to the Chamber of Deputies and in 1839 became a peer.

2121. LEGALLOIS, Julien Jean César. 1770-1814. (Physiology 1741)
While a medical student he took part in the drive to produce saltpetre and gunpowder during the wars of the Revolution. As a result he was later selected to study at the new Ecole de Santé, where he graduated in 1801. Thereafter, until his early death, he practised medicine in Paris while at the same time carrying on experimental investigations in physiology.

1810s

2122. ACADÉMIE DES SCIENCES. Re-established in 1816, its functions during the period 1795-1815 having been performed by the First Class of the Institute (2099). Through much of the nineteenth century the Academy retained its traditional importance in French science, and with the publication of its *Comptes Rendus* (2146) made a major contribution to the dissemination of science on the international scene. In the latter part of the century, however, for a variety of reasons both internal and external (especially the increase in the number of specialized societies and the vast growth in the size and complexity of the scientific movement) its importance rapidly declined and by the end of the century it was little more than an honorific body.

2123. CAUCHY, Augustin Louis. 1789-1857. (Mathematics 213. Mechanics 549) Graduated at the Ecole Polytechnique in 1807 and at the Ecole des Ponts et Chaussées two years later. After working in the provinces as an engineer he returned to Paris in 1813 and thereafter devoted himself to mathematics. He became a member of the Academy in 1816 and at about the same time was appointed to chairs at the Ecole Polytechnique and the Faculté des Sciences.

Following the revolution of 1830 Cauchy refused to take the oath of allegiance to the new régime and went into exile. He spent most of the 1830s in Prague, at the court of the exiled Charles X, but in 1838 returned to Paris. He was able to resume his place in the Academy but did not get back his chair at the Faculty until after the revolution of 1848.

In that episode, as in much else, his behaviour was highly quixotic. But, despite his eccentricities, he was one of the greatest mathematicians of the nineteenth century.

2124. CORIOLIS, (G.) Gustave. 1792–1843. (Mechanics 551, 554)
      Graduated at the Ecole Polytechnique about 1810 and then at
      the Ecole des Ponts et Chaussées. After practising as an
      engineer he was appointed to a junior teaching post at the
      Ecole Polytechnique in 1816. From 1829 he was also professor
      of mechanics at the recently-established Ecole Centrale des
      Arts et Manufactures, and in 1836 he succeeded Navier as
      professor of applied mechanics at the Ecole des Ponts et
      Chaussées. He was elected to the Academy in 1836 and became
      director of studies at the Ecole Polytechnique two years later.

2125. DULONG, Pierre Louis. 1785–1838. (Physics 647) Began the
      course at the Ecole Polytechnique in 1801 but was unable to
      finish it because of ill health. After some vicissitudes
      he became an assistant in Thenard's laboratory and then in
      Berthollet's private laboratory at Arcueil, subsequently
      becoming a member of the Société d'Arcueil. From 1811 he held
      some minor teaching posts and in 1820 was appointed professor
      of chemistry at the Paris Faculté des Sciences. From 1820 he
      was also professor of physics at the Ecole Polytechnique and
      from 1830 director of studies there. He was elected to the
      Academy in 1823.

2126. FRESNEL, Augustin Jean. 1788–1827. (Physics 646) Entered the
      Ecole Polytechnique in 1804 and after the two-year course
      there went on to the three-year course at the Ecole des Ponts
      et Chaussées. He began his investigations in optics while
      working as a civil engineer in various parts of France, and
      after being assigned to Paris was able to pursue them more
      intensively. In 1823 he was elected to the Academy and just
      before his death he was awarded the Rumford medal of the Royal
      Society of London.
          In addition to his work in pure science Fresnel made
      a major contribution with the special lenses he invented for
      lighthouses in the early 1820s. Simplified versions of Fresnel
      lenses are now used in innumerable applications.

2127. CHEVREUL, Michel Eugène. 1786–1889. (Chemistry 875) From 1803
      he studied chemistry at the Muséum d'Histoire Naturelle under
      Vauquelin. After several junior teaching positions in various
      institutions he became professor of chemistry at the Museum in
      1830. He also became director of dyeing at the Gobelins in
      1824 (in succession to Berthollet) and taught chemistry there
      as well as at the Museum for almost sixty years. He was elected
      to the Academy in 1826 and was director of the Museum from 1864
      to 1879.
          In addition to his various other interests Chevreul
      was a notable historian of chemistry and also wrote on phil-
      osophy and scientific method. He was best known for his work
      on the perception of colours which was more a contribution to
      the arts than to the sciences and was highly influential. He
      was one of the most esteemed scientists of the late nineteenth
      century and his centenary in 1886 was celebrated as a national
      event.

2128. PELLETIER, Pierre Joseph. 1788–1842. (Chemistry 862) Came
      from a family of pharmacists. Qualified at the Ecole de

Pharmacie in 1810 and two years later gained the degree of
docteur ès sciences. He became assistant professor at the
Ecole de Pharmacie in 1815 and full professor in 1825; he also
ran his own pharmacy and a chemical manufacturing business.
He was elected to the Académie de Médecine in 1820 and to the
Académie des Sciences in 1840.

2129. DUTROCHET, (R.J.) Henri. 1776-1847. (Plant physiology 1363)
Graduated in medicine at Paris in 1806. After serving in the
army as a medical officer he gave up medicine in 1809 and,
being now of independent means, devoted himself to research.
He was elected to the Académie de Médecine in 1823 and to the
Adadémie des Sciences in 1831.

2130. MAGENDIE, François. 1783-1855. (Physiology 1742) Studied
medicine in Paris hospitals and at the Ecole de Médecine
where he graduated in 1808. In 1811 he was appointed demonstrator in anatomy at the Paris Faculté de Médecine but resigned
in 1813 and went into private practice; he also began giving a
private course in physiology. His course, illustrated by
numerous experimental demonstrations, proved very successful
with the medical public and formed the basis of his influential
Précis Elémentaire de Physiologie of 1816-17. His researches,
carried on at the same time, were highly fruitful and led to
his election in 1821 to both the Académie des Sciences and the
Académie de Médecine. In the same year he founded the *Journal
de Physiologie Expérimentale*, the first periodical of its kind.
    In 1831 Magendie became professor of medicine at the
Collège de France where his teaching--incorporating the many
experimental techniques he had developed--was very successful
and attracted many capable students.

1820s

2131. PONCELET, Jean Victor. 1788-1867. (Mathematics 214. Mechanics
548) Graduated at the Ecole Polytechnique in 1810 and became
a military engineer. From 1825 to 1835 he taught engineering
mechanics at the military Ecole d'Application at Metz, and
from 1838 to 1848 he gave a similar course at the Faculté des
Sciences in Paris. He was elected to the Academy in 1834.
Following the revolution of 1848 he took some part in politics
and was commandant of the Ecole Polytechnique until his retirement in 1850.

2132. LAMÉ, Gabriel. 1795-1870. (Mathematics 232) Graduated at the
Ecole Polytechnique in 1817 and at the Ecole des Mines three
years later. From 1820 to 1832 he was in Russia where he was
engaged in the teaching and practice of civil engineering.
After his return to Paris he was professor of physics at the
Ecole Polytechnique from 1832 to 1844 while concurrently acting
as a consulting engineer. He was elected to the Academy in
1843 and in the following year became an examiner at the
Sorbonne and was appointed to a chair in mathematical physics
there in 1851.

2133. STURM, Charles François. 1803-1855. (Mathematics 222, 224.
Physics 659) Born in Geneva and educated at the Geneva Academy.

In 1825 he settled in Paris where he carried on research while earning his living by teaching and editing; he was able to make the acquaintance of several of the leading French scientists, especially Ampère, Arago, and Fourier. In 1837 his publications won him election to the Academy and in the following year he became a *répétiteur* at the Ecole Polytechnique, and in 1840 a professor. In that year he was also appointed professor of mechanics at the Paris Faculté des Sciences.

2134. CARNOT, Sadi. 1796-1832. (Physics 657) Son of Lazare Carnot. After graduating at the Ecole Polytechnique in 1814 he went on to the two-year course at the Ecole du Génie at Metz. Thereafter he served on garrison duty in various parts of France. In 1819 he went into the army reserve on half pay and settled in Paris where he was able to pursue his scientific interests.

2135. CLAPEYRON, Benoit Pierre Emile. 1799-1864. (Physics 672) Graduated at the Ecole Polytechnique in 1818 and at the Ecole des Mines two years later. He worked as a civil engineer in Russia from 1820 to 1830, and after returning to France specialized in the design and construction of steam locomotives; from 1844 he gave a course on the steam engine at the Ecole des Ponts et Chaussées. He was elected to the Academy in 1848.

2136. DUMAS, Jean Baptiste André. 1800-1884. (Chemistry 877, 886. See also 1664) Born in France but studied pharmacy and chemistry in Geneva. In 1823 he went to Paris and was appointed Thenard's *répétiteur* at the Ecole Polytechnique. He was much concerned with applied chemistry and in 1829 was one of the founders of the Ecole Centrale des Arts et Manufactures where he lectured until 1852. In 1835 he succeeded Thenard at the Ecole Polytechnique and from 1839 was also professor of organic chemistry at the Ecole de Médecine. He was elected to the Academy in 1832 and in the same year was appointed assistant professor at the Sorbonne, becoming full professor in 1841; this post he held until 1868 when he became permanent secretary of the Academy.

Dumas was the most prominent French chemist of his time and a brilliant teacher. He is said to have been the first in France to give practical laboratory instruction in chemistry to his students—at the Ecole Polytechnique from 1832 (partly at his own expense). With the revolution of 1848 he became involved in politics. He was briefly a minister and was later made a senator by Napoleon III. In 1859 he became president of the Paris Municipal Council and collaborated with Haussmann in the transformation of the city.

2137. BALARD, Antoine Jérome. 1802-1876. (Chemistry 878) Graduated at the Ecole de Pharmacie in his native Montpellier in 1826. After holding minor teaching posts in Montpellier and then in Paris he became professor of chemistry at the Sorbonne in 1842. He was elected to the Academy two years later and was appointed professor of chemistry at the Collège de France in 1851. He did much to further the careers of his two most outstanding students, Pasteur and Berthelot.

2138. ÉLIE DE BEAUMONT, (J.B.A.L.) Léonce. 1798-1874. (Geology 1155,

1169) Graduated at the Ecole Polytechnique in 1819 and at the Ecole des Mines in 1822. In the following year he and another young mining engineer, P.A. Dufrénoy, were sent to England to inspect its mining industry and to collect information about methods of geological surveying. In 1825 the two of them were commissioned to carry out surveys to collect data for the long-delayed project of a geological map of France. The task took them fifteen years, working half-time, and in parallel with that activity Elie de Beaumont lectured on geology at the Ecole des Mines from 1827 and at the Collège de France from 1832.

He was one of the founders of the Société Géologique de France in 1830 and was elected to the Academy in 1835, becoming one of its secretaries in 1853. From 1861 he was chairman of the Conseil Général des Mines, and from 1865 director of the newly-established Service de la Carte Géologique (the national geological survey).

2139. DESHAYES, (G.) Paul. 1797-1875. (Palaeontology 1151) Began studying medicine but gave it up for natural history, especially geology and malacology. Being financially independent he was able to devote himself to research. He became well known in Parisian scientific circles and was one of the founders of the Société Géologique de France in 1830 and later its president. In 1869 he was appointed professor of conchology at the Muséum d'Histoire Naturelle.

2140. BRONGNIART, Adolphe Théodore. 1801-1876. (Palaeontology 1154. Botany 1359) Son of Alexandre Brongniart who gave him his basic training in the natural history sciences and in methods of research. He began the study of medicine in 1818 (and qualified ten years later) but at the same time carried on original research, publishing his first paper in 1820; by the late 1820s he was producing results of great importance. With two associates he founded the important periodical *Annales des Sciences Naturelles* in 1824. In 1831 he became assistant to Desfontaines, professor of botany at the Muséum d'Histoire Naturelle, and two years later succeeded him; he held the chair for the rest of his life. He was elected to the Academy in 1834 and became inspector-general of the Université de France in 1852.

2141. BOUSSINGAULT, Jean Baptiste. 1802-1887. (Plant chemistry 1374) Spent ten years in South America as a mining engineer. His researches there on geology and other earth sciences won the commendation of Humboldt and gained him a scientific reputation. On returning to France in 1832 he received the degree of *docteur ès sciences* and was appointed dean of the Faculté des Sciences at Lyons. Thenceforward his chief scientific interest was the application of chemical principles to agriculture. In 1839 he was elected to the Academy and in the same year was appointed professor of chemistry at the Sorbonne and professor of agriculture at the Conservatoire des Arts et Métiers.

2142. GEOFFROY SAINT-HILAIRE, Isidore. 1805-1861. (Zoology 1478) Son of Etienne Geoffroy Saint-Hilaire. He became his father's laboratory assistant in 1824 and thereafter published a wide range of researches on mammals and birds. By 1830 he was attracting attention for his original ideas (notably on what

is now called ecology) and was elected to the Academy in 1833.
He succeeded his father as professor of comparative anatomy at
the Faculté des Sciences in 1837, and as professor of zoology
at the Muséum d'Histoire Naturelle in 1841.

Geoffroy was active in the administration of higher
education, organizing the Faculté des Sciences at Bordeaux in
1838 and serving in the demanding post of inspector-general
of education from 1844 to 1850. Another of his activities was
the promotion of the acclimatization of useful animals: he
established the Société d'Acclimatation in 1854 and the Jardin
d'Acclimatation in Paris. In 1847 he published a very capable
biography of his father.

2143. SOCIÉTÉ ANATOMIQUE, Paris. Its periodical began in 1826.

2144. FLOURENS, (M.J.) Pierre. 1794-1867. (Physiology 1743) Graduated in medicine at Montpellier in 1813. He then went to Paris,
decided on a research career, and became a protégé of Cuvier.
In the early 1820s he twice received the Academy's Montyon
Prize and in 1828 was elected a member. After acting as deputy
for Cuvier at the Collège de France he became professor there
in 1832. In the following year he succeeded Cuvier as secretary of the Academy and ably continued the tradition of compos-
ing *éloges* of deceased members. As a writer of some distinction
he was elected to the Académie Française in 1840.

2145. POUCHET, Félix Archimède. 1800-1872. (Microbiology 1806)
Studied medicine in his native Rouen and then in Paris, graduating in 1827. Soon afterwards he was appointed director of
the Muséum d'Histoire Naturelle at Rouen, a post he retained
for the rest of his career. He also held teaching positions
in medicine and science in Rouen, and was the author of some
successful popularizations of biological themes. He became a
well-known and respected naturalist, a member of many learned
societies, and a correspondent of the Paris Academy. "In view
of Pouchet's wide-ranging contributions to biology, it is
unfortunate that he is often remembered only as the defeated
adversary of Pasteur...." (*DSB*)

1830s

2146. *COMPTES RENDUS de l'Académie des Sciences.* A new periodical of
the Academy containing notes or short papers which was begun
in 1835 in order to circumvent the long delays in the publication of the Academy's *Mémoires*. It was issued weekly and—a
highly important innovation—included items by non-members
(after approval by official committees of the Academy). It
proved extraordinarily successful and soon became the biggest
of all scientific periodicals (and still remains one of the
biggest today).

2147. CONGRÈS SCIENTIFIQUE DE FRANCE. Held annual meetings in different
cities, beginning in 1833. (cf. 2192) Its periodical ceased
publication in 1879.

2148. CHASLES, Michel. 1793-1880. (Mathematics 235, 256, 271) Came
from a wealthy family. Graduated at the Ecole Polytechnique
in 1816 and subsequently qualified as an engineer but soon
gave up the profession. After a period in business he devoted

himself entirely to scholarly activities, chiefly mathematics
and its history. His publications, beginning in 1829, led to
his appointment in 1841 as a professor of applied mathematics
at the Ecole Polytechnique (a post which he gave up in 1851).
In 1846 the Sorbonne created a chair of higher geometry for
him which he held for the rest of his life. He was elected to
the Academy in 1851 and awarded the Copley Medal of the Royal
Society of London in 1865.

2149. LIOUVILLE, Joseph. 1809-1882. (Mathematics 231, 260) Graduated
at the Ecole Polytechnique in 1827 and at the Ecole des Ponts
et Chaussées three years later. His publications, beginning
in 1828, led to his appointment as a *répétiteur* at the Ecole
Polytechnique in 1831; he acceded to one of the mathematics
chairs there in 1838. He resigned the chair in 1851 when he
was appointed professor of mathematics at the Collège de France,
a post he held until his death. From 1857 he also occupied a
chair at the Paris Faculté des Sciences.

      Liouville was elected to the Academy in 1839 and in
following year became a member of the Bureau des Longitudes.
In 1836 he founded the *Journal de Mathématiques Pures et
Appliquées* which he continued to edit for almost forty years.
The *Journal de Liouville*, as it was called, soon became one
of the leading mathematical periodicals; through his editorship of it, as well as through his teaching, he exerted a
strong influence on mathematics in France.

2150. GALOIS, Evariste. 1811-1832. (Mathematics 246) A mathematical
prodigy. In 1829 he was failed in the (oral) entrance examination to the Ecole Polytechnique because of a clash with the
examiner. Despite this and several other misfortunes he was
beginning to acquire a reputation in French mathematical
circles when the July Revolution of 1830 broke out and he
became heavily involved in politics. Two years later he was
killed in a duel.

      Despite the brevity and tragic character of his life
Galois is now regarded as one of the most brilliant and original
mathematicians of the nineteenth century.

2151. LE VERRIER, Urbain Jean Joseph. 1811-1877. (Astronomy 443)
Graduated at the Ecole Polytechnique in 1835 and two years
later was appointed *répétiteur d'astronomie* there. His work
in celestial mechanics led to his election to the Academy in
January 1846--before his discovery of Neptune, which was made
in August/September 1846. Shortly after the discovery a chair
in celestial mechanics was created for him at the Paris Faculté
des Sciences.

      Le Verrier's successful prediction of the existence of
Neptune made a deep impression on the general public as well
as on the scientific world, and congratulations and honours
were showered upon him. When the earlier prediction of J.C.
Adams (442) was made public a sensational priority dispute
ensued--carried on mainly by chauvinist journalists, French
and English. Le Verrier and Adams took little part in it and
were on quite good terms.

      Le Verrier entered politics in 1849 and two years later
became a senator and a member of the Conseil Supérieur de

l'Instruction Publique. In 1854 he was appointed director of
the Paris Observatory (in succession to Arago) but so alienated
his staff that he was dismissed in 1870. On the death of his
successor three years later he was reinstated but with reduced
powers.

2152. REGNAULT, Henri Victor. 1810–1878. (Chemistry 901. Physics
675) Graduated at the Ecole Polytechnique in 1832 and at the
Ecole des Mines two years later. After working briefly with
Liebig at Giessen and with Boussingault at Lyons he was appointed assistant to Gay-Lussac at the Ecole Polytechnique in 1836
and succeeded him in the chair of chemistry in 1840. He was
elected to the Academy in the same year.

Hitherto Regnault's researches had been chiefly in
organic chemistry but in 1840 he turned to topics in physics
and was appointed professor of physics at the Collège de France
in the following year. His reputation as a very skilful experimenter led to a commission from the Ministry of Public Works
to determine (or redetermine) all the physical constants
involved in the design of steam engines, and it was in this
connection that he did the work on the thermal properties of
gases for which he is best known.

2153. LAURENT, Auguste. 1807–1853. (Chemistry 887) Graduated at the
Ecole des Mines in 1830. In 1831 he was assistant to Dumas at
the Ecole des Arts et Manufactures and for the next few years
worked in a succession of technical jobs. In 1837 he gained
the degree of *docteur ès sciences* and in the following year
was appointed professor of chemistry at the Faculté des Sciences
at Bordeaux. He visited Liebig at Giessen in 1843 and in the
next year began the collaboration with Gerhardt that was to
continue until his death (see 2164).

Leaving Bordeaux in 1845, he worked in various laboratories in Paris until 1848 when he gained the position of assayer at the Mint. In 1851 he was unsuccessful (possibly, it seems,
because of his political radicalism) in his candidature for
the chair of chemistry at the Collège de France. This dis-
appointment, added to his ill-health and financial difficulties,
helped to bring on his death.

Laurent never received the recognition or the institutional appointment that his contributions to chemistry warranted.
While at Bordeaux he was elected a corresponding member of the
Academy but after his return to Paris full membership was not
accorded to him. He was however elected a foreign member of
the Royal Society of London in 1849 and of the Chemical Society
in the following year.

2154. BRAVAIS, Auguste. 1811–1863. (Crystallography 1099, 1101)
Graduated at the Ecole Polytechnique in 1831 and joined the
navy. He took a prominent part in a naval scientific expedition in 1839–40 and, while remaining in the navy, became
professor of astronomy at the Faculté des Sciences at Lyons
in 1841. Four years later he was appointed professor of physics
at the Ecole Polytechnique. He was elected to the Academy in
1854.

2155. SOCIÉTÉ GÉOLOGIQUE DE FRANCE. Founded in 1830.

2156. SOCIÉTÉ ENTOMOLOGIQUE DE FRANCE. Its publications began in 1832.

2157. MILNE-EDWARDS, Henri. 1800-1885. (Zoology 1479, 1492) Born in Bruges of English parentage. Studied medicine and then zoology (under Cuvier) in Paris. In 1832 he became professor of hygiene and natural history at the Ecole Centrale des Arts et Manufactures. His researches won him election to the Academy in 1838 and appointment three years later to the chair of entomology at the Muséum d'Histoire Naturelle. Twenty years later he transferred to the chair of mammalogy. He also held a chair at the Faculté des Sciences and was for many years one of the most outstanding zoologists in France.

2158. DUJARDIN, Félix. 1801-1860. (Microbiology 1804) His early career was disorganized but he picked up an education in science during the 1830s and did some significant research in zoology, botany, and palaeontology. This led to his appointment in 1840 as professor of botany and zoology at the newly-established Faculté des Sciences at Rennes, a post he held for the rest of his career.

2159. DAVAINE, Casimir Joseph. 1812-1882. (Microbiology 1808) M.D., Paris, 1837. He practised medicine in Paris while carrying on research on a variety of biological and medical topics, especially under his mentor, Pierre Rayer, professor of medicine. In association with Rayer he was one of the group which established the Société de Biologie in 1848. He was elected to the Académie de Médecine in 1868.

1840s

2160. HERMITE, Charles. 1822-1901. (Mathematics 241, 264, 284) Began publishing original work in mathematics in 1842. He studied at the Ecole Polytechnique and was appointed to a junior post there in 1848. He was elected to the Academy in 1856. In 1869 he was appointed to the chair of analysis at the Ecole Polytechnique and at the same time to the corresponding chair at the Faculté des Sciences. He was an outstanding and influential mathematician and in his later years received many honours.

2161. ROCHE, Edouard Albert. 1820-1883. (Mathematics 266. Astronomy 446, 463) Came from an academic family in Montpellier where he spent nearly all his life. After gaining the degree of *docteur ès sciences* in 1844 he spent three years in Paris in association with such luminaries as Cauchy, Le Verrier, and Arago. In 1849 he became *chargé de cours* at the Faculté des Sciences of Montpellier and in 1852 was appointed professor of mathematics. He was elected a corresponding member of the Paris Academy in 1873 (being ineligible for full membership because he did not live in Paris).

2162. FOUCAULT, (J.B.) Léon. 1819-1868. (Physics 684, 694, 707. Mechanics 557) Began the study of medicine but abandoned it for science. He was able to devote himself very largely to research, supporting himself by scientific journalism and occasional teaching. His brilliant experimental work earned him the doctorate in 1853 and an appointment as physicist at

the Paris Observatory. In 1862 he was made a member of the
Bureau des Longitudes. He was elected to the Academy in 1865
and received numerous other honours.

2163. FIZEAU, (A.) Hippolyte (L.). 1819-1896. (Physics 683, 693)
Came from a wealthy family. Studied science informally by
attending Regnault's lectures at the Collège de France, Arago's
at the Observatory, and following (through his brother) those
at the Ecole Polytechnique. Being of independent means he was
able to devote himself to research. His brilliant experimental
work won him a prize from the Academy in 1856 and a medal from
the Royal Society of London in 1866, and led to his election
to the Academy in 1860 and to the Bureau des Longitudes in 1878.

2164. GERHARDT, Charles Frédéric. 1816-1856. (Chemistry 897) Studied
under Liebig at Giessen for six months (1836/37) and then under
Dumas at the Sorbonne (1838-41), finally becoming Dumas' assistant. In 1841 he obtained his doctorate and was appointed to
the staff of the Faculté des Sciences at Montpellier (initially
as *chargé de cours* and three years later as professor). Since
there were no facilities for research and no encouragement of
it at Montpellier he spent as much time as he could in Paris,
and finally resigned in 1851. For a time he ran a private
school of chemistry in Paris but it was financially unsuccessful. In 1855 he was appointed professor of chemistry at Strasbourg but less than two years later he died suddenly.
         In 1844 Gerhardt had formed a close friendship with
Laurent who had similar ideas about the need for a new approach
to the theory of organic chemistry and a similarly unfortunate
career. Until Laurent's death in 1853 the two collaborated
closely, influencing each other. In 1845 they launched a new
journal to publish their research reports and other writings
but it lasted only until 1851. Despite the brevity and the
difficulties of their careers they both had a deep influence
on the development of organic chemistry.

2165. WURTZ, Charles Adolphe. 1817-1884. (Chemistry 904, 914, 941,
955, 972) During his medical course in his native Strasbourg
he took a year off (in 1842) to study chemistry under Liebig
at Giessen. After graduating in medicine in the following
year he went to Paris and became an assistant to Dumas. In
1849 he obtained a teaching post at the Ecole de Médecine and
four years later succeeded Dumas as professor there; he later
became dean and proved to be a very capable administrator. In
1874 he transferred to a chair of organic chemistry that had
been created for him at the Sorbonne.
         Wurtz was an enthusiastic and inspiring teacher who
(together with Berthelot) created a school of chemistry in
Paris which, in the number and calibre of the young men it
attracted, was comparable with the best schools in the German
universities at the time. He was the leading figure in the
formation in 1858 of the Société Chimique de France and one of
its most prominent members. He was also one of the founders
of the Association Française pour l'Avancement des Sciences in
1872 and later its president. He was elected to the Académie
de Médecine in 1856 and to the Académie des Sciences in 1867.
In his later years he received many scholarly and civil honours.

2166. DEVILLE, Henri Etienne Sainte-Claire. 1818-1881. (Chemistry
915, 938) M.D., Paris, 1843. Before graduating he had already
devoted himself to chemistry, and his early researches led to
his appointment (on Thenard's recommendation) as professor of
chemistry at the newly-established Faculté des Sciences at
Besançon. The quality of the research he managed to do there
won him the chair of chemistry at the Ecole Normale Supérieure
in Paris in 1851. From 1853 he was also Dumas' deputy at the
Sorbonne and he became titular professor there in 1866. As
well as being very active in research, Deville was an important
teacher and trained a number of outstanding chemists.

2167. DAUBRÉE, (G.) Auguste. 1814-1896. (Geology 1184) Graduated at
the Ecole Polytechnique and then at the Ecole des Mines. As
a mining engineer he was stationed in Alsace in 1840, and later
became professor of mineralogy and geology at Strasbourg; there
he established a laboratory for the experimental study of mineralogical and geological processes. In 1861 he was appointed
professor-administrator at the Muséum d'Histoire Naturelle and
in the same year was elected to the Academy. From 1872 he was
director of the Ecole des Mines and from 1875 a member of the
committee controlling the national geological survey.

2168. THURET, Gustave Adolphe. 1817-1875. (Botany 1365) Graduated
in law at Paris in 1838. He began a diplomatic career but gave
it up in order to devote himself to botany, which he studied
under J. Decaisne. Being very wealthy he never sought an
institutional position. For his researches, which began in
1840, he engaged an assistant, E. Bornet, who became his lifelong collaborator and friend.

2169. SOCIÉTÉ DE BIOLOGIE, Paris. Founded in 1848. Its *Comptes Rendus*,
modelled on the Academy's, became a very big periodical in the
latter part of the century.

2170. BERNARD, Claude. 1813-1878. (Physiology 1758, 1767, 1776)
M.D., Paris, 1843. Earlier, as an interne, he had met Magendie
and from 1841 to 1844 was his research assistant at the Collège
de France. A wealthy marriage, in 1845, enabled him to devote
himself to research and the following ten years were his most
brilliant period. In 1853 he was awarded the doctorate for
his celebrated work on the glycogenetic function of the liver.
From 1847 Bernard was Magendie's deputy at the Collège
de France and finally succeeded him there in 1855. In 1854 a
chair of general physiology was created for him at the Faculté
des Sciences and in the same year he was elected to the Académie
des Sciences. His election to the Académie de Médecine followed
in 1861. In 1869 he had the distinction of being elected also
to the Académie Française. In his last years he received many
other honours, scholarly and civil.

2171. BROWN-SÉQUARD, Charles Edouard. 1817-1894. (Physiology 1763)
Of American-French parentage. M.D., Paris, 1846. Thereafter
he lived in the United States, practising medicine, teaching,
and carrying on his researches. He made numerous visits to
Paris, however, and his papers were published in French periodicals. In 1858 he founded the *Journal de la Physiologie de
l'Homme et des Animaux* (later *Archives de Physiologie*). He

then lived for a few years in London, becoming a member of the
Royal Society in 1861. In 1878 he succeeded Claude Bernard as
professor of medicine at the Collège de France, a post he held
until his death.

2172. PASTEUR, Louis. 1822-1895. (Microbiology 1805, 1807, 1812.
Chemistry 903) Graduated at the Ecole Normale Supérieure in
1845 and won the *doctorat ès sciences* two years later. In
1849 he became professor of chemistry at the Faculté des
Sciences in Strasbourg, in 1854 dean of the new Faculté des
Sciences in Lille, and in 1857 director of scientific studies
at the Ecole Normale Supérieure. He was elected to the Academy
in 1862. In 1867 he was enabled to give up his administrative
duties at the Ecole Normale and become director of the new
laboratory of physiological chemistry which had been created
for him by order of Napoleon III.

Pasteur was elected to the Académie de Médecine in
1873 and to the Académie Française in 1882. In 1888 the
Institut Pasteur was inaugurated and he directed it until his
death. In his later years he was one of the most famous men
in the world and honours of all kinds, French and foreign,
were showered upon him.

1850s

2173. JANSSEN, Pierre Jules César. 1824-1907. (Astronomy 458, 459,
465) Obtained the *licence ès sciences* at Paris in 1852 and
the doctorate in 1860. He held various minor teaching posts
while carrying on important research in astrophysics and in
1876 was appointed director of the new astrophysical observatory at Meudon, near Paris. He was elected to the Academy in
1873 and to the Bureau des Longitudes in 1875. In the 1890s
he established a temporary observatory on the summit of Mont
Blanc (4,800 metres).

2174. CAILLETET, Louis Paul. 1832-1913. (Physics 729) After graduating at the Ecole des Mines he took up a position in his
father's ironworks, later becoming manager. His early investigations in metallurgy led him to experimental physics and
for his achievements in the field he was elected to the Academy
in 1884.

2175. SOCIÉTÉ CHIMIQUE DE PARIS. Founded in 1855. It was renamed
the Société Chimique de France in 1907.

2176. BERTHELOT, (P.E.) Marcellin. 1827-1907. (Chemistry 925, 956.
Various writings 63) Graduated at the Paris Faculté des
Sciences in 1849 and soon afterwards became a demonstrator in
chemistry at the Collège de France. In 1854 he obtained his
doctorate. He also studied at the Ecole de Pharmacie where
he graduated in 1858; in the following year he was appointed
to the new chair of organic chemistry there which he held
until 1876. In 1865 a chair of organic chemistry was created
for him at the Collège de France and he held it until his
death; he had several outstanding students.

Berthelot was elected to the Académie de Médecine in
1863 and to the Académie des Sciences in 1873, becoming its
permanent secretary in 1889. At the height of his fame, in

1901, he was elected to the Académie Française. His involvement in politics stemmed from the notable part he played in the defence of Paris during the siege of 1870. He became a senator in 1871 (elected for life in 1881), inspector of higher education in 1876, and briefly minister of education (1886/87) and foreign minister (1895).

In addition to a large amount of high-quality research in several branches of chemistry Berthelot was the author of important historical works on ancient and medieval alchemy and of a distinguished biography of Lavoisier.

2177. RAOULT, François Marie. 1830-1901. (Physical chemistry 963, 983) Studied at the Paris Faculté des Sciences but was unable to complete his course for financial reasons; he published his first paper however in 1853. He became a *lycée* teacher but managed to gain scientific qualifications, culminating in the *doctorat ès sciences physiques* in 1863. In 1867 he was appointed to the staff of the Faculté des Sciences at Grenoble and was professor of chemistry there from 1870 until his death. He was elected a correspondent of the Paris Academy in 1883 and later received other honours and awards.

2178. FRIEDEL, Charles. 1832-1899. (Chemistry 939, 954) Studied at Strasbourg (under Pasteur) and at the Sorbonne (under Wurtz). In 1856 he became the curator of the mineral collections at the Ecole des Mines and in 1876 professor of mineralogy at the Sorbonne. He was elected to the Academy in 1878. On the death of Wurtz in 1884 Friedel succeeded him as professor of organic chemistry at the Sorbonne. During the period 1889-92 he was president of an international committee set up to standardize the nomenclature of organic chemistry. In 1899 he was the chief founder and editor of the *Revue Générale de Chimie*.

2179. SOCIÉTÉ MÉTÉOROLOGIQUE DE FRANCE. Its periodical began in 1853.

2180. SOCIÉTÉ BOTANIQUE DE FRANCE. Its periodical began in 1854.

2181. FABRE, Jean Henri. 1823-1915. (Entomology 1491) After several years as a *lycée* teacher in the south of France he gained the *doctorat ès sciences naturelles* at Paris in 1854. He then returned to his native Provence and devoted himself to entomology while earning a slender living by teaching and writing textbooks and popularizations. His entomological works were highly regarded by the general public for both their style and content. In 1887 he was elected a corresponding member of the Paris Academy.

2182. GAUDRY, Albert Jean. 1827-1908. (Evolutionary palaeontology 1527) His career was assisted by his brother-in-law, the palaeontologist, Alcide d'Orbigny. In 1853, after completing research at the Muséum d'Histoire Naturelle for his doctorate, he became assistant to d'Orbigny who had just been appointed to the Museum's new chair of palaeontology. Gaudry succeeded to the chair in 1872.

2183. BROCA, Pierre Paul. 1824-1880. (Anatomy 1639) M.D., Paris, 1849. In 1853 he became assistant professor at the Paris Faculté de Médecine and subsequently held senior posts in various Paris hospitals. He was appointed professor of clinical

surgery at the Faculty in 1868. Throughout his medical career
he also carried on extensive researches in anthropology (especially craniology) and was one of the chief pioneers of the
subject. He was a prominent member of the Société d'Anthropol-
ogie (which he helped to establish in 1859) as well as of the
Académie de Médecine. He received many honours.

1860s

2184. ÉCOLE NORMALE SUPÉRIEURE. Its origins went back to 1808 but
its importance as a teacher-training college dated from the
1830s. In 1857 Pasteur was appointed director of scientific
studies and during the next few years considerably improved
the Ecole's scientific section, with the result that it was
subsequently able to compete with the Ecole Polytechnique.
In 1864 Pasteur initiated the periodical *Annales Scientifiques
de l'Ecole Normale Supérieure*.

2185. JORDAN, Camille. 1838–1921. (Mathematics 278, 293) Graduated
at the Ecole Polytechnique about 1858 and obtained the doctorate
at the Sorbonne in 1860. For the next fifteen years he worked
as an engineer but managed to publish numerous mathematical
papers and in 1870 a major treatise which won the Academy's
Poncelet Prize. In 1876 he was appointed professor at the
Ecole Polytechnique and *suppléant* at the Collège de France;
he held both posts until his retirement in 1912. He was
elected to the Academy in 1881.

2186. DARBOUX, (J.) Gaston. 1842–1917. (Mathematics 285, 301, 315)
In 1861 he came first in the entrance examinations to both the
Ecole Polytechnique and the Ecole Normale Supérieure and, to
everyone's surprise, chose the latter. He graduated there in
1864 and then gained his doctorate at the Sorbonne two years
later. In 1870 he was instrumental in establishing the *Bulletin
des Sciences Mathématiques* which he edited for many years.
After some secondary school teaching he was appointed in 1872
maître de conférences at the Ecole Normale, a post he held
until 1881. From 1873 to 1878 he was also *suppléant* professor
of rational mechanics at the Sorbonne; in the latter year he
became professor of higher geometry there, a post he held for
the rest of his life.
        Darboux was elected to the Academy in 1884 and became
its secretary in 1900 (a collection of his *éloges* was published
in 1912). He was a prominent figure in the scientific community
in Paris and, being a capable administrator, was asked to serve
on numerous commissions of various kinds. He is said to have
been a member or honorary member of over a hundred learned
societies.

2187. TISSERAND, François Félix. 1845–1896. (Astronomy 483) Graduated
at the Ecole Normale Supérieure in 1866 and in the same year
was appointed to a post in the Paris Observatory. He obtained
his doctorate in 1868 and five years later was appointed director of the Toulouse Observatory. In 1878 he returned to Paris
to take up a position in the Faculté des Sciences teaching
rational mechanics (and from 1883, celestial mechanics). He
was elected to the Academy in 1878 and to the Bureau des Longitudes the next year. In 1892 he became director of the Paris
Observatory.

2188. BOISBAUDRAN, (P.E.) (*called* François) Lecoq de. 1838–1912. (Chemistry 950) A wine merchant who was self-taught in science. His publications dated from 1866. For his discoveries of new elements he received several prizes and awards and was elected a corresponding member of the Academy (in 1878) and a foreign member of the Chemical Society of London.

2189. SERVICE DE LA CARTE GÉOLOGIQUE. The national geological survey, established in 1868. The geological exploration and mapping of the country went back much earlier, of course (cf. 1113, 1149, 1169).

2190. MAREY, Etienne Jules. 1830–1904. (Physiology 1771, 1789) M.D., Paris, 1859. Practised medicine and at the same time carried on research in physiology in a laboratory of his own. In 1868 he was appointed professor of natural history at the Collège de France, a post he held until his death. He was elected to the Academy in 1878 and became its president in 1895.

2191. BERT, Paul. 1833–1886. (Physiology 1780) After qualifying in law he took up medicine and science, obtaining the M.D. in 1863 and the doctorate in science three years later; during those three years he was *préparateur* for Claude Bernard at the Collège de France. After two years at the Bordeaux Faculté des Sciences he was appointed to the chair of comparative physiology at the Muséum d'Histoire Naturelle in 1868 (in succession to Flourens) and to the chair of physiology at the Sorbonne in the following year (in succession to Bernard). He was elected to the Academy and was a prominent member of the Société de Biologie.

  Following the Franco-Prussian War Bert became deeply involved in politics. From 1872 until his death he was a member of the Chamber of Deputies and during 1881–82 he held the post of minister of education.

1870s

2192. ASSOCIATION FRANÇAISE POUR L'AVANCEMENT DES SCIENCES. Held annual meetings in different cities, beginning in 1872. (cf. 2147)

2193. SOCIÉTÉ MATHÉMATIQUE DE FRANCE. Founded in 1872.

2194. POINCARÉ, (J.) Henri. 1854–1912. (Mathematics 286. Celestial mechanics 487) Graduated at the Ecole Polytechnique in 1875 and then at the Ecole des Mines. In 1879 he gained the *doctorat ès sciences*. After teaching for a brief period at Caen he was appointed a professor at the Sorbonne in 1881; he held the post until his death. He was a famous teacher and changed his lecture course every year, reviewing in turn many fields of mathematics, especially applied fields. Many of these lecture courses were published by his students.

  Poincaré was elected to the Academy in 1887. The literary and philosophical distinction of his widely-read works—*La Science et l'Hypothèse* (1903), *La Valeur de la Science* (1905), and *Science et méthode* (1908)—won him appointment to the Académie Française in 1909. As one of the greatest mathematicians of the age he received many other honours, French and foreign.

2195. PICARD, (C.) Emile. 1856–1941. (Mathematics 287, 308) In the entrance examinations of 1874 he came first in the list for

the Ecole Normale Supérieure and second in the list for the
Ecole Polytechnique. His choice of the Ecole Normale received
much publicity. He graduated there in 1877 (having already
begun to publish original work) and soon afterwards received
the *doctorat ès sciences*. After a short period as an assistant
at the Ecole Normale he was appointed to a chair at Toulouse
in 1879. Two years later he returned to Paris to take up
concurrent lectureships at the Sorbonne and the Ecole Normale.
In 1885 he was appointed to one of the mathematical chairs at
the Sorbonne.

Picard was elected to the Academy in 1889 and was one
of its permanent secretaries from 1917 until his death. A
capable administrator, he served on numerous committees and
government bodies, including the Bureau des Longitudes. As
an outstanding mathematician he was awarded many honours.

2196. DESLANDRES, Henri. 1853–1948. (Astrophysics 485. Physics 745)
Graduated at the Ecole Polytechnique in 1874, served in the
army until 1881, and then did research in spectroscopy at the
Ecole Polytechnique and the Sorbonne. In 1889 he was appointed
to the Paris Observatory and in 1897 to the astrophysical
observatory at Meudon. He was elected to the Academy in 1902
and in 1908 became director of the Meudon Observatory.

2197. SOCIÉTÉ FRANÇAISE DE PHYSIQUE. Founded in 1873.

2198. BECQUEREL, Henri. 1852–1908. (Physics 759) His father was
professor of physics at the Muséum d'Histoire Naturelle (as
his grandfather had been, and as both he and his son were to
be). He graduated at the Ecole Polytechnique in 1874 and at
the Ecole des Ponts et Chaussées three years later, becoming
an engineer in the Corps des Ponts et Chaussées. His career
as an engineer, in which he reached high rank, continued in
parallel with his scientific career.

The latter began in 1876 with a post as *répétiteur* at
the Ecole Polytechnique, followed two years later by a junior
appointment at the Museum. His early researches won him a
doctorate in 1888 and election to the Academy in the following
year. After the death of his father in 1891 he succeeded to
the chair of physics at the Museum (and also to his father's
other chair at the Conservatoire des Arts et Métiers). He was
still teaching at the Ecole Polytechnique and in 1895 became
professor of physics there in addition to his other appointments.

When the significance of his discovery of radioactivity
became recognized Becquerel received many honours, including
the 1903 Nobel Prize, shared with the Curies.

2199. LE BEL, Joseph Achille. 1847–1930. (Chemistry 949) Studied at
the Ecole Polytechnique and was later an assistant to Balard
(professor of chemistry at the Collège de France and the
Sorbonne). From 1873 he was working in Wurtz' laboratory where
van't Hoff was also working. Their acquaintance was apparently
only slight and they evidently did not influence each other.

Le Bel never held any academic position. He came from
a wealthy family which owned an oil well and refinery which he
managed for some years. After 1889 he sold his share in it and
retired to Paris where he continued his chemical researches.
He was president of the Société Chimique in 1892.

2200. LE CHÂTELIER, Henry Louis. 1850-1936. (Physical chemistry 966)
Graduated at the Ecole Polytechnique and then at the Ecole des
Mines in 1873. After serving as a mining engineer he became
professor of chemistry at the Ecole des Mines in 1877 and held
the post until 1919. He was also professor of chemistry at the
Collège de France from 1887 to 1908, and at the Sorbonne from
1907 to 1925; at the latter he had many research students.
        Apart from the work in chemical thermodynamics for
which he is best known, most of Le Châtelier's many researches
were in applied chemistry and metallurgy. He was one of the
most outstanding chemists in France in his time and an influential adviser to the government on scientific and technical
matters. He was elected to the Academy in 1907 and in his
later years received many honours, both French and foreign.

2201. SOCIÉTÉ MINÉRALOGIQUE DE FRANCE. Its periodical began in 1878.

2202. BERTRAND, Marcel Alexandre. 1847-1907. (Geology 1206) Son of
the mathematician, Joseph Bertrand. Graduated at the Ecole
Polytechnique and then at the Ecole des Mines and was appointed
to the Geological Survey. In 1886 he became a professor at the
Ecole des Mines and in 1896 was elected to the Academy.

2203. SOCIÉTÉ ZOOLOLOGIQUE DE FRANCE. Its publications began in 1876.

2204. RICHET, Charles Robert. 1850-1935. (Physiology 1786, 1798.
His work on anaphylaxis was done after 1900 and won the Nobel
Prize in 1913) Graduated in medicine at Paris in 1875 and
thereafter worked in the laboratories of several of the leading
French physiologists. In 1878 he became *professeur agrégé* at
the Faculty of Medicine and was appointed to the chair of
physiology in 1887.

2205. ARSONVAL, Arsène d'. 1851-1940. (Biophysics 1788) Studied
medicine at Paris in the early 1870s and was led into physiological research by the influence of Bernard. He became
Bernard's *préparateur* and took his doctorate in 1876. Subsequently he was assistant to Brown-Séquard at the Collège de
France, eventually succeeding him in the chair of medicine.
From 1882 he was director of the College's new laboratory for
biophysics.
        From the advent of the electrical age about 1880
d'Arsonval was a leader in encouraging teaching and research
in the big new field of electricity and its applications. He
was best known to the public for his development of electro-
therapy and his other contributions to medicine.

1880s

2206. HADAMARD, Jacques. 1865-1963. (Mathematics 305) Graduated at
the Ecole Normale Supérieure in 1888 and became a *lycée* teacher.
He gained the *doctorat ès sciences* in 1892 and in the following
year was appointed a lecturer at the Faculté des Sciences at
Bordeaux. From 1897 he was a lecturer at the Sorbonne and then,
from 1909, professor at the Collège de France until his retirement in 1937. Concurrently he was also professor at the Ecole
Polytechnique from 1912 and at the Ecole Centrale des Arts et
Manufactures from 1920. He was elected to the Academy in 1912.

2207. SOCIÉTÉ ASTRONOMIQUE DE FRANCE. Its periodical began in 1887.

2208. CURIE, Pierre. 1859-1906. (Physics 733, 753) Graduated at the
        Paris Faculté des Sciences in 1877 and in the following year
        was appointed assistant in the physics laboratory. In 1882 he
        became director of laboratory work at the newly-founded Ecole
        Municipale de Physique et de Chimie Industrielle. His important
        researches on magnetism gained him the doctorate in 1895.
                Following the great discoveries in radioactivity made
        jointly with his wife, Marie (1867-1934), Curie was appointed
        a lecturer at the Sorbonne in 1900 and a professor in 1904.
        The Academy awarded him a prize in 1901 and elected him a
        member in 1905. Other honours, shared with his wife, included
        the Davy Medal of the Royal Society of London in 1903 and later
        in the same year the supreme honour of the Nobel Prize, which
        they shared with Becquerel.
                Following his accidental death in 1906 his wife succeed-
        ed him in the chair at the Sorbonne (becoming the first woman
        to teach there).

2209. MOISSAN, (F.F.) Henri. 1852-1907. (Chemistry 967, 984) Grad-
        uated at the Ecole de Pharmacie in 1879 and in the following
        year gained his doctorate at the Sorbonne. After holding junior
        teaching posts at the Ecole de Pharmacie he became professor
        of toxicology there in 1886 and professor of inorganic chemistry
        in 1899. In 1900 he was also made professor of inorganic chem-
        istry at the Sorbonne where his research school became the
        leading centre for the subject in France. He was awarded the
        Nobel Prize in 1906.

2210. SABATIER, Paul. 1854-1941. (Chemistry 993) Graduated at the
        Ecole Normale Supérieure in 1877 and then worked with Berthelot
        at the Collège de France, gaining his doctorate in 1880. In
        1882 he was appointed to the staff of the Faculté des Sciences
        at Toulouse, becoming professor of chemistry there two years
        later. He remained at Toulouse until his retirement in 1930,
        declining offers to succeed Moissan at the Sorbonne and
        Berthelot at the Collège de France.
                Sabatier was awarded the Nobel Prize (shared with
        Grignard) in 1912 and became the first non-resident member of
        the Academy in the following year. "Both by personal example
        and by administrative action, Sabatier was throughout his life
        an important influence in steps toward the decentralization of
        scientific institutions in France." (*DSB*)

2211. MARGERIE, Emmanuel (M.P.M.J.) de. 1862-1953. (Geology 1207)
        Came from a wealthy and cultured family of aristocratic lineage.
        He was educated privately and though he attended various high-
        level courses he sought no qualification. From an early age
        however he was active in the geological community in Paris.
        His only institutional appointment was as director of the
        geological survey of Alsace-Lorraine from 1918 to 1933. He
        was a member of many learned societies and received numerous
        honours.

2212. INSTITUT PASTEUR. Founded in 1888 for research in microbiology.
        It was directed initially by Pasteur and after his death in
        1895 by Emile Duclaux.

1890s

2213. BERTRAND, Gabriel. 1867–1962. (Biochemistry 992) Graduated at the Ecole de Pharmacie in 1890 and became a *préparateur* in chemistry at the Muséum d'Histoire Naturelle. In 1900 he was appointed to the staff of the new biochemistry department of the Institut Pasteur. He obtained his doctorate in 1904 and four years later was appointed professor of biochemistry at the Faculté des Sciences (in addition to his post at the Institut Pasteur where his courses were given).

2214. ASSOCIATION DES ANATOMISTES. Held annual meetings in different French cities. Its periodical began in 1899.

2.03 BRITAIN

1510s

2215. TUNSTALL, Cuthbert. 1474-1559. (Mathematics 81) Studied at
both Oxford and Cambridge and in 1499 went to Padua where he
became a doctor of laws. He returned to England a few years
later and pursued an ecclesiastical career which culminated
in his appointment as Bishop of London in 1522.

1530s

2216. TURNER, William. 1508-1568. (Botany 1288. Zoology 1411)
M.A., Cambridge, 1533. For some years he was a college fellow
but because of his involvement in the religious turmoil of the
time he was obliged to go into exile for the period 1540-46.
During that period he studied medicine in Italy (botany under
Ghini) and obtained an M.D. He also travelled widely, meeting
Gesner and other naturalists. After his return to England he
became dean of Wells cathedral but had to go into exile again
in 1553-58, and resumed his acquaintance with the Continental
naturalists and their work. After his second return to England
he got back his deanery at Wells and was able to devote his
time to botany.

2217. WOTTON, Edward. 1492-1555. (Zoology 1415) After holding a
college fellowship at Oxford he studied medicine at Padua,
obtaining his M.D. in 1526. Thereafter he was a physician in
London, becoming eminent in the profession and a prominent
member of the Royal College of Physicians.

1540s

2218. RECORDE, Robert. ca. 1510-1558. (Mathematics 88, 95. Astronomy
332) Graduated in medicine at Oxford about 1533. After practising as a physician he worked for the government as "compt-
roller of mints and monies." He also taught mathematics and
was a skilful textbook writer, choosing to write in English
instead of Latin. His vernacular textbooks gave a big impetus
to the development of the English school of mathematical
practitioners who were to play a key part in the navigational
and other practical concerns of the age.

1550s

2219. DIGGES, Leonard. ca. 1520-1559(?). (Mathematics 100) Younger
son of a landed family. Admitted to Lincoln's Inn (law school
in London) in 1537.

2220. DEE, John. 1527-1608. (Mathematics 99) After taking his M.A.
at Cambridge in 1548 he studied in Louvain and Paris with Gemma
Frisius and Gerhardus Mercator. He became an expert on navigation and navigational instruments and for many years was an
adviser to the numerous expeditions that were then being sent
out from England. His standing in court circles however was
due to his fame as an astrologer.

1570s

2221. NAPIER, John. 1550-1617. (Mathematics 119) A Scottish noble.
Apparently his education was largely theological but he was
interested in mathematics from an early age.

2222. DIGGES, Thomas. 1546(?)-1595. (Astronomy 336) Son of Leonard
Digges (2219). He learnt mathematics and its uses from his
father and from John Dee, and became a military engineer and
ballistics expert. From 1572 he was a member of parliament
and active in public affairs.

2223. GILBERT, William. 1544-1603. (Physics 576) M.D., Cambridge,
1569. He became one of the leading physicians of London and
for a time was president of the Royal College of Physicians.

2224. NORMAN, Robert. fl. 1560-96. (Physics 575) A maker of navigational instruments in London.

1580s

2225. MOFFETT, Thomas. 1553-1604. (Zoology 1420) Studied medicine
at Cambridge and then at Basel where he gained his M.D. in
1578 and became an advocate of Paracelsian medicine. After
some travelling he settled in London as a medical practitioner
and in 1588 became a fellow of the Royal College of Physicians.

1590s

2226. GRESHAM COLLEGE. Founded in 1596 and financially endowed by the
bequest of a leading London merchant, Sir Thomas Gresham, who
saw the need for an institution of higher education in London.
Its staff included professors of mathematics, astronomy, and
medicine (as well as other subjects) and through the seventeenth
century it provided valuable institutional backing for the
scientific movement in London. The Royal Society held its
meetings at the College from its foundation there in 1660 until
1711 and the two institutions were closely associated. By the
end of the seventeenth century however the College was in a
state of decline and thereafter it led only a tenuous existence.

2227. HARRIOT, Thomas. ca. 1560-1621. (Mathematics 126) B.A., Oxford,
1579. A protégé initially of Sir Walter Ralegh and then of
the Earl of Northumberland who gave him an annual pension and
living quarters in London.
   From the early 1590s until he was overtaken by ill
health in 1618 Harriot carried on researches of the first
importance in mathematics and most of the physical sciences
but never published any of them. The only work to appear was

a posthumous edition of some of his results in algebra published by one of his friends in 1631. His manuscripts still exist.

2228. BRIGGS, Henry. 1561-1630. (Mathematics 120, 123, 130) M.A.,
Cambridge, 1585. In 1596, after a few years teaching at
Cambridge, he became professor of geometry at the newly-established Gresham College in London. From 1619 he was Savilian
professor of geometry at Oxford.

1600s

2229. BACON, Francis. 1561-1626. (Philosophy of science 11) Educated
at Cambridge and studied law at Gray's Inn (law school in
London), becoming a barrister in 1582. From 1618 to 1621 he
was Lord Chancellor. His writings relevant to science date
from 1605 onward.

2230. FLUDD, Robert. 1574-1637. (Hermeticism etc. 10) M.D., Oxford,
1605. A successful medical practitioner in London and a
fellow of the Royal College of Physicians.

2231. TOPSELL, Edward. 1572-1625(?). (Zoology 1419) M.A., Cambridge,
ca. 1594. Held various Church appointments in the provinces
and in London.

2232. HARVEY, William. 1578-1657. (Physiology 1712. Embryology 1655)
Studied medicine at Cambridge and then at Padua (anatomy under
Fabrici) where he received the M.D. in 1602. Practised medicine
in London very successfully and became a fellow of the Royal
College of Physicians in whose affairs he was very active; for
many years he was the College's lecturer in surgery. Physician
to James I and Charles I.

1610s

2233. GUNTER, Edmund. 1581-1626. (Mathematics 121) M.A., Oxford,
1605; B.D., 1616. Rector of a parish in Southwark and from
1619 concurrently professor of astronomy at Gresham College.

1620s

2234. OUGHTRED, William. 1575-1660. (Mathematics 127) M.A., Cambridge,
1600. He was ordained in 1603 and five years later became
rector of Albury, in Surrey, where he remained until his death.

2235. GELLIBRAND, Henry. 1597-1636. (Terrestrial magnetism 1013)
M.A., Oxford, 1623. Became professor of astronomy at Gresham
College in 1627.

2236. BOTANICAL GARDEN, OXFORD. Founded in 1621 by a private benefactor
(Lord Danby) on the initiative of the professor of medicine.
It was largely created by the first keeper, Jacob Bobart, who
published a catalogue of the plants in 1648 and was succeeded
in the post by his son of the same name. The latter collaborated with Morison (2253) in the 1670s.

1630s

2237. GASCOIGNE, William. ca. 1612-1644. (Astronomy 359) He evidently

studied at Oxford. By 1640 he was engaged in scholarly correspondence but in 1644 he was killed in the Civil War.

2238. HORROCKS, Jeremiah. 1618-1641. (Astronomy 367, 375) Studied
at Cambridge in 1632-35 but did not take a degree. Self-taught
in astronomy and an ardent admirer of Kepler. His extraordinary achievements in almost every aspect of the astronomy of
his time, observational and theoretical, were not published
in his lifetime.

2239. GLISSON, Francis. 1597(?)-1677. (Anatomy 1590, 1602) M.D.,
Cambridge, 1634. Two years later he was appointed regius prof-
essor of medicine but he left Cambridge during the Civil War.
He later settled in London where he was a successful practition-
er and an active member of the Royal College of Physicians.
He was one of the founding fellows of the Royal Society.

2240. ENT, George. 1604-1689. (Physiology 1714) M.A., Cambridge,
1631. M.D., Padua, 1636. Became a successful practitioner
in London and a prominent member of the Royal College of
Physicians. He was one of the foundation members of the
Royal Society.

1640s

2241. HAAK, Theodore. 1605-1690. (Not in Part 1) Born and reared
in Germany but studied theology and mathematics at Oxford in
the late 1620s. After returning to Germany he settled in
England in 1638. During the Commonwealth he served the government in important ways for which he was awarded a pension.
       Haak's significance was as an "intelligencer" and
stimulator of the scientific movement in England during the
1640s and 1650s. He corresponded with Mersenne at intervals
during the period 1639-48 and apparently was the initiator of
an informal group which held meetings in London from about
1645 and was a precursor of the Royal Society. When the
Society was established in 1660 Haak was one of the original
members and he continued to be active in the Society until
his death.

2242. WALLIS, John. 1616-1703. (Mathematics 143, 165. Mechanics
513) M.A., Cambridge, 1640. Became a chaplain and minister
in London where he was a member of the group that met weekly
at Gresham College and elsewhere to discuss the new scientific
developments. In 1649 he was appointed Savilian professor of
mathematics at Oxford, a post he held until his death. He was
one of the founders of the Royal Society and a frequent contrib-
utor to the *Philosophical Transactions*. His collected mathematical works fill three large volumes.

2243. WILKINS, John. 1614-1672. (Astronomy 358. Mechanics 508)
M.A., Oxford, 1634; became a chaplain to various nobles. In
London in the mid-1640s he was a member of the group formed
by Theodore Haak to discuss developments in "the new experimental philosophy." In 1648 he was appointed Warden of Wadham
College, Oxford, and during the next ten years he both reinvig-
orated the College (after the stresses of the Civil War) and
gathered around him a brilliant group of young scientists,
several of whom were later to become of first importance.

This Oxford group, together with the above-mentioned London
group, constituted the nucleus of the Royal Society when it
came into existence in 1660. Wilkins was a key member of the
Society until 1668 when he became Bishop of Chester. He is
now regarded as the chief stimulus to the upsurge of science
in England around the middle of the century, both through his
popular writings and his personal example and inspiration to
others.

2244. WARD, Seth. 1617-1689. (Astronomy 363) M.A., Cambridge, 1640.
He began an academic career, first at Cambridge and then at
Oxford, but became a victim of the political reversals of the
time and turned to an ecclesiastical career. He was an original
fellow of the Royal Society.

2245. HIGHMORE, Nathaniel. 1613-1685. (Anatomy 1586. Embryology
1656) M.D., Oxford, 1643. Thereafter he was a successful
practitioner in Dorset. Despite his provincial isolation he
was able to keep in touch with scientific developments and
contribute articles to the *Philosophical Transactions*.

2246. CHARLETON, Walter. 1620-1707. (Physiology 1716. Natural philosophy 20) Studied at Oxford, where he was introduced to the
new science by John Wilkins, and took his M.D. in 1643. Became
a successful practitioner in London and a fellow of the Royal
College of Physicians. He was one of the foundation members
of the Royal Society.

1650s

2247. OLDENBURG, Henry. ca. 1618-1677. (Not in Part 1) Ranks with
Mersenne as the greatest of the "intelligencers" and organizers
of the scientific movement. He was also the initiator of the
first scientific periodical.

Born and educated in Bremen, and studied at the University of Utrecht in the early 1640s. For many years he was
a private tutor to upper-class youths, an occupation that
involved a good deal of travelling, enabling him to acquire a
good knowledge of western Europe and a remarkable command of
languages. He settled in England in 1653 and became acquainted
with a number of important people such as Milton and Hobbes
and the Boyle family; Robert Boyle remained always a firm
friend. He was at Oxford in 1656 where he met Wilkins and
doubtless others in the group of experimenters there.

In 1657 Oldenburg took one of his young English pupils
on a Continental tour during which he mixed in learned circles,
especially in Paris. On his return to England in 1660 he
became a foundation member of the Royal Society and two years
later was elected its secretary. Thereafter he served the
Society assiduously until his death and was a major factor in
its early success.

As secretary, Oldenburg handled the Society's correspondence which soon became very large. In 1665, as a means of
disseminating the information that flowed to him, he took the
step of printings selections from it in regular bulletins under
the name of *Philosophical Transactions*; the venture was an
immediate and lasting success. Though it was not the first
periodical--the *Journal des Sçavans* preceded it by two months--

the *Philosophical Transactions* was sufficiently distinctive in
character to merit the title of the first scientific periodical.

2248. MERCATOR, Nicolaus. ca. 1619–1687. (Mathematics 153. Physical
geography 1016) A native of Denmark. Graduated at the University of Rostock and a few years later was on the faculty of
the University of Copenhagen. He settled in England about 1654
as a private teacher of mathematics. He became a fellow of the
Royal Society in 1666 and was acquainted with Newton.

2249. STREETE, Thomas. 1622–1689. (Astronomy 366) A minor public
servant in London. He was associated with Gresham College
and was acquainted with many leading astronomers, often assisting them in their observations. He published many ephemerides
which were well received.

2250. ASHMOLE, Elias. 1617–1692. (Chemistry 791) Studied law in
London and established a practice there in 1638. At the Restoration he was granted some lucrative crown offices and became
very wealthy. His interests in antiquarianism and astrology
led him to alchemy, and his main publication (791) was part of
a larger plan to restore English alchemy. He was a founding
fellow of the Royal Society and regarded its programme of
"experimental philosophy" as complementary to the Hermetic
philosophy in which he was interested. In 1675 he donated his
large collection of antiquities and "curiosities" to Oxford
University which erected a special building (now known as the
Old Ashmolean Museum) to house them—the first public museum
in England.

2251. BOYLE, Robert. 1627–1691. (Chemistry 794. Physics 589. Natural
philosophy 21) A younger son of a wealthy aristocratic family;
educated at Eton, then by private tutors. An interest in
medical matters led him to chemistry which he taught himself.
From 1654 he lived in Oxford where he was an active member of
Wilkins' circle. He was one of the founders of the Royal
Society and for many years its most outstanding member (he
settled in London in 1668). A prolific (and prolix) author,
he wrote much on theology as well as on science.

2252. MERRETT, Christopher. 1614–1695. (Natural history 1229) M.D.,
Oxford, 1643. Medical practitioner in London and fellow of
the Royal College of Physicians. One of the original members
of the Royal Society.

2253. MORISON, Robert. 1620–1683. (Botany 1317) A native of Scotland.
M.A., Aberdeen, 1638. Being a royalist, he was in exile in
France during the Interregnum and while there obtained an M.D.
at Angers and took up botany which he cultivated assiduously.
At the Restoration he became physician and botanist to Charles
II and in 1669 was appointed the first professor of botany at
Oxford.

2254. RAY, John. 1627–1705. (Botany 1311, 1313, 1319, 1321. Zoology
1427, 1433, 1434, 1437. Earth science 1023. Natural theology
24) M.A., Cambridge, 1651. Apparently he derived his interest
in natural history from his family background (he had no medical
training). At Cambridge he became a college fellow and taught
various subjects until 1662 when, as a result of his refusal

to take the oath required by the Act of Uniformity, he was
obliged to leave the University. Thereafter his work was
supported by the patronage of his young aristocratic collaborator, Francis Willughby (2267). He became a fellow of the
Royal Society in 1667. Willughby died in 1672 but left him
an annuity.

2255. WHARTON, Thomas. 1614-1673. (Anatomy 1591) M.D., Oxford, 1647.
Physician in London and fellow of the Royal College.

2256. WILLIS, Thomas. 1621-1675. (Anatomy 1596) M.A., Oxford, 1642;
B.Med., 1646. While practising medicine in Oxford in the
1650s he was a member of the group centered on John Wilkins at
Wadham College, and it was in this milieu that he published
his first scientific and medical works. In 1660, after the
Restoration, he took his M.D. and was appointed Sedleian professor of natural philosophy (and medicine). He was one of the
original members of the Royal Society. In 1667 he left Oxford
for a lucrative practice in London.

1660s

2257. ROYAL SOCIETY OF LONDON. Founded in November 1660, six months
after the restoration of the monarchy. (During the Interregnum
it had been preceded by informal groups in London and Oxford)
The newly-formed society was successful in obtaining the King's
approval--vital in the politically-tense atmosphere of the time
--and called itself "the Royal Society for Improving Natural
Knowledge." It was however a private body, not an organ of
the State like the Continental academies. Since it received
no support from the State it had to finance its activities
from its own resources, chiefly its members' subscriptions.
For this reason, and also because in the 1660s the virtuoso
interest in "natural curiosities" was at its height, the
Society became a large and widely-inclusive body. Though the
virtuoso interest subsequently declined, the Society continued
to be dominated by non-scientists until the mid-nineteenth
century (see 2366).
    In its golden age in the 1660s, however, that conse-
quence had not yet emerged and the scientists and the virtuosi
were united in their enthusiasm for the "new experimental philosophy." Among the reasons for the Society's resounding success
in its early years were the unflagging energy and many-sided
abilities of its first secretary, Henry Oldenburg (2247).

2258. *PHILOSOPHICAL TRANSACTIONS.* The first scientific periodical;
began 6 March 1665. It was founded by the secretary of the
Royal Society, Henry Oldenburg, who continued to edit it until
his death in 1677; thereafter it was edited by successive
secretaries of the Royal Society and it was not formally taken
over by the Society until 1753.

2259. COLLINS, John. 1625-1683. (Not in Part 1) After a meagre
education and some vicissitudes he managed to establish himself
as a teacher of mathematics. He published several books on
practical mathematics but his importance was as an "intelligencer" in the field of mathematics (cf. 2264). He carried on

an extensive correspondence with many of the most outstanding
mathematicians of the time, not only in Britain but also
(through Oldenburg) with several important Continental figures.
He was elected a fellow of the Royal Society in 1667.

2260. SPRAT, Thomas. 1635-1713. (Not in Part 1) M.A., Oxford, 1657.
While a student at Wadham College he was closely associated
with Wilkins and his circle. He was ordained in 1660 and held
a succession of ecclesiastical positions, finally a bishopric.
Though he became a member of the Royal Society in 1663 there
is no indication that he ever engaged in scientific work and
his claim to a place in the history of science rests solely on
his famous *History of the Royal Society*, published in 1667.
   Though this was not formally a publication of the
Society it was so in fact: the Council of the Society asked
Sprat to write it, supplied him with information, discussed
his drafts, and generally supervised the production of the book.
Its purpose was to explain the Society's character and aims to
the public, and to defend it against the accusations and innu-
endoes that were being made against it on political and religious grounds. The book led to further controversy however, and
the attacks on the Society continued for some years.

2261. GLANVILL, Joseph. 1636-1680. (Not in Part 1) M.A., Oxford,
1658. Ordained in 1660 and for most of his life was rector of
the Abbey Church in Bath. At Oxford he had probably been
associated with Wilkins' circle, and in 1664 he became a member
of the Royal Society. His contributions to science were very
slight and his significance for the history of science was as
an apologist for the Royal Society. In a series of books
(published in 1661-71), which attracted much attention, he
defended the Society and "the new experimental philosophy"
against a number of accusations, especially of impracticality
and of irreligion and scepticism. He was at the same time a
prominent apologist for latitudinarian Anglicanism.

2262. BARROW, Isaac. 1630-1677. (Mathematics 157, 161) M.A., Cambridge, 1652. An original fellow of the Royal Society. From
1663 to 1669 he was the first Lucasian professor of mathematics.
He then became royal chaplain and finally returned to Cambridge
as master of Trinity College.

2263. WREN, Christopher. 1632-1723. (Mathematics 154, 158. Mechanics
512) M.A., Oxford, 1654. At Oxford he was closely associated
with Wilkins and his circle. From 1657 he was professor of
astronomy at Gresham College and in that position took a leading part in the formation of the Royal Society. In 1661 he
became Savilian professor of mathematics at Oxford. Charles II
appointed him "Surveyor of the Royal Works" in 1669 and thereafter his energies were devoted to architecture. He did however
serve as president of the Royal Society in 1680-82 and continued
to support the scientific movement in various ways.

2264. GREGORY, James. 1638-1675. (Mathematics 152) Born and educated
in Aberdeen. In 1662 he went to London and then to Padua where
he studied mathematics and astronomy under Torricelli's disciples for three years. By the time he returned to London in
1668 he had published three books. He was elected to the Royal

Society and appointed professor of mathematics at the University of St. Andrews (in Scotland). He was able to keep in touch with current developments through the well-known "intelligencer", John Collins.

2265. NEWTON, Isaac. 1642–1727. (Mathematics 155, 170. Mechanics 518. Physics 593, 601. Astronomy 372, 382) B.A., Cambridge, 1665; M.A., 1668. (The years 1665 and 1666 were his most fertile period, when most of his great discoveries were made, at least in essence) In 1669 he was appointed Lucasian professor of mathematics, in succession to Barrow, and he became a fellow of the Royal Society three years later.

At the end of the 1680s Newton apparently grew weary of academic life and made repeated attempts to obtain a post in London. He was not successful in doing so until 1696 when he was appointed Warden of the Mint (three years later he became Master of the Mint). He was knighted in 1705 and served as president of the Royal Society from 1703 until his death. He was elected a foreign member of the Paris Academy in 1699.

2266. HOOKE, Robert. 1635–1702. (Mechanics 516. Geology 1028. Microscopy 1257) M.A., Oxford, 1663. While a student he was a member of Wilkins' circle at Oxford and became Boyle's assistant, constructing his famous air-pump in 1658. On the establishment of the Royal Society he was appointed its "curator of experiments", a paid position which he held from 1662 until 1684. (He was elected a fellow in 1663) The brilliant way in which he carried out this demanding task was very beneficial to the Society and he also served it in various other ways for many years. From 1665 he was also professor of geometry at Gresham College.

After the great fire of London in 1666 he was employed for many years as a surveyor and architect in the task of rebuilding, in which role he was closely associated with Wren.

2267. WILLUGHBY, Francis. 1635–1672. (Not in Part 1) A member of the landed gentry. Graduated at Cambridge in 1656 and continued his scientific studies there in association with John Ray, who was then a college tutor. He became Ray's patron and from 1660 to 1665 they went on extensive travels, first in Britain and then on the Continent, studying natural history and collecting specimens. Because of his early death his two works on zoology were completed and published by Ray (see 1427 and 1433).

2268. LOWER, Richard. 1631–1691. (Physiology 1719) Graduated in medicine at Oxford in 1665. While he was a student there in the 1650s he was associated with Wilkins' circle and assisted Willis in his research. From 1666 he practised medicine in London, becoming a fellow of the Royal College of Physicians. He was also a fellow of the Royal Society.

2269. CROONE, William. 1633–1684. (Physiology 1718) B.A., Cambridge, 1650. From 1659 until 1670 he was professor of rhetoric at Gresham College and, with other members of the College, took part in the formation of the Royal Society in 1660; he continued to be an active member of the Society until his death. In 1662 he was granted the M.D. degree by royal warrant and subsequently had a successful practice in London, becoming a fellow of the Royal College of Physicians in 1674 (which he later endowed with the Croonian Lectures). He was also lecturer in anatomy at Surgeons' Hall from 1670.

2270. MAYOW, John. 1641-1679. (Physiology 1721) Graduated in law at Oxford in 1670 and also studied medicine. (His first book, published in 1668, was on medical subjects) He became a medical practitioner in Bath but also spent much time in London. He was elected to the Royal Society in 1678.

1670s

2271. GREENWICH OBSERVATORY. In 1675 Charles II agreed to the establishment of a national observatory at Greenwich and to the appointment of Flamsteed as astronomer royal. A year later the building was completed and Flamsteed began work.

2272. MOORE, Jonas. 1617-1679. (Not in Part 1. His importance was as a patron of science) A surveyor and military engineer. As a result of his work on fortifications he gained the royal favour, was knighted and appointed surveyor-general of ordnance. He was the author of several books on aspects of practical mathematics and was elected to the Royal Society in 1674.

In 1675 Moore used his influence with the King to obtain the royal assent to the establishment of Greenwich Observatory and the appointment of his protégé, Flamsteed, as astronomer royal. Since the King would not pay for the observatory's instruments Moore did so. The establishment of a mathematics school at Christ's Hospital (the Blue Coat School), for training navigation officers for the navy, was also due to his influence with the King.

2273. FLAMSTEED, John. 1646-1719. (Astronomy 381, 390) Was unable to attend a university and was self-taught in astronomy. In 1670 he became a protégé of Sir Jonas Moore (2272) who procured his appointment as astronomer royal at Greenwich. Since the King would pay for no more than the construction of the observatory building Flamsteed had to support himself (he was ordained and had a church living, and also gave private lessons in mathematics) and even pay for technical assistance and some of the instruments. He became a fellow of the Royal Society in 1677 and often served on its council.

2274. HALLEY, Edmond. 1656(?)-1743. (Astronomy 379, 386, 387. Earth science 1021) Came from a wealthy background. He was interested in astronomy and had a valuable collection of instruments even before he began his studies at Oxford in 1673. He was acquainted with Flamsteed and sometimes visited him at Greenwich. In 1676, while still a student, he went to the island of St. Helena and compiled a catalogue of southern stars. He then returned to Oxford, took his M.A. in 1678 and was elected to the Royal Society in the following year.

In the early 1680s Halley persuaded Newton to write the *Principia* and gave him much assistance and support while he was doing so. From 1686 to 1699 he was the clerk of the Royal Society, an important and influential position. In 1704 he became Savilian professor of geometry at Oxford and in 1720 succeeded Flamsteed as astronomer royal at Greenwich.

2275. BOTANICAL GARDEN, EDINBURGH. Established in several stages in the 1670s; a catalogue of the plants was published in 1683. The garden proved very successful in its several functions and was transferred to a larger site in 1763. It was transferred to its present, much larger, site in 1823.

2276. GREW, Nehemiah. 1641–1712. (Botany 1315, 1318) B.A., Cambridge, 1661. M.A., Leiden. Became a medical practitioner in Coventry. He took up the study of plant anatomy and in 1671 became a member of the Royal Society; in the following year he settled in London at the urging of Wilkins and other members of the Society who were impressed with his results. He was secretary (with Hooke) of the Society in 1677–79 and later served for many years on its council. Besides his books on plant anatomy he was author of *Musaeum Regalis Societatis; or, A catalogue and description of the natural and artificial rarities belonging to the Royal Society* (London, 1681).

2277. LISTER, Martin. 1639–1712. (Zoology 1428, 1432) M.A., Cambridge, 1662; then studied medicine at Montpellier. From 1669 he practised medicine in York where much of his research was done in rather isolated circumstances; he was a member of the Royal Society from 1671. In 1684 he moved to London where he became a fellow of the Royal College of Physicians.

1680s

2278. SLOANE, Hans. 1660–1753. (Not in Part 1. His importance was as a patron of science) Studied medicine in London around 1680 and was acquainted with Robert Boyle and John Ray. In 1683 he went on a tour of Europe in the course of which he acquired an M.D. and studied at Montpellier under Magnol and others. He returned to London and established himself as a successful physician, becoming a member of the Royal Society (1685) and of the Royal College of Physicians (1687). In 1687–89 he was a member of an expedition to Jamaica and subsequently published descriptions of its flora and fauna.

As a result of his lucrative practice, a fortunate marriage, and successful investments, Sloane became rich and influential. He was personal physician to Queen Anne and was knighted by George I in 1716. He became president of the Royal College of Physicians in 1719 and of the Royal Society (in succession to Newton) from 1727 to 1741. At his death he left to the nation his enormous collection of natural history specimens and antiquities, together with his large library and art collection. They became the nucleus of the British Museum, founded in 1753 as a result of his bequest.

2279. GREGORY, David. 1659–1708. (Astronomy 385) Nephew of James Gregory (2264). Took his M.A. at Edinburgh in 1683 and in the same year became professor of mathematics there. In 1691 he was appointed Savilian professor of astronomy at Oxford on the recommendation of Newton and Flamsteed. He joined the Royal Society in the following year.

Gregory's original work was not outstanding and his chief importance was in initiating the dissemination of Newtonianism through his lectures and his textbook.

2280. BURNET, Thomas. ca. 1635-1715. (Geology 1020) M.A., Cambridge, 1658. He remained there until the 1670s and was closely assoc- iated with the Cambridge Platonists. After various ecclesias- tical appointments he became chaplain to William III.

2281. TYSON, Edward. 1650/51-1798. (Zoology 1429, 1435) M.B., Oxford, 1677. Became a physician in London and lecturer in anatomy at Surgeons' Hall. He was elected to the Royal Society in 1679 and to the Royal College of Physicians in 1680.

2282. HAVERS, Clopton. ca. 1655-1702. (Anatomy 1607) Studied medicine at Cambridge and then at Utrecht where he graduated in 1685. He practised in London and joined the Royal Society in 1686.

1690s

2283. MOIVRE, Abraham de. 1667-1754. (Mathematics 176) Born and educated in France but migrated to England in the mid-1680s as a Huguenot refugee and became a private teacher of mathe- matics. He made the acquaintance of Halley who arranged for his election to the Royal Society in 1697. Though he became well known and respected as a mathematician (foreign member of the Berlin Academy in 1735 and of the Paris Academy in 1754) he was never able to obtain any sort of institutional appoint- ment and had to eke out a precarious living as a private tutor and a consultant on annuities and gambling odds.

2284. KEILL, John. 1671-1721. (Mechanics 524) Graduated M.A. at Edinburgh where he studied under David Gregory, the first teacher of Newtonianism. In the early 1690s he went to Oxford where he did some assistant teaching. He was elected to the Royal Society in 1700 and in 1712 became Savilian professor of astronomy at Oxford.

Keill was a fervent disciple of Newton and his book (524) did much to spread a knowledge of the Newtonian system in the scientific community; his use of experimental demonstra- tions was especially important (cf. 2293). In the famous controversy about the invention of the calculus he was Newton's "avowed champion", charging in 1708 that Leibniz had derived it from Newton's writings.

2285. GRAY, Stephen. 1666-1736. (Physics 606) Followed his father's craft as a dyer. Apparently he was self-educated in science but he knew Flamsteed and had a minor contact with the Royal Society. From 1696 he published numerous papers in the *Philo- sophical Transactions* on various subjects, chiefly astronomy, and became well known in London scientific circles, though he was not elected to the Royal Society until 1732. His experi- ments on electricity were conducted intermittently from about 1707 but his chief work was done in the early 1730s.

2286. WOODWARD, John. 1665-1728. (Geology 1024, 1031) During the 1680s he studied medicine under a leading physician in London whose influence helped him to get the position of professor of medicine at Gresham College in 1692. He was granted a Cambridge M.D. by special dispensation in 1696 and subsequently built up a medical practice in addition to his post at Gresham College. He became a fellow of the Royal Society in 1693.

2287. WHISTON, William. 1667-1752. (Geology 1025) M.A., Cambridge, 1693. Became assistant lecturer to Newton whom he succeeded in 1703 as Lucasian professor of mathematics (on Newton's recommendation). He wrote a number of theological--as well as mathematical and astronomical--works, and in 1710 was deprived of his chair because of his heretical opinions. Thereafter he lived in London and continued to write theological and scientific books, some of the latter being popularizations of Newtonianism.

1700s

2288. COTES, Roger. 1682-1716. (Mathematics 173) M.A., Cambridge, 1706; in the same year he was appointed Plumian professor of astronomy. He became a fellow of the Royal Society in 1711. For over three years he collaborated closely with Newton on the preparation of the second edition of the *Principia*. It was published in 1713 (with a preface by Cotes defending Newton's ideas against the still dominant Cartesianism). Newton's remark after his early death--"Had Cotes lived we might have known something"--appears to have been well justified.

2289. HAUKSBEE, Francis. ca. 1666-1713. (Physics 602) Originally a draper; apparently he later became an instrument-maker. In 1704 he was appointed curator (or demonstrator) of experiments for the Royal Society, becoming a fellow of the Society the next year. He continued in the post (which was a paid one) until his death. From 1704 he also gave a successful public lecture course on experimental physics, illustrated with demonstrations--the first such course to be given in London.

2290. FREIND, John. 1675-1728. (Chemistry 804) M.D., Oxford, 1707. F.R.S., 1712. He became a successful physician in London and a fellow of the Royal College. His *History of Physik* (1725) was well known.

1710s

2291. TAYLOR, Brook. 1685-1731. (Mathematics 175) Came from a wealthy family. Graduated in law at Cambridge in 1709. F.R.S., 1712.

2292. MACLAURIN, Colin. 1698-1746. (Mathematics 177, 182) A child prodigy. M.A., Glasgow, 1715. In 1717 he was appointed professor of mathematics at Marischal College, Aberdeen. Two years later he visited London where he met Newton and became a fellow of the Royal Society. During 1722-24 he travelled in France and won the prize offered by the Paris Academy for an essay on the percussion of bodies. On the recommendation of Newton he was appointed professor of mathematics at Edinburgh in 1725. In 1740 he shared another of the Paris Academy's prizes with Euler and Daniel Bernoulli.

2293. DESAGULIERS, John Theophilus. 1683-1744. (Not in Part 1) Born in France but taken to England as a child by his parents who were Huguenot refugees. M.A., Oxford, 1712. At Oxford he acquired the art of lecturing with the aid of experimental demonstrations from John Keill (cf. 524). He began public lectures in London in 1714 and in the same year became curator of experiments for the Royal Society (in succession to Hauksbee) and a fellow of the Society.

Desaguliers' many papers on mechanics and physics do
not contain much of importance and his chief book, *A Course of
Experimental Philosophy*, was of significance chiefly for practical mechanics and engineering. His main importance for
science was in the success of his public lecture course on
experimental physics which he gave over 120 times in the period
1714-34. In particular he devised demonstrations to make
Newtonian science understandable and convincing without the
use of much mathematics. He was thus one of the most important
of the early popularizers of Newtonianism.

2294. SMITH, Robert. 1689-1768. (Physics 608) M.A., Cambridge, 1715.
Cousin of Cotes (2288) with whom he worked at Cambridge and
whom he succeeded as Plumian professor of astronomy in 1716.
F.R.S., 1718. Later master of Trinity College and vice-chancellor of the University.

2295. HALES, Stephen. 1677-1761. (Plant physiology 1327. Physiology
1728. Chemistry 807) M.A., Cambridge, 1703. He was ordained
in 1709 and spent the rest of his life as curate of Teddington,
a village on the Thames not far west of London.

At Cambridge Hales obtained a good grounding in most
aspects of contemporary science, partly through the lectures
of Whiston, Cotes, and others, and partly through his own
efforts. Though Newton had left the University in the year
that Hales arrived there, he was profoundly influenced by
Newton through the Cambridge tradition and through his reading
of Newton's works, especially the *Opticks*. His "statical way
of inquiry" (i.e. quantitative experimentation), which he used
so fruitfully in plant and animal physiology, was the outstanding result of the Newtonian influence.

Hales became a fellow of the Royal Society in 1718 and
a foreign member of the Paris Academy in 1753.

2296. DILLENIUS, Johann Jacob. 1687-1747. (Botany 1331) Born and
educated in Germany. Graduated in medicine at Giessen in 1713
and became the town physician. He took up botany and became
a member of the Leopoldina Academy to which he contributed
several significant papers on cryptogams. In 1721 he went to
England to serve as assistant to the wealthy botanist, William
Sherard. He became a fellow of the Royal Society in 1724 and
ten years later was appointed professor of botany at Oxford.

1720s

2297. BRADLEY, James. 1693-1762. (Astronomy 392, 394, 396) M.A.,
Oxford, 1717. Ordained 1719. He had been instructed in observational astronomy by his uncle, a skilled amateur astronomer,
who introduced him to Halley. He did some observing for Halley
who was impressed by his ability and arranged for his election
to the Royal Society in 1718. Three years later Bradley was
appointed Savilian professor of astronomy at Oxford. In 1742
he succeeded Halley as astronomer royal. His work won him an
international reputation and he was elected a foreign member
of the academies of Paris, Berlin, Bologna, and St. Petersburg.

2298. COLLINSON, Peter. 1693/4-1768. (Not in Part 1. A naturalist
whose significance was entirely in his social role) He was
largely self-taught and derived his living from a haberdashery
business in London. Because of his expertise as a gardener,
especially in domesticating foreign plants, he became acquainted
with Sir Hans Sloane who arranged his election to the Royal
Society in 1728. As well as being a diligent member of the
Society he carried on an extraordinarily extensive activity
as an "intelligencer", especially with American naturalists
(partly because of his business contacts in America).
　　　　"At once gadfly, middleman, and entrepreneur ... Collinson was at the center of a network of scientific intelligence
which reached from Peking to Philadelphia ... He informed his
correspondents about those things which interested his friends
in the Royal Society, transmitted their collections to English
and European naturalists, and either read their letters and
formal papers to the Royal Society or secured their publication." (*DSB*) Nearly all the main American naturalists of the
time benefited from his activities and he is even credited with
starting Benjamin Franklin and his associates on their study
of electricity. He was widely respected for his useful activities and was elected to foreign membership of the Berlin and
Uppsala academies.

2299. MONRO, Alexander (*Primus*). 1697-1767. (Not in Part 1) Son of
a surgeon (who had studied at Leiden); educated in Edinburgh
and apprenticed to his father in 1713. In 1717-18 he visited
London, Paris, and Leiden, and in 1720 was appointed professor
of anatomy at Edinburgh. F.R.S., 1723. He was a leading
figure in a small society which began about 1730 as a purely
medical society but later became the Philosophical Society
(and in 1783 evolved into the Royal Society of Edinburgh).
　　　　Monro's importance was as a great teacher (his textbook
on osteology went through nineteen editions by 1828) who played
a major part in building up Edinburgh's medical school to the
stage where it could begin to rival Leiden's.

1740s

2300. NEEDHAM, John Turberville. 1713-1781. (Microscopy 1262. Embryology 1659) Ordained a Catholic priest in 1738 and became a
tutor to young men of English Catholic families, especially on
their grand tour of the Continent. Following the publication
of his book (1262) he was elected to the Royal Society in 1747
and soon afterwards, while in Paris, made the acquaintance of
Buffon with whom he maintained close relations for many years.
In 1768 he settled in Brussels and took an important part in
the establishment of the Brussels Academy in 1772, becoming
its director. He was well known for his theological and apolo-
getic writings as well as for his scientific work.

2301. WHYTT, Robert. 1714-1766. (Physiology 1730) Studied medicine
at Edinburgh from 1730 to 1734, then briefly at London, Paris,
and Leiden, and took an M.D. at Rheims in 1736. On returning
to Edinburgh he became a fellow of the Royal College of Physicians and set up in practice. In 1747 he was appointed professor

of medicine; his lectures on physiology were notable for their
experimental demonstrations. He was also a distinguished
physician. He became a fellow of the Royal Society in 1761.

1750s

2302. BLACK, Joseph. 1728-1799. (Chemistry 816. Physics 620)
      Studied medicine first at Glasgow and then at Edinburgh where
      he took his M.D. in 1754. (His dissertation for the degree
      was the starting point for his classical research on "fixed
      air") In 1756 he succeeded Cullen as professor of chemistry
      at Glasgow (where he also carried on a medical practice) and
      ten years later succeeded him again at Edinburgh, also as
      professor of chemistry.
           Black was an exceedingly popular teacher--one of the
      greatest of the eighteenth century, an age of great teachers.
      Some of his many students later became important chemists,
      including a number of Continentals who had been attracted to
      Edinburgh by the high reputation of its medical school.

2303. BRITISH MUSEUM. Established by Act of Parliament in 1753 in
      consequence of the bequest of Sir Hans Sloane (2278) who left
      to the nation his vast collection of natural history specimens
      and antiquities together with his rich library and art collec-
      tion. Thus the Museum was from its beginning a natural history
      museum combined with a collection of antiquities and *objets
      d'art* and what soon became a great library.
           The most outstanding of the early keepers of the Natural
      History Department was Daniel Solander (2310). In 1837 the
      Department was divided into three branches (later called depart-
      ments)--Botany (initially under the direction of Robert Brown;
      see 2341), Zoology, and Mineralogy and Geology. For later
      developments see 2464.

2304. MONRO, Alexander (*Secundus*). 1733-1817. (Anatomy 1618, 1622.
      Zoology 1454) Son of Alexander *primus* (2299). M.D., Edinburgh,
      1755. He then spent about a year on the Continent, studying
      with eminent anatomists. From 1758 he assisted his father as
      conjoint professor of anatomy at Edinburgh and continued teach-
      ing there successfully until his retirement in 1808. In 1798
      he arranged to have his own son, Alexander *tertius*, appointed
      as conjoint professor with him. He was a prominent member of
      the Royal College of Physicians of Edinburgh and of the Royal
      Society of Edinburgh.

1760s

2305. LUNAR SOCIETY OF BIRMINGHAM. An informal group, comprising
      hardly more than a dozen members, which gradually came into
      existence from the mid-1760s and gradually went out of exist-
      ence during the 1790s. It was significant because of the high
      calibre of its membership and because its concerns demonstrated
      the interest in science and its technological potentialities
      on the part of some of the leading figures of the Industrial
      Revolution. It included Joseph Priestley, James Watt, the
      famous industrialists Matthew Boulton and Josiah Wedgwood,
      inventors like Erasmus Darwin, and others with similar interests
      ranging over the physical sciences and technology of the time.

2306. CAVENDISH, Henry. 1731–1810. (Chemistry 817, 829. Geophysics 1043) A member of an aristocratic family. His father, a prominent fellow of the Royal Society, encouraged his scientific interests. He studied at Cambridge from 1749 to 1753 and acquired a good education in mathematics and physics but (like most aristocrats) did not take a degree. Being of independent means he was able to devote himself entirely to research. Most of his findings, including many of first-rate importance (especially in the study of electricity and heat) were never published and remained buried in his papers without any influence on the science of his time. Much of what he did publish, however, was outstanding and he was known to his contemporaries as a very able chemist and physicist.

    Because of his high social position and great wealth Cavendish held a distinguished place in the scientific life of London: as well as being active on the council of the Royal Society and its committees he was a trustee of the British Museum, a manager of the Royal Institution, and a prominent member of various societies.

2307. PRIESTLEY, Joseph. 1733–1804. (Chemistry 820. Physics 621) Educated for the ministry at a Dissenting Academy (a very good education which included some science). From 1755 onward he served as a minister in several parts of England and also taught for a few years at the Dissenting Academy in Warrington. While writing his first scientific work, his *History of Electricity*, in 1766 he became a fellow of the Royal Society. In 1773 he joined the staff of a prominent politician, the Earl of Shelburne, and during this period did his fundamental work in chemistry. He left Shelburne's employ in 1780 (but Shelburne continued to pay him a generous pension until his death) and settled as a preacher in Birmingham where he continued his chemical experiments and became a member of the Lunar Society. In 1791, because of his sympathy with the French Revolution, his house was wrecked by a rioting mob, and three years later continuing animosity drove him to America where he spent the last years of his life.

    Radical in politics and highly unorthodox in religion, Priestley wrote innumerable articles, pamphlets, and books on these and many other subjects. In his own time he was better known for these writings than for his scientific achievements.

2308. HUTTON, James. 1726–1797. (Geology 1124) Came from a wealthy merchant family in Edinburgh. Studied medicine at Edinburgh and Paris and took his M.D. at Leiden in 1749. He did not practise medicine however but took up farming. During the years he spent in the countryside as a scientifically-minded farmer he became interested in geology and was led to study it intensively. About 1768 he gave up farming and, with a good income from the rent of his farm and from a business interest, he settled in Edinburgh. In the city's flourishing intellectual community he had many distinguished friends and was an active supporter of the Royal Society of Edinburgh from its inception.

2309. MASKELYNE, Nevil. 1732–1811. (Geophysics 1040) Graduated at Cambridge in 1754 with honours in mathematics and soon afterwards made the acquaintance of Bradley, the astronomer royal.

F.R.S., 1759. In the early 1760s he was sent by the British
government on two ocean voyages to make astronomical observations and to investigate the reliability of proposed methods
for determining longitude. Following Bradley's death he was
appointed astronomer royal in 1765 and soon afterwards proposed
to the Board of Longitude the establishment of a national
ephemeris. As a result the *Nautical Almanac* began in 1766 and
he supervised its production until his death. In addition to
his various duties associated with practical improvements in
navigation he continued the tradition of accurate observation
which Bradley had established at Greenwich and made some
improvement in instrumental methods.

2310. SOLANDER, Daniel Carl. 1733-1782. (Not in Part 1. A botanist
who was not outstanding for his original work but was important
institutionally) Born and educated in Sweden. Studied at
Uppsala under Linnaeus and became his assistant. Linnaeus had
a high regard for him and commissioned him to publicize the
Linnean system in England; this he did very effectively from
his arrival in the country in 1760 at the invitation of some
English naturalists. Because of his expertise in Linnean
taxonomy he was given a post in the newly-established British
Museum in 1763 and proceeded to organize the natural history
collection. F.R.S., 1764. About that time he began a close
association with Joseph Banks who was then at the beginning of
his career. The two of them sailed with Captain James Cook on
the voyage of the *Endeavour* in 1768-71 and brought back a
wealth of specimens.

Solander was promoted to the position of keeper at the
British Museum in 1773 and continued to organize and enlarge
the natural history collection. At the same time he was in
charge of Banks' extensive collection and assisted in the
taxonomic work at Kew Gardens. In these and many other ways
he took a prominent part in the affairs of British naturalists.

2311. BANKS, Joseph. 1743-1820. (Not in Part 1. A wealthy patron of
science) Came from a prosperous family of landed gentry.
Studied at Oxford for a few years but did not take a degree.
After a voyage of botanical exploration to Labrador and Newfound-
land in 1766 he was elected to the Royal Society. In 1768-71,
together with Solander, he sailed with Captain James Cook on
the famous voyage of the *Endeavour*, thereby inaugurating the
practice--important for the future--of naturalists accompaning
naval voyages of exploration. His participation in the voyage,
and the specimens and data he brought back, made his reputation
and soon after his return he became friendly with George III.
He persuaded the King to add a research function to Kew Gardens
and in his capacity as honorary director of the Gardens he had
plants collected from all over the world and did much to make
the Gardens a centre for the dissemination of plants of economic importance.

Banks' friendship with the King was also a factor in
his election to the presidency of the Royal Society in 1778,
a position he held until his death. During his long tenure of
the office he was assiduous in carrying out his duties and did
much to enhance the prestige of the Society, though he did not

reform it. He was also one of the founders of the Linnean Society and the Royal Institution. His herbarium, one of the largest in existence, and his library, a major collection of works on natural history, both passed to the British Museum after his death.

2312. HUNTER, John. 1728-1793. (Anatomy 1621, 1624) Learnt anatomy and surgery during the 1750s by assisting his elder brother, William, who was a teacher of anatomy in London (and later a prominent obstetrician), and by attending classes at London hospitals. He set up in practice in London and his interests in natural history and comparative anatomy led to his election to the Royal Society in 1767. His practice flourished and he became well known and wealthy. From the early 1770s until 1790 he also gave private lectures on anatomy and surgery which had a wide influence; many of his students later became outstanding surgeons. His famous anatomical museum was created to advance the teaching of the subject; it eventually contained several thousand specimens and was the best of its kind in existence. After his death it was purchased by the British government who gave it to the Royal College of Surgeons.

Though he was primarily a surgeon (and a great one who did much for his profession) Hunter had a wealth of knowledge of comparative as well as human anatomy and a deep concern with, and insight into, the biological issues of his time. His contribution to science was more through the spoken word and personal example than through his published writings.

1770s

2313. RADCLIFFE OBSERVATORY, Oxford. Established in 1772 on the initiative of the Savilian professor of astronomy, Thomas Hornsby. It was one of the most outstanding of the numerous observatories constructed in universities and elsewhere in the eighteenth century.

2314. HERSCHEL, William. 1738-1822. (Astronomy 416. Physics 635) A native of Hanover. Son of a regimental bandmaster whose band he joined as an oboist in 1753. When Hanover was occupied by the French army in 1757, during the Seven Years War, he escaped to England (then linked politically with Hanover) and settled there. He earned his living as a musician and music-teacher, eventually achieving considerable distinction and popularity. His interest in the theory of music led him to the study of mathematics which in turn led him to astronomy and optics. During the early 1770s he taught himself telescope-making, becoming very skilful, and by the late 1770s he was making reflecting telescopes with unusually large mirrors.

Herschel's telescopes, though unsuitable for the exact positional astronomy which was the preoccupation of most astronomers of the time, far outstripped all others in their light-gathering power and suitability for the investigation of very faint objects. It was for this reason that, unlike nearly all other astronomers of the time, his interests lay in what may be called the natural history of stars.

His first triumph, however, was in planetary astronomy

--the discovery of Uranus, the first planet to be discovered
since prehistoric times. He immediately became world-famous
and was promptly elected to the Royal Society and awarded its
Copley Medal. George III granted him a pension which enabled
him to give up music and devote himself entirely to astronomy.
He continued to make telescopes for sale, and bigger and better
ones for his own use.

2315. BOTANICAL GARDENS, KEW. After a prehistory extending back to
the seventeenth century the ornamental gardens at Kew were
developed as botanical gardens by Princess Augusta, especially
with the appointment of William Aiton as superintendent in
1759. After the death of the Princess in 1772 the gardens
came into the possession of George III who enlarged them and
had them landscaped by the famous 'Capability' Brown. He also
appointed Joseph Banks as his scientific adviser, and Banks
enthusiastically built up the collection with plants gathered
from all over the world by expeditions which he organized.
After Banks' death in 1820 the gardens declined but in 1840
their formal ownership was transferred from the monarchy to
Parliament, and the Royal Botanic Gardens were constituted as
a national institution (see 2342).

2316. INGENHOUSZ, Jan. 1730-1799. (Botany 1336) A native of Holland.
M.D., Louvain, 1753. He later studied at Leiden and elsewhere,
and then practised medicine at Breda. In 1764 he migrated to
England where, as a hospital physician, he developed an expertise in smallpox inoculation. In 1768 he took his skills to
Vienna and was so successful at the imperial court that he
remained there for twenty years, with occasional short trips
back to England. It was in England, on one of these visits,
that he made his famous experiments on photosynthesis and
wrote his book on the subject. He returned permanently to
England in 1789.

       A person of varied scientific interests, Ingenhousz
also achieved some significant results in physics. He became
a member of the Royal Society in 1771.

2317. ROYAL PHYSICAL SOCIETY OF EDINBURGH. Founded in 1771 "for the
promotion of zoology and other branches of natural history."
It apparently did not publish a periodical until 1854.

1780s

2318. MANCHESTER LITERARY AND PHILOSOPHICAL SOCIETY. Founded in 1781.
Apart from the special case of the Lunar Society of Birmingham
(2305) it was the first and most important of the many learned
societies--often named "Literary and Philosophical"--which were
established in provincial cities, especially in the industrial
areas, during the following forty years. Their contributions
to science were generally slight (with some notable exceptions,
especially the case of Dalton in the Manchester Society) but
they did a great deal to disseminate an interest in science,
both for its own sake and for its possibilities for industry.

2319. ROYAL SOCIETY OF EDINBURGH. Founded in 1783. (For its ante-
cedents see 2299)

2320. ROYAL IRISH ACADEMY. Founded in Dublin in 1785.

2321. GOODRICKE, John. 1764–1786. (Astronomy 413) Came from the landed gentry. Though he was deaf and dumb from childhood he was able to acquire a good education which he completed at Warrington Academy (in 1778–81) where he had an able mathematics teacher who introduced him to astronomy. For his discovery of stellar variability the Royal Society awarded him its Copley Medal in 1783. Following his subsequent discoveries he was elected a fellow of the Society in 1786. Two weeks later he died "in the consequence of a cold from exposure to night air in astronomical observations."

2322. THOMPSON, Benjamin (*Count* RUMFORD). 1753–1814. (Physics 631) An American who was loyal to the British crown during the Revolution and was obliged to flee to London in 1776. He later commanded a British regiment in the war and received a knighthood from George III. His interest in physics, especially heat, arose out of his involvement with cannons and gunpowder. An investigation in ballistics in 1781 led to his election to the Royal Society of which he became an active member.

In 1785 Thompson joined the court of the Elector of Bavaria and became head of the army and one of the most powerful men in the country. He introduced numerous social reforms as well as many technological innovations which brought him international fame. He was made a count of the Holy Roman Empire in 1793, taking the title of von Rumford.

He returned to London in 1798 and was the chief mover in the establishment of the Royal Institution in the following year. In 1802 he settled in Paris where he was an active member of the Institut National. He received many honours and was elected to numerous academies and societies.

2323. TENNANT, Smithson. 1761–1815. (Chemistry 841, 848) Studied medicine at Edinburgh (chemistry under Black) and Cambridge, receiving the M.B. at Cambridge in 1788 (and the M.D. in 1796). F.R.S., 1785. He was a substantial landowner and a prominent agriculturist. In 1799 he was one of the founders of the Askesian Society (which became the Geological Society in 1807). For his achievements in chemistry he received the Royal Society's Copley Medal in 1804, and in 1813 he became professor of chemistry at Cambridge.

2324. KIRWAN, Richard. 1733(?)–1812. (Chemistry 827. Mineralogy 1086. Geology 1126) A member of an Irish landed family. From 1766 he was a barrister but gave up the profession after a few years and lived in London where he cultivated science, especially chemistry. F.R.S., 1780. He returned to Ireland in 1787 and was one of the foundation members of the Royal Irish Academy, of which he was president from 1799 until his death. He was well known in the scientific community in Britain and elsewhere.

2325. PLAYFAIR, John. 1748–1819. (Geology 1131) Studied at the universities of St. Andrews and Edinburgh, and exhibited considerable mathematical ability. He was ordained a minister and from 1772 to 1782 was engaged in parish duties. In 1785 he became professor of mathematics at Edinburgh, and in 1805 professor

of natural philosophy (i.e. mechanics and physics). For many
years he edited the *Transactions* of the Royal Society of Edinburgh and was for a time the Society's secretary and later its
president. Though most of his publications were in mathematics
and physics he became best known for his work in geology, a
subject to which he had been introduced by his friend, Hutton.

2326. HALL, James. 1761-1832. (Geology 1128) Came from a wealthy
Scottish landed family. He spent two years at Cambridge and
two years at Edinburgh (where he attended Black's chemistry
lectures) without taking a degree. From 1783 to 1786 he was
travelling in Europe where he developed his interest in chemistry and geology, and met many scientists. Thereafter he
settled on his country estate with periods of residence in
Edinburgh, becoming a fellow of the Royal Society of Edinburgh
in 1784 (president in 1812) and of the Royal Society of London
in 1806.

2327. LINNEAN SOCIETY OF LONDON. Founded in 1788; see 2328. It became
a well-established body and through the nineteenth century was
the premier natural history society in Britain.

2328. SMITH, James Edward. 1759-1828. (Not in Part 1. A botanist
whose original work--chiefly in taxonomy--was of no great
importance but who had a significant role institutionally)
Studied medicine at Edinburgh and took his M.D. at Leiden in
1786. In 1783 he was in London and in touch with Joseph Banks;
at Banks' suggestion he bought the library, manuscripts, herbarium, and specimens of Linnaeus from the latter's executors.
Thereafter his career was shaped by his ownership of the
collection. It led to his election to the Royal Society in
1785 and to the foundation in 1788 of the Linnean Society of
which he was president until his death. He published translations of some of Linnaeus' works and an edition of his
correspondence. After his death the Linnaean collection was
acquired by the Society.

1790s

2329. ROYAL INSTITUTION. Established in 1799 by Count Rumford (Benjamin
Thompson) with the support of Sir Joseph Banks and other wealthy
and influential persons with philanthropic concerns. It was
created, in the founders' intentions, "for diffusing the knowledge, and facilitating the general introduction, of useful
mechanical inventions and improvements; and for teaching, by
courses of philosophical lectures and experiments, the applications of science to the common purposes of life."

The character of the institution was however largely
determined in its early years by the interests and capabilities
of its staff. The first professor (of physics and chemistry),
Thomas Garnett, was an accomplished lecturer and began to
attract large and fashionable audiences to his public lectures.
His successor, Humphry Davy (appointed professor in 1802) developed this initiative with spectacular success. The tradition
of excellent public lectures, with demonstrations and experiments, that was thus established has continued to the present
day, and the attendance fees have contributed substantially to

the financial stability of the institution. With no less
success Davy also inaugurated the Royal Institution's other
leading activity--research--which had been implicit, if not
explicit, in the founders' intentions.

Following Davy, the luminary of the Royal Institution
was Michael Faraday who began there as a humble laboratory
assistant and twenty years later, in 1833, was appointed to
the newly-endowed Fullerian professorship of chemistry. Other
outstanding figures who ably continued the institution's trad-
itions in the late nineteenth century were John Tyndall (1853-
87). James Dewar (1877-1923), and Lord Rayleigh (1887-1905).

2330. YOUNG, Thomas. 1773-1829. (Physics 636. Physiology 1739)
A child prodigy; largely self-taught in mathematics and science
(and eastern languages). Studied medicine at London, Edinburgh,
and Göttingen where he was awarded the M.D. in 1796. His
researches, initially on physiological optics, were published
from 1791 onward and in 1794 he was elected to the Royal Society.
In 1801 he was appointed to the newly-created Royal Institution
but his lectures were unsuccessful and he was obliged to resign
in 1803.

Young continued practising medicine in London and in
1811 was appointed physician to St. George's Hospital, a post
he held for the rest of his life in addition to sundry other
activities, especially journalism and consulting. From 1804
he was foreign secretary of the Royal Society and from 1818
secretary of the Board of Longitude and superintendent of the
Nautical Almanac. His important work on the partial decipher-
ment of the Egyptian hieroglyphics on the Rosetta stone was
mostly done in the period 1813-19 and laid the foundation for
the subsequent work of Champollion.

2331. DALTON, John. 1766-1844. (Chemistry 851) Son of a weaver.
Educated at a Quaker school in the north of England and became
a teacher in a similar school. From 1787 he acquired an interest in meteorology and it was from this that his later scientific work (including his atomic theory) derived. In 1800 he
established his own school in Manchester where he taught mathematics, physics, and chemistry. From this occupation he
derived an adequate income for the rest of his life. After he
had become established scientifically he gave numerous courses
of public lectures in several major cities and these added
substantially to his income.

Dalton became a member of the Manchester Literary and
Philosophical Society in 1794, its secretary in 1800, and its
president from 1817 until his death. In the early stages of
his career he derived much stimulus and assistance from the
Society (he had his laboratory on its premises) and he later
brought it considerable prestige. He was an active member of
the British Association from its establishment in 1831 but he
was not interested in joining the Royal Society (possibly
because its mainly upper-class membership and metropolitan
style were so different from his provincial, Quaker, lower-
class background). He was however quite willing to accept
election as a corresponding member of the Paris Academy in
1816 and as a foreign member in 1830. In his late years he
received many other honours.

2332. SMITH, William. 1769-1839. (Geology 1127, 1141) Son of a
      village blacksmith. In 1787 he became assistant to a surveyor
      and learnt the trade. From the early 1790s he established
      himself as what would now be called a civil engineer, superin-
      tending the construction of canals (an activity then at its
      height in England). In the course of his work he travelled
      over much of the country, and from his constant observations
      was able to gain a general view of the distribution and success-
      ion of strata in a large part of England. His study of the
      individual strata was greatly assisted by his supervision of
      excavations for canals and other earthworks; he thus made the
      great discovery that the strata could be distinguished by their
      characteristic fossils.
           By the 1820s Smith was recognized as "the father of
      stratigraphical geology." In 1831 his achievements were form-
      ally acknowledged by the award of a medal by the Geological
      Society, and from 1832 he received a government pension.

2333. KNIGHT, Thomas Andrew. 1759-1838. (Botany 1348) Came from a
      wealthy family of landowners; graduated at Oxford but spent
      his life as a gentleman farmer, managing the family estate.
      He corresponded with Joseph Banks about his scientific experi-
      ments and was elected to the Royal Society in 1805. He was a
      founding member of the Horticultural Society and its president
      from 1811. In addition to his numerous experiments in plant
      physiology and plant breeding he made improvements in agricul-
      ture by application of scientific principles.

1800s

2334. BRISBANE, Thomas Macdougall. 1773-1860. (Not in Part 1. Though
      a capable astronomer his significance was as a patron of astron-
      omy) Came from the Scottish landed gentry. His studies at
      Edinburgh University and elsewhere included mathematics and
      astronomy. He served in the army from 1789, reaching the rank
      of general in 1841. In 1808 he established a well-equipped
      observatory at his residence in Scotland and two years later
      became a fellow of the Royal Society. He was subsequently a
      prominent member of the Astronomical Society and of the Royal
      Society of Edinburgh.
           When he became governor of New South Wales in 1821
      Brisbane established at his own expense an observatory at
      Parramatta (near Sydney) in order to advance the astronomy of
      the southern hemisphere; it eventually produced the "Brisbane
      Catalogue", listing over seven thousand stars. On his return
      to Scotland in 1826 he established another observatory at
      Makerstoun, to which in 1841 he added a geomagnetic observatory.
      The magnetic observations made there, as part of the internat-
      ional effort inspired by Humboldt (1049), were especially
      valuable because of the observatory's location in the far
      northwest of Europe.
           In his later years Brisbane received many scholarly
      and civil honours.

2335. BREWSTER, David. 1781-1868. (Physics 641) M.A., Edinburgh,
      1800. His interest in science, dating from his youth, had to

remain an avocation because there were no professorships or
other suitable posts available to him; nevertheless he managed
to publish much research. In the early part of his career he
earned his living chiefly through his literary and journalistic
activities, writing many books and articles and editing various
periodicals, both scientific and non-scientific. The former
included *The Edinburgh (New) Philosophical Journal* (1819-64)
and *The Edinburgh Journal of Science* (1824-32); he was also
for many years a member of the editorial board of *The Philosophical Magazine*.

Brewster's researches in optics began about 1798 and
he was elected to the Royal Society in 1815. His invention of
the kaleidoscope the next year made him well known to the public
(and brought him much-needed royalties). He also became well
known for his many writings, notably his biography of Newton.
He took a leading part in the establishment of the British
Association in 1831 and throughout his career agitated for
greater public recognition and support for science and scientific education.

His political connections, especially with Henry
Brougham, the Lord Chancellor, led to a knighthood in 1832 and
the post of principal of the University of St. Andrews a few
years later. In 1859 he became principal of the University of
Edinburgh. He received many honours and awards.

2336. WOLLASTON, William Hyde. 1766-1828. (Chemistry 847, 865. Crystallography 1090, 1092. Physics 638) Came from a distinguished
family; his father was a fellow of the Royal Society. Graduated
in medicine at Cambridge in 1787 and after completing his training in London began practising in 1792. He gave up medicine
in 1800, and subsequently derived a large income from his
process for making platinum malleable. He was an active member
of the Royal Society (president in 1820) and was also from 1818
a member of the Board of Longitude (because of his interest in
optical instruments). At his death he left substantial amounts
of money to the Royal Society and the Geological Society (of
which he had been a member since 1812).

2337. THOMSON, Thomas. 1773-1852. (Chemistry 853) M.D., Edinburgh,
1799. During his medical course he attended Black's lectures
on chemistry which inspired him to devote himself to the subject
and he never practised medicine. In his early career he earned
his living by teaching chemistry privately, consulting, and
publishing. F.R.S., 1811; in the next year he brought out his
History of the Royal Society. In 1813 he founded the *Annals
of Philosophy* which until its cessation in 1827 had a prominent
place in British science.

Thomson became professor of chemistry at Glasgow in
1818 and introduced teaching by means of practical laboratory
work, an innovation still quite uncommon everywhere. With
some of his best students he built up a small research school
which can be regarded as a forerunner of Liebig's more famous
school. By 1830 he was one of the leading scientists of Scotland and his reputation was further enhanced by his *History of
Chemistry* (1830-31).

2338. HENRY, William. 1774–1836. (Chemistry 850) M.D., Edinburgh,
      1807. He had joined the Manchester Literary and Philosophical
      Society (of which his father was a prominent member) in 1796
      and began research in chemistry about that time. In 1808 his
      publications brought him election to the Royal Society and
      award of its Copley Medal. As well as practising medicine he
      directed the chemical works (manufacturing magnesia, etc.)
      that had been established by his father. He was a close friend
      of Dalton and his work on the solubility of gases probably
      played a part in the genesis of Dalton's atomic theory.

2339. DAVY, Humphry. 1778–1829. (Chemistry 845, 861) Apprenticed
      to an apothecary–surgeon in 1798; taught himself chemistry and
      became assistant to the chemist and physician, Thomas Beddoes,
      who was investigating the medical effects of the newly–discov-
      ered gases. The work he did at Beddoes' institute in Bristol
      led to his appointment in 1801 as assistant lecturer at the
      recently established Royal Institution in London, and in the
      following year he became professor. His public lectures were
      highly popular, attracting large audiences and becoming quite
      fashionable in London society. Their success gained consider-
      able prestige for the Royal Institution and for science gener-
      ally.
           Davy became a fellow of the Royal Society in 1803, its
      secretary in 1807–12, its president in 1820, and received three
      of its medals (1805, 1816, 1827). He was knighted in 1812 and
      after his invention of the safety lamp was made a baronet.

2340. GEOLOGICAL SOCIETY OF LONDON. Founded in 1807. The first
      national geological society.

2341. BROWN, Robert. 1773–1858. (Botany 1350, 1358, 1360) Graduated
      in medicine at Edinburgh and in 1795 joined the army as an
      assistant surgeon; he spent much of his spare time developing
      his knowledge of botany. While in London he got to know Joseph
      Banks who was impressed with his ability and in 1800 recommended
      him to the Admiralty for the post of naturalist aboard the
      Investigator, a ship which had been assigned the task of survey-
      ing the coasts of Australia. After four years of botanical
      exploration in Australia he returned to England in 1805 with
      some four thousand specimens.
           In 1806 Brown became librarian to the Linnean Society
      and in 1810 (in addition) librarian and curator in the service
      of Joseph Banks. When Banks died in 1820 he left his library
      and collection to Brown, stipulating that on the latter's death
      they were to pass to the British Museum. Brown however negoti-
      ated their transfer to the Museum in 1827 on the condition that
      it establish a botanical department with himself as head. He
      held the post until his death, declining three offers of profess-
      orships. He was held in the highest regard by the botanists
      of his time and had an important influence in Germany (which
      he often visited) as well as in Britain.

2342. HOOKER, William Jackson. 1785–1865. (Not in Part 1. Though he
      was an outstanding systematic botanist and a prolific author
      his chief importance was in his institutional role) As a young
      man he made the acquaintance of some prominent botanists, such

as J.E. Smith and later Joseph Banks, whom he impressed with his ability. Until 1820 he lived on his private income, writing numerous papers and books, building up a large herbarium, and establishing a solid reputation; in that year however, because of a decline in his income, he accepted the chair of botany at Glasgow (obtained for him by Banks). He was a successful teacher and during his stay at Glasgow he greatly improved the city's botanical garden; at the same time he managed to maintain an impressive output of publications.

In 1836, as the leading botanist in England, Hooker was knighted. In conjunction with other botanists, notably John Lindley, and with the support of the Duke of Bedford, he used his influence in high places to have the gardens at Kew--the private possession of the monarchy and lacking scientific supervision since the death of Banks in 1820--transferred to the nation. In 1841 he became the first director of the reconstituted Royal Botanic Gardens and within a few years greatly enlarged the institution and converted it into a leading botanical centre, a national showpiece, and--in everything to do with plants--the hub of the British Empire. After his death he was succeeded as director by his son, J.D. Hooker (2400).

2343. BELL, Charles. 1774-1842. (Anatomy 1630) Learnt anatomy and surgery by assisting his elder brother, a surgeon who gave private lessons, and by attending lectures at Edinburgh University. In 1799 he was admitted to the Royal College of Surgeons of Edinburgh. He moved to London in 1804 and built up a successful practice and also established his own school of anatomy. His *Essay on the Anatomy of Expression in Painting* of 1806 proved very popular and helped to establish his reputation. He was awarded a medal by the Royal Society in 1829 for his work in neuroanatomy, knighted in 1831, and appointed professor of surgery at Edinburgh in 1836.

1810s

2344. WHEWELL, William. 1794-1866. (Philosophy of science 41) Graduated at Cambridge with high honours in mathematics in 1816, became a college fellow, and was ordained in 1826. Until 1839 he was a college tutor and concurrently professor of mineralogy (1828-32) and then professor of moral philosophy (1837-55); from 1841 until his death he was master of Trinity College. He became an influential figure in the University and served as its vice-chancellor in 1842 and 1855.

Whewell's interests extended beyond science into various fields of scholarship--theology, ethics, Kantian philosophy, etc.--and he was the author of many essays, sermons, and other writings. Within science he made significant, if minor, contributions to the theory of tides, mineralogy, and meteorology. More important was his involvement during the 1820s and later in the movement to reform Cambridge mathematics by introducing the methods of the great Continental mathematicians; his use of these methods in several textbooks of applied mathematics that he published between 1819 and 1837 was an effective way of disseminating them. By his participation in this movement and by use of his weighty influence in later years he did much to strengthen the Cambridge tradition in mathematics and physics.

Whewell was a fellow of the Royal Society from 1820 and later became a member of most of the other leading British scientific societies and a central figure in the scientific community. From the mid-1830s, apart from his numerous extra-scientific activities, he was chiefly concerned with the history and philosophy of science. His three-volume *History of the Inductive Sciences* (1837) was the most outstanding work of its kind to appear before the twentieth century, while his *Philosophy of the Inductive Sciences* (1840) ranked with the contemporary works in the field by John Herschel and J.S. Mill.

2345. BABBAGE, Charles. 1792-1871. (Mathematics 212. Organization of science 37) Came from a wealthy family. Graduated at Cambridge in 1814. While they were still undergraduates, he and John Herschel and George Peacock founded in 1812 the (short-lived) Analytical Society with the object of introducing Continental methods into Cambridge mathematics. He joined the Royal Society in 1816 and from 1827 to 1839 held the Lucasian chair of mathematics at Cambridge but without exercising its functions. He was a prominent figure in the scientific community in London and took part in establishing the Astronomical Society, the British Association, and the Statistical Society.

In addition to papers on pure and applied mathematics Babbage wrote on an extraordinary variety of subjects, often with remarkable insight. The great objective of his life, however, was his computational "engine" (see 212).

2346. HERSCHEL, John (F.W.). 1792-1871. (Astronomy 428, 433, 445. Philosophy of science 36. Various writings 43, 46, 52) Son of William Herschel. Graduated at Cambridge in 1813 with the highest honours in mathematics. In the same year he was elected to the Royal Society by virtue of the mathematical papers he had already published. After some hesitation about the choice of a career he decided in 1816 to assist his famous father in his astronomical researches. This apprenticeship gave him a unique training in astronomy and shaped his future achievements in the field. He was able to devote himself entirely to research because of his large private income (the source of which is not now known).

Throughout his career as an astronomer Herschel continued the kind of research that had been pioneered by his father. During the period 1834-38 he worked at the Cape of Good Hope, applying the Herschel methods and approach to the skies of the southern hemisphere. (He refused an offer of governmental financial assistance towards the cost of the expedition) On his return to England he was made a baronet by Queen Victoria and lionized by the public as well as by the scientific community. Around mid-century he was probably the best-known scientist in Britain and most of the chief European academies and societies made him a member. He served on numerous government bodies, took a leading place in the affairs of the scientific community and--with his numerous essays, lectures, and popular articles on many topics--became a notable figure in the cultural life of the times. When he died he was mourned by the whole country and was buried close to Newton in Westminster Abbey.

Herschel was one of the last great scientists to spread
his activities over several fields. Besides his work in astronomy (which included theoretical studies as well as the observational work for which he is best known) he made significant
contributions to mathematics and physics, especially optics,
while his ability in chemistry led him to make several valuable
innovations in the new art of photography.

2347. DANIELL, John Frederic. 1790–1845. (Chemistry 888. Meteorology
1045) Son of a prosperous lawyer; educated privately. During
the 1810s and 1820s he was involved in business affairs while
at the same time carrying on research in chemistry. F.R.S.,
1813. In 1827 he was one of the founders of the Society for
Promoting Useful Knowledge and later wrote several of its
popularizations of scientific subjects. In 1831 he was appointed professor of chemistry at the newly-established King's
College, London, where his teaching was distinguished by his
lecture demonstrations and instruction in practical laboratory
work; it also gave rise to a successful textbook.

Daniell became prominent in the affairs of King's
College and in the scientific community generally in London,
and on various occasions served as an adviser to government
bodies. He was one of the founders of the Chemical Society
and throughout his career was active in the affairs of the
Royal Society. In 1837 he received its highest award, the
Copley Medal, for his invention of the Daniell cell.

2348. BUCKLAND, William. 1784–1856. (Geology 1150) Graduated at
Oxford in 1804; became a college fellow and was ordained in
1809. He was one of a group at Oxford who were developing an
interest in mineralogy and geology, and he made a number of
tours in Britain and later to many parts of the Continent. In
1813 he became reader in mineralogy at Oxford and soon afterwards was elected to the Geological Society and the Royal
Society. He was later prominent in the affairs of both societies and the newly-formed British Association.

A follower and admirer of Cuvier, Buckland became
during the 1820s the most outstanding geologist in England and,
to the general public, one of the best-known scientists. Many
of the younger geologists (including Lyell) were his pupils or
had been introduced to geology through his influence. Especially important was the support he gave to De La Beche, a long-
time friend, in the latter's moves to establish the Geological
Survey and the Museum of Practical Geology.

2349. SEDGWICK, Adam. 1785–1873. (Geology 1157, 1164, 1180) Studied
theology and mathematics at Cambridge and graduated in 1808;
became a college fellow and was ordained in 1817. In the next
year he was appointed Woodwardian professor of geology, though
he knew little about the subject. He soon turned out however
to be the first distinguished (indeed the first active) occupant
of the chair since its establishment under the will of John
Woodward (2286) nearly a century earlier. He became a fellow
of the Geological Society in 1818 and of the Royal Society
three years later, and was a prime mover in the founding of
the Cambridge Philosophical Society in 1819. He was a fine
teacher with a very engaging manner and his lecture course

(which he continued to give until 1870) was highly popular even though, in the Cambridge system, it was extracurricular.

Woodward's bequest to Cambridge had included his mineral collection and on this foundation Sedgwick gradually assembled one of the best geological museums in the world. A new building that the University provided for it in 1841 became inadequate by the time of his death and the present Sedgwick Museum, opened in 1903, was built as a memorial to him.

2350. CONYBEARE, William Daniel. 1787-1857. (Geology 1147) Graduated at Oxford about 1810 and was ordained soon afterwards. After serving as a parish priest in various parts of England and Wales he became dean of Llandaff in 1845. He became a member of the Geological Society in 1811 and of the Royal Society in 1832.

2351. DE LA BECHE, Henry Thomas. 1796-1855. (Not in Part 1. A capable geologist who made many solid contributions though none of any great significance; his importance was chiefly in his institutional role) Educated for the army but left it after the peace of 1815. Being of independent means he was able to devote himself to his geological interests, joining the Geological Society in 1817 and the Royal Society two years later. During the next ten years he spent much time travelling over large areas of the Continent, and he published numerous papers on aspects of Continental and British geology.

In the early 1830s De La Beche conceived the idea of making a geological survey of Devonshire (an area he was familiar with) using the topographical maps recently published by the Ordnance Survey. He managed to obtain a government grant for the purpose as officialdom was beginning to perceive the potential economic value of geological information. On completion of the task he proposed that similar surveys should be undertaken in other parts of the country: the result was the establishment by the government in 1835 of the Geological Survey of Great Britain, the first institution of its kind.

De La Beche was appointed director and throughout the rest of his career was successful in expanding the Survey and convincing the government of the value of its activities. In 1842 he was knighted and in 1851 the Museum of Practical Geology was established under his control as an adjunct to the Survey.

2352. SABINE, Edward. 1788-1883. (Geodesy 1046. Geomagnetism 1051, 1059, 1060) Educated at the Royal Military Academy and served in the army. After the end of the Napoleonic Wars he took up scientific investigations while retaining his commission (he rose to the rank of major-general by 1859) and became a fellow of the Royal Society in 1818. He eventually became an influential figure in the British scientific community and was prominent in the Royal Society (president 1861-71) and the British Association (president 1852). He was knighted in 1869.

1820s

2353. HAMILTON, William Rowan. 1805-1865. (Mathematics 242. Mechanics 553. Physics 660) A prodigy in mathematics and languages.

While still a student at Trinity College, Dublin, he published such outstanding papers that in 1827, even before he had received his degree, he was appointed professor of astronomy and astronomer royal of Ireland. These posts he retained for the rest of his life, leaving the astronomical work of the observatory to assistants and devoting himself to research in mathematics, pure and applied. He was elected to the Royal Irish Academy in 1832, later serving as its president, and was prominent in the British Association. He was knighted in 1835.

Hamilton acquired a great reputation as a mathematician and received many honours and awards from scientific bodies.

2354. DE MORGAN, Augustus. 1806-1871. (Mathematics 228, 248) Graduated at Cambridge in 1827 with honours in mathematics. In the following year he was appointed professor of mathematics at the recently-established University College, London, where he remained (with an intermission in 1831-35) until his retirement. He became a fellow of the Astronomical Society in 1828, later serving as its secretary for many years, and subsequently a fellow of the Royal Society. In 1865 he was one of the founders of the London Mathematical Society and its first president.

Brilliant, original, and eccentric, De Morgan was a prolific writer and popularizer. From 1826 he was a leading member of the Society for the Diffusion of Useful Knowledge and he wrote innumerable articles for encyclopaedias and popular periodicals. He had a considerable influence on the teaching of mathematics in Britain both through his own pupils and his many textbooks. One of his chief interests was the history of mathematics to which he made many contributions.

2355. ASTRONOMICAL SOCIETY OF LONDON. (Later named the Royal Astronomical Society) Founded in 1820. The first national astronomical society.

2356. PARSONS, William, *3rd Earl of* ROSSE. 1800-1867. (Astronomy 444) A member of the Anglo-Irish nobility. Graduated at Oxford in 1822 with distinction in mathematics. In later life he capably fulfilled the duties of a person of his social rank—lord lieutenant of his county, member of the House of Lords, etc. His contributions to science stemmed from his determination to construct the most powerful telescope possible at the time. His experiments began about 1826 and soon led him to decide on a reflector rather than a refractor. After many trials he succeeded in casting and shaping a metal mirror of 72-inch diameter; the telescope incorporating it, fifty-four feet in length, was completed in 1845. No bigger telescope was built until 1917.

Rosse served at various times as president of the Royal Society and of the British Association. Among other public offices he was a member of the board of visitors of Greenwich Observatory.

2357. AIRY, George Biddell. 1801-1892. (Not in Part 1. A capable but not outstanding astronomer whose significance lay in his organizational role) Graduated at Cambridge in 1823 with the highest honours in mathematics. In 1826 he became Lucasian

professor of mathematics and in 1828 Plumian professor of
astronomy and director of the university observatory. He left
Cambridge for Greenwich in 1835 when he was appointed astronomer royal.

Airy completely re-equipped Greenwich Observatory and,
with great energy, converted it into a highly efficient organization for its official purpose of amassing large amounts of
accurate observational data. His methods of organizing and
running the observatory set a standard which was emulated in
other countries. Since the observatory was under the control
of the Admiralty, Airy was a government scientist--perhaps the
most outstanding one of the time. In addition to his obliga-
tions to the Admiralty he served on innumerable government
bodies and commissions dealing with all kinds of scientific
and technical matters.

A prominent member of the scientific community, Airy
was at various times president of the Royal Society, the Royal
Astronomical Society, and the British Association. He was
knighted in 1872 and received numerous honours and awards from
scientific bodies, foreign as well as British.

2358. FARADAY, Michael. 1791-1867. (Physics 665, 687. Chemistry 872,
885) Came from humble origins and grew up in poverty. He had
only a rudimentary education and at the age of fourteen was
apprenticed to a bookbinder. His omniverous reading and attendance at public lectures gained him an education and a special
interest in science. He came into contact with Davy who was
impressed by him and gave him a position as laboratory assistant
at the Royal Institution. He learnt much from Davy who took
him as his secretary and assistant on a two-year visit to the
Continent during which he met many important scientists.

On his return in 1815 he was given a better position
at the Royal Institution and began to undertake research,
becoming director of the laboratory in 1825 on Davy's recommendation; he had been elected to the Royal Society in the previous
year. He instituted the Friday evening discourses at the Royal
Institution--a series of lectures on popular science which were
highly successful--and in 1833 was appointed to the newly-
endowed Fullerian professorship of chemistry. His career of
research continued until his retirement in 1858.

2359. MURCHISON, Roderick Impey. 1792-1871. (Geology 1163, 1164, 1174)
Came from a wealthy Scottish landed family. After active
service in the Peninsular War he led a rather aimless life. In
the early 1820s, however, when he had settled in London, he
became interested in science, partly through meeting Davy and
attending the lectures at the Royal Institution. The influence
of Buckland led him to geology and he became a member of the
Geological Society in 1825 and of the Royal Society the next
year. After five years of geological exploration in Scotland,
France, and the Alps--sometimes with Lyell and sometimes with
Sedgwick--he began in 1831 his important researches on the
Lower Palaeozoic rocks of south Wales. In 1840-41 he travelled
in Russia where there was a sequence of strata very relevant
to his investigations.

Murchison was knighted in 1846 and appointed Director-

General of the Geological Survey (in succession to De La Beche) in 1855. His wealth and high social position, combined with his achievements in geology, made him a leading member of the British scientific community; at various times he was president of the Geological Society, the Geographical Society, and the British Association. He received many honours.

2360. LYELL, Charles. 1797–1875. (Geology 1156) Graduated at Oxford in 1819, his studies including geology under Buckland. He then studied law in London and qualified in 1825 but practised for only a few years, his private income enabling him to devote himself to geology. He had joined the Geological Society in 1819 and his study of law had been interspersed with geological tours in Britain and the Continent. In 1828 he resumed his travels, going via Auvergne and the Italian peninsula to Sicily where his numerous discoveries served to crystallize ideas that had been forming in his mind. He returned to England to write his great *Principles of Geology*.

   In 1831 Lyell was appointed professor of geology at King's College, London, but he gave up the post two years later as too time-consuming. Through the rest of his life he gave much effort to keeping his *Principles* up to date (with twelve editions up to his death) as well as writing other books and numerous papers. He also continued to make many geological tours, including several to the Continent and four to North America. From 1836 he was a close friend of Charles Darwin who had been greatly influenced by his *Principles* and whose geological observations in South America and elsewhere were very useful to him. However Lyell's extreme uniformitarianism (which tended to deny any progressive development in the course of Nature) made him doubtful of Darwin's theory and he did not fully accept it until 1864; he then revised his *Principles* (in the tenth edition) to bring it into line with the theory.

   Lyell became a member of most of the important European academies, was accorded the highest scientific honours as well as a knighthood and a baronetcy from Queen Victoria, and received the ultimate accolade of burial in Westminster Abbey.

2361. SCROPE, George Julius Poulett. 1797–1876. (Geology 1153) Came from a wealthy family. Studied at both Oxford (1815–16) and Cambridge (1816–21). His interest in geology, and specifically in volcanoes, was stimulated during visits to Italy in 1817–19 by the spectacle of Vesuvius in eruption. During the period 1819–23 he studied Stromboli and Etna in Sicily and the extinct volcanic regions of the Auvergne in France and the Eifel in Germany. He joined the Geological Society in 1824 and the Royal Society two years later.

   After 1830 Scrope's original work in geology largely ceased and he devoted his energies to political and social issues. He retained an interest in the subject however and in later life published revisions of his two books as well as various articles.

2362. LISTER, Joseph Jackson. 1786–1869. (Microscopy 1266) A London wine merchant with an amateur interest in optical instruments. As a result of his important paper of 1830 on the design of microscope lenses he was elected to the Royal Society early in

1832. He later did significant work in biological microscopy and was one of the founders of the Microscopical Society in 1839. He was the father of the famous surgeon.

2363. BENTHAM, George. 1800-1884. (Botany 1383) Came from a wealthy background. Qualified as a barrister in 1831 but on inheriting a fortune two years later gave up law and devoted himself to botany. His publications began in 1826 and he became a fellow of the Linnean Society two years later. From 1829 to 1840 he was a capable secretary of the Royal Horticultural Society, and from 1863 to 1874 an energetic president of the Linnean Society. The Royal Society awarded him one of its medals in 1859 and elected him a fellow three years later.

Bentham was a close friend of W.J. Hooker, director of Kew Gardens, and donated his library and extensive herbarium (containing over 100,000 specimens) to Kew in 1854. Thereafter he worked at Kew for the rest of his life, his greatest achievement (1383) being the product of many years' collaboration with J.D. Hooker (director of Kew from 1865).

2364. ZOOLOGICAL SOCIETY OF LONDON. Founded in 1826. The first national society for zoology. Its aims were "the advancement of Zoology and Animal Physiology and the introduction of new and curious subjects of the Animal Kingdom." The latter aim was well served by the foundation two years later of the London zoological garden which, besides its more popular functions, served as a research institution for the Society. In this respect, as in others, the Society provided a model for zoological societies in other countries.

2365. OWEN, Richard. 1804-1892. (Zoology 1477, 1482, 1495, 1504. Palaeontology 1166, 1185) Qualified as a member of the Royal College of Surgeons in 1826 and went into private practice. In the following year, while continuing his practice, he was appointed by the College to assist in preparing a catalogue of John Hunter's museum (see 1477b) which had earlier come into the possession of the College (cf. 2312); he was later put in charge of the museum. He was a member of the Zoological Society from its formation and in 1831 he went to Paris to visit Cuvier at the Muséum d'Histoire Naturelle; he later said that his contact with Cuvier was a major influence.

Owen was elected to the Royal Society in 1834 and three years later became professor of anatomy and physiology at the Royal College of Surgeons. He continued his private practice however until 1856 when he was appointed superintendent of the natural history departments of the British Museum. He was chiefly responsible for the creation of the new Natural History Museum at South Kensington during the 1870s (see 2464) and continued as its superintendent until his retirement in 1884. He received a knighthood and many scholarly honours.

1830s

2366. THE ROYAL SOCIETY. The domination of the Society by non-scientists (see 2257) was a major reason for its unimpressive record during the eighteenth century and the early decades of the nineteenth. Around 1830 several publications appeared, including a notable book by Babbage (see 37), which were sharply critical of its composition, management, and general performance. For some time the minority of scientist members had been attempting reforms and at the presidential election of 1830 the distinguished astronomer, John Herschel, stood as a candidate; the majority of the Society however showed that they preferred social eminence to scientific achievement by electing the Duke of Sussex. (The discouraging result of this hotly contested election may have been a factor leading to the formation of the British Association in the following year)

During the 1840s, however, scientists were in a majority on the Society's council and, led by W.R. Grove, were successful in 1847 in passing statutes which had the effect of severely limiting the number of persons to be admitted to the Society each year, as well as ensuring a proper scrutiny of their qualifications. By 1860, as a result of this strategic measure, scientists formed the majority of the Society which thereafter became the distinguished institution that it has ever since been.

2367. THE BRITISH ASSOCIATION FOR THE ADVANCEMENT OF SCIENCE. Formed in 1831 in order--as its constitution proclaimed--"To give a stronger impulse and a more systematic direction to scientific inquiry; to promote the intercourse of those who cultivate Science...; to obtain more general attention for the objects of Science and the removal of any disadvantages of a public kind which impede its progress."

It was modelled on the Gesellschaft Deutscher Naturforscher und Ärzte (2589) and, like it, held annual meetings in different cities around the country, with ample publicity, encouragement to the general public to participate in the main sessions, addresses by leading scientists, and the reading and discussion of research reports in specialist sections--a formula that proved very successful and was widely imitated in other countries.

2368. *THE PHILOSOPHICAL MAGAZINE.* Though by the end of the nineteenth century it was one of the best known and most highly respected journals in the scientific world, it came originally from humble origins. It was begun as a semi-popular journal in 1798 by Alexander Tilloch who, though hardly a scientist, was a capable journalist, and during the period from 1814 to 1832 it absorbed three other journals of a similar kind but of higher quality. On Tilloch's death in 1825 it was taken over by Richard Taylor, the founder of the publishing firm of Taylor and Francis which continues to publish it to the present day. During the 1830s Taylor established it as a well-regarded journal, and through the middle and latter part of the century he and his partner and editorial successor, William Francis, made good use of scientific advisers who included such figures as Brewster, Tyndall, and Lord Kelvin.

Though the scope of the journal originally covered the whole of science it gradually became restricted to the physical sciences and by the end of the century very largely to physics. Such specialization was essential for the survival of a non-institutional (or proprietary) journal in the late nineteenth century.

2369. SYLVESTER, James Joseph. 1814–1897. (Mathematics 249) Graduated
      at Cambridge in 1837 with honours in mathematics. His first
      publications led to his election to the Royal Society in 1839.
      From 1838 he was professor of natural philosophy at University
      College, London, but in 1841 he went to the United States and
      for a brief period had a post at the University of Virginia.
      Returning to England in 1843, he took up the study of law and
      was called to the bar in 1850. He continued his mathematical
      researches however, and from 1855 to 1870 was professor at the
      Royal Military Academy, Woolwich.
             Sylvester returned to the United States in 1876 to
      become professor of mathematics at the newly-founded Johns
      Hopkins University. During his stay there he did much to
      stimulate the cultivation of pure mathematics in the country,
      including the founding of the *American Journal of Mathematics*
      in 1878. In 1883 he returned to England to take up his appoint-
      ment as Savilian professor of mathematics at Oxford.

2370. WHEATSTONE, Charles. 1802–1875. (Physics 670) His family back-
      ground of musical instrument makers and dealers gave rise to
      his early researches in acoustics and, more broadly, to his
      inclination to science; he had no formal scientific education.
      As a result of his work in acoustics he was appointed professor
      of physics ("experimental philosophy") at the newly-established
      King's College, London, in 1834. His subsequent work was chiefly
      in electricity and optics. F.R.S., 1836.
             From the early 1830s Wheatstone was experimenting with
      an electric telegraph, and he and his entrepreneur partner,
      W.F. Cooke, obtained their first patent in 1837. For many
      years thereafter he was engaged in developmental work on the
      invention, concurrently with his researches in pure science,
      and he was also the originator of several other inventions
      (notably the use of electromagnets in dynamos). For his achieve-
      ments he received many honours, including a knighthood in 1868.

2371. FORBES, James David. 1809–1868. (Physics 673) Studied law at
      Edinburgh but gave it up for science. Following some early
      research he was elected to the Royal Society of Edinburgh in
      1830 and to the Royal Society of London in 1832. He took an
      active part in the formation of the British Association in 1831.
      He became professor of natural philosophy (i.e. physics) at
      Edinburgh in 1833.

2372. GROVE, William Robert. 1811–1896. (Physics 689. Chemistry 890,
      900) M.A., Oxford, 1835. Became a barrister but soon gave up
      his legal career and devoted himself to scientific research.
      He became a fellow of the Royal Society in 1840 and, as a
      member of its council during the late 1840s, was the leading
      figure in bringing about the reform of the Society (see 2366).
      He was one of the original members of the Chemical Society and
      from 1841 to 1846 was professor of "experimental philosophy"
      at the London Institution. For financial reasons he resumed
      his legal career, while continuing his interest in science.
      He became a judge in 1871.

2373. GRAHAM, Thomas. 1805–1869. (Chemistry 883, 905) M.A., Glasgow,
      1826; the lectures of Thomas Thomson inspired him to take up

chemistry. After several minor teaching posts in Glasgow he
became professor of chemistry at University College, London,
in 1837. His early researches had led to his election to the
Royal Society in 1834. He was one of the founders of the
Chemical Society in 1841 and became its first president. Like
other British chemists at the time he did much consulting work.
He resigned his chair in 1854 to become Master of the Mint (in
succession to John Herschel) but continued to publish important
research up to his death.

2374. ANDREWS, Thomas. 1813-1885. (Physical chemistry 928) After an
early involvement with chemistry he studied medicine at Dublin
and Edinburgh, receiving the M.D. at the latter in 1835. He
then went into medical practice in Belfast, while at the same
time lecturing in chemistry at the Belfast Academical Institution. In 1845 he became vice-president of the newly-established
Queen's College, Belfast, and from 1849 until his retirement
in 1879 he was its professor of chemistry.

2375. GEOLOGICAL SURVEY OF GREAT BRITAIN. The first national geological
survey. Begun in 1835 through the initiative of H.T. De La
Beche (see 2351).

2376. MALLET, Robert. 1810-1881. (Geology 1178) B.A., Trinity College,
Dublin, 1830. Became a successful engineer in Ireland and in
1861 settled in London as a consulting engineer. In addition
to his numerous activities in engineering he had wide scientific
interests and was a member of several scientific societies in
Dublin and London, including the Royal Society to which he was
elected in 1854. He was awarded the Geological Society's
Wollaston Medal in 1877.

2377. (ROYAL) MICROSCOPICAL SOCIETY. Founded in 1839.

2378. BOTANICAL SOCIETY OF EDINBURGH. One of the earliest botanical
societies. It was established in 1836 by a number of former
pupils of Robert Graham, professor of botany at the University
and keeper of the Royal Botanic Garden.

2379. (ROYAL) ENTOMOLOGICAL SOCIETY OF LONDON. Founded in 1833.

2380. DARWIN, Charles Robert. 1809-1882. (Evolution 1525, 1534, 1537.
Botany 1382, 1393, 1399. Zoology 1489, 1506, 1514. Geology
1170) Began the study of medicine at Edinburgh in 1825 but
two years later abandoned it and went to Cambridge to do the
arts course. Though he neglected the required studies--taking
a poor degree in 1831--he got to know some important scientists
at Cambridge, especially the professor of botany, J.S. Henslow,
who stimulated his interest in natural history and had a strong
influence on him.

     Soon after he left Cambridge in 1831 Darwin was invited
by the Admiralty, on Henslow's recommendation, to join the
survey ship, H.M.S. *Beagle*, as (unpaid) naturalist on its
voyage to the coasts of South America and then around the world.
The voyage, which lasted five years, gave him the opportunity
to develop into an experienced naturalist and shaped the
direction of his future work.

     Two years after his return Darwin was elected to the
Royal Society. He married and because of chronic ill-health

(due to a debilitating disease probably contracted in South America) retired in 1842 to a country house in Kent where he lived on the income from his investments (both he and his wife were quite wealthy) for the rest of his life.

2381. HALL, Marshall. 1790-1857. (Physiology 1749) M.D., Edinburgh, 1812. After travels on the Continent he went into private practice in Nottingham and then, from 1826, in London. He was elected to the Royal Society in 1832.

1840s

2382. PLAYFAIR, Lyon. 1818-1898. (Not in Part 1. A capable chemist but his importance was in his political role) Graduated in medicine but turned to chemistry. From 1839 to 1841 he studied under Liebig at Giessen; he later acted as Liebig's agent in England and translated several of his books. He became professor of chemistry at the Royal Institution, Manchester, from 1842 and then, from 1845, at the Royal School of Mines in London. He was elected to the Royal Society in 1848 and became prominent in the Chemical Society and the British Association.

From the early 1840s Playfair cultivated connections with the government, serving on some important commissions as a scientific expert. He became Prince Albert's chief adviser concerning the Great Exhibition of 1851 and had much to do with its success. From 1853 to 1858 he was secretary of the government's new Department of Science and Art, taking a leading part in shaping the Royal College of Science and the South Kensington Museum. From 1858 to 1869 he was professor of chemistry at Edinburgh but his tenure of the post was undistinguished chemically, probably because of his continuing involvement in governmental and political affairs. He was elected to Parliament in 1868 and subsequently held some high offices. He was knighted in 1883 and raised to the peerage in 1892.

Like Joseph Banks earlier, Playfair was a highly influential statesman of science. He had connections at Court, especially with Prince Albert, and was on close terms with many important politicians, as well as with the leaders of the scientific community. In contrast to nearly all the British politicians and industrialists of the time he knew the value of scientific technology for British industry, and did what he could to change the prevailing attitude and to introduce science into education.

2383. BOOLE, George. 1815-1864. (Mathematics 258) Came from a poor family and was largely self-educated in mathematics and other subjects; from the age of sixteen he taught in village schools. He began publishing papers in mathematics in 1840 and was awarded a medal by the Royal Society in 1844. In 1849 he was appointed professor of mathematics at the newly-founded Queen's College, Cork, where he remained until his death. F.R.S., 1857.

2384. CAYLEY, Arthur. 1821-1895. (Mathematics 239, 249) Graduated at Cambridge in 1842 with the highest honours in mathematics. Because there were no suitable university openings available he took up the study of law and was admitted to the bar in 1849. Over the next fourteen years he divided his time between

the practice of law and mathematical research, managing to
publish about 250 papers during the period. In 1863 he was
appointed to the new Sadlerian chair of mathematics at Cambridge,
a position he held for the rest of his life.

One of the most outstanding mathematicians of the nineteenth century, Cayley has been compared with Euler for his
versatility and analytical power, with Cauchy for the clarity
and elegance of his analysis, and with both for his immense
output (nearly a thousand papers and notes). He was a member
of most of the leading academies, president at various times
of the London Mathematical Society and the Royal Astronomical
Society, and recipient of two medals from the Royal Society
and of honorary degrees from many universities.

2385. ADAMS, John Couch. 1819–1892. (Astronomy 442) Graduated at
Cambridge in 1843 with the highest honours in mathematics and
was appointed a college fellow. For his part in the discovery
of Neptune in 1846 (see 2151 concerning Le Verrier's part and
the ensuing priority dispute) he was offered a knighthood the
next year but declined it. The Royal Society awarded him its
Copley Medal in 1848 and he was elected president of the Royal
Astronomical Society in 1851. In 1859 he was appointed Lowndean
professor of astronomy at Cambridge and soon afterwards director
of the university observatory.

2386. CARRINGTON, Richard Christopher. 1826–1875. (Astronomy 447)
Came from a wealthy family. Graduated at Cambridge in 1848
and in the following year was appointed to the staff of the
observatory of the University of Durham. He resigned the post
in 1852 and established a private observatory of his own. He
had earlier joined the Royal Astronomical Society and in 1860
was elected to the Royal Society. After the early 1860s ill-
health and the pressure of his business interests prevented
him continuing research.

2387. STOKES, George Gabriel. 1819–1903. (Mechanics 555. Physics
676, 696. Geodesy 1056) Graduated at Cambridge in 1841 with
the highest honours in mathematics and was appointed a college
fellow. In 1849 he became Lucasian professor of mathematics,
a post he held until his death. He had an important influence
on many students (including Maxwell) and in the early part of
his career was active in spreading a knowledge of the French
tradition in mathematical physics. He was elected to the
Royal Society in 1851 and served as its secretary (1854–85)
and president (1885–90), and was awarded two of its medals.
He was knighted in 1889 and received many other honours.

2388. JOULE, James Prescott. 1818–1889. (Physics 681) Came from a
wealthy family in Manchester and was educated privately—in
science by John Dalton who had a significant influence on him.
Being of independent means he was able to devote himself to
research. As a result of his work on the mechanical equivalent
of heat he was elected to the Royal Society in 1850 and became
a much respected figure in the British scientific community.

2389. THOMSON, William (*Lord* KELVIN). 1824–1907. (Physics 679, 692,
712. Geology 1186) His father was professor of engineering
at Belfast, and then, from 1832, at Glasgow. The son studied

at Glasgow and from 1841 to 1845 at Cambridge. He then spent
several months in Paris working in Regnault's laboratory in
order to gain experience in experimental work; he also made
himself conversant with the French tradition of mathematical
physics. In 1846 he was appointed to the chair of natural
philosophy (i.e. physics) at Glasgow where he remained for the
rest of his career, despite some blandishments from Cambridge.
At Glasgow he introduced laboratory teaching in physics for
the first time in Britain. F.R.S., 1851.

As well as being a theoretical physicist of the first
rank Thomson was also deeply interested in devising very
sensitive and accurate measuring instruments. This interest
led to numerous inventions as well as to contacts with the
worlds of industry, navigation, and commerce--the most outstanding case being his involvement with the design, laying, and
operation of the Atlantic telegraph cable in the 1850s and
1860s. His technical advice and the sensitive instruments he
invented were responsible for making the cable effective, and
as a result he became something of a national hero, receiving
a knighthood in 1866. He also became very wealthy through his
consulting and his many patents, especially those to do with
cable telegraphy.

Thomson published over six hundred papers on all
aspects of physics and in his later years was acclaimed as one
of the world's leading scientists as well as a famous inventor.
He was raised to the peerage in 1892.

2390. CHEMICAL SOCIETY OF LONDON. Founded in 1841. The first national
society for chemistry.

2391. ROYAL COLLEGE OF CHEMISTRY. Established by private initiative
in 1845 in a wave of enthusiasm for agricultural and industrial
chemistry prompted by Liebig's visit to England in 1842. In
a few years however financial support for it fell off and as
a result it was united with the Government School of Mines in
1853, becoming in effect its department of chemistry. From
the College's beginning in 1845 until 1865 it was very capably
directed by the young German chemist, A.W. Hofmann (see 2637).

2392. WILLIAMSON, Alexander William. 1824-1904. (Chemistry 912)
Born in England but lived on the Continent for many years.
Began the study of medicine at Heidelberg but changed to chemistry under Leopold Gmelin. From 1844 to 1846 he worked with
Liebig at Giessen, and from 1846 to 1849 he had a private
laboratory in Paris where he was on close terms with the leading
French chemists. In 1849 he was appointed to the chair of
practical chemistry at University College, London. His important work was done in the next five years; after 1855, for a
combination of reasons, he published little. He long remained
however one of the leaders of chemistry in Britain and an
important teacher.

2393. FRANKLAND, Edward. 1825-1899. (Chemistry 907) Initially a
druggist's apprentice. From 1845 he was employed in Playfair's
laboratory at the School of Mines. There he met Kolbe who had
come over from Germany to assist Playfair in gas analysis. In
1847 Frankland went with Kolbe to Marburg where he worked with

Bunsen, obtaining the D.Phil. in 1849. After working for a
short time with Liebig at Giessen he returned to England and
in 1851 became professor of chemistry at the newly-established
Owens College, Manchester. He resigned the post in 1857
however because the College seemed to be a failure (cf. 2417).
After teaching chemistry at a medical school for a few years
he was professor of chemistry at the Royal Institution from
1863, and finally at the Royal School of Mines from 1865 until
his retirement in 1885.

Frankland was elected to the Royal Society in 1853 and
was later awarded two of its medals. He was prominent in the
Chemical Society and the British Association and was a member
of the informal group of influential scientists known as the
X-Club. For his public services (on governmental commissions,
etc.) he was knighted in 1897.

2394. PALAEONTOGRAPHICAL SOCIETY, London. From its foundation in 1847
into the twentieth century it supported the publication of
numerous palaeontological monographs.

2395. FORBES, Edward. 1815-1854. (Geology 1175. Oceanography 1063)
Studied medicine at Edinburgh but gave it up in 1836 for botany
and zoology. In the following year he went to Paris and attended the lectures at the Muséum d'Histoire Naturelle. He also
went on natural history journeys in several countries and on
dredging expeditions in the Irish Sea. In 1841-42 he served
as a naturalist on a naval vessel surveying the coastal waters
of the Aegean and was able to do much dredging, bringing many
new species to light. After holding the (badly-paid) post of
professor of botany at King's College, London, for two years
he was a palaeontologist with the Geological Survey from 1844.
He became a member of the Geological Society in that year and
of the Royal Society the next year. He was appointed to the
chair of natural history at Edinburgh in 1854 but died soon
afterwards.

2396. WILLIAMSON, William Crawford. 1816-1895. (Palaeobotany 1192)
Though he had a rather fragmented education he acquired a good
knowledge of natural history from his father, an accomplished
naturalist, and from his father's acquaintances who included
some of the leading naturalists and geologists in Britain. He
qualified in medicine at University College, London, in 1841
and thereafter practised in Manchester, at the same time
carrying on research in zoology. In 1851 he was appointed
professor of natural history and geology at the newly-founded
Owens College, Manchester, and by 1880 was able to concentrate
entirely on botany. He held the chair of botany until his
retirement in 1892.

Williamson was elected to the Royal Society in 1854
for his early work in zoology. For his subsequent pioneering
work in palaeobotany he received many honours and awards.

2397. SORBY, Henry Clifton. 1826-1908. (Geology 1177. Microscopy
1272) Educated privately, his tutor being of some scientific
ability. Being of independent means he was able to devote
himself to research in a laboratory in his house in Sheffield.
He was an active member of the local Literary and Philosophical

Society and also attended meetings of the British Association, thereby becoming acquainted with many leading scientists.

Initially Sorby was interested in agricultural chemistry, in which field he published his first paper in 1847, but two years later he turned to geology. Besides making some significant contributions to various aspects of geology during the 1850s he took the important step of developing methods for using the microscope in the study of rocks—a step that was later to earn him the title of "the father of microscopical petrography." From rocks he went on in 1863 to meteorites, studying their crystalline structure microscopically, and then to steel—thereby becoming the founder of metallography as well. After 1864 he turned to the new field of spectrum analysis and later to marine biology.

Sorby became a fellow of the Royal Society in 1857 and was subsequently a member of its council. At various times in the 1870s he was president of the Royal Microscopical Society, the Mineralogical Society, and the Geological Society. He took an active part in establishing what later became the University of Sheffield and was awarded an honorary degree by Cambridge.

2398. KEW OBSERVATORY. Originally known as the King's Observatory. The use of it was acquired from the government in 1842 by the British Association on the instigation of Edward Sabine and, with the help of private endowments, it was converted into a geophysical laboratory. Until 1871, when control of it was transferred to the Royal Society, it was managed by a committee of the British Association on which Sabine was a major figure. It became a leading centre for geophysical research and in 1867 was designated as the central meteorological laboratory of Great Britain, the government thereafter providing an annual grant.

2399. RAY SOCIETY, London. From its foundation in 1844 into the twentieth century it supported the publication of numerous monographs on topics in botany and zoology.

2400. HOOKER, Joseph Dalton. 1817-1911. (Botany 1370, 1383, 1391) Son of W.J. Hooker. M.D., Glasgow, 1839. His father got him the position of assistant surgeon and naturalist on the Antarctic expedition of H.M.SS. *Erebus* and *Terror* during 1839-43. His subsequent account of his botanical findings during the expedition, published in six volumes in 1844-60, established his reputation as an outstanding taxonomist and plant geographer. In 1847-50, with the aid of a government grant, he explored the botany (and topography, etc.) of northeast India. Smaller expeditions that he undertook later were to Syria and Palestine (1860), the Atlas Mountains in Morocco (1871), and the western United States in company with Asa Gray (1877).

In 1855 he was appointed assistant director of the Royal Botanic Gardens, Kew, and ten years later succeeded his father as director. He continued the improvement of the gardens and built up the institution as an international centre for botanical research. The important series *Index Kewensis* was begun under his supervision.

Hooker first met Darwin in 1839 and they became lifelong friends. He was acquainted with Darwin's ideas as they developed but his gradual acceptance of the concept of evolution through natural selection was based on his own work, especially on the geographical distribution of plants. On the publication of Darwin's *Origin of Species* he was one of the first to support it and he became a vigorous advocate of the theory, demonstrating its importance to botany in general and plant geography in particular.

Hooker retired from his post at Kew in 1885 and in his last years received many honours, including a knighthood.

2401. GALTON, Francis. 1822-1911. (Heredity 1535, 1542. Meteorology 1064) A cousin of Charles Darwin. Though intellectually precocious he took only a poor degree at Cambridge. He then began the study of medicine but dropped it when his father died, leaving him a fortune. After adventurous explorations in Africa during the period 1845-52 he settled in London where he lived privately for the rest of his life. For his geographical discoveries he was awarded a medal by the Geographical Society and elected to the Royal Society in 1856. He was knighted in 1909 and received many scholarly honours.

2402. WALLACE, Alfred Russel. 1823-1913. (Evolution 1524, 1536, 1540. Zoology 1499, 1509, 1512) Received only a rudimentary education and was self-taught in science. He shared an interest in natural history with his friend, H.W. Bates (later an important naturalist--see 2404), and their enthusiasm for collecting, together with their reading of travel books, inspired them to go to South America in 1848 to collect specimens (which they could sell in Europe and so support themselves). Wallace returned to England in 1852 (Bates remained longer) and acquired a scientific reputation for his publications on the natural history of the Amazon basin and also established himself with the general public as a writer of travel books.

From 1854 to 1862 he was exploring in the Malay Archipelago and the many publications that resulted from his discoveries there enhanced his scientific reputation while his popular writings (which included some of the best natural history travel books ever written) increased his fame with the public. In subsequent years he continued the writing and lecturing by which he earned his living, and in his old age was awarded a government pension as well as numerous honours.

2403. HUXLEY, Thomas Henry. 1825-1895. (Zoology 1488, 1497. Evolution 1528. Various writings 55) Received only two years of formal schooling and was largely self-educated. With the aid of a scholarship he qualified in medicine in 1845. He then joined the navy and from 1846 to 1850 was surgeon on H.M.S. *Rattlesnake* during its survey of the coast of New Guinea. During the voyage he did much research on marine organisms and sent some important papers to the Royal Society. On his return in 1850 he was elected to the Society and two years later was awarded its Royal Medal.

He left the navy in 1854 and became lecturer in natural history at the School of Mines. He was also appointed naturalist to the Geological Survey and helped to organize the Museum of Practical Geology. The School of Mines gave rise to the Normal School of Science, later known as the Royal College of

Science; Huxley remained with the institution in its successive incarnations throughout his career, despite some attractive offers from Oxford and elsewhere after he had become famous. His innovative teaching methods and approach had a wide influence through his disciples, notably Michael Foster and E. Ray Lankester, and through his ten textbooks.

From 1855 Huxley regularly gave a series of lectures to workingmen which were well received. He came to be much in demand as a lecturer and public speaker, especially on educational matters. As is well known, he was Darwin's main supporter and--in innumerable lectures, debates, and articles--did more than anyone else to overcome the opposition to the theory of evolution.

As one of the most outstanding and influential members of the British scientific community, Huxley served on many government commissions, his contributions to the reform of education--elementary as well as advanced--being especially important. He was a prominent member of numerous scientific and other societies and in his last years received worldwide recognition. He was offered a peerage but characteristically declined it on principle.

2404. BATES, Henry Walter. 1825-1892. (Zoology 1498) Received only a rudimentary education and was self-taught in science. With his friend A.R. Wallace (see 2402) he went to South America in 1848 to collect specimens. He remained there until 1859 and estimated that he had collected some fifteen thousand species (mostly insects) of which eight thousand were previously unknown. On his return to England he began work on his collections and also in 1863 produced a popular account of his travels, *The Naturalist on the River Amazon*, which went through many editions and translations.

In 1864 Bates was appointed assistant secretary of the Royal Geographical Society, a post of some importance which he held with distinction for the rest of his life. He was a leading systematic entomologist and became a member of the Entomological Society (1861), the Zoological Society (1863), the Linnean Society (1871), and the Royal Society (1881).

2405. BOWMAN, William. 1816-1892. (Anatomy 1635. Physiology 1757) Became an assistant surgeon at King's College Hospital in 1840 and surgeon in 1856. His histological researches led to his election to the Royal Society in 1841 and the award of its Royal Medal in the following year. He subsequently became a leading ophthalmic surgeon and was knighted in 1884.

1850s

2406. SOUTH KENSINGTON MUSEUM. Founded in 1857 as a museum for science and art. In 1899 it was divided into the present Science Museum and the Victoria and Albert Museum (of art).

2407. SMITH, Henry John Stanley. 1826-1883. (Mathematics 252) Graduated at Oxford in 1849 with first class honours in both mathematics and *literae humaniores* and became a college fellow. In 1860 he was appointed Savilian professor of geometry. He was elected to the Royal Society in 1861 and was prominent in the British Association. Much of his time was devoted to administration and reform at Oxford and to serving on various Royal Commissions.

2408. TAIT, Peter Guthrie. 1831-1901. (Mathematics 273. Physics 712) Graduated at Cambridge in 1852 with the highest honours in mathematics. He became professor of mathematics at Queen's College, Belfast, in 1854, then professor of natural philosophy at Edinburgh in 1860.

2409. HUGGINS, William. 1824-1910. (Astronomy 456, 462) Had an early interest in science but was prevented from going to Cambridge, as he had wished, because of his responsibility for the family business. In 1854 he was able to sell the business and devote himself to astronomy, setting up a private observatory and joining the Royal Astronomical Society. He was elected to the Royal Society in 1865, later being awarded three of its medals and becoming its president in 1900. At various times he was also president of the Royal Astronomical Society and the British Association. He was knighted in 1897 and received the Order of Merit in 1902.

2410. RANKINE, William John Macquorn. 1820-1872. (Physics 697) After studying at Edinburgh University he qualified as a civil engineer by apprenticeship and in 1842 went into practice. By 1850 he was carrying on research in both science and engineering, and in that year was elected to the Royal Society of Edinburgh and in 1853 to the Royal Society of London. In 1855 he was appointed professor of civil engineering and mechanics at Glasgow. His several textbooks of engineering were very successful and had a major influence in shaping engineering education in Britain and elsewhere.

2411. TYNDALL, John. 1820-1893. (Physics 709, 719. Microbiology 1810. Various writings 56) Initially a surveyor and engineer and then a teacher of mathematics. In 1848 he went to Germany and two years later gained a D.Phil. at Marburg; he then worked there with K.H. Knoblauch on diamagnetism, and in 1851 with G. Magnus in Berlin. After his return to England he was elected to the Royal Society in 1852 and in the following year became professor of natural philosophy at the Royal Institution. He derived much benefit from his association with Faraday whom he succeeded as superintendent of the Royal Institution and whom he emulated as a skilful and successful lecturer. His many essays and popularizations were no less successful.

Tyndall's central institutional position as well as his character and abilities made him one of the most outstanding figures in the British scientific community--a member of the X-Club, influential in the British Association and other scientific bodies, one of the founders of the journal *Nature*, science journalist and controversialist in the popular press, and adviser to publishers and government bodies. He was a forthright protagonist for national support of research and education in science, and acquired considerable notoriety by his advocacy of the ideological scientism of the time.

2412. STEWART, Balfour. 1828-1887. (Physics 703. Geophysics 1074) Studied at the universities of St. Andrews and Edinburgh and

then spent ten years in the business world. In the mid-1850s
however he turned to a career in science and became assistant
to J.D. Forbes, professor of physics at Edinburgh. His successful researches led in 1859 to his appointment as director of
the Kew geophysical and meteorological laboratory (2398). In
1870 he became professor of natural philosophy at Owens College,
Manchester, where he established a teaching laboratory (one of
his students was J.J.Thomson).

    Stewart was a fellow of the Royal Society, receiving
its Rumford Medal for his work on radiant heat, and became
president of the Physical Society. His textbooks and popularizations were widely influential.

2413. MAXWELL, James Clerk. 1831-1870. (Physics 700, 705, 706, 720.
Astronomy 449. Physiology 1764) Graduated at Cambridge in
1854. Became professor of natural philosophy at Marischal
College, Aberdeen, in 1856, then at King's College, London,
in 1860. In 1865 he retired to his estate in Scotland and
spent his time in research and writing; he also made regular
visits to Cambridge where he acted as an examiner. In 1871
he accepted an invitation to become Cambridge's first professor
of experimental physics and to plan the Cavendish Laboratory
and recruit its staff. Much of his time and energy in the
last years of his short life were devoted to developing the
Laboratory.

2414. ODLING, William. 1829-1921. (Chemistry 909) After studying
medicine at Guy's Hospital he took London University's M.D.
degree in 1851, being one of the first to do so. During his
medical training he had cultivated an interest in chemistry,
joining the Chemical Society in 1848, and from 1850 he was
demonstrator in chemistry at Guy's Hospital, later holding
other teaching positions there. He became acquainted with
many of the leading chemists, especially Williamson through
whom he met Kekulé when the latter was in London. In 1854 he
spent a few months in Paris with Gerhardt. F.R.S., 1859.
In 1867 he became Fullerian professor of chemistry at the Royal
Institution and in 1872 Waynflete professor at Oxford.

2415. COUPER, Archibald Scott. 1831-1892. (Chemistry 919) Studied
classics and philosophy at Glasgow and Edinburgh in the early
1850s. In 1854-55 he was in Berlin studying chemistry, and
in 1856 he went to Paris to work in Wurtz' laboratory, publishing his first paper the next year. The publication of his
famous paper (919) on the theory of chemical structure in 1858
was delayed by Wurtz, and as a result Kekulé's paper (918)
appeared first. When Couper's paper appeared about two months
later Kekulé attacked it, insisting on his priority. Couper
complained bitterly to Wurtz about the delay and Wurtz responded
by dismissing him from the laboratory. He returned to Edinburgh
where he worked for a short time as assistant to the professor
of chemistry but then suffered a severe mental breakdown from
which he never recovered.

2416. CROOKES, William. 1832-1919. (Chemistry 927. Physics 727)
Studied at the Royal College of Chemistry and was Hofmann's
assistant from 1850 to 1854. In 1856 he inherited a fortune

and, apart from attending to his business interests, devoted himself to research in his private laboratory. He also did some consulting and edited several photographic and scientific journals, notably *Chemical News* which he founded in 1859. He was elected to the Royal Society in 1863 and was later prominent in the affairs of the Chemical Society and the Photographic Society. He received various honours and awards from scientific societies and was knighted in 1897.

2417. ROSCOE, Henry Enfield. 1833-1915. (A capable chemist--cf. 906 and 953--but important chiefly for his institutional role) Graduated in 1853 at University College, London, where he studied under Graham and Williamson. Over the next few years he worked intermittently at Heidelberg with Bunsen while earning a precarious living in London. In 1857 he was appointed professor of chemistry at Owens College, Manchester, in succession to Frankland who had left because the College (founded in 1851 by John Owens, a wealthy merchant) appeared to be a failure and was almost defunct. Together with some other new appointees, however, Roscoe was able to rejuvenate it and redirect it along lines more suitable to its social and industrial setting. Over the next thirty years he built up its chemistry department into the leading school of chemistry in Britain. Its graduates filled many positions in academia and industry and its research tradition continued into the twentieth century (when the College became the University of Manchester).

An excellent teacher, Roscoe wrote several successful textbooks and was active in the promotion of scientific and technical education. He was elected to the Royal Society in 1863 and was a prominent figure in the scientific community, nationally as well as on the Manchester scene. In 1872 he declined the offer of a chair at Oxford. He was knighted in 1884 for his services to technical education.

2418. PERKIN, William Henry. 1838-1907. (Chemistry 916) Entered the Royal College of Chemistry in 1853 (at the age of fifteen) to study under Hofmann and became his assistant two years later. In 1856 he accidently discovered a valuable dye, later called mauve, and decided to manufacture it. His enterprise, which initiated the synthetic dye industry, was highly successful-- so much so that in 1874 (at the age of thirty-six) he was able to sell out and retire. Thereafter he devoted himself to continuing full-time the research in pure chemistry he had already been carrying on to some extent during his industrial career. He published numerous papers and made some substantial contributions.

Perkin was elected to the Royal Society in 1866 and was awarded two of its medals. He was knighted in 1906 during the celebration of the fiftieth anniversary of the discovery of mauve.

2419. SCHOOL OF MINES. Established together with the Museum of Practical Geology in 1851. The two institutions were closely linked with the Geological Survey and were created by the government on the initiative of its director, De La Beche. The School of Mines was the British government's first investment in scientific and technical education; it initially had the title "The

Government School of Mines and of Science applied to the Arts"
and was later known as the Royal School of Mines. As well as
teachers of geology, mining, and metallurgy, it had teachers
of chemistry (initially Lyon Playfair) and natural history
(initially E. Forbes, succeeded in 1854 by T.H. Huxley). In
1853 it absorbed the Royal College of Chemistry (2391), thereby
acquiring a strong chemistry department.

> For many years there was a strong difference of opinion
about future policy for the School, one party maintaining that
it should concentrate on the restricted role of a technical
college for mining and geology, and the opposing party, led by
Huxley, urging that it should serve a much wider scientific
and technological purpose. In 1872 the Huxley faction was
successful in having the departments of physics, chemistry,
and natural history split off and relocated in new buildings
at South Kensington. These departments later formed the
nucleus of the Normal School of Science (see 2457).

2420. GEIKIE, Archibald. 1835-1924. (Geology 1203, 1215) Born and
educated in Edinburgh. Though self-taught in geology his
acquaintance with established geologists such as J.D. Forbes
and Hugh Miller led to his appointment to the Geological Survey
in 1855. He was elected to the Royal Society of Edinburgh in
1861 and to the Royal Society of London in 1865. In 1867 he
was appointed director of the newly-formed Scottish branch of
the Geological Survey and from 1871 he also held the Murchison
chair of geology at Edinburgh University. He had to resign
the chair in 1882 when he became head of the Geological Survey
and moved to London. In London he took a prominent part in the
affairs of the Geological Society and the Royal Society, became
a trustee of the British Museum, and served in various official
roles. He was knighted in 1891 and received many other honours.

> Geikie was a man of wide interests (he was president
of the Classical Association in 1910) and a notable writer.
His numerous books included several well-known biographical
works bearing on the history of geology and some successful
popularizations.

2421. (BRITISH) (ROYAL) METEOROLOGICAL SOCIETY. Founded in 1850.

2422. THOMSON, Charles Wyville. 1830-1882. (Oceanography 1065, 1068)
Studied at the University of Edinburgh and in 1851 became
lecturer in botany at Aberdeen. In 1854 he was appointed professor of mineralogy and geology at Queen's College, Belfast,
six years later exchanging the chair for that of zoology and
botany. He finally became regius professor of natural history
at Edinburgh in 1870.

> The success of Thomson's oceanographic cruises won him
election to the Royal Society in 1869 and appointment as head
of the scientific staff of the *Challenger* Expedition (1068).
His part in this famous expedition gained him a knighthood in
1876. He began the editing of the *Challenger Report* but died
before the massive task could be completed.

2423. LUBBOCK, John (From 1900 Lord AVEBURY). 1834-1913. (Zoology
1515) Came from a wealthy, upper-class family which happened
to live close to Darwin's home in Kent; from his early years

he was a friend of Darwin. He joined the staff of his family's
bank in his youth (becoming a partner in 1856) but, with Darwin's
support, continued his self-education in natural history.
Through Darwin, and by virtue of the quality of his early
researches, he became friendly with some of the leading English
scientists, being proposed for the Geological Society by Lyell
in 1855; three years later, with Darwin's backing, he was
elected to the Royal Society. Not surprisingly, he was a
vigorous advocate of Darwinism and took a prominent part in
the great debate over the *Origin of Species*.

 During the 1860s Lubbock was one of the pioneers of
human prehistory, setting it in an evolutionary perspective;
his books on the subject were important and influential. In
1870 he was elected to Parliament and began a long career as
a politician. His scientific researches—ranging over zoology,
botany, and geology—continued in parallel with his public
career. A prolific author and prominent in many scientific
societies, he was an outstanding figure in the British scientific community.

2424. TURNER, William. 1832-1916. (Not in Part 1. Though he published
over two hundred papers on comparative anatomy and anthropology
he made no outstanding discoveries. He had much institutional
significance however) Admitted to the Royal College of Surgeons
in 1853. In the following year he became senior demonstrator
in anatomy at the University of Edinburgh (which then had the
best school of anatomy in Britain). He was elected to the
Royal Society of Edinburgh in 1861 and to the Royal Society of
London in 1877. In 1867 he became professor of anatomy at
Edinburgh and in the same year he and G.M. Humphrey, professor
of anatomy at Cambridge, began the *Journal of Anatomy and
Physiology* which he edited for many years. He and Humphrey
also established the Anatomical Society in 1887.

 Turner's teaching at Edinburgh had a big influence on
anatomy in Britain and many of his students later gained chairs.
He was also a central figure in the reform of British medical
education in the 1880s. He was knighted in 1886 and was a
prominent figure in medical and scientific societies.

1860s

2425. LONDON MATHEMATICAL SOCIETY. Founded in 1865.

2426. LOCKYER, Joseph Norman. 1836-1920. (Astronomy 459, 460, 466)
After an ordinary education he became a clerk in the War Office
in 1857. His remarkable achievements as an amateur astronomer
led to his election to the Royal Society in 1869 and to his
appointment (still as a public servant) as secretary to the
Devonshire Commission—the Royal Commission on "scientific
instruction and the advancement of science", 1870-75. In line
with some recommendations of the Commission he was transferred
in 1875 from the War Office to the Science and Art Department
where he vigorously supported research and teaching in astronomy,
along with numerous other activities such as building up the
South Kensington Museum. When the Royal College of Science
was constituted in 1890 he was appointed professor of astron-

omical physics, a post he held until his retirement in 1913.
He was knighted in 1897.

A successful author and a popular lecturer, Lockyer
contributed in many ways to the scientific life of the time--
most of all in the establishment in 1869 of the journal *Nature*
which he made into a widely respected organ of the international
scientific community and edited for fifty years.

2427. REYNOLDS, Osborne. 1842-1912. (Mechanics 564) After an apprenticeship in mechanical engineering he went to Cambridge where
he graduated in 1867. In the following year he was appointed
to the new chair of engineering at Owens College, Manchester.
He held the position until his retirement in 1905 and was a
successful and influential teacher. He was elected to the
Royal Society in 1877 and awarded its Royal Medal in 1888.

Besides his contributions to mechanics and physics
Reynolds did much valuable work on several aspects of engineering science. He is now regarded as the most outstanding academic exponent of mechanical engineering in Britain in the
nineteenth century.

2428. DEWAR, James. 1842-1923. (Physics 730) M.A., Edinburgh, ca.
1865. He then worked for several months with Kekulé at Ghent
and in 1867 became assistant to the professor of chemistry at
Edinburgh. In 1875 he was appointed Jacksonian professor of
"natural experimental philosophy" at Cambridge, and in 1877
Fullerian professor of chemistry at the Royal Institution; he
held both chairs until his death, most of his research being
done at the Royal Institution. He was elected to the Royal
Society in 1877 and was prominent in several scientific
societies. As well as doing consulting work for industry he
acted as a government adviser, especially concerning explosives;
with F.A. Abel he invented cordite in 1889.

2429. GEIKIE, James. 1839-1915. (Geology 1195) Born and educated in
Edinburgh. Younger brother of Archibald Geikie (2420) who
procured him a post in the Geological Survey in 1861. He
joined the Geological Society in 1873 and was elected to the
Royal Societies of Edinburgh and London a few years later.
In 1882 he succeeded his brother as Murchison professor of
geology at Edinburgh. He was an able teacher and wrote some
successful textbooks.

2430. BUCHAN, Alexander. 1829-1907. (Meteorology 1067, 1078) M.A.,
Edinburgh, 1848. He then became a schoolteacher but in 1860
was appointed secretary of the Scottish Meteorological Society
(which had been founded, with government support, in 1855).
In that position, which he held for the rest of his life, he
supervised the operations of a network of meteorological
stations and the compilation of statistics, and was active in
research. He was a prominent member of the Royal Society of
Edinburgh and in 1887 was appointed a member of the Meteorological Council which administered the Parliamentary grant for
meteorology. He was elected to the Royal Society of London
in 1898 and received several honours and awards.

2431. FOSTER, Michael. 1836-1907. (Physiology 1785. Embryology 1677.
Not remarkable for his researches but highly important for his

institutional role) M.D., University College, London, 1859.
After practising medicine for a few years he became lecturer
in physiology and histology at University College in 1867 and
professor two years later; he established there the first
laboratory course in physiology in England. In 1870 he was
appointed to the newly-established position of praelector in
physiology at Trinity College, Cambridge. (Both the position
and his appointment to it were largely due to the influence
of his former teacher, T.H. Huxley, who in other respects also
had a big effect on his early career) His courses at Trinity
College were open to all Cambridge students and the numbers
attending them grew rapidly. In 1883 he became the first
professor of physiology at Cambridge.

 Highly successful as a teacher and research director,
Foster attracted a large number of talented research students
with interests in most fields of experimental biology, chiefly
physiology; many of them later gained chairs in other British
universities and some became world leaders. Before 1870 experimental biology in Britain was quite minor; after 1870 it rapidly
expanded and the catalyst of its expansion was Foster.

 He was also an important figure in organizational
affairs--a leader in the establishment of the Physiological
Society in 1876, founder of the *Journal of Physiology* in 1878
and for many years its editor, one of the secretaries of the
Royal Society, prominent in the British Association and other
societies, a major figure in the founding of the International
Physiological Congresses and other international undertakings,
and a member of various government committees.

1870s

2432. GILL, David. 1843-1914. (Astronomy 489) After two years at
the University of Aberdeen he went to Switzerland to learn
clockmaking and subsequently took over his family's clockmaking
business. His growing interest in astronomy led to his appointment in 1872 as private astronomer to Lord Lindsay who was
then building a fine observatory at Dun Echt, near Aberdeen.
While in this post Gill made some important observations and
in 1879 he was appointed director of the Royal Observatory at
Cape Town.

 By the time of his retirement in 1906 Gill had made
the Cape Observatory one of the best in the world and had
acquired a reputation as one of the most outstanding practical
astronomers of the time. He was knighted in 1900.

2433. DARWIN, George Howard. 1845-1912. (Astronomy 472) Second son
of Charles Darwin. Graduated at Cambridge in 1868. He then
studied law and qualified as a barrister in 1874 but never
practised; instead he returned to Cambridge as a college
fellow and devoted himself to research in applied mathematics.
He was elected to the Royal Society in 1879 and four years
later was appointed Plumian professor of astronomy and experimental philosophy, which post he held for the rest of his life.
He was knighted in 1905 and at various times served as president
of the Royal Astronomical Society and the British Association.

2434. LAMB, Horace. 1849–1934. (Mechanics 561) Graduated at Cambridge
      in 1872 and was appointed a college fellow. From 1875 he was
      professor of mathematics at the University of Adelaide (in
      Australia), returning to England in 1885 to become professor
      of mathematics at Owens College, Manchester; he held the post
      until his retirement in 1920.
           Lamb had a wide influence through his teaching and his
      excellent textbooks, and made important contributions in many
      fields of applied mathematics (his notable work in geophysics
      was mostly done after 1900). He was elected to the Royal
      Society in 1884 and in his later years received many scholarly
      honours. He was knighted in 1931.

2435. CAVENDISH LABORATORY. Cambridge University's celebrated labora-
      tory for experimental physics. Established in 1871, the
      finance being provided by the Duke of Devonshire (a descendant
      of Henry Cavendish). The early directors were Maxwell (1871–79),
      Strutt (1879–84), and J.J. Thomson (1884–1919). Until 1895 the
      Laboratory could be used only by Cambridge graduates (and under-
      graduates) but in that year the University decided to accept
      graduates from other universities as candidates for its higher
      degrees. As a result the Laboratory's leading position as a
      research school was much enhanced.

2436. PHYSICAL SOCIETY OF LONDON. Founded in 1874. In its early years
      it was not a learned society comparable to those that existed
      in London for the other main sciences. Its members were mostly
      teachers, in schools or technical colleges, and it was concerned
      with experimental rather than theoretical physics.

2437. STRUTT, John William (*Lord* Rayleigh). 1842–1919. (Physics 718.
      Chemistry 988. Some of his chief work was done after 1900)
      Graduated at Cambridge in 1865. He then set up a laboratory
      in the family mansion in Essex and most of his researches were
      done there. On the death of his father in 1873 he succeeded
      to the title of Baron Rayleigh. In 1879 he accepted the offer
      of the Cavendish chair of experimental physics at Cambridge
      where he continued the tradition of teaching and research that
      had been established by his predecessor, Maxwell. He resigned
      the chair in 1884 and returned to his private laboratory.
           Strutt's scientific ability and sound judgement, coupled
      with his high social position and diplomatic manner, made him
      a leading member of the British scientific community. He was
      prominent in the affairs of the Royal Society and the British
      Association, and was also active in government circles as an
      adviser and chairman of various committees on technical matters;
      thus he played a leading part in the establishment of the
      National Physical Laboratory in 1900. He was awarded the Nobel
      Prize in 1904 and is today regarded as the last of the great
      classical physicists of nineteenth-century Britain.

2438. HEAVISIDE, Oliver. 1850–1925. (Physics 723. His well-known
      prediction of the existence of the ionosphere was made in 1902)
      Had no higher education and was almost entirely self-taught in
      mathematics and science. In 1870 he became a telegraph operator
      but was forced to retire in 1874 by increasing deafness. There-
      after he devoted himself to study and research in theoretical

physics with (rather slender) financial support from his brother and others, and later from a government pension.

In the late 1880s Heaviside's reputation became established by the importance of his publications, and Kelvin's public commendation of his work helped to make him prominent. He was elected to the Royal Society in 1891 and awarded an honorary doctorate by the University of Göttingen. In his later years he became quite famous.

2439. FITZGERALD, George Francis. 1851-1901. (Physics 747) Graduated at Trinity College, Dublin, in 1871. After some years of further study of theoretical physics he became a fellow of Trinity College, and in 1881 professor of "natural and experimental philosophy", a post he held until his death. He was elected to the Royal Society in 1883 and later received one of its medals for his contributions to theoretical physics.

Fitzgerald's main work was in the development of Maxwell's electromagnetic theory. The 'Fitzgerald contraction' (747) for which he is now known was probably of little importance to him. It was characteristic of him to throw out ideas and leave them to be taken up by others.

2440. LODGE, Oliver Joseph. 1851-1940. (Physics 746, 752) D.Sc., University College, London, 1877. Four years later he was appointed professor of physics at the new University College, Liverpool (later Liverpool University). F.R.S., 1887. In 1900 he became principal of the newly-constituted University of Birmingham, a post he held until his retirement in 1919. He was knighted in 1902 and received many honours and awards from scientific bodies.

2441. BOYS, Charles Vernon. 1855-1944. (Physics 748) Graduated at the Royal School of Mines in 1876 and became a demonstrator and then an assistant professor at the Normal School of Science. He resigned the post in 1897 and set up as a consultant in applied physics, developing a lucrative practice as an expert witness, mainly in patent cases. He also patented many inventions of his own. He was a prominent member of the Physical Society and was elected to the Royal Society in 1888 and later awarded two of its medals. He was knighted in 1935.

2442. RAMSAY, William. 1852-1916. (Chemistry 989) After early studies of chemistry in his native Glasgow he went to Germany to work in organic chemistry under Fittig at Tübingen, gaining the D.Phil. there in 1872. After his return he obtained a junior post at Glasgow and in 1880 became professor of chemistry at University College, Bristol. Seven years later he was appointed to the chair at University College, London, where he remained until his retirement in 1912. F.R.S., 1888.

Ramsay had the unique distinction of discovering a whole family of elements whose existence was previously unsuspected, namely the rare gases. For this achievement he was knighted in 1902 and awarded the Nobel Prize in 1904.

2443. MINERALOGICAL SOCIETY of Great Britain and Ireland. Its periodical began in 1876.

2444. LAPWORTH, Charles. 1842-1920. (Geology 1199) Became a school-teacher in Scotland in 1864 and developed a strong interest in

geology. His important researches in the late 1870s on the
stratigraphy of the Lower Palaezoic brought him recognition,
and in 1881 he was appointed to the new chair of geology at
Mason College (later the University of Birmingham). In 1888
he was elected to the Royal Society which awarded him its
Royal Medal in 1891. In 1899 he received the Geological
Society's Wollaston Medal.

2445. MILNE, John. 1850-1913. (Seismology 1201) Studied at the
Royal School of Mines and then worked as a mining engineer
and field geologist. In 1875 he was appointed professor of
geology and mining at the Imperial College of Engineering,
Tokyo. The prevalence of earthquakes in Japan led him to
take up the study of seismology and when he returned to England
in 1895 he continued to do important research in the subject
until his death. The value of his work was recognized by the
award of medals by the Geological Society and the Royal Society.

2446. OLDHAM, Richard Dixon. 1858-1936. (Seismology 1216) Graduated
at the Royal School of Mines and in 1879 was appointed to the
Geological Survey of India, eventually becoming its superintendent. He returned to England in 1903 and three years later
succeeded in demonstrating from seismic records that the earth
has a central core. He was awarded a medal by the Geological
Society in 1908 and was elected to the Royal Society in 1911.

2447. MURRAY, John. 1841-1914. (Oceanography 1079) Studied at the
University of Edinburgh but did not take a degree. His enthusiasm for studies of the ocean (which in 1868 had led him to
visit the Arctic in a whaling ship) resulted in his appointment
in 1872 to the scientific staff of the *Challenger* Expedition.
After the return of the expedition in 1876 he continued as
assistant to its scientific leader, C.W. Thomson, in the immense
task of preparing the *Challenger Report*, and when Thomson died
in 1882 Murray was appointed to succeed him. The *Report* was
finally published in fifty volumes in 1890-95 and in 1898 he
was awarded a knighthood.
    Murray's financial independence enabled him to continue
his oceanographic researches, and in 1912 (in collaboration
with the Norwegian scientist, J. Hjort) he published a classic
work, *The Depths of the Ocean*, which was for many years the
leading textbook of oceanography.

2448. CONCHOLOGICAL SOCIETY of Great Britain and Ireland. Its periodical began in 1874.

2449. LANKESTER, Edwin Ray. 1847-1929. (Zoology 1505) Son of the
zoologist and microscopist, Edwin Lankester. Graduated in
science at Oxford in 1868, and then studied at Vienna and
Leipzig and at the Zoological Station at Naples. He was
appointed professor of zoology at University College, London,
in 1872 and elected to the Royal Society three years later.
In 1884 he was one of the chief movers in the establishment of
the Marine Biological Association and its important laboratory
at Plymouth. He became professor of comparative anatomy at
Oxford in 1891 and during his tenure of the post reorganized
the University's natural history museum. In 1898 he was
appointed director of the British Museum (Natural History)

and was the most outstanding figure in British zoology at the turn of the century. He was knighted when he retired in 1907.

For many years Lankester was editor of the *Quarterly Journal of Microscopical Science* which had been founded in 1853 by his father and which, from rather humble beginnings, he made into a professional journal of international standing. He was also editor of a multi-volume *Treatise on Zoology* and in his retirement became widely known for his successful popularizations of science.

2450. ROMANES, George John. 1848-1894. (Zoology 1516) Came from a wealthy family. Graduated in the natural science course at Cambridge in 1870 and then did research in physiology under Foster for three years. In 1874 he moved to London where he continued physiological research at University College. His wealth enabled him to devote himself to research during the twenty years until his death. He served as secretary of the Linnean Society and also of the Physiological Society, and was a friend of Darwin and an enthusiastic advocate of his ideas.

2451. BALFOUR, Francis Maitland. 1851-1882. (Embryology 1682) Graduated in the natural science course at Cambridge in 1871 and in the following year worked with Foster. From 1873 to 1878 he spent part of his time at the research station for marine biology at Naples. Concurrently he was lecturing on morphology and embryology at Trinity College, Cambridge, and from 1873 he was director of the University's morphological laboratory where he soon created a vigorous research group. He was elected to the Royal Society in 1878 and awarded its Royal Medal in 1881. In 1882 Foster persuaded the University to create a chair of animal morphology for him but later in the same year he was accidently killed.

2452. PHYSIOLOGICAL SOCIETY. Founded in 1876.

2453. FERRIER, David. 1843-1928. (Physiology 1784) M.D., Edinburgh, 1870. From 1872 he was a professor of medicine at King's College Hospital and Medical School. An original member of the Physiological Society, he was also one of the founders and editors of the neurological journal *Brain* which began in 1878. F.R.S., 1876. His important research on the brain brought him honours from scientific and medical societies and a knighthood in 1911.

2454. GASKELL, Walter Holbrook. 1847-1914. (Physiology 1791) Graduated at Cambridge in 1869 with honours in mathematics. He then decided to take up medicine and as a step in that direction studied physiology under Foster who, then and later, had a powerful influence on him. Although he qualified in medicine in 1874 he never practised; instead, under Foster's inspiration, he chose physiological research and went to Leipzig to work with Carl Ludwig for a year. After returning to England in 1875 he continued research at Cambridge and in 1883 was appointed lecturer in physiology, a post he held until his death. He was elected to the Royal Society in 1882 and was later awarded two of its medals.

2455. SHARPEY-SCHÄFER, Edward Albert (known as SCHÄFER until 1918).
1850-1935. (Physiology 1797) Graduated in medicine at University College, London, in 1874 and became assistant in physiology there in the same year. He was appointed to the chair in 1883 and was one of the founders and editors of the *Internationale Monatsschrift für Anatomie und Physiologie* which began in 1884. In 1899 he transferred to the chair of physiology at Edinburgh. In his later years he received many honours from scientific societies and universities and was knighted in 1913.

2456. LANGLEY, John Newport. 1852-1925. (Physiology 1783, 1794) Graduated in the natural science course at Cambridge in 1875, having studied biology under Foster. In the following year he became Foster's chief demonstrator and in 1877 he spent some months working with Kühne at Heidelberg. From 1883 he was lecturer in physiology at Cambridge and in 1903 succeeded Foster as professor of physiology; he successfully maintained the high reputation of the Cambridge school. In 1894 he rescued the esteemed *Journal of Physiology* from financial collapse and for the rest of his life continued to edit it in an exemplary manner.
Langley was a prominent member of the Neurological Society and was elected to the Royal Society in 1883. He received many honours and awards.

1880s

2457. NORMAL SCHOOL OF SCIENCE. Created by the government in 1881 out of the scientific departments at South Kensington which had earlier been detached from the School of Mines (see 2419). Its character was defined officially as "a Metropolitan School or College of Science, not specially devoted to Mining but to all Science applicable to Industry, and with a special organisation as a Training College for Teachers." The latter was its chief function and the reason for the term 'normal' in its name.
The conception of the School was very largely due to T.H. Huxley who had been professor of natural history at the School of Mines since 1854 and had developed far-reaching and influential ideas about education generally and scientific education in particular. Huxley (who had recently declined a chair at Oxford) was appointed Dean of the new institution and held the post until his retirement in 1885. In 1890 the School was renamed the Royal College of Science.

2458. LOVE, Augustus (E.H.). 1863-1940. (Mechanics 566. His major work--in theoretical geophysics--was done after 1900) Graduated in mathematics at Cambridge and became a college fellow in 1886. He was a prominent member of the London Mathematical Society and was elected to the Royal Society in 1894. In 1899 he was appointed to the Sedleian chair of natural philosophy at Oxford.

2459. THOMSON, Joseph John. 1856-1940. (Physics 762) After studying engineering at Owens College, Manchester, he won a scholarship to Cambridge where he graduated in 1880. In the following year he became a fellow of Trinity College and in 1884 was appointed professor of experimental physics and head of the Cavendish Laboratory (in succession to Lord Rayleigh). He was both an inspiring teacher and an able director of research, and under his leadership the Laboratory grew considerably in size. In 1895 Cambridge decided to accept graduates from other univer- sities as candidates for its higher degrees and thereafter he attracted a brilliant band of young research workers, many of whom later became distinguished physicists.

Thomson was awarded the Nobel Prize in 1906, a knight- hood in 1908, and many other honours. He resigned his chair in 1919 in favour of his most distinguished pupil, Rutherford, and ended his days as master of Trinity College.

2460. LARMOR, Joseph. 1857-1942. (Physics 756) Graduated at Cambridge in 1880. After five years as professor of natural philosophy at Queen's College, Galway, he returned to Cambridge in 1885 as lecturer at St. John's College. In 1903 he was appointed Lucasian professor of mathematics. He was prominent in the London Mathematical Society and served as secretary of the Royal Society from 1901 to 1912. He was awarded two of the Royal Society's medals, as well as other honours from British and foreign scientific bodies, and was knighted in 1909.

2461. PERKIN, William Henry, Jnr. 1860-1929. (Chemistry 969, 979) Son of W.H. Perkin, Snr. (2418). Graduated at the Royal College of Chemistry in 1880 and then went to Germany to work with Wislicenus at Würzburg (1880-82) and Baeyer at Munich (1882-86). On returning to Britain he was appointed to the chair of chem- istry at the new Heriot-Watt College in Edinburgh in 1887. He was elected to the Royal Society in 1890 and two years later became professor of organic chemistry at Owens College (soon to become the University of Manchester). He built on the strong tradition of chemistry at Manchester (cf. 2417) and created an outstanding research school. In 1912 he was appointed Waynflete professor of chemistry at Oxford.

Perkin was accomplished as a teacher as well as a research director and had a far-reaching influence on the development of organic chemistry in Britain. In his later years he received many honours and awards.

2462. WALKER, James. 1863-1935. (Physical chemistry 980) D.Sc., Edinburgh, 1886; he then went to Germany where he worked mostly with Ostwald at Leipzig. After his return to Britain he held research posts at Edinburgh (1889-92) and University College, London (1892-94), and in 1894 was appointed professor of chem- istry at Dundee. From 1908 until his retirement in 1928 he held the chair of chemistry at Edinburgh. He was elected to the Royal Society of London in 1900 and was knighted in 1921. Through his teaching and research he did much to establish the study of physical chemistry in Britain.

2463. BARLOW, William. 1845-1934. (Crystallography 1109) Came from a wealthy family and was privately educated. Because of his wealth he held no institutional position but he collaborated with W.J. Pope, professor of chemistry at Manchester and later Cambridge. He was elected to the Royal Society in 1908 and was president of the Mineralogical Society in 1915-18.

2464. BRITISH MUSEUM (NATURAL HISTORY). From the early nineteenth century the natural history departments of the British Museum (see 2303) had been suffering increasingly from lack of space

in the Museum's building in central London (the present building
in Bloomsbury). In 1856 Richard Owen (2365) was appointed
superintendent of the natural history departments and thenceforward continued to press for a solution to the problem. As
a result of his efforts Parliament approved the purchase of
twelve acres of land at South Kensington (part of the Exhibition
site) in 1863, and ten years later construction of the present
Natural History Museum began. It was opened in 1881 with the
official name of British Museum (Natural History).

2465. MARINE BIOLOGICAL ASSOCIATION. After its establishment in 1884
it set up its important laboratory on Plymouth Sound.

2466. MORGAN, (C.) Lloyd. 1852-1936. (Zoology 1518) Studied at the
institution later known as the Royal College of Science under
T.H. Huxley who influenced him greatly. From 1878 he taught
at a school in South Africa. He returned to England in 1883
and became professor of geology and zoology at University
College, Bristol, where he remained until his retirement in
1919. He was elected to the Royal Society in 1899.

2467. PEARSON, Karl. 1857-1936. (Biometrics 1544. Philosophy of
science 66) Graduated at Cambridge in 1879 with honours in
mathematics. He then took up the study of law and though he
qualified in it he never practised. Meanwhile he was publishing
papers in mathematics and in 1884 was appointed professor of
applied mathematics at University College, London. One of his
colleagues there was the zoologist, W.F.R. Weldon, who had a
determining influence on his future career. In 1901, in conjunction with Weldon, he founded the journal *Biometrika* which
he continued to edit until his death.
         Pearson was elected to the Royal Society in 1896 and
two years later was awarded its Darwin Medal. He gave up his
mathematics chair in 1911 to become the first Galton professor
of eugenics and head of a new department of applied statistics
at University College.

2468. WELDON, Walter Frank Raphael. 1860-1906. (Biometrics 1543)
Graduated at Cambridge in 1881 with honours in zoology. After
some research at the Naples Zoological Station and then at
Cambridge he became a college fellow there in 1884 as well as
university lecturer in invertebrate morphology. In 1890 he
was appointed professor of zoology at University College,
London, where one of his colleagues was the mathematician,
Karl Pearson. Their common interest in the work of Francis
Galton (1542) led to a highly fruitful collaboration which
initiated the subject of biometrics.
         Weldon was appointed to the Linacre chair of zoology
at Oxford in 1900. His early death put an end to a brilliant
career.

2469. BATESON, William. 1861-1926. (Heredity 1546) Graduated in
zoology at Cambridge in 1883. He continued research and was
appointed a college fellow in 1885. After holding various
junior teaching positions at Cambridge he was appointed reader
in zoology there in 1907 and in the next year professor of
genetics (the first in Britain). He gave up the chair in 1910
however and spent the rest of his life as director of the

John Innes Horticultural Institution which he made a leading centre for research in genetics. With his collaborator, R.C. Punnett, he began the *Journal of Genetics* in 1910.

Bateson was elected to the Royal Society in 1894 and later awarded two of its medals. Other honours included the presidency of the British Association in 1914 and appointment as a trustee of the British Museum in 1922.

2470. ANATOMICAL SOCIETY of Great Britain. Founded in 1887 by G.M. Humphrey, professor of anatomy at Cambridge, and W. Turner (2424), professor of anatomy at Edinburgh.

2471. SHERRINGTON, Charles Scott. 1857-1952. (Physiology 1792. Most of his famous researches in neurophysiology were done after 1900) His medical studies at St. Thomas' Hospital, London, were suspended between 1879 and 1883 when he took the natural sciences course at Cambridge and studied physiology in Foster's laboratory. He resumed his medical training, graduating in 1885, and then did graduate work in Germany under Virchow and Koch. On his return to England in 1887 he became a fellow of Caius College, Cambridge, and lecturer in physiology at St. Thomas' Hospital. He was appointed professor of physiology at Liverpool in 1895 and then at Oxford in 1913. In his later years he received many honours, civil and scholarly, including the Nobel Prize in 1932.

1890s

2472. BRITISH ASTRONOMICAL ASSOCIATION. Founded in 1890 for amateur astronomers.

2473. DIXON, Henry Horatio. 1869-1953. (Botany 1407) Graduated at Trinity College, Dublin, in 1891 and then studied at Bonn with Strasburger for two years. After junior appointments in the department of botany at Trinity College he became professor there in 1904. He was elected to the Royal Society in 1908.

2474. MALACOLOGICAL SOCIETY of London. Founded in 1893.

2.04 GERMANY

1480s

2475. STÖFFLER, Johannes. 1452–1531. (Astronomy 322, 324) Professor of mathematics at Tübingen University.

1490s

2476. KÖBEL, Jacob. ca. 1462–1533. (Mathematics 84) Graduated in arts and law at Heidelberg in 1491. Apparently he then studied mathematics at Cracow where he is said to have been a fellow student of Copernicus. He settled in Oppenheim and became a scholar of wide interests, author of many books, and a printer and publisher.

1500s

2477. DÜRER, Albrecht. 1471–1528. (Mathematics 82) The celebrated artist. During his lifetime he was as well known for his use of mathematics as for his art.

1510s

2478. WERNER, Johannes. 1468–1522. (Mathematics 80. Cartography 1002) Studied at the University of Ingolstadt in the 1480s and at Rome in the 1490s. From 1498 until his death he was a parish priest in Nuremberg and became a highly regarded scholar.

2479. RIES, Adam. 1492–1559. (Mathematics 79) Germany's most famous *Rechenmeister*. In addition to various other occupations he ran a school for teaching arithmetic, first at Erfurt from 1518 to 1523, and then for many years at Annaberg. His school became well known and he was honoured by the Duke of Saxony.

2480. HARTMANN, Georg. 1489–1564. (Physics 570) After studying at Cologne around 1510 he settled in Nuremberg in 1518 and became an outstanding instrument-maker. He was acquainted with several important figures, including Dürer and Rheticus.

1520s

2481. APIAN, Peter. 1495–1552. (Mathematics 85. Astronomy 327. Cosmography 1003) Studied mathematics and astronomy at Leipzig and Vienna. Became professor of mathematics at the University of Ingolstadt. Best known as a writer of popular works on astronomy and geography.

2482. PARACELSUS. ca. 1493–1541. (Not in Part 1) One of the most famous figures of sixteenth-century medicine and science. Apparently he studied at several Italian universities and he

may have taken a medical degree at Ferrara. Because of his
tempestuous personality he only had one academic appointment,
as professor of medicine at Basel in 1527, and that lasted
for only a year. Thereafter he wandered around central Europe
until his death.

    Of his many books only a few were published in his
lifetime. His importance lay in his initiation of the iatrochemical movement and in his powerful influence on his posthumous followers (cf. 773-776 and many others later).

2483. CORDUS, Euricius. 1486-1535. (Botany 1282) M.D., Ferrara,
      1521. Became a municipal physician in Brunswick, and later
      Bremen, and was for a few years professor of medicine at the
      new University of Marburg.

2484. BRUNFELS, Otto. ca. 1489-1534. (Botany 1280) Initially a
      Lutheran pastor. Studied medicine, gaining the M.D. at Basel
      in 1532, and becoming physician in Bern for a brief period
      before his death.

1530s

2485. STIFEL, Michael. ca. 1487-1567. (Mathematics 89) Lutheran
      pastor, graduate of the University of Wittenberg, and teacher
      of mathematics in various places, including the universities
      of Königsberg and Jena.

2486. REINHOLD, Erasmus. 1511-1553. (Astronomy 328, 331) Professor
      of astronomy at Wittenberg from 1536. He became a leading
      mathematical astronomer, comparable with Copernicus in whose
      work he was highly interested.

2487. RHETICUS, Georg Joachim. 1514-1574. (Astronomy 329. Mathematics
      110) M.A., Wittenberg, 1536. After spending a few years
      travelling and visiting important astronomers, including
      Copernicus (with whom he stayed for some months), he taught
      briefly at Wittenberg and then, in 1542, became professor of
      mathematics at Leipzig. He had to leave Leipzig in 1551
      because of a scandal and went to Prague where he studied medi-
      cine, finally settling in Cracow where he was a physician for
      the rest of his life.

2488. AGRICOLA, Georgius. 1494-1555. (Earth sciences 1006, 1007)
      Studied first at the University of Leipzig and later spent
      three years in Italy where he obtained his M.D. about 1525.
      From 1526 he was town physician in Joachimsthal, and later
      Chemnitz, in the mining area of Saxony.

2489. BOCK, Jerome. 1498-1554. (Botany 1284) From 1533 to 1550 he
      was a canon at a Benedictine church and he also acted as a
      physician.

2490. FUCHS, Leonhart. 1501-1566. (Botany 1285) M.D., Ingolstadt,
      1524. After practising medicine with distinction he became
      professor of medicine at Tübingen in 1535 and held the post
      for the rest of his life. In 1548 he declined an invitation
      from the Grand Duke of Tuscany to become director of the
      botanical garden at Pisa.

2491. DRYANDER (or EICHMANN), Johann. 1500-1560. (Anatomy 1560)
      Graduated in medicine at Mainz and about 1536 became professor

of medicine and mathematics at Marburg where he remained until
his death. Besides his work in anatomy he was also active as
an astronomer.

1540s

2492. CORDUS, Valerius. 1515-1544. (Botany 1287, 1292) Son of
Euricius Cordus (2483) from whom he learnt pharmacy and botany
at an early age. He studied at Wittenberg and on three occasions during the period 1540-43 lectured there on materia medica.
He then travelled to Italy where he met his untimely death.
His books were published after his death through the efforts
of Conrad Gesner.

1560s

2493. DASYPODIUS, Cunradus. ca. 1530-1600. (Not in Part 1) Educated
at the famous academy of Johannes Sturm in Strasbourg and was
a teacher of mathematics there from 1558 onward. From the
1560s until his death he published a series of mathematical
(including astronomical and mechanical) textbooks which illus-
trate the enlarged place of mathematics in some of the new
educational institutions of the time. He was also an important
editor and translator of the classics of Greek mathematics.

2494. WILHELM IV, *Landgrave of Hesse*. 1532-1592. (Astronomy 333)
His education at the court in Kassel had included some mathematics and astronomy, and in his early years he became a capable
astronomer. On becoming landgrave in 1567 he had to give up
observing but he continued to be an enthusiastic patron.

2495. RAUWOLF, Leonhard. 1535-1596. (Botany 1297) Educated at Wittenberg, then studied medicine at Montpellier (botany under
Rondelet), graduating in 1562. He settled in Augsburg, eventually becoming city physician; he maintained a botanical garden
and exchanged specimens with other botanists, including L'Obel.
In 1573-76 he went on a journey of botanical exploration to the
Middle East, as far as Baghdad, the journey being financed by
a firm of merchants which had connections in the region and
hoped to profit from discoveries of new drugs.

2496. BAUHIN, Jean. 1541-1613. (Botany 1310) A native of Basel,
brother of Gaspard Bauhin, and a pupil and friend of Conrad
Gesner. M.D., 1562 (at Montpellier?). After practising
medicine at Lyons and Geneva he became physician to the Duke
of Württemberg in 1571. He published several minor botanical
works in the 1590s and 1600s but his main work did not appear
until thirty-seven years after his death.

2497. COITER, Volcher. 1534-1576. (Anatomy 1570. Embryology 1653)
Born and educated in the northern Netherlands. Studied in
Tübingen (with Fuchs), Montpellier (briefly with Rondelet),
Padua (with Falloppio), Rome (with Eustachi), and Bologna
(with Aranzio and Aldrovandi). Graduated in medicine at
Bologna in 1562 and taught surgery there for a few years. He
apparently fell foul of the Inquisition and in 1566 left for
Germany, settling at Nuremberg as city physician.

1570s

2498. FLEISCHER, Johannes. 1539–1593. (Physics 571) After studying at Wittenberg he became a gymnasium teacher in Breslau in the late 1560s.

2499. ALBERTI, Salomon. 1540–1600. (Anatomy 1574) Graduated in medicine at Wittenberg in 1574 and was later professor of medicine there for many years.

1580s

2500. MÄSTLIN, Michael. 1550–1631. (Astronomy 340) M.A., Tübingen, 1571; he then studied theology and became a Lutheran pastor in 1576. He had earlier shown much ability in mathematics and astronomy, and had taught them briefly at Tübingen; as a result he was appointed professor of mathematics at Heidelberg in 1580. Four years later he returned to Tübingen as professor of mathematics and held the post until his death. An adherent of the Copernican system from 1578, he was an influential teacher and author of many books and tracts on astronomy. Perhaps his main contribution to the subject was to convince his favorite student, Kepler, of the superiority of the Copernican system.

2501. BOTANICAL GARDEN, LEIPZIG. Established by the University in 1580. The first in Europe outside Italy.

1590s

2502. PITISCUS, Bartholomeo. 1561–1613. (Mathematics 112) Studied theology at Heidelberg and was later court chaplain to the Elector of the Palatinate.

2503. LIBAVIUS, Andreas. ca. 1560–1616. (Chemistry 778) Studied at Wittenberg and Jena, and received his M.D. at Basel in 1588. After teaching and practising medicine for many years he became rector of a gymnasium in Coburg in 1607. He had wide-ranging interests and wrote many books on a variety of subjects.

2504. HARTMANN, Johannes. 1568–1631. (Chemistry 786) Studied at several German universities and in 1592 became professor of mathematics at Marburg as well as *mathematicus* to the Landgrave of Hesse-Kassel. He also studied medicine and received the M.D. in 1606. In 1609 he was appointed professor of "chymiatria" (iatrochemistry) at Marburg and he can be regarded as the first university professor of chemistry. Until he left the University in 1621 (to become court physician) he gave lectures and practical laboratory instruction in materia medica and the preparation of chemical medicines in the "laboratorium chymicum publicum" at Marburg.

1600s

2505. FAULHABER, Johann. 1580–1635. (Mathematics 125) Originally a weaver. Learnt mathematics from a private teacher and from his own efforts. In 1600 he established a school for mathematics in Ulm which was very successful and became largely a school for military engineering. Descartes, as a young man, studied with him in 1620 and had a high opinion of him.

2506. CHRISTMANN, Jacob. 1554-1613. (Astronomy 350) Studied oriental
languages at Heidelberg. From 1584 he was professor of Hebrew
there, and from 1608 professor of Arabic. During the 1590s he
edited some medieval Hebrew and Arabic astronomical texts and
developed an interest in the subject. About 1600 he acquired
some instruments and began making observations.

2507. BAYER, Johann. 1572-1625. (Astronomy 346) A lawyer in Augsburg.

2508. MAYR (or MARIUS), Simon. 1573-1624. (Astronomy 352) Spent four
months in 1601 as an assistant to Tycho Brahe in Prague. He
then went to Padua to study medicine, remaining there until
1605. On his return to Germany he became court astronomer
and mathematician to the Margrave of Brandenburg-Anspach (who
had financed his studies earlier).

2509. SENNERT, Daniel. 1572-1637. (Chemistry 784) Studied medicine
at several German universities and received the M.D. at Wittenberg in 1601. He became professor of medicine there the next
year and held the post for the rest of his life. He was the
author of many books on medicine and iatrochemistry.

2510. SALA, Angelo. 1576-1637. (Chemistry 785) Born and perhaps
educated in Italy but spent his adult life in Germany as a
physician in several cities.

1610s

2511. JUNGIUS, Joachim. 1587-1657. (Part 1 entry: see below) M.A.,
Giessen, 1608. Taught mathematics for a few years, then
studied medicine at Rostock and Padua, receiving the M.D. at
the latter in 1619. During the next ten years he practised
medicine and for brief periods taught at the universities of
Helmstedt (medicine) and Rostock (mathematics). He finally
became rector of the Akademisches Gymnasium in Hamburg.
    Jungius was a person of considerable ability and origin-
ality who achieved much less than he might have done, possibly
because of adverse circumstances and possibly because he spread
his energies over too many fields--principally logic, mathematics, chemistry, and botany, but other scientific and non-
scientific fields as well. He did not publish much of importance and his influence on his contemporaries proceeded chiefly
from his correspondence and the many dissertations he wrote
for his students. For posthumous editions of some of his
lecture notes on botany see 1312.

2512. SCHEINER, Christoph. 1573-1650. (Astronomy 351) Jesuit. From
1610 to 1616 he was professor of Hebrew and mathematics at
Ingolstadt. He was then at the court of the Archduke of Tyrol
as a mathematician (and presumably astronomer) until the early
1620s. Thereafter he held administrative and teaching posts
in Rome and several places in Germany.

2513. CYSAT, Johann Baptist. ca. 1586-1657. (Astronomy 353) Jesuit.
Studied at Ingolstadt (astronomy under Scheiner). From 1618
to 1623 he was professor of mathematics there, and was later
rector of various Jesuit colleges.

1620s

2514. ROLFINCK, Guerner (or Werner). 1599-1673. (Not in Part 1)
Studied medicine at Wittenberg and Leiden, and continued his
education at Oxford and Paris; received the M.D. at Padua in
1625. In 1629 he became professor of anatomy, surgery, and
botany at Jena where he built an anatomical theatre and founded
the botanical garden. He also set up a chemical laboratory in
1638 and a few years later became professor of chemistry.

1630s

2515. BOTANICAL GARDEN, JENA. Founded in 1631 by G. Rolfinck.

1640s

2516. GUERICKE, Otto von. 1602-1686. (Physics 587) Came from a
family of high social standing; educated at the universities
of Leipzig and Helmstedt. In the early 1620s he studied law
at Jena and then at Leiden where he also studied mathematics
and engineering. He served his city, Magdeburg, in various
capacities, finally as mayor. On his official journeys on
behalf of his city he was able to collect information about
scientific developments.

1650s

2517. LEOPOLDINA ACADEMY. Began about 1652 as the Academia Naturae
Curiosorum, a small local society founded by the town physician
of Schweinfurt in Franconia. It achieved little until the
1660s when it was invigorated by the inspiration and example
of the Royal Society of London. From 1670 it issued its
Miscellanea Curiosa Medico-physica, a periodical modelled on
the *Philosophical Transactions* but with a larger medical content.
(The *Miscellanea* was quite substantial until the early eighteenth
century but thereafter it declined)

  The Academy sought the patronage of the Emperor Leopold
I and succeeded in getting it in 1687, whereupon it styled
itself the Academia Caesareo-Leopoldina Naturae Curiosorum,
claiming to be the academy of the Holy Roman Empire (though
what this actually meant is uncertain). It had no fixed place
of abode, moving from one city to another in central Europe
until the late nineteenth century when it settled in Halle.
(From about the beginning of the nineteenth century it was
known as the Kaiserlich Leopoldinisch-Carolinische Academie
der Naturforscher and it continues today as the Deutsche
Akademie der Naturforscher)

2518. SCHOTT, Gaspar. 1608-1666. (Not in Part 1) Jesuit. Born and
educated in Germany. Taught at Palermo until 1652 when he
went to Rome to work with Kircher (1883). In 1655 he returned
to Germany and taught mathematics and physics at Mainz and
then at Würzburg until his death.

  During his period in Germany from 1655 to 1666 Schott
acted as a clearing-house for scientific information sent to
him by his fellow Jesuits and many others. He first spread

the news of Guericke's experiments and later did much to make Boyle's work in pneumatics widely known in Germany. In the same period he published eleven books which, like those of his mentor, Kircher, contained little that was original but were useful in disseminating scientific and technical knowledge.

1660s

2519. KUNCKEL, Johann. 1630(or 1638)–1702. (Chemistry 800) Learnt chemistry from his father who was an alchemist; he had no university education. Served several German princes as apothecary, alchemist, and technologist. In 1688 he became minister of mines in Sweden and was ennobled in 1693; in that year he also became a member of the Leopoldina Academy.

2520. BECHER, Johann Joachim. 1635–1682. (Chemistry 797) M.D., Mainz, 1661. After having been physician to some German princes he was appointed in 1666 as economic councillor to the Emperor in Vienna and became an outstanding proponent of mercantilism. He was, however, dismissed in 1678 and went to Holland and then to England where he was involved in mining affairs until his death.

2521. KERCKRING, (T.) Theodor. 1640–1693. (Anatomy 1599) Born in Hamburg but brought up in Amsterdam. He gained the M.D. at Leiden in the early 1660s and became a medical practitioner in Amsterdam, at the same time carrying on anatomical research in conjunction with Ruysch. He left Amsterdam, apparently in 1675, and after a few years in Italy settled in Hamburg as "resident" (ambassador) of the Grand Duke of Tuscany. He was a foreign member of the Royal Society of London.

1670s

2522. LEIBNIZ, Gottfried Wilhelm. 1646–1716. (Mathematics 162. Mechanics 519) Studied law at the universities of Leipzig, Jena, and Altdorf where he graduated in 1666. From about 1670 he was engaged in legal and diplomatic activities in the course of which he visited Paris and London in 1672–76 and became acquainted with many of the leading scientific figures. From 1676 he was in the service of the Duke of Brunswick-Lüneburg as adviser, librarian, historian, etc., and in the course of his duties travelled over much of Europe. In 1700 the Berlin Academy was founded on his recommendation and he was appointed its president. He was elected a foreign member of the Paris Academy in 1700.

1680s

2523. *ACTA ERUDITORUM.* The first German periodical, begun in 1682 and published from Leipzig. It covered the whole range of learning and included a substantial amount of mathematics and science until the early decades of the eighteenth century. It ceased publication in 1782.

2524. TSCHIRNHAUS, Ehrenfried Walther. 1651–1708. (Mathematics 163) Came from a wealthy land-owning family. Studied at Leiden around 1670 (mathematics under van Schooten). Visited Paris

and London in 1675 and met a number of mathematicians. After
further travels he settled on his estates where he spent much
of his time in research in mathematics and science; he also
carried on an extensive correspondence with Leibniz and others.
On a visit to Paris in 1682 he was elected a foreign member of
the Academy.

1690s

2525. STAHL, Georg Ernst. 1660–1734. (Chemistry 803. Physiology
       1724) M.D., Jena, 1684. After a few years as court physician
       at Weimar he was appointed professor of theoretical medicine
       at Halle in 1694. (The professor of practical medicine was
       Hoffmann) In 1715 he became court physician in Berlin. He
       was a member of the Leopoldina Academy.

2526. HOFFMANN, Friedrich. 1660–1742. (Chemistry 801. Physiology
       1727) M.D., Jena, 1681. After travelling in the Low Countries
       and England (where he met Boyle) he became a physician in
       Saxony. In 1693 he became the first professor of medicine at
       the new University of Halle; he was a very successful and
       influential teacher in both medicine and chemistry. Member of
       the Berlin, Leopoldina, and St. Petersburg academies and of
       the Royal Society of London.

2527. RIVINUS, Augustus Quirinus. 1652–1723. (Botany 1322) Studied
       medicine at Leipzig and took his degree at Helmstedt in 1676.
       Became professor of medicine at Leipzig and during the 1690s
       was also professor of botany.

2528. CAMERARIUS, Rudolph Jakob. 1665–1721. (Botany 1324) After
       travelling through much of western Europe he studied medicine
       at Tübingen (where his father was professor of medicine) and
       graduated in 1687. In the following year he was appointed
       extraordinarius in medicine at Tübingen and director of the
       botanical garden; he later succeeded his father as *ordinarius*.

2529. VALENTINI, Michael Bernhard. 1657–1729. (Zoology 1438) M.D.,
       Giessen, 1686; in the following year he became professor of
       medicine there. He was an outstanding physician and also
       published several works on various branches of science, espec-
       ially natural history. Member of the Leopoldina and Berlin
       academies and of the Royal Society of London.

1700s

2530. BERLIN ACADEMY. Founded in 1700 by the Elector Frederick III of
       Brandenburg (who made himself King of Prussia in 1701) and
       inaugurated in 1711 as the Societas Regia Scientiarum. The
       Academy owed its existence to Leibniz' influence on his former
       pupil, the Electress Sophia, and it was he who drew up its
       constitution and became its first president. It was given an
       economic base by the grant of a monopoly on the publication of
       calendars and astronomical almanacs, in connection with which
       it acquired a small observatory.
          The Academy had many difficulties in its early years,
       especially during the reign of the barbarous Frederick William
       I (1717–40), but it managed to bring out eight volumes of its
       Miscellanea Berolinensia. For its subsequent fortunes see 2536.

2531. LEUPOLD, Jacob. 1674-1727. (Mechanics 529) An engineer in Leipzig. He became a correspondent of the Berlin Academy.

1720s

2532. JUNCKER, Johann. 1679-1759. (Chemistry 809) M.D., Erfurt, 1717. Became a distinguished physician in Halle and in 1729 professor of medicine at the University.

1730s

2533. MARGGRAF, Andreas Sigismund. 1709-1782. (Chemistry 812) Son of the apothecary to the royal court in Berlin. Studied chemistry with apothecaries in several cities and at the universities of Strasbourg and Halle (but apparently did not take a degree) and then worked with his father. In 1738 he was appointed to the Berlin Academy and in 1753 made director of its chemical laboratory. He was later elected a foreign member of the Paris Academy and an honorary member of the Mainz (Erfurt) Academy.

2534. CRAMER, Johann Andreas. 1710-1777. (Chemistry 811) His family had interests in metallurgy. From 1734 he studied chemistry at the University of Helmstedt, then travelled in Holland, England, and Saxony to extend his knowledge of metallurgy. In 1743 he was appointed director of the Brunswick administration of mining and smelting in the Harz Mountains.

2535. KLEIN, Jacob Theodor. 1685-1759. (Zoology 1439) A lawyer who was a high official of the city of Danzig. He was the founder and director of the local naturalists' society and established the city's botanical garden. He published many books and papers, chiefly on zoological topics, and was a foreign member of the Royal Society of London and the St. Petersburg Academy.

..... HALLER, A. von. Professor of medicine at Göttingen, 1736-53. See 2896.

1740s

2536. BERLIN ACADEMY. (For its early years see 2530) In 1740 the Academy entered on a new era with the accession of Frederick II ('the Great'). He completely re-organized it, divided it into four classes (mathematics, science, philosophy, and literature), and renamed it the Académie Royale des Sciences et des Belles-Lettres, its official language being changed from Latin to the new international language, and its periodical becoming the *Histoire et Mémoires*.

    Thereafter the Academy flourished, despite Frederick's interference and his restrictions on its autonomy (cf. 2537). After his death in 1786 it was not so well treated, and there was resentment against the non-Prussian members (cf. 2548). The Academy was well enough established however to ride out the difficulties. For its later period see 2569.

2537. EULER, Leonhard. 1707-1783. (For entries in Part 1 and for his earlier career in St. Petersburg see 2973) His invitation to join the Berlin Academy in 1741 was a consequence of Frederick the Great's policy to restore and enlarge the Academy. The reorganization, in which Euler played an important part, was completed in 1744. For the next twenty years Euler, now the leading mathematician in Europe, was the Academy's brightest luminary. For most of the period he ran the Academy, though without the title of president, and his administrative and advisory duties were numerous; nevertheless he was able to maintain his customary spate of publications.

Relations between Euler and Frederick deteriorated however--in their personalities and values the two men were incompatible--and the tension increased to the point where Euler was constrained to leave. Fortunately he had maintained ties with the St. Petersburg Academy (he had been an honorary member since he left, and he had kept up a correspondence with the Academy and had carried out various assignments for it) and when it became known in St. Petersburg that he wished to return, an invitation was speedily forthcoming. He left Berlin in 1766. For his second period in St. Petersburg see 2976.

2538. MAYER, (J.) Tobias. 1723-1762. (Astronomy 399, 411) Of humble origin; self-taught in mathematics and science. From 1744 to 1750 he worked as a cartographer for publishing firms in Augsburg and Nuremberg and in this capacity made many astronomical observations. He was an active member of the Nuremberg Cosmographical Society which provided him with a scientific milieu and valuable stimulus. The astronomical papers which he published in its periodical, together with his reputation as a cartographer, led to his appointment in 1751 as professor of mathematics at Göttingen. Three years later he was put in charge of the University's observatory of which he made good use until his untimely death.

1750s

2539. GÖTTINGEN SOCIETY. Societas Regia Scientiarum Gottingensis (known from the early nineteenth century as the Königliche Gesellschaft der Wissenschaften in Göttingen) Founded in 1751 by the Hanoverian government with the advice of Albrecht von Haller, the renowned professor of medicine at Göttingen University, who became its first president. (The term 'academy' was not used in its name because at that time in Germany the term could also mean a *Hochschule* or university, and so in Göttingen would have caused confusion) The Society functioned in intimate association with the University and was intended by its founders to enhance the reputation and quality of the still new University. Its *Commentarii* began in 1751 (and was continued from 1838 as *Abhandlungen*).

2540. ERFURT ACADEMY. Academia Electoralis Moguntina Scientiarum Utilium quae Erfurti est (known from the end of the eighteenth century as the Kurfürstlich-Mainzische Akademie nützlicher Wissenschaften zu Erfurt). Founded--to a large extent on the model of the Göttingen Society--in 1754 under the patronage of the Elector of Mainz in whose territory the old university city of Erfurt then was. The Academy's *Acta* began in 1757 (and ceased in 1826, perhaps as a result of the abolition of the University in 1816).

2541. MUNICH ACADEMY. Founded (on the model of the Berlin Academy) in 1759 as the Churfürstliche (later Königliche) Baierische Akademie der Wissenschaften. Its *Abhandlungen* began in 1763.

2542. LAMBERT, Johann Heinrich. 1728-1777. (Mathematics 185, 188. Astronomy 404. Physics 617, 624. Cartography 1039) Born in Alsace of humble parentage; after an elementary education he was entirely self-taught. From 1748 he was for ten years a private tutor to a noble Swiss family in Chur. During this period he carried out astronomical and meteorological observations and various physical experiments using instruments and apparatus that he made himself; he also became a member of the Societas Physico-Medica at Basel, publishing several papers in its *Acta Helvetica*. After two years of travel with some of his young pupils through much of western Europe--during which he made many scientific contacts--he gave up his tutorship in 1759. For the next five years he led an unsettled life in Switzerland and Germany but in 1765 he was appointed to the Berlin Academy and he remained in Berlin until his death.

     Despite his rather fragmented life, Lambert was a very prolific writer on a wide variety of subjects, including philosophy and psychology.

2543. AEPINUS, Franz Ulrich Theodosius. 1724-1802. (Physics 616) Came from a family of theologians and professors. M.A., Rostock, 1747; he then became a junior lecturer in mathematics at the University. In 1755 his publications in mathematics led to his appointment to the Berlin Academy. Two years later (perhaps because of the threatening political situation) he accepted an invitation to join the St. Petersburg Academy, where he remained for the rest of his career. (His important book on electricity was mostly written in Berlin but published in St. Petersburg in 1759) After the late 1760s he became involved in court appointments and his scientific career ceased.

2544. LEHMANN, Johann Gottlob. 1719-1767. (Geology 1114) M.D., Wittenberg, 1741. During the 1740s he practised medicine in Dresden but became involved in the study of mining and metallurgy. In 1750 he went to Berlin where he was appointed to a position in the Prussian mining administration, becoming a *Bergrat* in 1754; in the same year he was elected to the Berlin Academy. He went to Russia in 1761 at the invitation of the St. Petersburg Academy which appointed him professor of chemistry and director of its natural history collection.

2545. FÜCHSEL, Georg Christian. 1722-1773. (Geology 1116) Studied medicine at Jena and Leipzig and took the M.D. at Erfurt. For the rest of his life he was a physician in Rudolstadt.

2546. WOLFF, Caspar Friedrich. 1734-1794. (Embryology 1660) M.D., Halle, 1759 (his *Theoria Generationis* was his doctoral dissertation). After service as a surgeon in the Prussian army he settled in Berlin in 1763 and gave private lectures in medicine. In 1766, through the influence of Euler, he was offered a place in the St. Petersburg Academy as an anatomist, and remained an academician there for the rest of his life.

1760s

2547. MANNHEIM ACADEMY. Founded in 1765 as the Academia Electoralis
       Scientiarum et Elegantiarum Litterarum Theodoro-Palatina. It
       was created and richly endowed by the Elector of the Palatinate,
       Carl Theodor, who was an enthusiastic patron of science and
       the arts. He also established an associated meteorological
       society (Societas Meteorologica Palatina) which organized and
       financed a remarkable international project (see 1041). Both
       institutions came to an end in 1795 as a result of the French
       invasion and subsequent political changes.

2548. LAGRANGE, Joseph Louis. 1736-1813. (For entries in Part 1 and
       for his earlier career in Turin see 1921) He arrived in Berlin
       in 1766, taking up the position in the Academy that had been
       vacated by Euler's departure for St. Petersburg. During his
       twenty-year period in Berlin his mathematical productivity was
       enormous and he repeatedly won prizes in the international
       competitions in celestial mechanics, the 'glamour science' of
       the age. Otherwise his life was uneventful: he was of retiring
       disposition, living only for his researches and having as
       little as possible to do with Academy politics and the parties
       and intrigues around the King. When Frederick died in 1786,
       however, the position of the foreign academicians became diffi-
       cult and he was glad to accept the invitation of Louis XVI to
       join the Paris Academy. For his career in Paris see 2085.

2549. BERGAKADEMIE FREIBERG. The mining school at Freiberg in Saxony,
       established in 1765. It was made famous by the teaching of
       the great mineralogist and geologist, A.G. Werner, who was on
       its staff from 1775 until his death in 1817. Though the Frei-
       berg school was the first of its kind, less formal training in
       mining technology had been going on in Germany and Sweden for
       many years earlier.

2550. GAERTNER, Joseph. 1732-1791. (Botany 1340) M.D., Tübingen,
       1753; he then travelled in Italy, France, England, and Holland,
       studying scientific subjects. Later he became professor of
       anatomy at Tübingen and then professor of botany at St. Peters-
       burg. In 1770 he returned to his native town of Calw where he
       wrote his great book, in which task he was aided by the gifts
       of specimens he received from many botanists.

2551. KOELREUTER, (J.) Gottlieb. 1733-1806. (Botany 1334) M.D.,
       Tübingen, 1755. For the next six years he was custodian of
       the natural history collection of the St. Petersburg Academy
       and it was there that he began his experiments on pollination.
       He returned to Germany and from 1764 to 1786 was professor of
       natural history and director of the gardens at Karlsruhe.

2552. BLOCH, Marcus Eliezer. 1723-1799. (Zoology 1455) Graduated in
       medicine at Frankfurt a.d. Oder, and practised in Berlin. He
       became a naturalist of considerable standing.

1770s

2553. BODE, Johann Elert. 1747-1826. (Astronomy 408, 410, 414, 419)
       Self-taught in astronomy, becoming a very capable teacher of

mathematics and astronomy. In 1772 he was appointed to a post
in the observatory of the Berlin Academy and two years later
he initiated the successful *Astronomisches Jahrbuch*, an almanac
or collection of ephemerides which also included astronomical
articles, observations, and news items; it was thus one of the
first astronomical periodicals (and one of the first periodicals
for an individual science). He continued to edit it until his
death. In 1786 he became a member of the Berlin Academy and
director of the observatory.

2554. WERNER, Abraham Gottlob. 1749-1817. (Mineralogy 1083, 1087.
Geology 1122, 1125) Came from a family which had been associated with mining and metallurgy for generations. Studied for
two years at the new Mining Academy at Freiberg and then for
three years at the University of Leipzig. As a result of the
success of his first book (1083) he was appointed in 1775 to
the staff of the Freiberg Mining Academy where he remained for
the rest of his life. It was largely due to his teaching and
researches that the Academy became one of the leading scientific
and technical schools of the eighteenth century. His students
included many of the chief geologists of the next generation.
He was made a *Bergrath* (a position of honour) by the government
of Saxony and was elected to membership of many of the leading
scientific bodies of Europe.

2555. GESELLSCHAFT NATURFORSCHENDER FREUNDE, Berlin. The first important society for natural history. Established in 1773 as a
private society with official approval. It soon became a large
and active body.

2556. HEDWIG, Johann. 1730-1799. (Botany 1337, 1345) M.D., Leipzig,
1759. While a medical practitioner, first in Chemnitz and then
in Leipzig, he gave much of his time to botany and published
some important work. In 1786 he became professor of botany at
Leipzig. In his later years he was elected to several academies
and to the Royal Society of London.

2557. BLUMENBACH, Johann Friedrich. 1752-1840. (Natural history 1239.
Zoology 1461. The founder of physical anthropology) Came from
a wealthy and cultured family. Studied medicine at Jena and
then at Göttingen where he gained the M.D. in 1775 (his dissertation later became famous as the pioneering work on anthropology). In 1776 he was appointed professor of medicine at Göttingen and curator of the natural history collection. He was
very successful as a teacher, as well as an author, and many
of his students were later outstanding.

   Blumenbach became so eminent that he was elected to
all the major academies and societies of Europe, as well as to
dozens of minor ones. From 1812 he was the permanent secretary
of the Göttingen Society.

2558. SÖMMERRING, Samuel Thomas. 1755-1830. (Anatomy 1623, 1627)
M.D., Göttingen, 1778; he then travelled for a year in Holland
and Britain. He was professor of anatomy at the Collegium
Carolinum in Kassel from 1779, and then at the University of
Mainz from 1784. In 1797 he lost his post at Mainz because of
the French invasion and had to go into medical practice in
Frankfurt to maintain himself. He was by then the most eminent

anatomist in Germany and when conditions improved he received
offers from several universities. In 1805 he accepted an offer
of membership of the Munich Academy. In his last years he
received many honours, civil and scholarly.

1780s

2559. OLBERS, (H.) Wilhelm (M.). 1758-1840. (Astronomy 417, 420, 423,
      430) M.D., Göttingen, 1781; he then became a medical practit-
      ioner in Bremen. From an early age he had been interested in
      astronomy which he had taught himself, and in 1780 he discovered
      his first comet. When he became a successful physician he was
      able to set up a well-equipped observatory in his own house
      an amass an excellent astronomical library. He managed to
      combine an active career as an astronomer with his medical
      practice and published nearly two hundred papers.

2560. CHLADNI, Ernst. 1756-1827. (Physics 629) Graduated in law at
      Leipzig in 1782 but soon took up science, specifically acoustics
      because of his interest in music--he also constructed musical
      instruments and was the inventor of the euphonium. He earned
      his living by writing books in these fields and by travelling
      around Europe giving musical performances and lectures illus-
      trated with acoustical demonstrations.

2561. KLAPROTH, Martin Heinrich. 1743-1817. (Chemistry 832) Qualified
      as an apothecary in 1764 by apprenticeship. After working in
      several towns as a journeyman he settled in Berlin in 1771 and
      a few years later acquired his own shop. In 1785 he began
      publishing his researches in analytical chemistry and was
      elected to the Berlin Academy in 1788. From 1792 he played a
      leading part in obtaining an acceptance of Lavoisier's new
      theory in Germany. In addition to his other activities he
      gave courses in chemistry, and after a succession of teaching
      posts in various medical and technical colleges he became
      professor of chemistry at the University of Berlin when it was
      established in 1810.

2562. SPRENGEL, Christian Konrad. 1750-1816. (Botany 1343) A Lutheran
      pastor. From 1780 he was headmaster of the town school at
      Spandau where he taught natural science and languages. He gave
      so much of his time to his botanical research that he lost his
      position in 1794.

1790s

2563. RICHTER, Jeremias Benjamin. 1762-1807. (Chemistry 840) Gained
      the D.Phil. at Königsberg in 1789 with a dissertation *De usu
      matheseos in chemia*--the theme of his subsequent work. He
      continued research in chemistry and in 1795 became an assayer
      in the department of mines in Silesia. Three years later he
      was appointed a chemist in the Royal Porcelain Works in Berlin.

2564. SCHLOTHEIM, Ernst Friedrich von. 1765-1832. (Palaeontology
      1133) Came from an upper-class family in Gotha. Studied
      public administration at Göttingen and also some science
      (natural history under Blumenbach). Being interested in miner-
      alogy and geology he went (about 1790-91) to the Freiberg

Mining Academy to study under Werner (while there he became friendly with a fellow student, A. von Humboldt). In 1792 he joined the public service of the duchy of Gotha in which he spent his career, rising to high rank.

2565. HOFF, Karl Ernst Adolf von. 1771–1837. (Geology 1146) Studied law, diplomacy, and history at Jena and Göttingen, but extended his studies to include the natural sciences (at Göttingen under Blumenbach). In 1791 he joined the civil service of the duchy of Gotha and became an important diplomat during and after the Napoleonic Wars. The Duke of Gotha encouraged scholarly activities by his staff and von Hoff was able to use his official journeys and other opportunities to further his interest in geology.

2566. BUCH, Leopold von. 1774–1853. (Geology 1132, 1136, 1152) Came from a wealthy Prussian landed family. From 1790 to 1793 he studied at the Freiberg Mining Academy under Werner (and formed a lifelong friendship with his fellow student, A. von Humboldt). After subsequent study of law and government at Halle and Göttingen he joined the Prussian civil service in 1796 as a mining inspector but resigned the next year in order to devote himself fully to research (which his wealth enabled him to do). In 1798 he set out on his travels, first to the Alps and then to Italy and Switzerland for lengthy periods. In 1802 he was in the Auvergne where his study of the famous volcanoes helped to weaken his attachment to Wernerian Neptunism. Later travels took him to Scandinavia (1806–08), the Canary Islands (1815), and Scotland (1817).

      Von Buch was elected to the Berlin Academy in 1806 and when not travelling he worked in Berlin. He assiduously attended scientific congresses, such as those of the Gesellschaft Deutscher Naturforscher und Ärzte, and until his death some part of every year was spent in travel--often to the Alps, the formation of which was one of his chief interests. In the latter part of his life he was regarded by many as the greatest geologist of his time.

2567. HUMBOLDT, Alexander von. 1769–1859. (Earth sciences generally 1049, 1052. Geology 1173. Natural history 1242, 1249. Botany 1347) Came from a wealthy family of the minor Prussian nobility; educated privately. After false starts in economics and engineering he became interested in botany and spent a year (1789/90) at Göttingen studying science. Having become especially interested in mineralogy and geology he then spent two years at the Freiberg Mining Academy under Werner. From 1792 he served the Prussian government as a mining inspector but resigned the post in 1796 when he inherited his mother's fortune. By that time he had decided that his chief aim in life was scientific exploration (a decision partly due to his friendship with the naturalist and explorer, Georg Forster).

      The next two years were spent in making preparations to join projected expeditions which had to be cancelled because of the Napoleonic War. Finally Humboldt obtained permission from the Spanish government to visit the Spanish colonies in South and Central America. In July 1799 he and his companion, the French botanist, Aimé Bonpland, arrived in South America and

began five years of arduous and adventurous exploration, in
the course of which they collected many thousands of plant
specimens and amassed vast amounts of data in all the earth
sciences.

They returned to Europe in 1804 and Humboldt settled
in Paris, then the unrivalled metropolis of science, as the
place to compile and publish his findings. Having already
acquired a high reputation and being a gregarious person he
promptly became an outstanding figure in the French scientific
community and in Parisian society. The golden years in Paris
came to an end in 1827 with the exhaustion of his fortune in
consequence of the cost of his expedition and the ensuing
massive publications.

Humboldt returned to Berlin, as the King of Prussia
had wanted him to do, and was appointed royal chamberlain, a
well-paid post that allowed him ample leisure for his scientific
pursuits. He used his great influence to do whatever he could
to stimulate scientific life in Berlin and to popularize science
at all levels of Prussian society. He also used his influence
at court to arrange an invitation from the Russian government
to undertake a long-planned expedition to Central Asia during
1829, as a result of which he made important additions to
knowledge in the earth sciences.

The last three decades of his life were spent in revis-
ing his earlier publications and writing his crowning work,
Kosmos, one of the most ambitious scientific popularizations
ever written and one of the most successful.

2568. BAYERISCHE BOTANISCHE GESELLSCHAFT, Regensburg. Founded in 1790.
The first major botanical society. It was very active in the
early nineteenth century and its chief periodical, *Flora*,
continued to be one of the main German periodicals for botany.

1800s

2569. BERLIN ACADEMY. (For its earlier period see 2536) In 1804 its
name became Königliche Akademie der Wissenschaften zu Berlin
(later Königliche Preussische Akademie der Wissenschaften).
Its *Abhandlungen*, begun at that time, continued through the
century as a small periodical and the big expansion took place
with its *Bericht* which began in 1836. After the University of
Berlin was established in 1810 the Academy became closely
associated with it.

2570. GAUSS, Carl Friedrich. 1777-1855. (Mathematics 202. Astronomy
422, 424. Physics 668, 677. Geomagnetism 1050. Geodesy 1053)
One of the greatest of all mathematicians. Born in Brunswick
of poor and uneducated parents. His ability was so impressive
that from the age of fourteen (until he was thirty) he was
granted a pension by the Duke of Brunswick to enable him to
develop his talents. He studied at Göttingen but graduated at
Helmstedt in 1799. In 1801 his reputation was established by
the publication of his monumental *Disquisitiones Arithmeticae*
and by some brilliant astronomical calculations which made
possible the spectacular rediscovery of an asteroid. In 1807,
after declining an offer of a post at the St. Petersburg Academy,

he became professor of astronomy and director of the observatory at Göttingen, a position that he retained for the rest of his life despite offers from Berlin and elsewhere.

2571. RITTER, Johann Wilhelm. 1776-1810. (Physics 637. Chemistry 844) After an apprenticeship as an apothecary he began the medical course at Jena in 1796. He happened to meet A. von Humboldt who encouraged him to undertake research in the new field of galvanism. His publications, from 1798 onward, soon gained him a reputation and in 1801 he was appointed to a post in the ducal court of Gotha. In 1805 he became a member of the Munich Academy.

Ritter was much influenced by the philosophical ideas of Romanticism and the allied *Naturphilosophie* which was becoming a powerful force in Germany in his time. In some respects its emphasis on unities and polarities was stimulating and suggestive for his scientific work, but the extravagant speculations he began to indulge in during the late 1800s did harm to his scientific reputation.

2572. WEISS, Christian Samuel. 1780-1856. (Crystallography 1091) D.Phil. in chemistry and physics, Leipzig, 1800. He then spent two years working on mineral analysis with Klaproth in Berlin and another year studying under Werner at the Freiberg Mining Academy. From 1803 he was a *Privatdozent* at Leipzig and, after travels in France (where he met Haüy and other notables) and elsewhere, he was appointed professor of physics there in 1808. When the University of Berlin was established in 1810 he was invited to become its professor of mineralogy, a position he held for the rest of his life. Weiss was an able teacher and several of his students later became important scientists. He was elected to the Berlin Academy in 1815.

2573. LEONHARD, Karl Cäsar von. 1779-1862. (Geology 1148, 1159) Studied at Göttingen where he was introduced to mineralogy by Blumenbach. He entered the Hessian public service in 1800 but studied mineralogy and geology in his spare time and did much travelling in central Europe. In 1807 he founded the *Taschenbuch für die gesammte Mineralogie* which he edited until his death; it soon became a leading periodical and has continued to the present day (under the title *Neues Jahrbuch für Mineralogie, Geologie und Paläontologie*). Leonhard became a senior official in the Frankfurt bureau of mines in 1809 and (after some disruptions to his career in consequence of the political changes of the time) in 1818 he was appointed professor of mineralogy and geognosy at Heidelberg.

Besides numerous works on mineralogy and geology Leonhard's publications included several popularizations of the geological sciences which contributed to the growth of interest in them.

2574. GAERTNER, Karl Friedrich von. 1772-1850. (Botany 1368) Son of J. Gaertner (2550). M.D., Tübingen, 1796. Practised medicine in Calw while at the same time carrying on his botanical researches. Became a member of the Leopoldina Academy in 1826.

2575. OKEN, Lorenz. 1779-1851. (Zoology 1462) M.D., Freiburg, 1804. *Privatdozent* at Göttingen from 1805. *Extraordinarius* in medi-

cine at Jena from 1807, then professor of natural history there
from 1812. In 1817 he founded the journal *Isis* which, during
the thirty years of its existence, was important in disseminating the biological ideas of the time and stimulating research
and discussion. Oken also provided an organizational focus
for German science generally by his adoption of the idea (pioneered by the Swiss) of national scientific congresses. His
efforts in this direction bore fruit in 1822 with the first
meeting of the Gesellschaft Deutscher Naturforscher und Ärzte.
        Oken lost his chair at Jena in 1819 for political
reasons and did not have an academic appointment until 1828
when he became professor of physiology at Munich but there too
he had difficulties with the authorities. He was finally professor of natural history at Zurich from 1832 until his death.

2576. MECKEL, Johann Friedrich. 1781-1833. (Zoology 1471) M.D.,
      Halle, 1802; he then spent a year or two in Paris where he
      worked with Cuvier, E. Geoffroy Saint-Hilaire, and A. von
      Humboldt. When the University of Halle was re-established in
      1808 after the Napoleonic invasion he was appointed professor
      of anatomy, surgery, and obstetrics, a post he held for life.
            Meckel was a popular teacher, and was influential also
      in his role as editor--from 1815 until his death--of the
      Deutsches Archiv für die Physiologie (later *Archiv für Anatomie
      und Physiologie*), the first journal of its kind. He was dubbed
      by his contemporaries "the German Cuvier" and his most import-
      ant achievement lay in the impetus he gave to the study of
      comparative anatomy in Germany. He also performed a valuable
      service for embryology by reviving the forgotten work of Wolff
      (1660).

1810s

2577. MÖBIUS, August Ferdinand. 1790-1868. (Mathematics 220, 263)
      Studied at Leipzig, Göttingen (under Gauss), and Halle, and
      took his doctorate at Leipzig in 1814. In 1816 he was appointed
      an *extraordinarius* in astronomy at Leipzig and though he receiv-
      ed offers from elsewhere (which he declined because he did not
      want to leave Leipzig) he did not become full professor until
      1844. Though his appointment was in astronomy, as were many
      of his publications, his chief work was done in mathematics.

2578. BESSEL, Friedrich Wilhelm. 1784-1846. (Astronomy 425, 431, 439.
      Mathematics 216) At the age of fifteen he became a clerk and
      for the next few years spent much of his time educating himself
      in various subjects. Becoming especially interested in astron-
      omy he made the acquaintance of Olbers who was impressed by his
      ability and in 1806 recommended him for the post of assistant
      in the well-known private observatory of J.H. Schröter at
      Lilienthal. In 1809, on the recommendation of Humboldt, Bessel
      was appointed director of Prussia's new observatory at Königs-
      berg (the first big German observatory) and professor of astron-
      omy at Königsberg University; since a doctorate was a necessary
      qualification for the latter post he was promptly awarded one,
      on Gauss' recommendation, by Göttingen University.
            In 1811 Bessel was awarded the Lalande Prize by the
      Institut de France and in the following year was elected to

the Berlin Academy. He remained at Königsberg until his death,
working at the observatory and teaching at the university.
Several of his students later became prominent astronomers.

2579. ENCKE, Johann Franz. 1791-1865. (Astronomy 426, 429, 432)
Graduated in 1815 at Göttingen where he was greatly influenced
by Gauss. In 1816, on Gauss' recommendation, he was appointed
assistant at the small observatory of the Duke of Mecklenburg
at Seeberg, near Gotha; by 1822 he had become its director.
In 1825 he was elected to the Berlin Academy and appointed
director of the Berlin Observatory; from 1841 he was also
professor of astronomy at the University. With the powerful
support of Humboldt he was able to procure the establishment
of a new and much better observatory which was completed in
1835. From 1828 until his death he was also the editor of the
important *Berliner Astronomisches Jahrbuch*.

2580. FRAUENHOFER, Joseph. 1787-1826. (Physics 644, 653) Came from
a background of craftsmen associated with the glass and optical
trade; he was apprenticed in the trade and also taught himself
theoretical optics. In 1806 he joined a Munich firm of scientific instrument-makers and gained valuable experience in the
making of high-quality optical glass. His ability brought him
rapid promotion and by 1811 he was one of the partners of the
firm which was in process of becoming the leading manufacturer
of precision optical instruments in Germany. From 1819 he was
an active member of the Munich Academy and in 1823 he was made
director of its "physics cabinet" with the honorary title of
Royal Bavarian professor. In his last years he received various
honours and was elected to several scientific bodies.

2581. GMELIN, Leopold. 1788-1853. (Chemistry 868) Son of J.F. Gmelin
(professor of chemistry at Tübingen). M.D., Göttingen, 1812.
Travelled in Italy and France, and in 1814 became professor of
medicine and chemistry at Heidelberg; from 1818 he was able
to concentrate on chemistry.

2582. BISCHOF, (C.) Gustav (C.). 1792-1870. (Geochemistry 1176)
D.Phil., Erlangen, ca. 1814; then *Privatdozent* in chemistry
and physics. In 1819 he became professor of chemistry and
technology at Bonn, a post he held for the rest of his career.

2583. BREITHAUPT, Johann (F.A.). 1791-1873. (Mineralogy 1093, 1100)
Studied at Jena and then at the Freiberg Mining Academy under
Werner, graduating there in 1813. He was appointed assistant
teacher at the Mining Academy and in 1826 became its professor
of mineralogy (in succession to Mohs). He was an excellent
teacher and had students from many countries.

2584. ALBERTI, Friedrich August von. 1795-1878. (Geology 1162)
Trained as a mining engineer in Stuttgart and from 1815 was
employed at various saltworks, eventually becoming manager of
a major installation.

2585. BURDACH, Karl Friedrich. 1776-1847. (Anatomy 1631. Physiology
1747) M.D., Leipzig, 1799. After practising medicine for ten
years his writings gained him a medical chair at Dorpat in 1811.
Three years later he was appointed professor of anatomy at
Königsberg, a position he held for the rest of his career. In

1817 he succeeded in inducing the University to establish an anatomy institute for research as well as teaching.

2586. BAER, Karl Ernst von. 1792–1876. (Embryology 1667) Born in Estonia (then part of the Russian Empire) of Prussian descent. After gaining his M.D. at Dorpat in 1814 he continued his medical and biological studies at Berlin, Vienna, and Würzburg. In 1817 he was appointed a prosector in anatomy at Königsberg, becoming an *extraordinarius* in 1819, and full professor of zoology in 1826. His best work was done at Königsberg but in 1834 he resigned his chair there and became a member of the St. Petersburg Academy where he remained for the rest of his career, his work there being mostly in anthropology and natural history. He became a foreign member of the Royal Society of London and the Paris Academy, and received awards from both as well as various honours from German bodies.

2587. PANDER, Christian Heinrich. 1794–1865. (Embryology 1663) Born in Latvia (then part of the Russian Empire) of a wealthy German family. Studied medicine at Dorpat, Göttingen, and Würzburg where he took his M.D. in 1817. After some years travelling in western Europe, in the course of which he accumulated many biological and geological observations, he was appointed a member of the St. Petersburg Academy in 1821. He resigned from the Academy in 1827, however, and retired to his estate near Riga. He later published works on geology.

2588. WEBER, Ernst Heinrich. 1795–1878. (Physiology 1745, 1762, 1765) Studied medicine at Wittenberg and Leipzig, and took his M.D. at Wittenberg in 1815. He became a *Privatdozent* and then an *extraordinarius* at Leipzig, and in 1821 was appointed professor of anatomy (and later also of physiology).

1820s

2589. GESELLSCHAFT DEUTSCHER NATURFORSCHER UND ÄRZTE. Founded in 1822 by L. Oken on the model of the Swiss society (2904) as a national body holding annual meetings in different cities. Like the similar societies established later in Britain (2367) and other countries it was highly successful. Moreover it served the important purpose of providing a national focus for the scientific movement in a politically divided country.

2590. STEINER, Jacob. 1796–1863. (Mathematics 229) Came from a poor family and had a very irregular education which did not lead to a degree, though he studied at the universities of Heidelberg and Berlin in the early 1820s. The brilliance of his publications however earned him an honorary doctorate from Königsberg in 1833, election to the Berlin Academy in 1834 and, in the same year, appointment to a special chair in geometry at the University of Berlin which he held for life.

2591. STAUDT, (K.G.) Christian von. 1798–1867. (Mathematics 247, 259) Studied at Göttingen under Gauss and took his doctorate at Erlangen in 1822. He then taught mathematics in secondary schools and from 1827 at the Polytechnische Schule in Nuremberg. In 1835 he was appointed professor of mathematics at Erlangen.

2592. PLÜCKER, Julius. 1801–1868. (Mathematics 221, 233, 245. Physics 702) Studied at several universities and took his doctorate at Marburg in 1824. In the following year he became a *Privatdozent* at Bonn, and in 1828 an *extraordinarius*. After a brief sojourn at Berlin (where he clashed with Steiner) he was appointed professor of mathematics at Halle in 1834. Two years later he returned to Bonn where he was professor of mathematics until 1847 and then professor of physics until his death.

2593. JACOBI, Carl Gustav Jacob. 1804–1851. (Mathematics 223, 230, 238. Mechanics 559) D.Phil., Berlin, 1825. In 1826 he became a *Privatdozent* in mathematics at Königsberg, in 1827 an *extraordinarius*, and in 1832 *ordinarius*. He was a dynamic teacher and created a flourishing school. In 1833, in conjunction with his physicist colleague, F.E. Neumann, he introduced the important institution of the seminar into university mathematics. In later years Jacobi was second only to Gauss as the most outstanding mathematician in Germany.

2594. DIRICHLET, Gustav Peter Lejeune (sometimes known as LEJEUNE-DIRICHLET). 1805–1859. (Mathematics 219, 253, 268. Mechanics 556) Born and educated in Germany. Studied mathematics in Paris and was especially influenced by Fourier. On his return to Germany he became a *Privatdozent* at Breslau, in 1828 an *extraordinarius* at Berlin, and in 1839 *ordinarius* there. He was elected a member of the Berlin Academy in 1831.

Dirichlet was an excellent teacher and through his many students had an important influence on the development of mathematics in Germany. In 1855 he accepted an attractive offer from Göttingen to succeed Gauss.

2595. HANSEN, Peter Andreas. 1795–1874. (Astronomy 438) Became a skilled clockmaker by apprenticeship and educated himself in mathematics and other subjects in his spare time. He made the acquaintance of the Danish astronomer, H.C. Schumacher, who was impressed by his ability and made him his assistant. In his five years with Schumacher Hansen became an able astronomer and in 1825 was appointed director (in succession to Encke) of the small observatory of the Duke of Mecklenburg at Seeberg, near Gotha; he held the post until his death.

Partly because of the inadequacy of the observatory's equipment and partly because of his mathematical ability Hansen concentrated on theoretical astronomy. His achievements in the field won him a high reputation among contemporary astronomers.

2596. ARGELANDER, Friedrich (W.A.). 1799–1875. (Astronomy 437, 441, 450) D.Phil., Königsberg, 1822; two years earlier he had become Bessel's assistant at the Königsberg Observatory. In 1823, on Bessel's recommendation, he was appointed director of the observatory at Åbo (now Turku) in Finland; he transferred to the new observatory at Helsinki in 1832. Four years later he became professor of astronomy at Bonn where he was later provided with a new, well-equipped observatory. In 1863 he was one of the founders of the Astronomische Gesellschaft.

Through his famous star-catalogue Argelander became

one of the best-known astronomers of the nineteenth century.
He was made a member of most of the important academies and
astronomical societies of the world.

2597. OHM, Georg Simon. 1789–1854. (Physics 658) D.Phil., Erlangen,
1811. He became a teacher of mathematics and physics at secondary schools, first in Bamberg and then in Cologne until 1827.
From 1828 he taught at military colleges in Berlin, and in
1833 became professor of physics at the Polytechnische Schule
in Nuremberg. Following the belated recognition of the significance of the electrical researches he had published in 1826–27
(he received the Copley Medal of the Royal Society of London
in 1841) he was appointed a member of the Munich Academy in
1845, becoming the curator of its "physical cabinet" and an
extraordinarius at the University of Munich in 1849; he was
made *ordinarius* three years later.

2598. POGGENDORFF, Johann Christian. 1796–1877. (Physics 652. Bio-
bibliography 48) Originally trained as an apothecary but went
to Berlin in 1820 to study physics and chemistry. In 1823 he
was assisting L.W. Gilbert in editing the *Annalen der Physik
und Chemie* and when Gilbert died the next year he took over
the editorship. By the early 1830s his success as an editor,
together with his researches in experimental physics, had made
him an important figure in the scientific community in Berlin.
He was appointed an *extraordinarius* at the University in 1834
and elected to the Academy in 1839. He received offers of
chairs from several universities but declined them because of
his commitment to his editorial work.

     He continued as editor of *Poggendorff's Annalen*, as it
was invariably known, until his death, and throughout the
long period of his editorship the journal was universally
acknowledged as the leader in its field. In this and other
ways Poggendorff made a major contribution to the development
of the physical sciences in Germany in the second quarter of
the century. His best-known work, his *Handwörterbuch* (48),
was a product of his editorial and bibliographical abilities
combined with his long-standing interest in the history of the
physical sciences.

2599. NEUMANN, Franz Ernst. 1798–1895. (Physics 685) D.Phil., Berlin,
1825. He then became a *Privatdozent* at Königsberg and in 1829
professor of mineralogy and physics there. His early work was
in crystallography but at Königsberg he was influenced by his
colleagues--Bessel, Dove, and Jacobi--and turned to mathematical
physics, becoming the first of Germany's great theoretical
physicists. He was a very influential teacher and many of his
students became important physicists. In 1833, in conjunction
with Jacobi, he initiated the first seminar in mathematical
physics in the German universities.

2600. WEBER, Wilhelm Eduard. 1804–1891. (Physics 688, 699. Physiology 1752) D.Phil., Halle, 1826. After holding junior teaching
positions at Halle he became professor of physics at Göttingen
in 1831. There he did important work on magnetism in collaboration with his senior colleague, Gauss (see 1050a). He was
one of the "Göttingen Seven" who were dismissed from their

chairs by the King of Hanover in 1837 because of their liberal
stance, and was unable to get another university position until
1843 when he became professor of physics at Leipzig. After
the political events of 1848 he was able to return to his
position at Göttingen which he held for the rest of his career.

2601. MITSCHERLICH, Eilhard. 1794-1863. (Chemistry 873. Crystallography 1094) Gained the D.Phil. (Göttingen, 1817?) for work in
oriental languages but then took up chemistry. As a research
student at Berlin he discovered isomorphism in 1819. This
attracted the interest of Berzelius who happened to be passing
through Berlin at the time and Mitscherlich went to Stockholm
to work with him for two years.

On his return to Berlin in 1822 he was appointed professor of chemistry on Berzelius' strong recommendation. He
became a member of the Berlin Academy in the same year and used
its laboratory for practical teaching as well as for his own
research: at that time the Prussian authorities were unwilling
to provide a laboratory in the University for practical teaching or research. He spent much money of his own on apparatus
and the improvement of the laboratory.

Mitscherlich was also professor of chemistry and physics
at the Berlin Military College and, as part of his duties, an
adviser to the Prussian government. Besides his initial work
on isomorphism he did much significant research in inorganic
and organic chemistry and wrote a successful *Lehrbuch der Chemie*.

2602. WÖHLER, Friedrich. 1800-1882. (Chemistry 879, 884) M.D., Heidelberg, 1823. He never practised medicine but took up chemistry,
having studied it under Leopold Gmelin during his medical course.
On Gmelin's advice he spent a year in Stockholm working with
Berzelius. (He later played a major part in disseminating
Berzelius' influence through his translations of most of the
great Swede's works into German--three editions of his massive
textbook and nearly all his annual reports) In 1825 he was
appointed to the new Technical School in Berlin and in 1831 he
transferred to the (also new) Technical School in Cassel. In
1836 he became professor of chemistry at Göttingen.

Wöhler was an outstanding teacher and many of his
students later became important chemists. Though his fame did
grow so rapidly as that of his friend, Liebig, Göttingen became
increasingly important as a centre for chemistry. Wöhler's
own estimates of the number of students attending his lectures
were: 1845-52, 1750; 1853-59, 2950; 1860-66, 3550.

2603. LIEBIG, Justus. 1803-1873. (Chemistry 882, 884, 893. Philosophy
of science 49) Son of a druggist and initially apprenticed to
an apothecary; he was however able to study at Bonn and then
Erlangen where he received the D.Phil. in 1822. With the help
of a grant from the government of Hesse he then went to Paris
where he worked in Gay-Lussac's laboratory (a rare privilege,
made possible by the influence of Humboldt) and absorbed a more
advanced tradition of chemistry than yet existed in Germany.

In 1824, again through the influence of Humboldt, he
was appointed professor of chemistry at the small university
of Giessen (in Hesse). There he established a school for
training students in research which soon became famous and

attracted students from many countries. A large proportion of
the leading chemists of the next generation were his students.
The extraordinary success of the school was due not only to
Liebig's ability, leadership, and teaching methods, but also
to the favorable state of development of chemistry at the time,
socially as well as cognitively. The Giessen school set the
pattern for the teaching of chemistry and other laboratory
sciences in the German universities, and was a major factor
underlying the great expansion of chemistry in Germany.

Liebig was made a baron in 1845 and, after declining
several offers of chairs elsewhere, accepted the chair at
Munich in 1852. There he was chiefly occupied in writing and
lecturing. His popularizations of chemistry and his works on
agricultural chemistry, food chemistry, and nutrition made him
well known to a wide public, and at the time of his death he
was one of the most famous scientists in the world. The *Annalen
der Pharmacie* (later *Annalen der Chemie und Pharmacie*) which
he had founded in 1832 and edited for many years became one of
the chief chemical journals and after his death was renamed
Justus Liebig's Annalen der Chemie.

2604. NAUMANN, Karl Friedrich. 1797-1873. (Crystallography 1096.
Mineralogy 1098. Geology 1179) Studied at the Freiberg
Mining Academy under Werner in 1816-17 and then at Leipzig and
Jena, receiving his doctorate at the latter in 1819. After
travelling in Scandinavia he became a *Privatdozent* at Jena in
1823 and then at Leipzig. In 1826 he was appointed professor
of crystallography at the Freiberg Academy and in 1842 professor
of mineralogy and geognosy at Leipzig. He was a successful
teacher and attracted many students to Leipzig. In 1865 he
was awarded a medal by the Geological Society of London.

2605. DOVE, Heinrich Wilhelm. 1803-1879. (Meteorology 1047) Studied
at Berlin and Breslau and took his doctorate at Königsberg in
1826. He taught physics at Königsberg until 1829 when he was
appointed *extraordinarius* at Berlin; he became full professor
there in 1844 and also taught at the Kriegsschule and the
Gewerbe-Institut. In 1837 he was elected to the Academy and
in the same year founded the *Repertorium der Physik*, a review
of developments in physics which he edited until 1849. He was
director of the Prussian meteorological organization from its
establishment in 1849.

2606. BRONN, Heinrich Georg. 1800-1862. (Palaeontology 1158. Zoology
1494. Evolution 1526) Graduated at Heidelberg, becoming a
Privatdozent there in 1821. Between 1824 and 1827 he travelled
in Italy and France, investigating topics in palaeontology and
geology. In 1828 he was appointed professor of natural science
and technology at Heidelberg and from 1830 he held the influential position of palaeontology editor of Leonhard's *Jahrbuch*
(see 2573). His writings won a valuable prize of the Paris
Academy in 1857 and the Wollaston Medal of the Geological
Society of London in 1861.

2607. RATHKE, Martin Heinrich. 1793-1860. (Embryology 1665) M.D.,
Berlin, 1818. He practised medicine in his native Danzig until
1829 when he was appointed professor of physiology and pathology

at Dorpat. In 1835 he succeeded von Baer as professor of
zoology and anatomy at Königsberg and remained there until
his death. In addition to the researches in embryology for
which he is best known he also did important work in marine
zoology. He was elected a foreign member of the Royal Society
of London in 1855.

2608. MOHL, Hugo von. 1805-1872. (Plant cytology 1690. Microscopy
      1269) M.D., Tübingen, 1828. After studying at Munich and a
      brief period as professor of physiology at Bern he became
      professor of botany at Tübingen in 1835 and held the post for
      the rest of his life. In 1843 he was one of the founders of
      the important journal *Botanische Zeitung*.

2609. PURKYNĚ, Jan Evangelista. 1787-1869. (Physiology 1744. Anatomy
      1634. Embryology 1666) M.D., Prague, 1819; then assistant in
      anatomy there. In 1823 he was appointed professor of physiology
      at Breslau. Reacting against the highly speculative treatment
      of physiology that was dominant in Germany at the time (especi-
      ally under the influence of *Naturphilosophie*), he emphasized
      the experimental approach and the value of practical instruction.
      By 1839, after much campaigning, he was able to persuade the
      University of Breslau to establish a modestly-equipped institute
      for physiology, the first of its kind. (Hitherto he had had
      to do his experimental work in his own home) A number of
      research students gathered around him, some of whom later
      became important physiologists, and his institute played a
      similar role--though on a much smaller scale--to that of
      Liebig's famous laboratory at Giessen.
           In 1850 Purkyně became professor of physiology at
      Prague and thereafter devoted his energies to the dissemination
      and promotion of science in the Czech language and culture.

2610. MÜLLER, Johannes. 1801-1858. (Physiology 1746, 1748. Anatomy
      1633. Zoology 1481) M.D., Bonn, 1822. He then spent three
      semesters at Berlin where he came under the influence of the
      anatomist, Carl Rudolphi, who helped to cure him of his infatu-
      ation with the *Naturphilosophie* of the time. He returned to
      Bonn where he became *Privatdozent*, then *extraordinarius*, and
      in 1830 full professor in the Faculty of Medicine. Following
      the death of Rudolphi he became professor of anatomy and
      physiology at Berlin in 1833.
           Müller was outstanding as a teacher and administrator
      as well as an investigator, and was the key figure in the
      integration of experimental biology into the German university
      system. At Berlin he attracted around him an extraordinarily
      brilliant group of disciples, several of whom became the leaders
      in experimental biology in the next generation--they included
      such names as Henle, Schwann, Virchow, Remak, Kölliker, Du Bois-
      Reymond, and Helmholtz. (Unlike many German professors, Müller
      encouraged his disciples to go their own way, having taught
      them his approach and methods) He also exerted a major influ-
      ence on the field through his *Archiv für Anatomie, Physiologie
      und wissenschaftliche Medicin* (universally known as *Müller's
      Archiv*) which he founded in 1834 and edited until his death.

2611. WAGNER, Rudolph. 1805-1864. (Physiology 1754, 1755. Embryology
      1668) M.D., Würzburg, 1826; he then spent eight months in

Paris studying comparative anatomy with Cuvier. On returning
to Germany he became prosector in anatomy at Erlangen, then
Privatdozent in 1829, and professor of comparative anatomy and
zoology in 1832. In 1840 he went to Göttingen where he succeeded Blumenbach as professor of physiology, comparative anatomy,
and general natural history. He also maintained the connection
with anthropology that Blumenbach had created.

2612. EHRENBERG, Christian Gottfried. 1795-1876. (Microbiology 1803.
Palaeontology 1165, 1183) M.D., Berlin, 1818. From 1820 to
1825, with the financial support of the Berlin Academy, he was
a member of a natural history expedition in Egypt, a valuable
experience which inspired much of his later zoological work.
In 1827 he became a member of the Berlin Academy and an *extraordinarius* at the University. Two years later he accompanied
Humboldt on an eight-month expedition to Siberia. Though he
became a full professor at Berlin in 1839 he did little teaching and concentrated on research which he carried out at the
Academy; he became secretary of the Academy's natural science
section in 1842. Much of his work was in invertebrate zoology
but he became best known for his research on the Infusoria.

1830s

2613. GRASSMANN, Hermann Günther. 1809-1877. (Mathematics 243)
Son of a teacher of mathematics and physics at the Stettin
Gymnasium who wrote textbooks of physics and did some research
in the subject. He studied theology and philology at Berlin
University in the late 1820s but after his return to Stettin
in 1830 the stimulus of his father's interests led him to
begin an intensive study of mathematics and physics. In 1836,
after some junior teaching positions, he was appointed to the
staff of a *Realschule* in Stettin where he taught mathematics,
physics, and other subjects. He continued to hold various
posts in the Stettin school system for the rest of his life.
    Grassmann was elected to the Leopoldina Academy in
1864--for his relatively minor work in physics, not for his
mathematics. Disappointed by the lack of response to his
magnum opus of 1844, he gradually gave up mathematics in the
1860s and turned to linguistics and philology. By the time of
his death he had become an eminent Sanskrit scholar.

2614. KUMMER, Ernst Eduard. 1810-1893. (Mathematics 225, 236, 269)
D.Phil. in mathematics, Halle, 1831. For the following ten
years he taught at a gymnasium while carrying on important
research in mathematics. His publications led to his appointment in 1842 as professor of mathematics at Breslau, and in
1855 he moved to the chair at Berlin. The seminar he established at Berlin became of major importance and, together with his
excellent teaching, attracted large numbers of students from
several countries. (From the mid-1850s mathematics at Berlin
was also strengthened by the presence of Weierstrass and
Kronecker)
    Kummer also taught at the Berlin Kriegsschule and was
one of the secretaries of the Academy. He was also a foreign
member of the Paris Academy, the Royal Society of London, and
other bodies.

2615. MAGNUS, Heinrich Gustav. 1802–1870. (Not in Part 1. A capable
investigator in chemistry and physics, with many publications
to his credit, but important mainly for his institutional role
in physics) D.Phil., Berlin, 1827. He then worked for a year
with Berzelius in Stockholm, and spent another year in Paris
where he worked for a time with Gay–Lussac. In 1831 he became
a *Privatdozent* at Berlin, in 1834 an *extraordinarius*, and in
1845 professor of technology and physics. He was elected to
the Academy in 1840 and was a leading figure in the establishment of the Physikalische Gesellschaft in 1843.

    Magnus was a popular and influential teacher of physics
who initiated at Berlin the sort of development that was already
taking place on a larger scale in chemistry at Giessen under
Liebig. He began by allowing the most promising of his students
(who included Clausius, Helmholtz, and the Englishman, Tyndall)
to work in his private laboratory in his own home. "As the
number of students increased, the private laboratory became
more and more inadequate; the university began to give financial
aid and the private establishment grew into a regular university
institution. By this process the private laboratory of Magnus
evolved into the physical laboratory of the University of Berlin,
which was opened in 1863." (F. Cajori) Magnus was succeeded
in 1871 by the famous Helmholtz for whom the University provided
an enlarged institute.

2616. BUNSEN, Robert Wilhelm. 1811–1899. (Chemistry 892, 906, 920)
Son of a Göttingen professor. D.Phil., Göttingen, 1830. With
financial assistance from the Hanoverian government he travelled
in several countries during 1830–33, meeting many leading
scientists. After junior appointments at Göttingen and Cassel,
he was professor of chemistry at Marburg from 1838. In 1852
he was appointed to the chair at Heidelberg where he remained
for the rest of his career. Being provided with adequate
facilities he was able to build up an important school of
research. In both teaching and research, however, he avoided
theory and concentrated on experiment and analysis.

2617. COTTA, (C.) Bernhard von. 1808–1879. (Geology 1184a) After
studying at the Freiberg Mining Academy he did research at
Heidelberg under von Leonhard, graduating in 1832. He then
taught at a forestry school while conducting important research
in geology, and in 1842 was appointed a professor at the Freiberg Academy. He remained there for the rest of his career
and was outstanding as a teacher and a writer of influential
treatises and textbooks.

    Cotta was one of the founders of the Deutsche Geolog-
ische Gesellschaft in 1848 and later became an influential
adviser to the government on mining affairs in Saxony and an
honorary member of various scientific societies.

2618. ENTOMOLOGISCHER VEREIN ZU STETTIN. Founded in 1837. Though a
provincial society it was quite important.

2619. SIEBOLD, Carl Theodor Ernst von. 1804–1885. (Zoology 1485, 1490.
Embryology 1673) M.D., Berlin, 1828. He then practised medicine while at the same time doing much zoological research. In
1841 his numerous publications, together with the support of

Humboldt, gained him the chair of zoology at Erlangen. He moved to Freiberg i.B. in 1845, to Breslau in 1850, and finally to Munich in 1853. In 1848, together with Kölliker, he founded the *Zeitschrift für wissenschaftliche Zoologie* which became one of the main journals for experimental biology.

2620. HENLE, (F.G.) Jakob. 1809–1885. (Anatomy 1636) M.D., Bonn, 1832. He had studied under Müller at Bonn and in 1833 became his assistant at Berlin and later a *Privatdozent*. The research he did at Berlin established him as a leader in the new field of histology, and his involvement in the editing of Müller's *Archiv* gave him a valuable vantage point in experimental biology generally. From 1840 he was professor of anatomy at Zurich, from 1844 at Heidelberg, and from 1852 until his death at Göttingen. He was one of the founders, in 1856, of the *Bericht* (later *Jahresberichte*) *über die Fortschritte der Anatomie und Physiologie*. In the latter part of his career his work was largely in pathology.

2621. BISCHOFF, Theodor Ludwig Wilhelm. 1807–1882. (Embryology 1670) M.D., Heidelberg, 1832. In the following year he became a *Privatdozent* in physiology at Bonn, and in 1843 was appointed professor of anatomy and physiology at Heidelberg.

2622. SCHLEIDEN, Jacob Mathias. 1804–1881. (Cell theory 1687. Botany 1366) After practising as a lawyer for a few years he gave up the profession and in 1833 began to study science, especially botany, at Göttingen. He then went to Berlin where he worked in Müller's laboratory and met Schwann. He obtained his D.Phil. at Jena in 1839 and, being wealthy, was able to devote himself to research. From the beginning of his career he was an enthusiastic microscopist and he did much to encourage the use of the microscope in biological research.

Schleiden became a highly successful author and lecturer in popular science, and several collections of his lectures in book form, such as his *Die Pflanze und ihr Leben*, went through many editions and translations. He declined an offer of a chair at Giessen in 1846 but four years later accepted a titular professorship at Jena. He left Jena in 1862 and continued his career as a writer and popular lecturer.

2623. SCHWANN, Theodor. 1810–1882. (Cell theory 1688. Physiology 1751) M.D., Berlin, 1834. For the next five years he was Müller's assistant at Berlin and during the period he achieved some brilliant results. After 1839, however, he did no more important research. During his Berlin period he became acquainted with Schleiden whose lead he followed in the formulation of the cell theory. He became professor of anatomy at Louvain from 1839 and then at Liège from 1848.

1840s

2624. HESSE, Ludwig Otto. 1811–1874. (Mathematics 267) Graduated at Königsberg in 1840 and became a *Privatdozent* in mathematics and in 1845 an *extraordinarius*. He was professor of mathematics at Heidelberg from 1856, and in 1868 accepted the chair at the newly-established Polytechnicum at Munich.

2625. WEIERSTRASS, Karl (T.W.) 1815-1897. (Mathematics 237) Qualified as a secondary teacher in 1841 and thereafter taught (mainly mathematics and physics) at various schools. In his spare time he worked at a project he had conceived for giving mathematical analysis the greatest possible rigour. A paper he published in 1854 created much interest and won him an honorary doctorate from the University of Königsberg and, together with subsequent papers, an appointment as professor at the Gewerbeinstitut in Berlin and concurrently *extraordinarius* at the University; in the same year he was elected to the Academy. He became *ordinarius* in 1864 and was a highly respected teacher. He was associated with Kummer in developing the important mathematical seminar at Berlin.

2626. EISENSTEIN, Ferdinand (G.M.). 1823-1852. (Mathematics 244) A precocius genius who came from a poor family and was a chronic invalid. In 1844, while a first-year student at the University of Berlin, he published twenty-five papers, many of them of high importance. His work was warmly praised by Gauss and he was befriended by Humboldt who arranged financial support for him. Early in 1845 he was awarded an honorary doctorate by the University of Breslau. From 1847 he was a *Privatdozent* at Berlin and in 1852, a few months before his death he was elected to the Academy.

2627. KRONECKER, Leopold. 1823-1891. (Mathematics 251) Studied at Berlin, Bonn, and Breslau, and took his doctorate at Berlin in 1845. For the next ten years he ran his family's business but continued his interest in mathematics. His financial abilities were such that in 1855 he was able to dispose of the business and return to academic life in Berlin as a private scholar of independent means. His numerous publications soon after his return led to his election to the Academy in 1861 and he later became active and influential in its affairs. Membership of the Academy carried with it the right to lecture at the University and in this capacity he taught there for twenty years, declining an offer of the prestigious chair at Göttingen. In 1883 he succeeded Kummer in the Berlin chair.

2628. SPÖRER, Gustav Friedrich Wilhelm. 1822-1895. (Astronomy 455) Studied mathematics and astronomy at Berlin, graduating in 1843. After working with Encke at the Berlin Observatory he became a gymnasium teacher in 1846 but continued astronomical research. In 1874 he was appointed an observer at the new astrophysical observatory at Potsdam, and later rose to a senior position there.

2629. PHYSIKALISCHE GESELLSCHAFT ZU BERLIN. Founded in 1843, developing out of a physics colloquium held by G.H. Magnus, professor of physics at Berlin. In 1877 it took over the famous *Annalen der Physik* (whose origins went back to 1790) and in 1899 it became the Deutsche Physikalische Gesellschaft.

2630. MAYER, Julius Robert. 1814-1878. (Physics 680. Physiology 1761) M.D., Tübingen, 1838. After some travelling he settled in his native Heilbronn as a medical practitioner. He received hardly any recognition for his 1842 statement of the conservation of energy until the 1860s. Thereafter however his priority

was acknowledged and he was universally acclaimed, being elected
a corresponding member of the Paris Academy in 1870 and receiving the Copley Medal of the Royal Society of London in 1871.

2631. HELMHOLTZ, Hermann von. 1821-1894. (Physics 690, 717. Astrophysics 448. Physiology 1768, 1774. Popularization 50)
M.D., Berlin, 1842. He had an early inclination to physics
and during his student years did much informal study in mathematics and physics. On graduation he became an army physician
and was stationed in Berlin where he was able to continue his
physiological research in Müller's school. He became professor
of physiology at Königsberg in 1849, at Bonn in 1855, and at
Heidelberg (where he was provided with an institute) in 1858.
After completing his great work in the physiology of the senses
he turned to physics and in 1871 was offered the chair of
physics at Berlin (with a new institute).

By the 1880s Helmholtz had become one of the leading
scientists in the world and was well known to the educated
public for his popular (but often profound) lectures, which
in book form were widely read, and for his philosophical writings. His career coincided with, and exemplified, the great
expansion of the German universities and their committment to
research. And he ended his career as the director of a new
kind of research institution—state-endowed but non-teaching—
the Physikalisch-Technische Reichsanstalt (2729).

2632. CLAUSIUS, Rudolf. 1822-1888. (Physics 691, 710) D.Phil., Halle,
1847. After teaching at a military college in Berlin for a
few years he was appointed in 1855 professor of mathematical
physics at the new Polytechnikum in Zurich. He became professor of physics at Würzburg in 1867 and at Bonn in 1869 where
he remained for the rest of his life. In his later years he
received many honours.

2633. KIRCHHOFF, Gustav Robert. 1824-1887. (Physics 686, 704) D.Phil.,
Königsberg, 1847; then a *Privatdozent* at Berlin, and from 1850
an *extraordinarius* at Breslau. In 1854 he became professor of
physics at Heidelberg where he remained until 1875 (by which
time ill-health was making experimental work difficult) when
he accepted the chair of theoretical physics at Berlin.

2634. HITTORF, Johann Wilhelm. 1824-1914. (Physical chemistry 913.
Physics 714) D.Phil., Berlin, 1846. After some junior positions he became professor of chemistry and physics at the
new University of Münster in 1852 (physics only from 1876).

2635. WILHELMY, Ludwig Ferdinand. 1812-1864. (Physical chemistry 910)
A pharmacist whose interest in science led him in 1843 to give
up his business (as his financial success enabled him to do)
and study physics and chemistry at Berlin, Giessen, and Heidelberg. He received the doctorate at Heidelberg in 1846 and
after working in Paris with Regnault he returned to Heidelberg
where he was a *Privatdozent* from 1849 to 1854. He then left
academic life but continued his interest in science.

2636. FRESENIUS, Carl Remegius. 1818-1897. (Chemistry 894) Initially
an apprentice apothecary. Studied at the University of Berlin
and while still a student devised the system of qualitative

analysis which was his great contribution to chemistry and the foundation of his career. He published it in 1841 and then went to Giessen to study under Liebig who promptly adopted his system. It gained him his doctorate in 1842 and a post as assistant to Liebig and *Privatdozent*.

In 1845 Fresenius became professor at the Wiesbaden Agricultural College where three years later he established a laboratory which gave training in practical chemistry, especially analysis, and also carried out analyses on request. It soon became highly successful in both roles. In 1862 he established the important *Zeitschrift für analytische Chemie* and his Wiesbaden laboratory eventually acquired an international reputation as a centre for analytical chemisty.

2637. HOFMANN, August Wilhelm. 1818-1892. (Chemistry 898, 911) Studied under Liebig at Giessen, gaining his doctorate in 1841 and becoming Liebig's assistant. In 1845, on Liebig's recommendation to Prince Albert, he became professor of chemistry at the Royal College of Chemistry in London. There he introduced the Liebig system of laboratory teaching and created a flourishing school of organic chemistry. Several of his English pupils became distinguished chemists--some of them the progenitors of the synthetic dyestuffs industry and the coal-tar distillation industry. Hofmann was a prominent member of the Chemical Society and was elected to the Royal Society in 1851.

He returned to Germany in 1865 to take up the chair of chemistry at Berlin where he built up a large research school. He became one of the leaders of German chemistry and the chief founder in 1868 of the Deutsche Chemische Gesellschaft and was its president for many years.

2638. KOLBE, (A.W.) Hermann. 1818-1884. (Chemistry 895, 924) Studied at Göttingen under Wöhler (through whom he met Berzelius who had a lifelong influence on his thinking). In 1842 he went to Marburg as Bunsen's assistant and took his doctorate there. During 1845-47 he was in England, acting as a consultant on Bunsen's method of gas analysis, and there he became a close friend of Frankland who also had an influence on his later theories. From 1851 he was professor of chemistry at Marburg (in succession to Bunsen) and then, from 1865 until his death, at Leipzig where he had the biggest and best-equipped chemical institute of the time.

As well as being an outstanding practical chemist and discoverer, Kolbe was an excellent teacher and wrote some successful textbooks. From 1870 he was editor of the important *Journal für praktische Chemie* and, in his editorial role, was notorious for the violence of his criticisms.

2639. DEUTSCHE GEOLOGISCHE GESELLSCHAFT. Founded in 1848. The seat of the Society was in Berlin but its annual congresses were held in different cities.

2640. NÄGELI, Karl Wilhelm von. 1817-1891. (Botany 1369, 1380, 1385. Cytology 1689, 1699. Microscopy 1273) Born and educated in Zurich. He began the study of medicine at Zurich (and attended Oken's lectures on zoology) but gave it up for botany which he studied at Geneva under Alphonse de Candolle, obtaining his doctorate in 1840. After working with Schleiden at Jena during 1842–44 he returned to Zurich where he became a *Privatdozent* and later an *extraordinarius*. He was appointed professor of botany at Freiburg i.B. in 1852 and five years later moved to Munich where he remained for the rest of his career.

One of the most outstanding botanists of the nineteenth century, Nägeli was remarkably versatile, making important contributions to plant anatomy and cytology as well as to morphology and systematics.

2641. PRINGSHEIM, Nathanael. 1823–1894. (Botany 1376, 1387) Studied at Breslau, Leipzig, and Berlin where he took his doctorate in 1848. After further study in Paris and London he became a *Privatdozent* at Berlin in 1851. In 1858 he founded the *Jahrbücher für wissenschaftliche Botanik* which he continued to edit until his death. He was elected to the Berlin Academy in 1860 and accepted the chair of botany at Jena in 1864 on condition that the University establish a new botanical institute, which it did. He resigned the post in 1868 when he inherited his father's fortune, and settled in Berlin where he continued his research in a private laboratory. He was one of the chief founders of the Deutsche Botanische Gesellschaft in 1882 and served as its president until his death.

2642. HOFMEISTER, Wilhelm (F.B.). 1824–1877. (Botany 1373, 1375, 1386) Had no university education. Learnt systematic botany from his father, a keen amateur botanist, but was completely self-taught in the field of the structure and function of plants in which his later work was done. He entered his father's business in 1841 and a few years later began research in his spare time. By 1851 his reputation was such that the University of Rostock awarded him an honorary doctorate. In 1863 he was appointed professor of botany at Heidelberg and in 1872 he moved to Tübingen. He was an excellent research director and many of his students later achieved distinction.

"The scientific career of Wilhelm Hofmeister is unique ... In Germany the amateur has taken a much less important place [than in England]. Yet Hofmeister, entirely without academic training, not only won for himself the leading position among German botanists, but also came to hold an important university chair. The consensus of expert opinion places him among the very greatest of modern botanists." (C.Singer)

2643. DEUTSCHE ORNITHOLOGEN-GESELLSCHAFT. Founded in 1845.

2644. LEUCKART, (K.G.F.) Rudolf. 1822–1898. (Zoology 1486, 1501) M.D., Göttingen, 1845. *Privatdozent* in zoology there, 1847. In 1850 he was appointed an *extraordinarius* in zoology at Giessen, becoming a full professor a few years later. He moved to Leipzig in 1869 and in 1880 was provided with a well-equipped institute which, together with his excellence as a teacher, made Leipzig an important centre for zoological research. Many of his students later held chairs in Germany and other countries. In 1888 he founded the journal *Bibliotheca zoologica* (later entitled *Zoologica*). In his later years Leuckart received many honours from state authorities and the scholarly world.

2645. KOELLIKER, Rudolf Albert. 1817–1905. (Anatomy 1638. Embryology 1669, 1672) Born and educated in Zurich. Studied medicine at Zurich, Bonn, and Berlin (at the last under Müller, Henle, and Remak). For some independent microscopical research he was awarded the D.Phil. at Zurich in 1841. Having completed his medical studies he took the M.D. at Heidelberg in 1842 and soon afterwards became assistant to Henle at Zurich. He became a *Privatdozent* in 1843 and an *extraordinarius* in the following year. In 1847 he was appointed professor of physiology and anatomy at Würzburg, a post he retained until 1902 (though he gave up the chair of physiology in 1864). In 1848 he and von Siebold founded the *Zeitschrift für wissenschaftliche Zoologie* which became a leading journal for experimental biology.

Koelliker was an outstanding teacher and many of his students later became important biologists (notably Haeckel and Gegenbauer). His treatises and textbooks were highly regarded and very influential. In addition to his books he published about three hundred papers and was still engaged in research in his eighties. He received many honours.

2646. REMAK, Robert. 1815–1865. (Embryology 1671) M.D., Berlin, 1838. Being a Jew he was ineligible for a teaching position so he maintained himself by practising medicine while doing part-time research in Müller's laboratory, and from 1843 in the laboratory of J.L. Schönlein, the professor of medicine. In 1847, as a result of appeals to the King of Prussia by Schönlein (who was also the King's physician) and Humboldt, he was made a *Dozent*--the first Jew to teach at the University of Berlin. Nevertheless his career lay with his medical practice, though --in a belated and inadequate recognition of his achievements-- he was made an *extraordinarius* in 1859.

2647. LUDWIG, Carl (F.W.). 1816–1895. (Physiology 1759, 1766) M.D., Marburg, 1840; then prosector in anatomy there and in 1846 *extraordinarius*. He became professor of anatomy and physiology at Zurich and then, in 1855, at the army medical academy in Vienna. In 1865 he was appointed to the new chair of physiology at Leipzig and there designed a famous physiological institute (completed in 1869) which served as a model for other universities. Ludwig was one of the greatest teachers and research directors in the history of physiology, and students from many countries were attracted to his institute. Nearly two hundred of them later became prominent. Through his own work and that of his students he was one of the chief founders of modern physiology.

2648. DU BOIS-REYMOND, Emil. 1818–1896. (Physiology 1756. Various writings 64) M.D., Berlin, 1843. From 1840 he was working with Müller and became a prominent member of the brilliant group which gathered around him. He was also a member of the new Physikalische Gesellschaft and his interest in physics had a determining influence on his approach to physiology. While continuing his research he held a minor teaching post in Berlin from 1848, and in 1855 he became an *extraordinarius*.

When Müller died in 1858 his chair of anatomy and physiology was divided into two--one chair for anatomy and one for physiology. (A similar division was made in several German universities in the late 1850s and the 1860s and constituted a major step in the institutionalization of physiology as a separate discipline) Du Bois-Reymond was appointed to the new chair of physiology (which he held for the rest of his life). It was not until 1877 however that adequate provision was made for research, with the erection of a new institute. Thereafter his research school flourished and many of his students later became important physiologists.

His other activities included the editorship of the important *Archiv für Physiologie* (a descendant of Müller's *Archiv*) and the duties arising from his appointment in 1876 as one of the secretaries of the Berlin Academy. From about 1870 he became a prominent and often controversial public figure with his lectures and eloquent speeches on the cultural relations of science and the philosophical and ideological issues of the time.

2649. COHN, Ferdinand Julius. 1828–1898. (Microbiology 1809) D.Phil. in botany, Berlin, 1847. *Privatdozent* at Breslau from 1850, *extraordinarius* from 1859, and *ordinarius* from 1872. In 1866, after much agitation, he managed to persuade the University of Breslau to establish an institute for plant physiology, the first in the world. He remained its director until his death.

Until the late 1860s Cohn's researches dealt mainly with primitive plants, especially microscopic algae and fungi, but from that time he concentrated on bacteriology (then regarded as part of botany). The journal *Beiträge zur Biologie der Pflanzen* which he founded in 1870 to publish the work coming from his institute contained many of the pioneering investigations in bacteriology, including Koch's classic paper of 1876. (Cohn was largely responsible for establishing Koch in his scientific career) In the 1880s, when bacteriology became a medical specialty, he returned to conventional botany.

1850s

2650. RIEMANN, (G.F.) Bernhard. 1826–1866. (Mathematics 255, 274) Studied at Göttingen (under Gauss and Weber) and Berlin (under Jacobi and Dirichlet) and took his doctorate at Göttingen in 1851. After being an assistant to H. Weber at Göttingen he became a *Privatdozent* there in 1854, an *extraordinarius* in 1857, and *ordinarius* (in succession to Dirichlet) in 1859. In the last four years of his tragically short life he was severely afflicted by illness. He is now generally regarded as one of the greatest and most creative of nineteenth-century mathematicians.

2651. CHRISTOFFEL, Elwin Bruno. 1829–1900. (Mathematics 262) D.Phil., Berlin, 1856; after further studies he became a *Privatdozent* there in 1859. He was appointed professor of mathematics at the Polytechnicum in Zurich in 1862, then at the Gewerbesakademie in Berlin in 1869, and finally at the newly-constituted University of Strasbourg in 1872.

2652. DEDEKIND, (J.W.) Richard. 1831–1916. (Mathematics 282, 303) Graduated at Göttingen (where he studied under Gauss) in 1852

and after two years of further study qualified as a *Privatdozent*.
He became professor of mathematics at the Polytechnicum in
Zurich in 1858 and then, in 1862, at the Polytechnicum in his
native city of Brunswick where he remained for the rest of his
life, being quite satisfied with the post. He was an eminent
mathematician of great originality and in his later years
received many honours.

2653. CLEBSCH, Rudolf (F.A.). 1833-1872. (Mathematics 272, 283)
Graduated at Königsberg in 1854. After further study at Berlin
he became a *Privatdozent* there in 1858; later in the same year
he was appointed professor of theoretical mechanics at the
Polytechnicum in Karlsruhe. From 1863 he was professor of
mathematics at Giessen and from 1868 until his death he held
the chair at Göttingen. In 1868, in conjunction with Carl
Neumann, he founded the important *Mathematische Annalen*.

2654. KEKULÉ, (F.) August. 1829-1896. (Chemistry 918, 937) Began
the study of architecture at Giessen but through Liebig's
influence was attracted to chemistry. After working with
Liebig for his doctorate he went to Paris for a year in 1851-52.
There he got to know the leading French chemists, especially
Gerhardt whose theories he studied. After a stay in Switzerland he went to London in late 1853 as assistant to the lecturer
in chemistry at St. Bartholomew's Hospital. He remained there
for two years and got to know many English chemists, especially
Odling and Williamson who, like Gerhardt, had a big influence
on the development of his ideas.
    Returning to Germany, Kekulé became a *Privatdozent* at
Heidelberg in 1855 where his association with the brilliant
young experimentalist, Adolf von Baeyer, was mutually profit-
able. He became professor of chemistry at Ghent in 1858 and
then, in 1867, at Bonn where he remained for the rest of his
life and created a flourishing school.

2655. MEYER, (J.) Lothar. 1830-1895. (Chemistry 934) M.D., Würzburg,
1854. Being interested in physiological chemistry he went to
Heidelberg to learn Bunsen's methods of gas analysis which he
then used to investigate the gases of the blood--a remarkable
piece of research which earned him a D.Phil. at Breslau in 1858.
After being a *Privatdozent* at Breslau from 1859 he became
professor of chemistry at the Karlsruhe Polytechnic in 1868
and then, from 1876 until his death, at Tübingen.

2656. BAEYER, (J.F.W.) Adolf von. 1835-1917. (Chemistry 921) D.Phil.,
Berlin, 1858. He then worked with Kekulé, initially at Heidelberg and then at Ghent. From 1860 he taught at the Gewerbe-
Schule in Berlin, where he created an important research school,
and then became professor of chemistry at Strasbourg in 1872.
From 1875 until the end of his career he held the chair at
Munich where he established a famous school of research. His
many students included several of the most important chemists
of the next generation. He was awarded the Nobel Prize in 1905.

2657. HOPPE-SEYLER, (E.) Felix (I.). 1825-1895. (Biochemistry 908,
952) M.D., Berlin, 1851. He practised medicine for a time
but about 1855 obtained a position in Virchow's new pathology
institute at Berlin. Virchow made him head of the chemical

laboratory and encouraged his researches in the new field of
physiological chemistry. In 1860 he was appointed an *extraordinarius* in the Berlin medical faculty and in the following
year professor of applied chemistry in the medical faculty at
Tübingen. From 1872 until his death he was professor of physiological chemistry at Strasbourg where he built up an important
research school.

      Hoppe-Seyler acquired an international reputation as
the outstanding advocate of physiological chemistry (now called
biochemistry) as an academic discipline in its own right,
distinct from the traditional field of physiology. One of his
chief and most effective efforts towards this end was his establishment in 1877 of the *Zeitschrift für physiologische Chemie*.
After his death it was renamed *Hoppe-Seyler's Zeitschrift....*

2658. SCHWENDENER, Simon. 1829-1919. (Botany 1388, 1392. Microscopy
     1273) Born and educated in Switzerland. D.Phil., Zurich, 1856.
     He then became Nägeli's assistant at Munich, qualifying as a
     Privatdozent there in 1860. He was appointed professor of
     botany at Basel in 1867, at Tübingen in 1877, and at Berlin in
     1878. His appointment to the Berlin Academy followed in 1880
     and two years later he was one of the founders of the Deutsche
     Botanische Gesellschaft.

2659. DE BARY, (H.) Anton. 1831-1888. (Botany 1377, 1384, 1395.
     Microbiology 1815) M.D., Berlin, 1853. He then became a
     Privatdozent in botany at Tübingen and was for a short time
     assistant to von Mohl. He was appointed professor of botany
     at Freiburg i.B. in 1855, at Halle in 1867, and at Strasbourg
     in 1872. He had many students, especially at Strasbourg where
     he was provided with a well-equipped institute.

2660. SACHS, Julius von. 1832-1897. (Plant physiology 1381, 1389,
     1402) Left an orphan in his youth in Breslau, he was befriended
     by Purkyně who, when he moved to Prague, took Sachs with him
     and made him his assistant. Sachs was thereby enabled to gain
     his D.Phil. at Prague in 1856, and in the following year he
     became a *Privatdozent* there in plant physiology. From 1859 he
     taught botany in agricultural colleges, and in 1868 became
     professor of botany at Würzburg where he remained for the rest
     of his life, despite attractive offers from elsewhere. He was
     a brilliant teacher and research director, and had many students.

      At the time when Sachs began his research plant physiol-
ogy had been neglected for many years. He revived the field
and initiated its subsequent rapid expansion. As its unrivalled
leader he achieved international renown, with many honours.

2661. ENTOMOLOGISCHER VEREIN, BERLIN. Founded in 1856.

2662. MÜLLER, Fritz (J.F.T.). 1822-1897. (Zoology 1511. Evolution
     1529) D.Phil., Berlin, 1844. He then studied medicine at
     Greifswald until 1849 but was not allowed to take the *Staats-
examen* because of his participation in the revolution of 1848.
     As a result of the harsh treatment he received he migrated to
     Brazil in 1852. After following various occupations there he
     was appointed travelling naturalist for the National Museum
     in Rio de Janeiro in 1876, a post he held until 1891.

2663. MÖBIUS, Karl August. 1825–1908. (Zoology 1503) Studied the biological sciences at Berlin University until 1853 and a few years later joined the staff of the Hamburg Museum of Natural History. He was one of the founders of the Hamburg zoo in 1863 and five years later was appointed professor of zoology at Kiel. In 1887 he became director of the new natural history museum in Berlin.

2664. GEGENBAUER, Carl. 1826–1903. (Zoology 1493, 1502, 1507, 1520) M.D., Würzburg (where he studied under Virchow, Leydig, and Koelliker), 1851; then a *Privatdozent* in zoology. In 1856 he was appointed an *extraordinarius* in zoology (in the medical faculty) at Jena; subsequently as a full professor he occupied various chairs there, finally that of anatomy. In 1873 he became professor of anatomy at Heidelberg where he remained for the rest of his career.
      Gegenbauer had many students and was the founder of a major school of research in the evolutionary interpretation of comparative anatomy. In 1875 he founded the *Morphologisches Jahrbuch* which became the leading journal in the subject.

2665. HIS, Wilhelm. 1831–1904. (Anatomy 1640, 1647. Embryology 1675, 1676, 1681) Born and educated in Basel. Studied medicine at Basel, Bern, Berlin (under Müller and Remak), and Würzburg (under Virchow, Leydig, and Koelliker), and took his M.D. at Basel in 1855. After visiting scientific centres in Vienna and Paris he was appointed professor of anatomy at Basel in 1857. In 1872 he moved to Leipzig where he was provided with a fine institute. As well as being a leading figure in the Leipzig Academy and the Gesellschaft Deutscher Naturforscher und Ärzte he was one of the founders of the Anatomische Gesellschaft in 1886 and later its president.

2666. SCHULTZE, Max (J.S.). 1825–1874. (Cytology 1692) M.D., Greifswald (where his father was professor of anatomy), 1849; then a *Privatdozent*. From 1854 he was an *extraordinarius* at Halle and in 1859 was appointed professor of anatomy and director of the anatomical institute at Bonn where he remained for the rest of his life. In 1865 he founded the *Archiv für mikroskopische Anatomie und Entwicklungsmechanik* which he edited until his death.

2667. FICK, Adolf (E.). 1829–1901. (Physiology 1769) Studied medicine at Marburg (physiology under Carl Ludwig) and Berlin (where he became friendly with Helmholtz and Du Bois-Reymond), and took his M.D. at Marburg in 1851. In the following year he went with Ludwig to Zurich where he was Ludwig's prosector; in 1855 he became an *extraordinarius* there, and in 1862 professor of physiology. From 1868 until his retirement in 1899 he was professor of physiology at Würzburg. He was provided with a new institute there in 1883.

2668. PFLÜGER, Eduard (F.W.). 1829–1910. (Physiology 1770, 1775) Studied medicine at Marburg and Berlin (physiology under Müller), graduating at Berlin in 1855. After further research with Müller and Du Bois-Reymond he was appointed professor of physiology at Bonn (in succession to Helmholtz) in 1859. In

1868 he founded the *Archiv für die gesammte Physiologie des Menschen und der Thiere* which he continued to edit until his death. It became a leading journal in the field and after his death was renamed *Pflügers Archiv*....

2669. VOIT, Carl von. 1831-1908. (Physiology 1773) Graduated in medicine at Munich in 1854. During his course he had attended Liebig's lectures on chemistry and his subsequent career was much influenced by Liebig's writings on "animal chemistry." After graduating he spent a year studying chemistry under Wöhler at Göttingen. In 1856 he became an assistant at the physiological institute at Munich, in 1859 a *Privatdozent*, and in 1863 professor of physiology and director of the institute. During the 1860s much of his research was done in collaboration with Max Pettenkofer, professor of medical chemistry at Munich, and in 1865 they founded the *Zeitschrift für Biologie*. The success of the methods they devised for the study of metabolism led to Voit's institute becoming the acknowledged centre for research in the field.

2670. KÜHNE, Wilhelm Friedrich. 1837-1900. (Physiology 1777. Biochemistry 922) D.Phil., Göttingen, 1856. He then worked with several distinguished physiologists at Jena, Berlin, Paris, and Vienna. In 1861 he was appointed to a post in the biochemical department of Virchow's pathology institute at Berlin. He was then professor of physiology at Amsterdam from 1868 and at Heidelberg from 1871 until his retirement in 1899.

1860s

2671. GORDAN, Paul Albert. 1837-1912. (Mathematics 272, 275) Studied at Breslau, Königsberg, and Berlin, and took his doctorate at Breslau in 1862. He then worked with Clebsch at Giessen and became professor of mathematics there in 1867 (in succession to Clebsch). From 1874 until his retirement he held the chair at Erlangen.

2672. PASCH, Moritz. 1843-1930. (Mathematics 294) D.Phil., Breslau, 1865. After further study at Berlin he became a *Privatdozent* at Giessen in 1870, an *extraordinarius* in 1873, and *ordinarius* in 1875. He remained at Giessen until his retirement in 1911.

2673. CANTOR, Georg. 1845-1918. (Mathematics 276, 296, 311) Studied at Zurich, Berlin (under Weierstrass, Kummer, and Kronecker), and Göttingen, and took his doctorate in 1867. He became a *Privatdozent* at Halle in 1869, an *extraordinarius* in 1872, and *ordinarius* in 1879. He remained at Halle for the rest of his life. Cantor is now regarded as one of the great mathematicians of the nineteenth century but for many years his work was controversial and its significance was not fully recognized until the turn of the century.

He was also notable for his efforts to promote the exchange of ideas among mathematicians: in 1890 he was the founder and first president of the Deutsche Mathematiker-Vereinigung, and he was the prime mover in the inauguration of the first international congress of mathematicians, held in Zurich in 1897.

2674. ASTRONOMISCHE GESELLSCHAFT. Founded in Leipzig in 1863.

2675. VOGEL, Hermann Carl. 1841-1907. (Astronomy 468, 482) Studied
at Leipzig and in 1865 became an assistant in the university
observatory. He took his doctorate at Jena in 1870 and in
the same year was appointed the director of a well-equipped
observatory belonging to a wealthy amateur astronomer. There
he had complete freedom and was able to do some valuable
research in the new field of astrophysics. In 1874 he was
appointed to the staff of the new Astrophysikalisches Observa-
torium at Potsdam, and in 1882 became its director. Until
his retirement he held this important post with distinction--
both as regards his own research and his administration and
development of the observatory.

2676. KUNDT, August Adolph. 1839-1894. (Physics 711, 726) D.Phil.,
Berlin, 1864. After some years as Magnus' assistant at Berlin
he became a *Privatdozent* there in 1867. He was appointed
professor of physics at the Zurich Polytechnic in 1868, at
Würzburg in 1869, at Strasbourg in 1872, and finally at Berlin
(in succession to Helmholtz) in 1888.

2677. KOHLRAUSCH, Friedrich (W.G.). 1840-1910. (Physics 715. Physical
chemistry 944) Son of R.H.A. Kohlrausch (professor of physics
at Marburg, then Erlangen). Studied at Marburg, Erlangen, and
Göttingen where he received his D.Phil. in 1863 and became an
extraordinarius three years later. He was professor of physics
at the Zurich Polytechnic from 1870, at the Darmstadt Polytechnic from 1871, at Würzburg from 1875, and at Strasbourg from
1888. In 1895 he was appointed director of the Physikalisch-
Technische Reichsanstalt (in succession to Helmholtz) and in
the same year was elected to the Berlin Academy.

2678. DEUTSCHE CHEMISCHE GESELLSCHAFT. Founded in Berlin in 1867.
By the late nineteenth century its *Berichte* ranked with the
Comptes Rendus of the Paris Academy as the biggest of all
scientific periodicals.

2679. WISLICENUS, Johannes. 1835-1902. (Chemistry 947) D.Phil.,
Halle, 1859. After some junior appointments he became professor of chemistry at Zurich (from 1867 at the University and
from 1870 at the Polytechnic as well), then at Würzburg from
1872, and finally at Leipzig from 1885.

2680. SOHNCKE, Leonhard. 1842-1897. (Crystallography 1104) Studied
mathematics and physics at Halle and Königsberg, taking his
doctorate at the latter in 1866. After some years as a *Privatdozent* at Königsberg he became professor of physics at the
Technische Hochschule in Karlsruhe from 1871, then at the
University of Jena from 1883, and finally at the Technische
Hochschule in Munich from 1886. He was a good teacher and
several of his students later became important physicists.

2681. ZIRKEL, Ferdinand. 1838-1912. (Geology 1188, 1193) D.Phil.
in geology, Bonn, 1861. During the following two years he
worked with Haidinger at the Geologische Reichsanstalt in
Vienna. He became professor of mineralogy and geology at
Lemberg in 1863, then at Kiel in 1868, and finally at Leipzig
in 1870.

2682. ZITTEL, Karl Alfred von. 1839–1904. (Palaeontology 1198, 1213. "The Linnaeus of palaeontology") D.Phil., Heidelberg, 1860. After some travels and a year in Vienna working at the Geologische Reichsanstalt he became professor of mineralogy at the Karlsruhe Technische Hochschule in 1863. Three years later he was appointed professor of geology and palaeontology at Munich where he further enlarged the already extensive palaeontological collections. These, together with his abilities as a teacher and research director, made Munich a leading centre for palaeontology. His *Geschichte der Geologie und Paläontologie* (1899; English trans., 1901) was highly regarded.

2683. HELMERT, Friedrich Robert. 1843–1917. (Geodesy 1072) D.Phil. in geodesy, Leipzig, 1867. After taking part in some triangulations in Saxony and working at the Hamburg astronomical observatory he was appointed professor of geodesy at the new Polytechnikum in Aachen in 1870. In 1877 he became a member of the advisory council of the Prussian Geodetic Institute and in 1886 he was appointed director of the Institute. In the following year he was also appointed professor of "higher geodesy" at the University of Berlin. He was elected to the Academy in 1900 and received many other honours.

2684. ABBE, Ernst. 1840–1905. (Microscopy 1274, 1275) Studied physics at Jena and then at Göttingen where he graduated in 1861. He became a *Privatdozent* in physics and astronomy at Jena in 1863 and an *extraordinarius* in 1870. In 1866 he began a collaboration with Carl Zeiss, the owner of a small Jena firm making optical instruments. Under Abbe's technical direction and Zeiss' commercial management the firm flourished, and in 1876 Abbe became a partner (retaining his position at the University on an honorary basis).

Continued technical developments, including important new kinds of optical glass, brought the firm to a position of world leadership. Following Zeiss' death, Abbe became sole owner in 1891 and established the Carl Zeiss Foundation, reconstituting the firm as a co-operative with the managers, workmen, and the University of Jena sharing the profits.

2685. BREFELD, (J.) Oscar. 1839–1925. (Botany 1390) D.Phil. in chemistry, Heidelberg, 1864. He soon gave up chemistry and turned to his other main interest, botany. From 1868 to 1870 he worked on fungi with De Bary at Halle, and after periods of further research at Munich and Würzburg went to Berlin in 1873, becoming a *Privatdozent* in botany there two years later. From 1878 he was professor of botany at a forestry academy and from 1884 head of the botanical gardens and institute at Münster. In 1898 he became professor of botany at Breslau.

2686. EICHLER, August Wilhelm. 1839–1887. (Botany 1394, 1403) D.Phil., Marburg, 1861. He then became a private assistant of the distinguished botanist, K.F. von Martius (formerly director of the Munich botanical gardens) whom he helped to edit his monumental *Flora Brasiliensis*; concurrently (from 1865) he was a *Privatdozent* at the University of Munich. In 1872 he became professor of botany at Kiel and six years later professor of systematic and morphological botany at Berlin and director of the University's herbarium and of the Royal Botanical Gardens at Schoeneberg. He was elected to the Academy in 1880 and received numerous other honours.

2687. ENGLER, (H.G.) Adolf. 1844-1930. (Botany 1397, 1404, 1405) D.Phil., Breslau, 1866. After a few years school teaching at Breslau he was appointed in 1871 curator of the collections at the State Botanical Institute in Munich, under the direction of Nägeli (from whom he learnt much). From 1878 he was professor of botany at Kiel, and from 1884 at Breslau. In 1880, while at Kiel, he founded the important journal *Botanische Jahrbücher für Systematik, Pflanzengeschichte und Pflanzengeographie* which he continued to edit for the rest of his life.

In 1889 he was appointed to the prestigious post of professor of botany and director of the botanical gardens at Berlin, a position he held with distinction until his retirement in 1921. The new botanical garden which he developed at Dahlem (near Berlin) became one of the most outstanding in the world, closely aligned (like its British exemplar at Kew) with imperial expansion and colonial development.

2688. PFEFFER, Wilhelm (F.P.). 1845-1920. (Plant physiology 1396, 1400) D.Phil. in chemistry and botany at Göttingen, 1865. After a period as a pharmacist he decided to go on with botany and in 1869 worked with Pringsheim at Berlin and then with Sachs at Würzburg. Sachs encouraged him to go into the expanding field of plant physiology (where his earlier training, which included chemistry and physics, would be helpful). He became a *Privatdozent* at Marburg in 1871, an *extraordinarius* at Bonn in 1873, and professor of botany at Basel (1877), then at Tübingen (1878), and finally at Leipzig (1887). He was a good teacher and had many students, foreign as well as German.

2689. DEUTSCHE MALAKOZOOLOGISCHE GESELLSCHAFT. Founded in 1868.

2690. HAECKEL, Ernst. 1834-1919. (Zoology 1500. Evolution 1531, 1549) Studied medicine at Würzburg (under Koelliker and Virchow), Berlin (under Müller who had a major influence on him), and Vienna, taking his M.D. at Berlin in 1857. After a zoological expedition to the Mediterranean he became a *Privatdozent* in comparative anatomy at Jena in 1861, an *extraordinarius* in zoology in the following year, and *ordinarius* and director of the zoological institute in 1865. This post he held until his retirement in 1909. Though he inspired many students he did not create a school.

Haeckel became well known as the outstanding German advocate of (or propagandist for) the Darwinian theory of evolution. His speculative elaborations of it tended to alienate his fellow scientists but had an immense appeal to the general public. His many popularizations had a vast influence throughout the civilized world, especially around the end of the century.

2691. DOHRN, (F.) Anton. 1840-1909. (Zoology 1508) Came from a wealthy and cultured family; his father was a distinguished entomologist. He studied at several universities, especially Jena where he worked under Haeckel and Gegenbauer and was habilitated in 1868. An academic career did not appeal to him however and in 1871 he settled permanently in Naples where three years later he established the subsequently famous Zoological Station for research in marine biology. The finance for its erection was donated by his father and by the Berlin Academy, while the running expenses and research costs came from the scientists who used it (or the universities, academies, etc., which supported them). The Naples Zoological Station became an international research institute of the greatest value and the prototype for later laboratories for marine biology in several other countries.

2692. HENSEN, (C.A.) Victor. 1835-1924. (Marine biology 1077) Studied medicine at Würzburg, Berlin, and Kiel where he graduated in 1859. He then became a *Privatdozent* in anatomy at Kiel, in 1864 an *extraordinarius* in physiology, and in 1868 *ordinarius*; he held the chair until his retirement in 1911.

Through most of his career Hensen carried on parallel researches in two different fields--physiology and marine biology. His contributions to the former were substantial though not remarkable, while his contributions to the latter were more original. From 1871 his work in the latter field (which initially was concerned with commercial fisheries) was carried on in association with the newly-formed Kommission zur Wissenschaftlichen Untersuchungen der Deutschen Meere, of which he later became president.

2693. WEISMANN, August. 1834-1914. (Evolution and heredity 1538, 1541, 1548) M.D., Göttingen, 1856. For some years he practised medicine while at the same time carrying on research in zoology, but in 1863 he gave up medicine and became a *Privatdozent* in zoology at Freiburg i.B. He was appointed an *extraordinarius* there in 1866, and *ordinarius* in 1874, retaining the position until his retirement in 1912. He was an able teacher and research director and his reputation attracted many students to his institute.

From the early 1860s Weismann was an advocate of the Darwinian theory of evolution and did much to spread a knowledge of it in Germany. Around the end of the century he was recognized as the most outstanding of the neo-Darwinians and a leading theorist of biology. He received many honours, from civil authorities as well as from the scholarly world.

2694. FLEMMING, Walther. 1843-1905. (Cytology 1697) Studied medicine at several universities and took his M.D. at Rostock in 1868. After researches in comparative anatomy in several places he became a *Privatdozent* at Rostock in 1872, and then moved to Prague and later to Königsberg. In 1876 he was appointed professor of anatomy and director of the anatomical institute at Kiel.

2695. STRASBURGER, Eduard Adolf. 1844-1912. (Botany 1406. Plant cytology 1694, 1698) Studied botany at Bonn and then at Jena as Pringsheim's assistant, graduating in 1866. At Jena he was much influenced by his association with Haeckel. He became *extraordinarius* in botany there in 1869, and *ordinarius* two years later. In 1880 he was appointed to the chair at Bonn which he held for the rest of his career. Under his direction

Bonn's botanical institute became the leading centre for plant cytology, attracting students from many countries. In his later years he was honoured by the Prussian government and received many awards from scientific bodies and universities.

2696. HERING, (K.) Ewald (K.). 1834-1918. (Physiology 1772, 1778)
M.D., Leipzig, 1860. He went into medical practice and from 1862 was concurrently a *Privatdozent* in physiology at Leipzig. In 1865 he was appointed professor of physiology and medical physics at the Medico-Surgical Academy in Vienna. From 1870 he was professor of physiology at Prague, and from 1895 at Leipzig.

2697. ENGELMANN, Theodor Wilhelm. 1843-1909. (Physiology 1782, 1796)
Studied medicine at several universities and graduated at Leipzig in 1867. He then went to Utrecht as an assistant to the professor of physiology, F.C. Donders; he became an *extra-ordinarius* there in 1871 and succeeded Donders in 1888, having declined offers from three other universities. In 1897 he became professor of physiology at Berlin (in succession to Du Bois-Reymond).

Engelmann's researches in animal physiology were numerous and varied, and at times in his career he also did notable work on the physiology of micro-organisms and on photosynthesis.

1870s

2698. MATHEMATISCHE GESELLSCHAFT IN HAMBURG. Its origins as a society for popular mathematics went back to 1690. Its *Mittheilungen* began in 1873.

2699. FREGE, (F.L.) Gottlob. 1848-1925. (Mathematics 289, 298, 309)
Studied at Jena and Göttingen, and took his doctorate at the latter in 1873. He became a *Privatdozent* in mathematics at Jena in 1874, an *extraordinarius* in 1879, and *ordinarius* in 1896. He taught in all branches of mathematics but his publications were confined almost entirely to mathematical logic. Their significance was not generally recognized until after his death.

2700. FROBENIUS, Georg Ferdinand. 1849-1917. (Mathematics 290, 304)
Studied at Göttingen and Berlin, and took his doctorate at the latter in 1870. After teaching in secondary schools he became an *extraordinarius* at Berlin in 1874 and in the follow-ing year professor of mathematics at the Polytechnikum in Zurich. In 1892 he was appointed to the chair at Berlin.

2701. KLEIN, (C.) Felix. 1849-1925. (Mathematics 280, 295, 306)
D.Phil., Bonn, 1868. After further study at Göttingen, Berlin, and Paris, he qualified as a *Privatdozent* at Göttingen in 1871. In the following year he was appointed to the chair of mathe-matics at Erlangen. From 1875 he was professor at the Tech-nische Hochschule in Munich, from 1880 at Leipzig, and from 1886 (until his retirement in 1913) at Göttingen.

As one of the leading mathematicians of the time, a stimulating teacher, and editor of the *Mathematische Annalen*, Klein did much from the late 1880s to strengthen Göttingen's position as a leading centre for the mathematical sciences.

From 1895 he took a prominent part in the production of the great *Encyklopädie der Mathematischen Wissenschaften* (316) and in the early twentieth century became widely known as an advocate of the modernization of the teaching of mathematics in schools.

2702. SEELIGER, Hugo. 1849–1924. (Astronomy 477) Studied at Heidelberg and Leipzig, taking his doctorate in astronomy at the latter in 1871. After holding positions at the observatories of Leipzig, Bonn, and Gotha, he was appointed professor of astronomy and director of the observatory at Munich in 1882. He served at various times as president of the Astronomische Gesellschaft and the Munich Academy.

2703. RÖNTGEN, Wilhelm Conrad. 1845–1923. (Physics 758) Born in Germany but educated in Holland and Switzerland. Graduated in mechanical engineering at the Zurich Polytechnikum in 1868, and in the following year gained a D.Phil. and became assistant to the professor of physics, A. Kundt. He later accompanied Kundt to Würzburg and then to Strasbourg. In 1879 he became professor of physics at Giessen and ten years later moved to Würzburg where he was provided with an institute. His sensational discovery of X-rays in 1895 brought him great fame and many honours, including the Nobel Prize in 1901. In 1900 he became professor of physics at Munich.

2704. GOLDSTEIN, Eugen. 1850–1930. (Physics 728, 742) Studied at Breslau and Berlin (under Helmholtz), taking his doctorate at the latter in 1881; by then he had already (in 1878) been appointed a physicist at the Potsdam Observatory. He spent his career at the Observatory, finally becoming head of its astrophysical section in 1927.

2705. WALLACH, Otto. 1847–1931. (Chemistry 958) D.Phil., Göttingen, 1869. After a brief period at Berlin he went to Bonn as Kekulé's assistant, becoming a *Privatdozent* there in 1873 and an *extraordinarius* three years later. In 1889 he was appointed professor of chemistry at Göttingen and remained there until his retirement in 1915. He received the Nobel Prize in 1910.

2706. MEYER, Victor. 1848–1897. (Chemistry 943, 986) D.Phil., Heidelberg, 1867. After a year as Bunsen's assistant there he spent three years with Baeyer in Berlin. He then became professor of chemistry, first at Stuttgart Polytechnic in 1871, then at Zurich Polytechnic in 1872, and finally (after a brief stay at Göttingen) at Heidelberg in 1888. A brilliant and popular teacher, he attracted many students and had large research schools at Zurich and Heidelberg.

2707. CLAISEN, Ludwig. 1851–1930. (Chemistry 951, 987) Studied at Bonn (under Kekulé) and Göttingen (under Wöhler), graduating at Bonn in 1874. His career consisted of a succession of short-term posts, mostly minor: 1875–81, assistant in the Bonn laboratory; 1882–85, lecturer at Owens College, Manchester; 1886–90, *Privatdozent* in Baeyer's institute in Munich; 1890–97, professor at the Aachen Technische Hochschule; 1897–1904, *extraordinarius* at Kiel. From 1904 to 1907 he was an honorary professor in Emil Fischer's institute in Berlin, and in the latter year he retired but continued research in a laboratory of his own until 1926.

2708. FISCHER, Emil (H.). 1852–1919. (Chemistry 964, 997) Studied
at Bonn (under Kekulé) and Strasbourg (under Baeyer), taking
his doctorate at the latter in 1874. In the following year
he accompanied Baeyer to Munich where he became a *Privatdozent*
in 1878 and an *extraordinarius* in the following year. He was
appointed professor of chemistry, first at Erlangen in 1882,
then at Würzburg in 1885, and finally at Berlin in 1892. He
had links with the dyestuffs industry and received several
offers of highly-paid posts, all of which he declined.
       Fischer received the Nobel Prize in 1902 and is consid-
ered by many to have been the greatest of all organic chemists.

2709. BECKMANN, Ernst Otto. 1853–1923. (Chemistry 973) D.Phil.,
Leipzig, 1878. After some minor teaching posts he became
professor of chemistry at Erlangen from 1892 and then at Leipzig
from 1897. In 1912 he was appointed director of the new Kaiser
Wilhelm Institut für Chemie in Dahlem.

2710. OSTWALD, (F.) Wilhelm. 1853–1932. (Physical chemistry 971,
977) Born in Riga (Latvia) of German descent. Graduated at
Dorpat in 1875 and became a *Privatdozent* there. From 1881 he
was professor of chemistry at the Riga Polytechnic and in 1887
was appointed to the chair of physical chemistry at Leipzig
(then the only chair of its kind). He was an inspiring teacher
as well as an able organizer and publicist, and the research
school which he created at Leipzig soon became the leading
centre for physical chemistry; many of his students later
gained chairs in the new discipline. The *Zeitschrift für
physikalische Chemie* which, in association with Van't Hoff,
he launched in 1887 was the standard bearer for the new subject.
       Ostwald retired in 1906 and devoted the rest of his
life to his philosophical and humanistic interests (including
the history of science). He was one of the most famous scientists of his time and was awarded the Nobel Prize in 1909.

2711. GROTH, Paul (H.). 1843–1927. (Mineralogy and crystallography
1103) Was introduced to mineralogy at the Freiberg Mining
Academy and then studied it and related subjects at Berlin,
obtaining his D.Phil. in 1868. After holding some junior
academic posts at Berlin he was professor of mineralogy at
Strasbourg from 1872 and at Munich from 1883 until his retirement in 1924. At Strasbourg he established an important
collection of minerals, and at Munich his academic appointment
also included the directorship of the long-standing Bavarian
State Collection which he enlarged.
       In 1877 Groth founded the *Zeitschrift für Kristallog-
raphie und Mineralogie* which he continued to edit until 1920.
He was a member of many academies and societies and received
numerous honours and awards.

2712. GOLDSCHMIDT, Victor. 1853–1933. (Crystallography 1106, 1110)
Graduated at the Freiberg Mining Academy in 1874 and for a few
years held a junior teaching position there. He then did
research at the universities of Munich, Prague, and Heidelberg,
obtaining his doctorate at the last in 1880. Being financially
independent he was able to continue full-time research, which
he did at Vienna from 1882 to 1887. In 1888 he settled in
Heidelberg where he became a *Privatdozent*, then an *extraordin-*

arius, and finally an honorary full professor. He created a
research school in crystallography at Heidelberg which attracted
students from several countries.

2713. ROSENBUSCH, (K.) Heinrich (F.). 1836-1914. (Geology 1194)
D.Phil., Freiburg, 1869; then *Privatdozent* there. In 1873 he
became an *extraordinarius* in mineralogy and petrography at
Strasbourg, and in 1878 professor of mineralogy at Heidelberg.
From 1888 until his retirement in 1908 he was also director
of the geological survey of Baden. His reputation as the
leading exponent of the new field of microscopical petrography
attracted students from many countries to his research school
at Heidelberg.

2714. PENCK, Albrecht. 1858-1945. (Geology 1202. Geomorphology
1212) After studying at Leipzig he took his D.Phil. at Vienna
in 1878. He then worked with von Zittel at Munich where he
became a *Privatdozent*. In 1885 he was appointed professor of
physical geography at Vienna where he built up a thriving
institute and had many students. From 1906 until his retirement in 1926 he held the prestigious chair at Berlin where he
was held in high public esteem. In his Berlin period he turned
from physical geography to regional, demographic, and political
geography--an area in which he became very influential.

2715. PREUSSISCHES GEODÄTISCHES INSTITUT, Potsdam. Its periodical
began in 1870.

2716. DEUTSCHE SEEWARTE, Hamburg. Its periodical began in 1878.

2717. SCHIMPER, Andreas Franz Wilhelm. 1856-1901. (Botany 1409)
D.Phil., Strasbourg, 1878. After working with Sachs at Würzburg he travelled in 1881-83 to the United States, the West
Indies, and South America--an experience which aroused his
interest in plant geography. In 1883 he became a *Privatdozent*
in plant physiology at Bonn, and in 1886 an *extraordinarius*.
During this period he was closely associated with Strasburger
at his famous institute in Bonn. In the late 1880s he made
further trips to tropical countries and in 1898 was a member
of the German scientific expedition on the *Valdivia* which
visited many parts of the world. In 1899 he became professor
of botany at Basel.

2718. ZOOLOGISCHE STATION ZU NEAPEL. The first research institute for
marine biology. It was established at Naples in 1874 by German
initiative and with German finance; see A. Dohrn (2691).

2719. WIEDERSHEIM, Robert. 1848-1923. (Zoology 1517) M.D., Würzburg,
1872; he became a *Privatdozent* in comparative anatomy there,
working with Koelliker until 1876. In that year he was appointed an *extraordinarius* in anatomy at Freiburg, and in 1883 he
became *ordinarius* and director of the anatomy institute there.
His renown as a teacher of comparative anatomy attracted many
students to his institute from several countries.

2720. HERTWIG, (W.A.) Oscar. 1849-1922. (Embryology 1678, 1680, 1683.
Cytology 1705) Brother of Richard Hertwig and with him studied
medicine at Jena, and especially zoology under Haeckel by whom
they were both greatly influenced. He graduated in 1872 and

after holding junior posts at Jena and Bonn was appointed an *extraordinarius* in anatomy at Jena in 1878 and *ordinarius* three years later. In 1888 he became the first professor of cytology and embryology at Berlin, and director of the new anatomical-biological institute.

2721. HERTWIG, (K.W.T.) Richard. 1850-1937. (Embryology 1680)
Brother of Oscar Hertwig and with him studied medicine at Jena, and especially zoology under Haeckel by whom they were both greatly influenced. He became a *Privatdozent* in zoology at Jena in 1872 and an *extraordinarius* in 1878. He was then appointed professor of zoology at Königsberg in 1881, at Bonn in 1883, and finally at Munich in 1885.

2722. ROUX, Wilhelm. 1850-1924. (Embryology 1684) M.D., Jena, 1878.
In the following year he became a *Privatdozent* in anatomy at Breslau and later an *extraordinarius*. He was appointed prof-essor of anatomy at Innsbruck in 1889, and in 1895 moved to Halle as director of the University's anatomical institute, a post he held for the rest of his career. Roux was a vigorous advocate of the new field of experimental embryology which he had initiated, and to further it he founded the *Archiv für Entwicklungsmechanik der Organismen* in 1894. In his later years he received many honours from the scholarly world.

2723. BÜTSCHLI, Otto. 1848-1920. (Cytology 1695) Studied mineralogy at the Karlsruhe Polytechnic and then at Heidelberg where he gained his D.Phil. in the subject in 1868. Subsequently, however, he decided to take up zoology and studied it at Leipzig during 1869. He then began research in the subject, first at Kiel for two years and then at Frankfurt for three years. In 1876 he became a *Privatdozent* at the Karlsruhe Polytechnic and two years later was appointed professor of zoology at Heidelberg.

2724. PHYSIOLOGISCHE GESELLSCHAFT ZU BERLIN. Founded in 1875.

2725. RUBNER, Max. 1854-1932. (Physiology 1790) M.D., Munich, 1878.
He then took part in nutritional research in Voit's institute as an unpaid assistant, and in 1880-81 worked in Ludwig's institute in Leipzig. He became a *Privatdozent* at Munich in 1883, an *extraordinarius* at Marburg in 1885, and professor of hygiene (which he regarded as applied physiology) at Berlin in 1891, transferring to the chair of physiology in 1909.

2726. KOCH, (H.H.) Robert. 1843-1910. (Microbiology 1811, 1813, 1818)
M.D., Göttingen, 1866. Practised medicine in several places-- from 1872 in a town in Prussian Poland where he did research in his own house. In 1876 he described his investigations of anthrax to the eminent botanist, Cohn, who was impressed with them and arranged for their publication (see 2649). Further important papers followed and in 1880 Koch was appointed to the new Imperial Department of Health in Berlin. There he was provided with a laboratory and assistants and commissioned to develop bacteriology in relation to public health. From that time onward he trained many pupils who, as his fame spread, flocked to his laboratory from many countries.
    In 1885 Koch accepted appointment to a new chair of

hygiene at Berlin University while retaining honorary membership of the Department of Health. In conjunction with Carl
Flügge (director of the hygiene institute in Göttingen) he
founded the *Zeitschrift für Hygiene* in 1886. A new institute
for infectious diseases was established for him in 1891 and he
continued as its director until 1904.

From the 1880s Koch was one of the most famous scientists in the world and the many honours he received were
crowned with the award of the Nobel Prize in 1905.

1880s

2727. HILBERT, David. 1862-1943. (Mathematics 300, 314. His work
in analysis, the foundations of mathematics, and mathematical
physics was done after 1900) D.Phil., Königsberg, 1885.
Privatdozent at Königsberg, 1886; *extraordinarius*, 1892;
ordinarius, 1893. In 1895 he was appointed to a chair at
Göttingen which he held until his retirement in 1930. As one
of the greatest mathematicians of the time he attracted many
students and created a strong school, further strengthening
the Göttingen mathematical tradition.

2728. MINKOWSKI, Hermann. 1864-1909. (Mathematics 312. His contributions to the theory of relativity were made after 1900)
D.Phil., Königsberg, 1885. After teaching at Bonn he returned
to Königsberg in 1894. From 1896 he was professor of mathematics at Zurich and in 1902 was appointed to a new chair at
Göttingen (which Hilbert had procured for him).

2729. PHYSIKALISCH-TECHNISCHE REICHSANSTALT. The first national research institute, founded in 1887 for pure and applied physics.
As a non-teaching establishment with a staff engaged in full-
time research using expensive equipment and facilities it
represented a new kind of institution, the significance of
which for the emerging science-based technologies became
widely recognized. The National Physical Laboratory in Britain
(founded in 1900) and the Bureau of Standards in the United
States (founded in 1901) were modelled upon it.

The chief originator was the technologist and indust-
rialist, Werner von Siemens, who donated the land at Charlottenburg (near Berlin) together with a substantial sum of money;
the rest of the capital cost and the running expenses were
provided by the German government. The first director of the
institute was the world-famous physicist, Helmholtz.

2730. HERTZ, Heinrich Rudolf. 1857-1894. (Physics 743, 744) D.Phil.,
Berlin, 1880; then assistant to Helmholtz for three years.
After being a *Privatdozent* at Kiel from 1883 to 1885 he became
professor of physics at the Technische Hochschule in Karlsruhe.
The work he did there made him famous and in 1888 he accepted
an offer of the chair of physics at Bonn (in succession to
Clausius). His tragic death put an untimely end to a brilliant
career.

2731. PLANCK, Max. 1858-1947. (Physics 763. His monumental achievement in establishing the quantum theory lies beyond the time
limit of the present chronology) D.Phil., Munich, 1879; then

a *Privatdozent* at Kiel and from 1885 an *extraordinarius*. In 1888 he was appointed professor of physics at Berlin (in succession to Kirchhoff) and director of the new institute for theoretical physics. He became a prominent member of the Academy as well as being a foreign member of other leading societies. He was awarded the Nobel Prize in 1918, and from 1930 to 1937 was president of the Kaiser Wilhelm Gesellschaft (now named the Max Planck Gesellschaft).

2732. LENARD, Philipp (E.A.). 1862-1947. (Physics 751) Studied physics at Heidelberg and Berlin, graduating at the former in 1886. After three years as assistant to Quincke at Heidelberg he became assistant to Hertz at Bonn in 1891 and *Privatdozent* there the next year. Following brief periods at Breslau and Aachen he became an *extraordinarius* at Heidelberg in 1896 and *ordinarius* at Kiel in 1898. In 1905 he was awarded the Nobel Prize and two years later was appointed professor of physics at Heidelberg and director of a fine new laboratory there.

2733. DRUDE, Paul Karl Ludwig. 1863-1906. (Physics 757. His electronic theory of metals was developed after 1900) Studied mathematics and physics at Göttingen, Freiburg, and Berlin, and took his doctorate in theoretical physics at Göttingen in 1887; there he was strongly influenced by Voigt, with whom he continued working until 1894; in that year he went to Leipzig as an *extraordinarius*. In 1900 he became professor of physics at Giessen and in the same year took over the editorship of the famous *Annalen der Physik*. He was appointed to the prestigious post of director of the physics institute at Berlin in 1905 but died suddenly a year later.

2734. WIEN, Wilhelm. 1864-1928. (Physics 754) Studied mathematics and physics at Berlin and Heidelberg, graduating at the former in 1886. In 1890 he became an assistant to Helmholtz at the Physikalisch-Technische Reichsanstalt. From 1896 he was an *extraordinarius* in physics at the Technische Hochschule in Aachen, and in 1899 (after a brief sojourn at Giessen) he became *ordinarius* at Würzburg. He was awarded the Nobel Prize in 1911 for his work on heat radiation. From 1920 he held the chair at Munich where he was provided with a new institute.

2735. PASCHEN, (L.C.H.) Friedrich. 1865-1947. (Physics 755. His work on infrared spectral series was done after 1900) Studied physics at Strasbourg and Berlin, graduating at the former in 1888. He was assistant to Hittorf at Münster until 1890 and then to Kayser at the Technische Hochschule at Hanover; there in 1895 he was appointed to a permanent lectureship (an unusual post which carried an adequate income and ample time for research). In 1901 he became professor of physics at Tübingen where, as one of the leading spectroscopists of the time, he built up a large and important research school.

2736. HANTZSCH, Arthur Rudolf. 1857-1935. (Chemistry 981) D.Phil., Würzburg, 1880. After being an assistant in the physical chemistry institute at Leipzig he became professor of chemistry at the Zurich Polytechnic in 1885. In 1893 he moved to Würzburg and in 1903 to Leipzig where he remained until his retirement in 1927.

2737. NERNST, (H.) Walter. 1864-1941. (Physical chemistry 976. His
       famous heat theorem dates from 1906) Studied physics at Zurich
       (under H. Weber), Berlin (under Helmholtz), Graz (under Boltz-
       mann), and Würzburg (under Kohlrausch), graduating at the last
       in 1887. He then became assistant to Ostwald at Leipzig, thus
       entering the new field of physical chemistry. In 1891 he was
       appointed an *extraordinarius* in physics at Göttingen and three
       years later he became Göttingen's first professor of physical
       chemistry; with a new institute for the subject he was able to
       create a flourishing school of research. In 1905 he became
       professor of physical chemistry at Berlin where--following the
       retirement of Ostwald in 1905--his school took the leadership
       in the field.
              In the first quarter of the twentieth century Nernst
       was a prominent and influential figure in the German scientific
       community. He had much to do with the establishment of the
       Kaiser Wilhelm Gesellschaft in 1911, and in 1913 he was instru-
       mental in bringing Einstein to Berlin and creating a post for
       him. He was awarded the Nobel Prize in 1920.

2738. LE BLANC, Max Julius Louis. 1865-1943. (Physical chemistry 982)
       Studied chemistry at Tübingen, Munich, and Berlin, graduating
       at the last in 1888. From 1891 he was an *extraordinarius* in
       Ostwald's institute in Leipzig, from 1896 director of the
       electrochemical division of the Höchst chemical firm, from 1901
       professor of physical chemistry at the Karlsruhe Technische
       Hochschule, and finally, from 1906, professor of physical chem-
       istry at Leipzig (in succession to Ostwald).

2739. KOSSEL, (K.M.L.) Albrecht. 1853-1927. (Biochemistry 959)
       Graduated in medicine at Strasbourg in 1878 and became assis-
       tant to Hoppe-Seyler. In 1883, as a *Privatdozent*, he was
       appointed director of the chemical section of Du Bois-Reymond's
       physiology institute at Berlin, becoming an *extraordinarius*
       four years later. He was professor of physiology at Marburg
       from 1895 and then at Heidelberg from 1901 until his retire-
       ment in 1924. After Hoppe-Seyler's death he edited the *Zeit-
       schrift für physiologische Chemie* for many years. He was
       awarded the Nobel Prize in 1910.

2740. BUCHNER, Eduard. 1860-1917. (Biochemistry 991) Studied under
       Baeyer at Munich and gained his D.Phil. in 1888; two years
       later he became Baeyer's assistant. As a student he had worked
       at Munich's institute for plant physiology under Nägeli and had
       become interested in the subject of alcoholic fermentation.
       With Baeyer however his researches were in classical organic
       chemistry. He resumed his interest in fermentation in his
       subsequent posts: *Privatdozent* and then *extraordinarius* in
       chemistry at Kiel from 1893, *extraordinarius* at Tübingen from
       1896, and full professor at the College of Agriculture in
       Berlin from 1898; concurrently with the latter appointment he
       was director of the Institute for the Fermentation Industry.
              Buchner received the Nobel Prize in 1907 and two years
       later was appointed to the chair of physiological chemistry at
       Breslau. In 1911 he moved to Würzburg.

2741. SCHOENFLIES, Arthur Moritz. 1853-1928. (Crystallography 1108)
       D.Phil. in mathematics, Berlin, 1877. After teaching in

secondary schools he became a *Privatdozent* at Göttingen in 1884 and eight years later an *extraordinarius* in applied mathematics. He was appointed professor of mathematics at Königsberg in 1899 and then at the new University of Frankfurt a.M. in 1914.

2742. PREUSSISCHE GEOLOGISCHE LANDESANSTALT. Its *Jahrbuch* began in 1880.

2743. GEOLOGICAL SURVEY OF BAVARIA. The Bavarian Department of Mines (Oberbergamt), an agency of long standing, began issuing its *Geognostische Jahreshefte* in 1888.

2744. DEUTSCHE METEOROLOGISCHE GESELLSCHAFT. Founded in 1883.

2745. DEUTSCHE BOTANISCHE GESELLSCHAFT. Founded in 1882.

2746. DEUTSCHE ENTOMOLOGISCHE GESELLSCHAFT. Founded in 1881 as an offshoot of the Berlin society (2661).

2747. ANATOMISCHE GESELLSCHAFT. Founded in 1886. Held meetings in different cities.

2748. BOVERI, Theodor. 1862-1915. (Cytology 1701, 1703) D.Phil. in anatomy, Munich, 1885. Through the influence of Richard Hertwig, who had just become director of the zoological institute at Munich, he took up research in cytology, becoming Hertwig's assistant in 1891. Two years later he was appointed professor of zoology and comparative anatomy at Würzburg and director of the zoological institute there. He held the post for the rest of his life, declining some attractive offers from elsewhere.

1890s

2749. DEUTSCHE MATHEMATIKER-VEREINIGUNG. Founded at Halle in 1890.

2750. VEREINIGUNG VON FREUNDEN DER ASTRONOMIE UND KOSMISCHEN PHYSIK, Berlin. Founded in 1891.

2751. RUBENS, Heinrich. 1865-1922. (Physics 760) Studied physics at Strasbourg and Berlin, receiving his doctorate at the latter in 1889. After a period as an assistant at Berlin he was appointed to the staff of the Charlottenburg Technische Hochschule in 1896, becoming full professor there in 1900. In 1906 he became professor of experimental physics at Berlin and director of the physics institute.

2752. DEUTSCHE ELEKTROCHEMISCHE GESELLSCHAFT. Its periodical began in 1894. In 1902 it became the Deutsche Bunsen-Gesellschaft.

2753. THIELE, (F.K.) Johannes. 1865-1918. (Chemistry 996) Studied at Breslau and Halle, obtaining his doctorate at the latter in 1890. In 1893 he was appointed an *extraordinarius* in Baeyer's institute in Munich where his best work was done. From 1902 until his death he was professor of chemistry at Strasbourg. In 1910 he was elected to the Munich Academy and in the same year he became the editor of *Justus Liebigs Annalen der Chemie*.

2754. DEUTSCHE ZOOLOGISCHE GESELLSCHAFT. Founded in 1890. Held its annual meetings in different cities.

2755. CORRENS, Carl (F.J.E.). 1864-1933. (Heredity 1550) Graduated in botany at Munich (having studied under Nägeli who had a

strong influence on him) and after further research became a
Privatdozent in botany at Tübingen in 1892. He was appointed
an *extraordinarius* at Leipzig in 1902 and *ordinarius* at Münster
in 1909. In 1913 he became director of the new Kaiser-Wilhelm
Institut für Biologie, a post he held for the rest of his life.

2756. DRIESCH, Hans (A.E.). 1867-1941. (Embryology 1686) Came from
a wealthy family. Studied zoology at Freiburg (under Weismann)
and Jena (under Haeckel), taking his doctorate at the latter
in 1889. Because of his wealth he did not need an academic
appointment and through the 1890s he travelled extensively.
It was in this period that he did his most important work--
chiefly at the Zoological Station at Naples. In 1900 he
settled in Heidelberg, continuing his embryological experiments
in his private laboratory until 1909 when he finally entered
the academic world--as a philosopher.

In 1912 Driesch became an *extraordinarius* in philosophy
at Heidelberg, and in 1919 he was appointed *ordinarius* in
philosophy at Cologne and then, from 1921, at Leipzig. He
gave public lectures in several countries and became a well-
known and influential figure.

2757. VERWORN, Max. 1863-1921. (Physiology 1799) D.Phil. in zoology,
Berlin, 1887. M.D., Jena, 1889. Until 1901 he remained in
the physiology institute at Jena, first as an assistant, then
from 1891 as a *Privatdozent*, and from 1895 as an *extraordinarius*.
He became professor of physiology at Göttingen in 1901 and then,
in 1910, at Bonn. At both universities he had an active research school, with many students from various countries. In
1902 he founded the *Zeitschrift für allgemeine Physiologie.*

2.05 NORTHERN NETHERLANDS

(HOLLAND)

1530s

2758. LONGOLIUS, Gisbert. 1507–1543. (Zoology 1412) Studied in Italy and obtained his M.D. there. Practised medicine in Deventer and became city physician. He was later in Cologne where he was physician to the Archbishop.

1580s

2759. CEULEN, Ludolph van. 1540–1610. (Mathematics 111) Teacher of mathematics in several Dutch towns, and finally, from 1600, at a new engineering school in Leiden.

2760. STEVIN, Simon. 1548–1620. (Mathematics 107. Mechanics 498) Born in Bruges and in his early years was employed in the financial world of Bruges and Antwerp. From 1581 he was living in Leiden. He matriculated at the University of Leiden in 1583 (at the age of thirty-five) but nothing is known of his studies there. Eventually he became a prominent engineer and organized a school for engineers in Leiden. From 1604 he was quartermaster-general of the Dutch army and he also served in some other senior administrative capacities. He was the author of many books on a variety of topics and wrote nearly all of them in Dutch, on principle.

2761. BOTANICAL GARDEN, LEIDEN. Established by the University in 1587. However it was not planted until 1594—after Clusius (L'Ecluse) had been appointed professor of botany.

1590s

2762. BLAEU, Willem Janszoon. 1571–1638. (Astronomy 357) Worked with Tycho Brahe in Uraniborg in 1595–96. He then settled in Amsterdam where he became a famous maker of maps and globes. From 1633 he was official cartographer to the Dutch East India Company.

..... L'ECLUSE, C. de. Professor of botany at Leiden from 1593. See 1978.

1600s

2763. SNEL (or SNELL), Willebrord. 1580–1626. (Physics 581) Travelled widely in his early years and worked for a short time with Tycho Brahe in Prague where he also met Kepler. He took an M.A. at Leiden in 1608 and three years later became professor of mathematics there.

1610s

2764. BEECKMAN, Isaac. 1588-1637. (Not in Part 1) Studied philosophy
and theology at Leiden and later graduated in medicine but
never practised it. He became a schoolteacher, finally the
rector of an important school in Dordrecht.
Beeckman was a thinker (and experimenter) of extraordinary insight into the issues that were central to the Scientific Revolution of the early seventeenth century, especially
in mechanics. He published nothing, however, being content
to jot down his ideas and experimental results in his journal
(or diary). Nevertheless he had a considerable effect on the
course of scientific events, above all through his influence
on Descartes. As a young man at the beginning of his career,
Descartes met him in 1618 and was introduced by him to a range
of problems in applied mathematics and, in general, to the
fruitfulness of a mathematical approach to physical problems.
Descartes retained a high regard for him and visited him again
in Dordrecht in the late 1620s, as did Gassendi and Mersenne.
Seven years after his death a selection of his notes was
published by his brother.

1620s

2765. GIRARD, Albert. 1595-1632. (Mathematics 124) A native of
Lorraine who settled in Holland and became a military engineer.

1630s

2766. LEIDEN OBSERVATORY. The first institutional astronomical observatory; established at the University of Leiden in 1632. Its
main instrument was a seven-foot quadrant formerly belonging
to Snel.

2767. GLAUBER, Johann Rudolph. 1604-1670. (Chemistry 789) Born in
Germany of humble origin. He did not attend a university and
was apparently largely self-taught. For several years he
travelled and worked in Germany and other countries as an
apothecary, chemical technologist, and alchemist. About 1639
he left Germany because of the Thirty Years War and settled in
Amsterdam where he had a large and well-equipped laboratory
and where most of his books were written. Except for a short
return to Germany after the end of the war he remained in
Amsterdam until his death.

1640s

2768. SCHOOTEN, Frans van. ca. 1615-1660. (Mathematics 145) Had a
good training in mathematics at Leiden University (where his
father was professor of mathematics) in the 1630s. Became
acquainted with Descartes and was greatly influenced by him.
In 1645 he was appointed professor of mathematics in the
engineering school at Leiden. He was a very successful teacher;
Huygens was one of his students and became a close friend. In
1646 he brought out the collected edition of Viète's works and
in 1649 the Latin edition of Descartes' *Géométrie* which was
the effective edition, outside France at least. He also carried
on an extensive correspondence with many mathematicians.

2769. GOEDAERT, Johannes. 1617-1668. (Zoology 1423) Did not attend a university and probably not even a secondary school. He apparently did not know Latin and his work on entomology was based entirely on his own observations. He spent all his life in Middelburg.

2770. SYLVIUS, Franciscus (or DELE BOË, François). 1614-1672. (Physiology 1717. Chemistry 792) Studied medicine at Leiden and, after short periods at some other universities, graduated at Basel in 1637. He became a distinguished physician in Amsterdam and in 1658 was appointed professor of medicine at Leiden. He was probably the most famous teacher of medicine in Europe in the third quarter of the century and attracted large numbers of students from many countries.

2771. VARENIUS, Bernhardus. 1622-1650. (Geography 1015) A native of Hanover. Gained his M.D. at Leiden and, because of the devastation of Hanover in the Thirty Years War, settled permanently in Amsterdam. There he practised medicine and wrote numerous books on geography and other subjects.

1650s

2772. HUYGENS, Christiaan. 1629-1695. (Mathematics 141, 156. Mechanics 509, 511, 515, 520, 525. Astronomy 364. Physics 598) A member of a wealthy, distinguished, and cultured family. Studied law and mathematics at the University of Leiden (mathematics under van Schooten) and elsewhere during the late 1640s; at this time he was already in touch with Mersenne and had met Descartes. During the period 1650-66 he lived at home, devoting himself to his researches and to his steadily increasing correspondence. He did however make some visits to Paris and London where he made the acquaintance of many of the leading scientific figures.

When the Académie Royale des Sciences was established Huygens was invited to become a member, and he moved to Paris in 1666. During the following fifteen years he was the most outstanding member of the Academy. He returned to Holland in 1681 because of serious illness (and could not go back to France later because of political developments). He continued his researches at home until his death.

2773. JONSTON, John. 1603-1675. (Zoology 1422) Born in Poland of Scottish extraction; attended St. Andrews, Cambridge, Frankfurt, and Leiden universities, and took his M.D. at Leiden in 1632. He practised medicine in Leiden for many years and gained a high reputation. He was offered the chair of medicine there in 1640 but declined.

2774. BLASIUS (or BLAES), Gerardus. 1625-1692. (Zoology 1430) M.D., Leiden, 1646. After practising for some years he became professor of medicine and director of the hospital at Amsterdam in 1660. He joined the Leopoldina Academy in 1682.

1660s

2775. SWAMMERDAM, Jan. 1637-1680. (Zoology 1425, 1441. Anatomy 1592, 1600) M.D., Leiden, 1667. Even as a student he impressed his

professors and others with his researches. In 1664-65 he was
in Paris where he was an active member of Thevenot's gatherings
(cf. 2012) and he later carried on a lifelong correspondence
with Thévenot. After returning to Amsterdam he devoted himself
to research until his early death. (It is doubtful if he ever
practised medicine)

2776. RUYSCH, Frederik. 1638-1731. (Anatomy 1597) M.D., Leiden,
1664. Practised medicine and from 1667 was praelector in
anatomy for the surgeons' guild in Amsterdam. From 1685 he
was also professor of botany at the Athaneum in Amsterdam and
supervisor of the botanical garden.

2777. STENSEN, Niels. 1638-1686. (For entries in Part 1 see 1900)
A native of Copenhagen where he first studied medicine. From
1660 to 1663 he studied it in Holland, first at Amsterdam but
chiefly in Leiden; during this period, while still a student,
he made several important anatomical discoveries and published
two books. He was awarded the M.D. at Leiden in 1664. In the
following year he was in Paris where he participated in the
meetings at Thévenot's mansion (cf. 2012). For his subsequent
career in Italy see 1900.

2778. GRAAF, Regnier de. 1641-1673. (Anatomy 1598, 1601. Physiology
1720) After studying medicine at Utrecht and Leiden he went
to France and obtained an M.D. at Angers in 1665. He became
a successful physician in Delft and is said to have declined
an offer of a university chair.

1670s

2779. HARTSOEKER, Nicolaas. 1656-1725. (Physics 599) Educated
informally; apparently largely self-taught in mathematics
and science. He learnt to make lenses and visited Leeuwenhoek
in 1672. He was then with Huygens in Paris and worked briefly
at the Paris Observatory. Thereafter he earned his living
making lenses, microscopes, and telescopes--initially in
Rotterdam, then from 1684 to 1696 in Paris, and subsequently
in Amsterdam. From 1704 to 1716 he was mathematician to the
Elector of the Palatinate and taught mathematics and physics
in Düsseldorf. He was a foreign member of the academies of
Paris and Berlin.

2780. LEEUWENHOEK, Antoni van. 1632-1723. (Microscopy 1258) Did not
know Latin or other languages, had no university education,
and was self-taught in science; was employed in various administrative posts by the municipality of Delft. He communicated
his numerous findings to the Royal Society of London and was
elected a foreign member of the Society in 1680. By the end of
the century his discoveries had won him international fame.

1680s

2781. BIDLOO, Govard. 1649-1713. (Anatomy 1606) Studied medicine in
Amsterdam and received the M.D. at the University of Franeker
in 1682. From 1688 he was professor of anatomy at The Hague,
and from 1694 professor of medicine and surgery at Leiden. He

became physician to William III of Orange. In 1701 he was
elected a foreign member of the Royal Society of London.

1700s

2782. BOERHAAVE, Hermann. 1668-1738. (Chemistry 808. Physiology
1725) Studied medicine at Leiden and took his M.D. at Harderwijk in 1693. In 1701 he was appointed lecturer in medicine
at Leiden, in 1709 professor of botany and medicine, and in
1718 professor of chemistry as well. In medicine he was the
most famous teacher of the age and his influence was widespread.
Crowds of students from many countries came to Leiden to study
under him, and the important reforms he introduced into the
medical curriculum spread to other universities, notably Edinburgh and Vienna.
     Boerhaave was also a popular and inspiring teacher of
chemistry. In botany too his influence was considerable, and
he greatly enlarged and improved the Leiden botanical garden,
of which he was director. Perhaps his chief contribution to
botany was the help he gave to Linnaeus at an early stage in
his career when he was in Holland. Boerhaave was elected a
foreign member of the Paris Academy in 1728 and of the Royal
Society of London in 1730.

1710s

2783. 'sGRAVESANDE, Willem Jacob. 1688-1742. (Mechanics 527) Had an
early interest in mathematics but studied law, graduating at
Leiden in 1707. He practised at The Hague until 1715 when he
went to England as secretary to the Dutch embassy. There he
became a fellow of the Royal Society and met Newton, Keill,
Desaguliers, and others. In 1717 he became professor of mathematics and astronomy at Leiden where he had many students.
His teaching of physics and astronomy, based on Newton's writings and illustrated by lecture demonstrations, formed the
basis of the successive editions of his highly influential
textbook.

1720s

2784. MUSSCHENBROEK, Petrus van. 1692-1761. (Physics 604, 610) Came
from a well-known family of instrument-makers (his elder brother
supplied him and 'sGravesande with many instruments). M.D.,
Leiden, 1715; D.Phil., 1719. About 1720 he visited London
where he met Desaguliers. In 1723 he became professor of
natural philosophy and mathematics at Utrecht, and in 1740 at
Leiden (succeeding 'sGravesande) where his students were very
numerous. He became famous for his lecture course on experimental physics, illustrated with many well-chosen demonstrations,
which (like 'sGravesande's teaching earlier) was effective in
disseminating Newtonianism on the Continent.

2785. ALBINUS, Bernard Siegfried. 1697-1770. (Anatomy 1616) Born in
Frankfurt a.d. Oder where his father, Bernard Albinus (1653-
1721), was professor of medicine. In 1702 his father was
appointed to one of the medical chairs at Leiden (and, with
Boerhaave, built up Leiden's fame in medicine). The son studied

at Leiden where he received the M.D. in 1719 and, chiefly on
Boerhaave's recommendation, succeeded to his father's chair
in 1721. He soon became one of the most outstanding anatomists
of the time and, together with Boerhaave, made Leiden the
leading medical university in Europe. He was the author of
many books and was highly esteemed by his contemporaries,
receiving many honours.

1730s

2786. LYONET, Pierre. 1706-1789. (Zoology 1447) Born in Holland of
Huguenot descent. Graduated in theology at Leiden in 1727,
having included some mathematics and science in his studies.
Though he became a pastor he subsequently studied law and
established himself in The Hague as a lawyer. His studies of
insects were carried on as an avocation and were largely
inspired by Réaumur's famous book on the subject.

1740s

2787. TREMBLEY, Abraham. 1710-1784. (Zoology 1443) Born and educated
in Geneva. From 1733 he was a private tutor to upper-class
families in Holland. His researches on the hydra, which made
him famous, were done in Leiden. His sensational findings
were first communicated in 1741 by letter to Réaumur who
checked their correctness and then read the letter to the
Paris Academy. The Royal Society of London was also informed
by letter; it elected him a foreign member in 1743. He returned
to Geneva in 1757.

1750s

2788. HAARLEM SOCIETY. Founded in 1752 as the Hollandsche Maatschappij
der Wetenschappen te Haarlem (Société Hollandaise des Sciences).
Similar societies were founded at Vlissingen (1765), Rotterdam
(1769), and Utrecht (1773).

2789. CAMPER, Peter. 1722-1789. (Zoology 1453) M.D., Leiden, 1746.
After travelling through much of western Europe he became
professor of medicine at the University of Franeker in 1751,
then at the Amsterdam medical school in 1755, and finally at
the University of Groningen in 1763. He became well know in
the scientific world and was a member of many learned societies.

1840s

2790. NEDERLANDSCHE ENTOMOLOGISCHE VEREENIGING. Its publications
began in 1845.

1850s

2791. AMSTERDAM ACADEMY. Kon. Akademie van Wetenschappen. Its origins
went back to 1808, if not earlier, but in 1853 it was reorganized and began issuing its *Verslagen* which by the late nineteenth century was a big periodical.

1860s

2792. RIJKSMUSEUM VAN NATUURLIJKE HISTORIE, Leiden. (Muséum d'Histoire Naturelle des Pays-Bas) Its publications began in 1862.

1870s

2793. STIELTJES, Thomas Jan. 1856-1894. (Mathematics 288) Graduated at the Polytechnic in Delft in 1877 and became an assistant at the Leiden Observatory. He was publishing original work in mathematics from 1878 onward, and in 1884 the University of Leiden awarded him an honorary doctorate; in the same year he was appointed professor of mathematics at the University of Groningen. In 1885 however he left Holland and went to live in Paris where he gained the *doctorat ès sciences* in 1886. In that year he was appointed professor at Toulouse where he remained until his untimely death.

2794. WAALS, Johannes Diderik van der. 1837-1923. (Physics 722) D.Phil., Leiden, 1873. In 1875 he became a member of the Amsterdam Academy and two years later was appointed professor of physics at the new University of Amsterdam. He was awarded the Nobel Prize in 1910.

2795. LORENTZ, Hendrik Antoon. 1853-1928. (Physics 750) D.Phil., Leiden, 1875. In 1877 he was appointed to the new chair of mathematical physics at Leiden. He resigned the chair in 1912 (but remained honorary professor) and became director of the science section of Teyler's Museum at Haarlem; there he gave popular lectures and had a laboratory at his disposal (on the model of the Royal Institution in London).

In 1902 he and his assistant, Pieter Zeeman, shared the Nobel Prize for physics. Lorentz was one of the leading theoretical physicists of the time and a central figure in the international community of physicists that took shape in the *belle époque* before World War I.

2796. VAN'T HOFF, Jacobus Henricus. 1852-1911. (Stereochemistry 948. Physical chemistry 968, 994) D.Phil., Utrecht, 1874. After a short period at Bonn with Kekulé he went to Paris to work with Wurtz; there he met Le Bel (see 2199). After returning to Holland he held some junior positions and then in 1878 became professor of chemistry at Amsterdam. His brilliant work in physical chemistry brought him into close touch with Ostwald, and in 1887 they jointly inaugurated the *Zeitschrift für physikalische Chemie*--an event often taken to mark the emergence of physical chemistry as a separate discipline.

Van't Hoff moved to Berlin in 1896, having been appointed to a post in the Berlin Academy which enabled him to give all his time to research, except for a few lectures at Berlin University. He was awarded the Nobel Prize in 1901 and received numerous other honours.

2797. VRIES, Hugo de. 1848-1935. (Plant physiology 1398. Heredity 1545; most of his work in the field was done after 1900) D.Phil. in botany, Leiden, 1870. He then worked with Sachs at Würzburg for several years part-time, while earning his

living in minor occupations. In 1877 he was appointed lecturer
in plant physiology at the new University of Amsterdam, becoming
an *extraordinarius* the next year and *ordinarius* in 1881. After
1900 his work in heredity and mutation gained him an international reputation.

2798. NEDERLANDSCHE DIERKUNDIGE VEREENIGING. [Dutch Zoological Associ-
      ation] Its periodical began in 1872.

..... ENGELMANN, T.W. Professor of physiology at Utrecht, 1871-97.
      See 2697.

2799. BEIJERINCK, Martinus Willem. 1851-1931. (Microbiology 1817,
      1819) After graduating in chemical engineering at the Delft
      Polytechnic he gained a D.Phil. in botany at Leiden in 1877.
      He then had a post in an agricultural school where he was
      able to do some botanical research. In 1884 he was appointed
      to a position as bacteriologist in the fermentation industry
      and in 1895 he became professor of microbiology at the Delft
      Polytechnic.

1880s

2800. NEDERLANDSCH NATUUR- EN GENEESKUNDIG CONGRES. [Dutch Congress
      for Science and Medicine] An association of the 'advancement
      of science' type, presumably based on the German model (2589).
      Its periodical began in 1887.

2801. ROOZEBOOM, Hendrik (W.B.). 1854-1907. (Physical chemistry 975)
      D.Phil., Leiden, 1884. After holding a junior post at Leiden
      he became lecturer in physical chemistry there in 1890. Six
      years later he was appointed professor of general chemistry
      at Amsterdam (in succession to Van't Hoff).

2802. RIJKS GEOLOGISCH-MINERALOGISCH MUSEUM, Leiden. Its periodical
      began in 1881.

1890s

2803. SOCIÉTÉ MATHÉMATIQUE D'AMSTERDAM. In 1893 it began its *Revue
      Semestrielle des Publications Mathématiques*.

2804. ZEEMAN, Pieter. 1865-1943. (Physics 761) D.Phil., Leiden,
      1893. From 1890 he had been Lorentz' assistant at Leiden and
      in 1894 he became a *Privatdozent* there. He was appointed to
      the teaching staff at Amsterdam in 1897 and three years later
      became professor of physics, succeeding van der Waals as
      director of the physics institute in 1908. He shared the
      Nobel Prize with Lorentz in 1902 and received many other
      honours and awards.

2.06   SOUTHERN NETHERLANDS

(BELGIUM)

1530s

2805. GEMMA FRISIUS, Reiner. 1508-1555. (Astronomical geography 1004)
A physician in Louvain and later a teacher in the University's
medical faculty. He was well known as a designer of globes
and astronomical instruments.

2806. MERCATOR, Gerardus. 1512-1594. (Cartography 1008) Graduated
at Louvain in 1532 and subsequently studied mathematics and
astronomy with the assistance of Gemma Frisius. He became a
skilled engraver and in 1536 established a business in Louvain
for producing maps, globes, and scientific instruments. He
soon became famous and was commissioned to make globes and
instruments for the Emperor. In 1552, because of religious
persecution, he migrated to Germany, becoming cosmographer
to the Duke of Cleves.

1540s

2807. DODOENS, Rembert. 1516-1585. (Botany 1290) After graduating
in medicine at Louvain in 1535 he travelled extensively and
then became a physician in his native city of Mechelen (now
Malines). From 1574 to 1580 he was in Vienna as physician to
the Emperor, and from 1582 until his death he was professor
of medicine at Louvain.

1580s

2808. ROOMEN, Adriaan van. 1561-1615. (Mathematics 109, 116) Native
of Louvain. Studied mathematics and philosophy at the Jesuit
college in Cologne (and perhaps also in Rome under Clavius).
About 1586 he became professor of medicine and mathematics at
Louvain and then, from 1593, at Würzburg. He was ordained a
priest in 1604.

1610s

2809. SAINT VINCENT, Gregorius (also known as GREGORIUS A SANCTO VIN-
CENTIO). 1584-1667. (Mathematics 139) Born and educated in
Bruges. Became a Jesuit and from 1617 onward taught mathematics
in various Jesuit colleges in the Spanish Netherlands and
briefly in Rome and Prague.

2810. HELMONT, Johannes Baptista van. 1579-1644. (Chemistry 790)
A member of the Flemish landed gentry. He took an M.D. at
Louvain in 1599 but apparently never practised medicine. He
travelled widely for several years and after his return he
carried on chemical research in his private laboratory.

473

1620s

2811. LAET, Jan de. 1593–1649. (Mineralogy 1014. Natural history
      1225, 1226) Born in Antwerp and studied (medicine?) at Louvain.
      Travelled widely in America and after his return to Antwerp
      became a director of the Dutch West India Company.

1640s

2812. TACQUET, Andreas. 1612–1660. (Mathematics 142) Born and
      educated in Antwerp. Became a Jesuit and from about 1645
      taught mathematics and other subjects at Jesuit colleges in
      Louvain and Antwerp.

1650s

2813. SLUSE, René François de. 1622–1685. (Mathematics 146) Came
      from a wealthy family in Liège. Educated first in Louvain
      and then in Rome where he obtained a doctorate in law in 1643.
      Subsequently he held a series of high ecclesiastical positions
      in Liège. He carried on an active correspondence with many
      mathematicians and also with the Royal Society of London,
      becoming one of its foreign members in 1674.

1770s

2814. BRUSSELS ACADEMY. Founded in 1772 as the Académie Impériale et
      Royale des Sciences et Belles-Lettres de Bruxelles. The
      English priest, J.T. Needham (2300), took a leading part in
      its foundation and early years. Brussels was then the capital
      of the Austrian Netherlands and the Academy was well supported
      by the Austrian monarchy.

1800s

2815. OMALIUS D'HALLOY, Jean Baptiste d'. 1783–1875. (Geology 1135,
      1149) Born in Liège of a wealthy, aristocratic family and
      educated for a career in public office. During a stay in Paris
      however he developed an interest in the natural history sciences
      which he studied intensively under the Parisian luminaries from
      1803 onward, with geological tours through northern France and
      his homeland. His geological career began auspiciously with
      a major publication in 1808 but was interrupted by political
      events, as a result of which he held a succession of high
      public offices from 1813 to 1830. The attainment of Belgian
      independence in 1830 finally enabled him to step down and
      resume his scientific career.

1830s

2816. STAS, Jean Servais. 1813–1891. (Chemistry 923) M.D., Louvain,
      1835. He never practised medicine but began a career of chem-
      ical research, and in 1837 went to Paris to work with Dumas.
      In 1840 he was appointed professor of chemistry at the Ecole
      Royale Militaire in Brussels where he spent the rest of his
      career.

1850s

2817. SOCIÉTÉ ENTOMOLOGIQUE BELGE. Its periodical began in 1857.

1860s

2818. SOCIÉTÉ ROYALE DE BOTANIQUE DE BELGIQUE. Its periodical began in 1862.

2819. SOCIÉTÉ MALACOLOGIQUE DE BELGIQUE. Its periodical began in 1863.

1870s

2820. SOCIÉTÉ GÉOLOGIQUE DE BELGIQUE. Its periodical began in 1874.

2821. BENEDEN, Edouard van. 1846-1910. (Cytology 1696) Appointed professor of zoology at Liège in 1870.

1890s

2822. VALLÉE-POUSSIN, Charles (J.G.N.) de la. 1866-1962. (Mathematics 313) Graduated in engineering but then concentrated on pure mathematics with the encouragement of L.P. Gilbert, professor of mathematics at Louvain. He succeeded Gilbert in 1892 and held the chair for the rest of his career.

2.07 SCANDINAVIA

1570s

2823. BRAHE, Tycho. 1546–1601. (Astronomy 335, 341, 342, 345)
A Danish noble. Educated at the University of Copenhagen,
then studied law at several German universities but became
interested in astronomy; travelled widely and visited many
astronomers. In 1576, with financial support from the King
of Denmark, he established his famous observatory at Uraniborg
(with ancillary laboratories, workshops, and a printing press)
which has been called the first research institute of modern
times. It attracted a number of young astronomers who worked
as his assistants and was visited by many scholars and travellers. In 1598, as a result of lack of support by the new king,
he left Denmark and settled in Prague under the patronage of
the Emperor Rudolf II. In his last years Kepler was one of
his assistants.

2824. SEVERINUS, Petrus (or SØRENSON, Peder). 1542–1602. (Chemistry
775) Studied first at the University of Copenhagen and then
(with financial support from it) at various German, Italian,
and French universities. He took his M.D. in France about
1571 and on his return home was appointed physician to the
King of Denmark.

1580s

2825. FINK, Thomas. 1561–1656. (Mathematics 106) A native of Denmark.
Studied at several German universities, and at Padua and Basel,
receiving the M.D. at Basel in 1587. After practising medicine
for a short period he was appointed professor at Copenhagen,
initially of mathematics and later of medicine.

1610s

2826. BARTHOLIN, Caspar. 1585–1629. (Anatomy 1580) After a preliminary education at Copenhagen he studied medicine and other
subjects at Wittenberg, Leiden, Basel, Padua, and Naples,
taking his M.D. at Basel in 1610. On his return to Denmark
in 1611 he was appointed to a post in the University of Copenhagen, becoming professor of medicine in 1613. In 1624 he
relinquished the post to become professor of theology.

1640s

2827. BARTHOLIN, Thomas. 1616–1680. (Anatomy 1584, 1588. Physiology
1715) Son of Caspar Bartholin and brother of Erasmus. Studied
medicine at Leiden and Padua, as well as visiting several other
universities in Italy and France. Took his M.D. at Basel in

1646, then returned to Copenhagen where he became professor of anatomy in 1649, later becoming very distinguished. In 1673 he began one of the earliest periodicals, *Acta Medica et Philosophica Hafniensia*, but it only lasted until 1679.

1650s

2828. BARTHOLIN, Erasmus. 1625-1698. (Physics 592) Son of Caspar Bartholin and brother of Thomas. Studied at Leiden, then went to Italy where he gained his M.D. at Padua in 1654. He travelled in France and England and returned home to become professor of mathematics at Copenhagen in 1656 and later professor of medicine.

2829. RUDBECK, Olof. 1630-1702. (Anatomy 1589) Studied medicine at Uppsala and then at Leiden where he graduated about 1654. Soon afterwards he became professor of medicine at Uppsala; he was also active as a botanist. He did much to upgrade the University, among other things building an anatomical theatre and establishing a botanical garden.

1660s

2830. BOTANICAL GARDEN, UPPSALA. Founded in the early 1660s by Olof Rudbeck. It became one of the best in Europe.

1680s

2831. LABORATORIUM CHYMICUM HOLMIENSE. Established in Stockholm in 1683 by the Swedish Bergskollegium, or Council of Mines, the government body for the development of mining and metallurgy (which had been instituted in the 1630s and was for generations the leading organization of its kind in Europe). The first director of the Laboratory was Urban Hiärne who initiated a programme of research in mineral chemistry. He was succeeded in 1727 by the important chemist, Georg Brandt (2833).

1710s

2832. UPPSALA SOCIETY. Founded in 1710 as a private society and reconstituted in 1728 as the Regia Societas Scientiarum Upsaliensis. It was concerned with the whole range of learning and functioned in intimate association with the University. Though it was less important for science than the Stockholm Academy its *Acta* continued to have a significant scientific content through the eighteenth and nineteenth centuries.

1720s

2833. BRANDT, Georg. 1694-1768. (Chemistry 810) Son of a mine-owner who was interested in chemistry and metallurgy. Studied medicine and chemistry at Leiden under Boerhaave during 1721-24 and took his M.D. at Rheims in 1726. On his way back to Sweden he studied mining and metallurgy in the Harz Mountains, and on his return was made director of the chemical laboratory (2831) belonging to the Council of Mines. He was appointed warden of

the Royal Mint in 1730 and became a member of the Council of
Mines in 1747. In 1750 he declined an offer of the new chair
of chemistry at Uppsala.

Brandt was one of the ablest chemists of his time, a
resolute opponent of alchemy, and one of the founders of the
distinguished Swedish tradition of mineral chemistry.

1730s

2834. STOCKHOLM ACADEMY. Founded by the Swedish government in 1739 as
Svenska Vetenskaps Academien. Its initiators included Linnaeus
and the physicist, M. Triewald, as well as industrialists and
politicians. Like the Berlin Academy it was given an economic
base through the grant of a monopoly on the publication of
calendars and astronomical almanacs.

The Academy was formed in pursuance of the government's
mercantilist policies, and its *Transactions*--published in
Swedish and intended to have an educational role in practical
matters--dealt largely with topics concerning agriculture and
industry. The emphasis on utility was not without benefit to
the sciences however: thus Linnaeus, as a powerful figure in
the Academy, made good use of his contacts with politicians
and industrialists.

The astronomer, P.W. Wargentin, was secretary from
1749 until his death in 1783 and during that period was the
Academy's main driving force. He established its observatory
in 1753 and did much to bring it into the international scientific community. Its standing is indicated by the fact that
its *Handlingar* [*Transactions*] was translated into German over
the period 1749-94 (on the initiative of A.G. Kaestner, professor of mathematics and physics at Göttingen).

2835. KLINGENSTIERNA, Samuel. 1698-1765. (Physics 614) Studied at
Uppsala and was awarded a travelling scholarship in 1727. His
studies with several leading mathematicians and scientists in
Marburg, Basel, Paris, and London provided him with a valuable
and stimulating training. On his return in 1731 he took up
the position of professor of geometry at Uppsala and in 1750
transferred to the newly-established chair of physics.

2836. LINNAEUS, Carl. 1707-1778. (Natural history 1232. Botany 1329,
1332) Studied medicine at Uppsala between 1728 and 1735 but
spent most of his time on botany. In 1735 he went to Holland
and took his M.D. at Harderwijk. He was greatly helped by the
patronage of the famous Boerhaave and stayed in Holland for
three very successful years. It was there that he published
his *Systema Naturae* and several other basic works which he had
largely completed before he left Sweden; there too that he
established connections with many leading botanists, making
brief trips to London and Paris. Returning to Sweden in 1738,
he practised as a physician in Stockholm and was one of the
founders of the Stockholm Academy and its first president.
In 1741 he became professor of medicine (including botany) at
Uppsala, and retained the post for the rest of his life.

Linnaeus was highly popular as a teacher and as a
supervisor of doctoral candidates; a remarkable number of his

students proceeded to the doctorate. Many foreign students
came to Uppsala to study under him and his ideas were disseminated widely through his students as well as through his
writings. In his later years he was perhaps the most famous
scientist in the world, and in botany and natural history
generally his influence was unparalleled.

2837. ARTEDI, Peter. 1705-1735. (Zoology 1442) Studied medicine at
       Uppsala in the late 1720s and was a fellow-student of Linnaeus
       with whom he formed a lifelong friendship. His studies became
       concentrated on zoology and in 1734 he spent a year in England
       on further study of the subject, with assistance from Sloane,
       the president of the Royal Society. In Amsterdam, on his way
       home, he was able to study a large collection of fishes, but
       there he met an accidental death. His manuscripts were saved
       and his book was published by Linnaeus.

1740s

2838. COPENHAGEN SOCIETY. Founded in 1742 as the Copenhagen Society
       of Friends of Learning and Science. In 1777 it became the
       Royal Danish Society of Sciences (K. Danske Videnskabernes
       Selskab).

2839. CRONSTEDT, Axel Frederik. 1722-1765. (Chemistry 815. Mineral-
       ogy 1082) Was associated with the Bergskollegium (Council of
       Mines) from 1742 and made a thorough investigation of the
       theory and practice of mining and metallurgy, in the course
       of which he spent much time on chemistry. In 1748 he was made
       the supervisor of a large mining area in central Sweden and,
       having a good laboratory at his disposal, he carried out some
       fruitful research in mineral chemistry. He became a member
       of the Stockholm Academy.

2840. WALLERIUS, Johan Gottschalk. 1709-1785. (Mineralogy 1081)
       M.D., Uppsala, 1735. In 1741 he was appointed assistant prof-
       essor of medicine at Uppsala and in 1750 professor of chemistry
       (the first in Sweden). His teaching of chemistry, which feat-
       ured lecture demonstrations and laboratory work by the students,
       was very successful. His writings included treatises on the
       application of chemistry to metallurgy and agriculture, the
       latter being especially influential. He became a member of
       the Leopoldina Academy, the Stockholm Academy, and the Uppsala
       Society.

2841. GEER, Charles de. 1720-1778. (Zoology 1446) Came from a very
       wealthy family. Educated in Holland, finally at the University
       of Utrecht where he became interested in scientific subjects.
       He returned to Sweden in 1739 and became a member of the Stock-
       holm Academy and later of several other learned societies. He
       was one of the richest men in the country and second only to
       Linnaeus in his standing as a naturalist.

1750s

2842. WILCKE, Johan Carl. 1732-1796. (Physics 623, 626) Studied at
       Uppsala and Rostock, graduating at the latter in 1757. He was

then in Berlin where he did significant research on electricity
in collaboration with his friend, Franz Aepinus. In 1759 he
was appointed lecturer (from 1770 professor) at the Military
Academy in Stockholm. He was secretary of the Stockholm Academy from 1784 until his death.

1760s

2843. TRONDHEIM SOCIETY. Founded in 1760 as the Trondheimske Selskab,
becoming in 1768 the K. Norske Videnskabers-Selskab [Royal
Norwegian Society of Sciences].

2844. BERGMAN, Torbern Olof. 1735-1784. (Chemistry 824) Studied
philosophy and science at Uppsala, graduating in 1758. He
became lecturer in mathematics there and was elected to the
Stockholm Academy in 1764. In 1767 he succeeded Wallerius as
professor of chemistry and pharmacy, subjects which at that
time he did not know much about, his research hitherto having
been in astronomy, physics, and geography. His point of entry
into his new field was the already existing Swedish tradition
of mineral analysis.
    Bergman became one of the leading chemists of his time
and was highly esteemed by his contemporaries. He was a foreign
member of several of the Continental academies and of the Royal
Society of London.

2845. MÜLLER, Otto Frederik. 1730-1784. (Zoology 1451) Studied law
and theology at Copenhagen in the early 1750s. For nearly
twenty years he was a private tutor to a noble family with
whom he travelled widely in western Europe. He became interested in natural history and used his travels to widen his
knowledge and make contacts with important naturalists. A
fortunate marriage in 1773 made him financially independent
and able to devote his full time to his investigations. He
became a member of the Leopoldina Academy, the Stockholm Academy, the Copenhagen Society, and several other learned bodies.

1770s

2846. SCHEELE, Carl Wilhelm. 1742-1786. (Chemistry 819) Qualified
as an apothecary by apprenticeship in 1763. As a journeyman
he worked in various towns, including Uppsala in the period
1770-75, and there met Gahn and Bergman who introduced him to
scientific circles and helped him generally. There also he
began his main lines of research, becoming elected to the
Stockholm Academy--an unprecedented honour for a junior apothecary. In 1775 he acquired his own business in the small town
of Köping where he remained for the rest of his life, declining
all offers of much more prestigious and lucrative positions.

2847. THUNBERG, Carl Peter. 1743-1828. (Botany 1339, 1349) Studied
medicine at Uppsala (botany under Linnaeus), graduating in
1770. He then went to Paris for further study, and through
the influence of Linnaeus and some Dutch connections received
an offer to go on a Dutch merchant ship to Japan and collect
garden plants there. (At that time Japan was closed to all
foreigners except those employed by the Dutch East India Comp-

any) On the way he stopped off at Cape Town for three years
(1772-75) to learn Dutch and to make extensive collections of
plants in Cape Colony. He was then in Japan for over a year
and despite the restrictions on travel in the country was able
to make a large collection of Japanese plants.

Thunberg returned to Sweden in 1779 and was appointed
demonstrator in botany at Uppsala, and in 1784 professor. He
held the post for the rest of his life and spent much of his
time working on his floras of Cape Colony and Japan. His
description of his travels went through many editions in Swedish, German, English, and French.

2848. FABRICIUS, Johann Christian. 1745-1808. (Zoology 1450) Studied
at Copenhagen and then spent two years at Uppsala as a student
of Linnaeus who had a high regard for him. From 1765 onward
he travelled widely and in 1775 was appointed professor of
natural history and economics at the University of Kiel (which
was then in Denmark).

1790s

2849. ACHARIUS, Erik. 1757-1819. (Botany 1351) Linnaeus' last student.
Studied medicine at Uppsala and then at Lund where he received
the M.D. in 1782. He practised medicine in a small town in
Ostergotland for the rest of his life.

2850. MOLDENHAWER, Johann Jacob Paul. 1766-1827. (Plant anatomy 1352)
Studied science, especially botany, at Copenhagen in the late
1780s. In 1792 he was appointed professor of botany and fruit-
tree culture at Kiel, a post he held until his death.

1800s

2851. OERSTED, Hans Christian. 1777-1851. (Physics 649) Son of an
apothecary. Graduated in pharmacy at the University of Copenhagen in 1797 and two years later was awarded a D.Phil. for a
thesis on Kantian natural philosophy. During the years 1801-04
he travelled in Germany and France, meeting many scientists
and surveying the scene in the physical sciences--and also in
philosophy where his predilection for the new *Naturphilosophie*
had some unfortunate consequences. In 1806 he was appointed
professor of physics at Copenhagen, and in 1829 was also made
director of the new Polytechnic Institute.

Oersted did much to strengthen the position of science
in Denmark by his excellent teaching, his energetic populariz-
ing, and his initiative in 1824 in establishing a society for
spreading a knowledge of science among the general public.

2852. BERZELIUS, Jöns Jacob. 1779-1848. (Chemistry 855, 860, 874)
M.D., Uppsala, 1802. During his medical course he developed
an interest in chemistry which he mostly taught himself. After
graduation he practised medicine in Stockholm and became
friendly with W. Hisinger, a wealthy mine-owner with an interest
in mineralogy and chemistry. He worked in Hisinger's private
laboratory and they jointly published numerous papers.

In 1807 Berzelius was appointed professor of medicine
and pharmacy at the Stockholm medical college and when this

institution was upgraded to the Karolinska Institutet in 1810 he taught only chemistry and pharmacy; he retained the post until 1832. In 1808 he became a member of the Stockholm Academy. He visited London in 1812 in order to meet the leading English chemists (especially Davy), and Paris in 1818 where he remained for a year, becoming well acquainted with the scientific scene in France; he travelled home through Germany and met many German chemists.

Berzelius became permanent secretary of the Stockholm Academy in 1818 and a dominant figure in Swedish science as well as the leading chemist in Europe. He was elected a foreign member of the Royal Society and the Paris Academy, and was made a baron by the King of Sweden.

1810s

2853. AGARDH, Carl Adolph. 1785-1859. (Botany 1356) Studied at Lund and became professor of botany and economics (largely agricultural) there in 1812. During the 1820s he travelled to Paris and Germany, and made botanical field trips to the shores of the Adriatic. In 1835 he resigned his chair and became a bishop.

2854. FRIES, Elias Magnus. 1794-1878. (Botany 1355, 1361) D.Phil. in botany, Lund, 1814. After holding junior teaching positions at Lund he became professor of botany at Uppsala in 1835. At that time the University was pervaded by *Naturphilosophie* and he made a stand in defence of empirical investigations, not without success. He was elected to the Stockholm Academy in 1847 and was generally regarded as one of the most distinguished Swedish botanists since Linnaeus.

1820s

2855. ABEL, Niels Henrik. 1802-1829. (Mathematics 217) A mathematical prodigy. Graduated at the University of Christiania (now Oslo) in 1822 and subsisted precariously while doing mathematical research of first-rate importance. In 1825 he was awarded a government grant to enable him to visit mathematicians in Berlin and Paris. He returned to Norway in 1827, sick and impoverished, and died two years later, unaware that an offer of a university post in Berlin was on the way.

2856. MOSANDER, Carl Gustaf. 1797-1858. (Chemistry 891) Initially a pharmacist, then studied medicine, graduating at Stockholm in 1825. He was a pupil and friend of Berzelius and, doubtless as a result, was appointed teacher of chemistry at the Karolinska Institutet. In 1832 he succeeded Berzelius as professor of chemistry and pharmacy there. He was also curator of the mineral collection of the Stockholm Academy.

2857. RETZIUS, Anders Adolf. 1796-1860. (Zoology 1470) Son of the professor of natural history at Lund. M.D., Lund, 1819. In 1824 he was appointed professor of anatomy at the Karolinska Medico-Kirurgiska Institutet. In the early years of the Institute he and Berzelius were its two outstanding figures.

Retzius was a pioneer of comparative anatomy in Sweden

and gave a substantial impetus to the development of the
biological sciences generally in the country. In the latter
part of his career he did fundamental work in physical anthropology and it is for this that he is best known.

1830s

2858. SKANDINAVISKA NATURFORSKARE OCH LÄKARE. [Scandinavian Scientists
      and Physicians] A society which held annual meetings in differ-
      ent cities, beginning in 1839. It was no doubt modelled on
      the German society (2589).

2859. AGARDH, Jacob Georg. 1813-1901. (Botany 1372) Son of C.A.
      Agardh (2853) whose line of work he continued. He taught
      botany at Lund from 1834 and was professor of the subject
      there from 1854 to 1879.

2860. SARS, Michael. 1805-1869. (Zoology 1475) From 1828 to 1854 he
      served as a pastor to communities on the Norwegian coast,
      generally travelling by boat and spending much of his time on
      studies of marine zoology. He had the opportunity to visit
      several countries and make contacts with leading zoologists,
      and he also visited sites on the Mediterranean and the Adriatic.
      In 1854 his publications gained him appointment as *professor
      extraordinarius* in zoology at the University of Christiania,
      a post he held until his death.

2861. LOVÉN, Sven Ludvig. 1809-1895. (Zoology 1481a) After gradua-
      ting at Lund in 1829 he studied for a year in Berlin. Later
      he took part in a number of excursions to the Arctic regions,
      studying aspects of marine zoology. In 1841 he was appointed
      curator of the invertebrate section of the Museum of Natural
      History in Stockholm, and he remained on the staff of the
      Museum for the rest of his career. He was a prominent member
      of the Stockholm Academy.

1840s

2862. SOCIETAS SCIENTIARUM FENNICA, Helsingfors (Helsinki). Its
      Commentationes began in 1840. The Society was later named
      Finska Vetenskaps-Societeten.

2863. ÅNGSTRÖM, Anders Jonas. 1814-1874. (Physics 698, 708) D.Phil.,
      Uppsala, 1839; then a *Privatdozent*. In 1843 he became assist-
      ant professor of astronomy at Uppsala Observatory, and in 1858
      professor of physics at the University. In the last years of
      his life his spectroscopic work won widespread recognition and
      he received many honours.

2864. COLDING, Ludvig August. 1815-1888. (Physics 682) Oersted was
      an old friend of his family and had a big influence on his
      education and early years. He graduated at the Copenhagen
      Polytechnic Institute (of which Oersted was the director) in
      1841 and subsequently held a succession of engineering and
      administrative posts. Concurrently however he managed to
      carry on a substantial amount of research on a variety of
      engineering and scientific topics. From 1869 he was a prof-
      essor at the Polytechnic Institute, in addition to other posts,

and in the latter part of his career he held a prominent place in the scientific and technical community in Copenhagen.

2865. THOMSEN, (H.P.J.) Julius. 1826-1909. (Physical chemistry 962) Graduated at the Copenhagen Polytechnic Institute in 1846. After junior teaching positions there and at the Danish Military College he was appointed to the staff of the University of Copenhagen in 1864, becoming professor of chemistry two years later. In 1869 he declined an offer of the chair of physical chemistry at Leipzig.

2866. STEENSTRUP, Japetus Smith. 1813-1897. (Zoology 1483) Educated at the University of Copenhagen. From 1841 he taught botany and zoology at the academy at Sorö. In 1846 his publications gained him appointment as professor of zoology at Copenhagen, a post he held for the rest of his career. Steenstrup had a strong influence on natural history in Denmark and was well known for his investigations--covering all aspects of natural history as well as archaeology--of the Danish peat-bogs. He was elected a foreign member of the Royal Society in 1863.

1850s

2867. GEOLOGICAL SURVEY OF NORWAY AND SWEDEN. Established in 1858.

1860s

2868. LIE, (M.) Sophus. 1842-1899. (Mathematics 279, 307) Graduated at Christiania in 1865. After further study he was awarded a travelling scholarship and spent the period 1869-71 very profitably in Berlin and Paris. He received his D.Phil. at Christiania in 1872 and was appointed an *extraordinarius* in mathematics. In 1886 he became professor of mathematics at Leipzig but his health broke down a few years later.

2869. GULDBERG, Cato Maximilian. 1836-1902. (Physical chemistry 935. Meteorology 1070) Studied mathematics and the physical sciences at Christiania, graduating in 1859. In 1861 he was awarded a travelling scholarship which enabled him to spend a year studying in France and Germany. From 1862 he was lecturer in mechanics at the Royal Military College in Christiania and concurrently from 1869 professor of applied mathematics at the University.

2870. MOHN, Henrik. 1835-1916. (Meteorology 1070) Graduated at Christiania in 1858 and became lecturer in astronomy there two years later. In 1866 he was appointed to the dual post of professor of meteorology and director of the newly-formed Norwegian meteorological organization. He was very active on both the national and international scene and became one of the leaders of European meteorology.

2871. BOTANISKE FORENING, Copenhagen. Its periodical began in 1866.

1870s

2872. MITTAG-LEFFLER, (M.) Gösta. 1846-1927. (Not in Part 1. An able but not outstanding mathematician whose importance was in his

organizational role) D.Phil., Uppsala, 1872; the next year he
was awarded a travelling scholarship to study in Paris, Berlin,
and Göttingen. He was appointed professor of mathematics at
Helsinki in 1877 and four years later at Stockholm where he
remained for the rest of his life. His work there did much to
advance the Scandinavian school of mathematics.

     In 1882, with the financial support of the King of
Sweden, Mittag-Leffler founded the international journal, *Acta
Mathematica*, which soon became of leading importance. As its
chief editor for many years he was in a highly influential
position in the mathematical world. (His support was of critical importance for Cantor, for example. He also obtained an
academic post for the outstanding woman mathematician, Sonya
Kovalevsky) He was prominent as an organizer of the international congresses of mathematicians, the first of which was
held in Zurich in 1897.

2873. GEOLOGISKA FÖRENING, Stockholm. Its periodical began in 1872.

2874. WARMING, (J.) Eugen (B.). 1841-1924. (Botany 1408) After
    studying at Munich and Bonn he received his D.Phil. at Copen-
    hagen in 1871. He was professor of botany at the Royal Instit-
    ute of Technology in Stockholm from 1882 and at the University
    of Copenhagen from 1885 to 1911.

2875. RETZIUS, Magnus Gustaf. 1842-1919. (Zoology 1513. Anatomy
    1643, 1648) Son of A.A. Retzius (2857). Studied medicine at
    Uppsala and Stockholm, and after some travels in other countries
    took his M.D. at Lund in 1871. In the same year he became a
    Dozent in anatomy at the Stockholm medical school (the Karolin-
    ske Institutet), in 1877 an *extraordinarius* in histology, and
    in 1889 full professor of anatomy. Being wealthy, he resigned
    the post in 1890 in order to devote himself to full-time
    research.

     In addition to several books, Retzius published over
    three hundred papers in several fields of biology and was best
    known for his work on the histology of the nervous system. He
    was a prominent member of the Swedish scientific community and
    received many honours from Swedish and foreign scholarly bodies.

2876. CARLSBERG LABORATORY. The successful Danish brewer and founder
    of the Carlsberg Breweries, J.C. Jacobsen, was so impressed
    with Pasteur's work that in 1876 he established a magnificent
    laboratory at the brewery, complete with a bust of Pasteur.
    The Carlsberg Laboratory soon became an important centre for
    research in microbiology and biochemistry, as well as producing
    results of fundamental value for the fermentation industries.
    Jacobsen also established the Carlsberg Foundation for the
    support of scientific research, and eventually transferred the
    ownership of the Carlsberg Breweries and the direction of the
    Foundation to the Royal Danish Society of Sciences.

1880s

2877. KEMISTS-SAMFUNDET [Chemists' Association], Stockholm. Its
    periodical began in 1887.

2878. ARRHENIUS, Svante. 1859-1927. (Physical chemistry 970) Studied

at Uppsala and Stockholm and in 1884 presented the first version
of his theory of electrolytic dissociation as his doctoral
dissertation at Uppsala. It was awarded only a fourth-class
degree--insufficient to qualify him for an academic post.
However he sent copies of his dissertation to several leading
chemists and physicists, eliciting a warm response from Ostwald
who offered him a post at Riga. As a result Arrhenius gained
an appointment as lecturer in physical chemistry at Uppsala at
the end of 1884.

In 1886 the Stockholm Academy awarded him a travel
grant which enabled him to spend the next four years working
at major centres--with Ostwald at Riga (later Leipzig), Kohlrausch at Würzburg, Boltzmann at Graz, and Van't Hoff at
Amsterdam. During these years he greatly improved his theory.
In 1891 he was offered a chair at Giessen but, preferring to
stay in Sweden, he accepted a lectureship at the Institute of
Technology in Stockholm, becoming professor of physics there
in 1895. In 1905 he was offered a prestigious chair at Berlin
and when he declined an attractive post was created for him as
director of the physical chemistry department of the new Nobel
Institute in Stockholm; he remained there until his death.

Around the turn of the century Arrhenius had an international reputation as one of the greatest chemists of the time.
He received the Nobel Prize in 1903.

2879. GEER, Gerhard Jakob de. 1858-1943. (Geology 1214. His well-
known work on varve geochronology was mostly done after 1900)
Graduated in geology at Uppsala in 1879 and was appointed to
the Swedish Geological Survey. In 1897 he became professor
of geology at the University of Stockholm.

2880. SEDERHOLM, Johannes Jakob. 1863-1934. (Geology 1211) After
gaining his B.A. at Helsinki in 1885 he studied mineralogy and
geology at Stockholm from 1886 to 1888 (receiving his doctorate
in 1892). In 1888 he was appointed to the staff of the Geological Commission (i.e. Survey) of Finland and in 1893 became
its director, a post he held with distinction until his death.

2881. ENTOMOLOGISKA FÖRENING, Stockholm. Its periodical began in 1880.

1890s

2882. DANMARKS GEOLOGISKE UNDERSØGELSE. [Geological Survey of Denmark]
Its publications began in 1890.

2.08 SWITZERLAND

1520s

2883. RUEFF, Jacob. 1500–1558. (Embryology 1650) A physician in
Zurich from 1525. In 1552 he became city physician and surgeon.

1540s

2884. GESNER, Conrad. 1516–1565. (Zoology 1414) M.D., Basel, 1541.
Settled in Zurich where he practised medicine, becoming chief
city physician in 1554. He was an assiduous traveller and
collector of botanical and zoological material. He planned a
survey of plant life which would have been comparable to his
great work on animals but died before he could complete it.
In his own time he was best known as a botanist and his correspondence (published in part posthumously) demonstrates his
ability in the field.
    In addition to his scientific interests Gesner was a
classical philologist and one of the founders of bibliography.

1560s

2885. BODENSTEIN, Adam von. 1528–1577. (Chemistry 773) Professor of
medicine at Basel and an early follower of Paracelsus, whose
doctrines he taught. His importance was as an editor and
interpreter of Paracelsus, his own writings being on alchemy.

2886. DORN, Gerard. fl. 1566–84. (Chemistry 774) A student of
Bodenstein and evidently a physician. He lived in Basel until
the late 1570s when he apparently went to Frankfurt.

2887. PLATTER, Felix. 1536–1614. (Anatomy 1575) Studied medicine at
Montpellier and took his M.D. at Basel in 1557. He later
taught medicine at the University of Basel and became the city's
chief physician. He was interested in botany and was well
known for his large herbarium.

1580s

2888. BOTANICAL GARDEN, BASEL. Established by the University in 1588.

2889. BAUHIN, Gaspard. 1560–1624. (Botany 1307. Anatomy 1579)
Brother of Jean Bauhin (2496). Educated at the University of
Basel, then studied at Padua (anatomy under Fabrizio and botany
at the garden), Bologna (anatomy under Aranzio), Montpellier,
and Tübingen. He returned to Basel, received the M.D. there
in 1581, and became a member of the Faculty of Medicine, teaching anatomy and botany. He was the prime mover in the establishment of an anatomical theatre and a botanical garden at
the University.

1660s

2890. GLASER, Johann Heinrich. 1629–1679. (Anatomy 1603) M.D., Basel, 1661. After establishing a high reputation as a physician he was appointed professor of anatomy and botany at Basel in 1667.

1680s

2891. BERNOULLI, Jakob, I. 1654–1705. (Mathematics 164) Graduated in theology in 1676 but turned to mathematics and astronomy; travelled in France, Holland, and England, meeting many mathematicians and scientists. He returned to Basel and in 1687 became professor of mathematics at the University. In 1699 he was elected a foreign member of the Paris Academy.

1690s

2892. BERNOULLI, Johann, I. 1667–1748. (Mathematics 168. Mechanics 526) Graduated in medicine at Basel in 1694 but then gave it up for mathematics which for several years previously he had been studying under his elder brother, Jakob (2891). From 1693 he was in correspondence with Leibniz and other mathematicians (and continued for the rest of his life to carry on an enormous correspondence). He was professor of mathematics at Groningen (in Holland) from 1695 until 1705 when, on the death of his brother, he succeeded him in the chair at Basel. A great admirer of Leibniz, he vigorously upheld his cause in the famous controversy with the Newtonians. He was a foreign member of the Paris, Berlin, and St. Petersburg academies, the Royal Society of London, and the Institute of Bologna.

1700s

2893. SCHEUCHZER, Johann Jakob. 1672–1733. (Earth science 1027) M.D., Utrecht, 1694. After a few years as a municipal physician in Zurich he became head of the city library (his own works included an enormous history of Switzerland) and also director of the natural history museum. His scientific explorations of the Alps made him very prominent in Zurich and he received a grant from the government for the purpose. He carried on an extensive correspondence with many scholars throughout Europe and became a member of the Leopoldina and Berlin academies and the Royal Society of London.

1720s

2894. CRAMER, Gabriel. 1704–1752. (Mathematics 183) Educated at the Academy of Geneva and became professor of mathematics there in 1724. He travelled extensively and was acquainted with most of the leading mathematicians of the time. Through his assiduous correspondence he did much to disseminate mathematical ideas and stimulate research. As well as being a member of several of the leading academies he was also a prominent citizen of Geneva.

2895. BERNOULLI, Daniel. 1700–1782. (Mechanics 530, 532) Son of

Johann Bernoulli (2892). Learnt mathematics from his father and elder brother but studied medicine (at Basel, Heidelberg, and Strasbourg), receiving the M.D. at Basel in 1721. In 1724 he published a collection of essays on various topics in pure and applied mathematics and as a result was invited to join the new academy being established at St. Petersburg. His years at St. Petersburg (1725-33), where he was a colleague of Euler, were a very creative period. For many years after he left Russia he kept up a correspondence with Euler and he continued to publish in the Academy's periodical for the rest of his life.

In 1733 Bernoulli returned to Basel where he had been appointed professor of anatomy and botany, though his interests and his researches were in the mathematical sciences (he did however make some contributions to physiology). Ten years later he was enabled to lecture in physiology instead of botany, and in 1750 he at last obtained the chair of physics. His lectures on experimental physics, illustrated by demonstrations, were highly popular.

Between 1725 and 1749 Bernoulli won ten of the Paris Academy's prizes for essays on various aspects of the physical sciences, technology, navigation, etc., and also shared the 1735 prize with his father. He became famous throughout Europe and was a member of all the leading academies.

1730s

2896. HALLER, Albrecht von. 1708-1777. (Anatomy 1615. Physiology 1729, 1731) Born and educated in Bern. Studied medicine first at Tübingen and then at Leiden (under Boerhaave) where he took his M.D. in 1727. After some travels he returned to Bern to practise medicine and also to continue his investigations in anatomy and botany. In 1736 he was appointed professor of medicine at the new University of Göttingen where he laid out a botanical garden and built an anatomical theatre. He was the chief founder and first president of the Göttingen Society (2539), and for several years was editor of the *Göttingische Zeitungen von gelehrten Sachen* which he made into a highly regarded journal. Despite his accomplishments in research at Göttingen he resigned his chair in 1753, to the surprise of his colleagues, and returned to Bern. There he held a number of important official posts in the city while at the same time continuing his scientific writing.

As well as being the leading physiologist of his time Haller was a man of wide interests, scientifically and otherwise --in his youth an active mountaineer and an accomplished poet, author of a major work on the Swiss flora, and a pioneer of scientific and medical bibliography. In addition to his many books he wrote several hundred periodical articles. He became a famous figure and was made a member of the leading academies.

1740s

2897. NATURFORSCHENDE GESELLSCHAFT IN ZÜRICH. A private society founded in 1746 on the model of the Royal Society of London. It became a well-established body though its publication record was fragmentary until the mid-nineteenth century.

2898. BONNET, Charles. 1720-1793. (Zoology 1444, 1448) Came from a
      wealthy family in Geneva; graduated in law in 1744 but did not
      practise. Being of independent means he was able to devote
      his time to his interests in biology. He was originally much
      influenced by Réaumur whose works he studied and with whom he
      corresponded. He subsequently built up an extensive corres-
      pondence with many scientists.

1750s

2899. SOCIETAS PHYSICO-MEDICA BASILIENSIS. Its periodical, *Acta
      Helvetica*, began in 1751 and finally ceased in 1787 after a
      long intermission. Presumably the Society was a predecessor
      of the Naturforschende Gesellschaft in Basel, whose *Bericht*
      began in 1835.

1770s

2900. SAUSSURE, Horace Bénédict de. 1740-1799. (Geology 1121) Came
      from a wealthy and cultured patrician family in Geneva. In
      his youth he was much influenced by his uncle, Charles Bonnet,
      and by Haller. He graduated in philosophy at Geneva in 1759
      and three years later, with Haller's support, was appointed to
      the chair of philosophy (including physics and natural history)
      at the Academy of Geneva.
              His alpine researches were carried out mainly in the
      period 1774-89 (during the 1780s he was also heavily involved
      in Genevan politics). His ascent of Mont Blanc in 1787 and
      the experiments he carried out at the summit won him an inter-
      national reputation and membership of the leading academies.
      In addition to his scientific achievements he was the pioneer
      of mountain-climbing and inspired an enthusiasm for the Alps
      in place of the former negative attitude.

2901. SENEBIER, Jean. 1742-1809. (Botany 1338. Philosophy of science
      31) Studied theology and was ordained a pastor in Geneva in
      1765. He had an early interest in science and during a year
      that he spent in Paris he became acquainted with a number of
      French scientists; in Geneva he was much influenced by Charles
      Bonnet and Abraham Trembley. After a few years as a pastor he
      became city librarian of Geneva in 1773 (his *Histoire Littér-
      aire de Genève* was of some importance). He was in close touch
      --as was Bonnet--with the Italian biologist, Spallanzani, and
      from 1777 onward translated most of his works into French.
      During the same period he published a great deal on the photo-
      synthetic process and related chemical investigations.

1790s

2902. SAUSSURE, Nicolas Théodore de. 1767-1845. (Chemistry 842, 859)
      Son of H.B. de Saussure whom he assisted in his researches in
      the 1780s; from 1790 he was a member of the Société de Physique
      et d'Histoire Naturelle de Genève. During the Revolution he
      lived in England, returning to Geneva in 1802. From that year
      he was an honorary professor at the Geneva Academy but never
      taught and lived on his private income. In 1815 he was a found-
      ing member of the Société Helvétique des Sciences Naturelles.

2903. CANDOLLE, Augustin de. 1778-1841. (Botany 1344, 1353, 1354)
Born and educated in Geneva. His early interest in botany
was greatly stimulated by his association with J.P. Vaucher,
J. Senebier, and H.B. de Saussure. In 1796 he went to Paris
to study medicine but soon gave it up for botany. He made
friends with Cuvier and Lamarck and in 1802 became Cuvier's
assistant at the Collège de France. A few years later Lamarck
deputed him to revise his *Flore Française* and he was commissioned by the government to carry out a botanical and agricultural survey of France. In 1808 he was appointed professor
of botany at the Faculté des Sciences at Montpellier.

The Academy of Geneva (a university in effect but not
in name) had earlier tried to attract him back, and when it
offered him a chair in natural history in 1816 he accepted.
He completely reorganized the city's botanical garden, established the Conservatoire Botanique (which became an important
herbarium), and took a leading part in the foundation of a
natural history museum. In addition to his scientific and
educational activities he was involved in local politics and
became a prominent figure in the city.

Candolle's standing as a botanist is illustrated by
the fact that over three hundred species, two genera, and one
family have been named after him.

1810s

2904. SCHWEIZERISCHE GESELLSCHAFT FÜR DIE GESAMMTEN NATURWISSENSCHAFTEN
(named also in French and Italian). Founded in 1815 as a
national society holding annual meetings in different Swiss
cities. It was the pioneer of such national societies, the
German one (2589) being modelled on it, the British one (2367)
being modelled on the German, and so on.

2905. CHARPENTIER, Johann (or Jean) de. 1786-1855. (Geology 1161)
Born and educated at Freiberg in Saxony. About 1806 he graduated at the Freiberg Mining Academy where his father, a geologist of some distinction, had been a professor. After working
as a mining engineer in Silesia and then in the Pyrenees (where
he did some excellent research on the geology of the range)
he was appointed in 1813 director of the important salt mines
in canton Vaud, where he spent the rest of his life. His
ability as a geologist and mining engineer was recognized by
an appointment to an honorary professorship at the Academy
of Lausanne.

1820s

2906. SOCIÉTÉ DE PHYSIQUE ET D'HISTOIRE NATURELLE DE GENÈVE. Its
Mémoires began in 1821.

2907. STUDER, Bernhard. 1794-1887. (Geology 1181) Graduated in
theology at the Academy of Bern and then studied science,
spending the period 1816-18 at Göttingen. After teaching
science at Bern he became professor of geology and mineralogy
at the newly-constituted University of Bern in 1834.

2908. CANDOLLE, Alphonse de. 1806-1893. (Botany 1378, 1401) Son of
Augustin de Candolle. Graduated in science at Geneva in 1825

and carried on research on taxonomy under his father whom he
succeeded in 1835 as professor of botany at the Academy of
Geneva. In addition to his own numerous publications (chiefly
on taxonomy) he continued two of his father's initiatives: the
massive *Prodromus* (1354b) and the standardization of botanical
nomenclature; he was one of the organizers of the First International Botanical Congress, held at Paris in 1867, and edited
the *Lois de Nomenclature Botanique* (1867) containing the
recommendations of the Congress.

2909. PREVOST, Jean Louis. 1790-1850. (Embryology 1664) After studies
in Geneva and Paris he obtained the M.D. at Edinburgh in 1818.
From 1820 he was in medical practice in Geneva while at the
same time cultivating his scientific interests in conjunction
with a group of collaborators. In addition to his work in
embryology (which won the Prix Montyon of the Paris Academy)
he also did significant research in physiology and biochemistry.

1830s

2910. AGASSIZ, Louis. 1807-1873. (Palaeontology and geology 1160,
1167. Zoology 1480, 1484, 1496) Born and educated in Switzerland; studied at Zurich and several German universities, his
teachers including Schelling, Oken, and Döllinger. D.Phil. in
zoology, Erlangen, 1829. M.D., Munich, 1830. After studying
in Paris under Cuvier he became (through the influence of A.
von Humboldt) professor of natural history at the recently-
established Collège de Neuchâtel in 1832. In 1847, following
a successful lecture tour in the United States, he accepted a
professorship in zoology at Harvard. For his subsequent
career see 3035.

1840s

..... NÄGELI, K.W. von. Born and educated in Zurich where he was an
extraordinarius until 1852. See 2640.

1850s

..... HIS, W. Born and educated in Switzerland. Professor of anatomy
at Basel, 1857-72. See 2665.

1860s

..... SCHWENDENER, S. Born and educated in Switzerland. Professor
of botany at Basel, 1867-77. See 2658.

2911. SCHWEIZERISCHE ENTOMOLOGISCHE GESELLSCHAFT. Its periodical
began in 1862.

1870s

..... NENCKI, M. Professor of medical chemistry at Bern from about
1875 until 1890. See 3003.

2912. SCHWEIZERISCHE PALÄONTOLOGISCHE GESELLSCHAFT. Its periodical
began in 1874.

2913. HEIM, Albert. 1849-1937. (Geology 1200) Graduated in geology
at the Zurich Polytechnic in 1869. After some travels in Italy

and Germany he became professor of geology at the Polytechnic in 1873 (succeeding A. Escher von der Linth by whom he had been greatly influenced and whose work on Alpine geology he continued). From 1875 he was also professor of geology at Zurich University. He was appointed to the Swiss Geological Commission (i.e. Survey) in 1888 and served as its president from 1894 to 1926.

2914. FOREL, François Alphonse. 1841-1912. (Limnology 1080) Graduated in science at the Academy of Geneva and then in medicine at the University of Würzburg. From 1870 he was professor of anatomy and physiology at the Academy of Lausanne.

2915. SCHWEIZERISCHER ORNITHOLOGISCHER VEREIN. Its periodical began in 1877.

2916. FOL, Hermann. 1845-1892. (Embryology 1679) A native of Geneva. Studied medicine at Jena (zoology under Gegenbauer and Haeckel), Heidelberg, Zurich, and Berlin, and took his M.D. in 1869. Being very wealthy he was able to devote himself to research. In 1878 he accepted a titular (unpaid) professorship at the University of Geneva but resigned in 1886. In 1880 he established, at his own expense, a laboratory for research in marine biology at Villefranche, near Naples, where he spent much of his time. He was the founder and editor of the *Recueil Zoologique Suisse*.

1880s

2917. SOCIÉTÉ GÉOLOGIQUE SUISSE. Its periodical began in 1888.

1890s

2918. WERNER, Alfred. 1866-1919. (Chemistry 985) A native of Alsace (which came under German rule during his childhood). Graduated at the Zurich Polytechnic in 1889 and was awarded a doctorate the next year for his classic research, with A.R. Hantzsch (981a), on the stereochemistry of nitrogen compounds. After working for several months with Berthelot in Paris he returned to Zurich in 1892 as a *Privatdozent* at the Polytechnic. In the following year he became an *extraordinarius* at the University of Zurich and in 1895 *ordinarius*. His co-ordination theory brought him great fame and in 1913 he was awarded the Nobel Prize.

2919. GEOLOGICAL SURVEY OF SWITZERLAND. Its publications began in 1899. (The first geological map of the country had been published in 1853: see 1181. The Swiss Geological Commission was in existence in 1888, if not earlier: see 2913)

2.09 SPAIN AND PORTUGAL

1500s

2920. LAX, Gaspar. 1487–1560. (Mathematics 78) After studying at the University of Zaragoza he went to France and taught in various colleges of the University of Paris from 1507 to 1523. Returning to Spain, he taught mathematics and philosophy at Zaragoza from 1525, remaining there until his death.

1510s

2921. ORTEGA, Juan de. ca. 1480–1568. (Mathematics 77) Dominican. Teacher of mathematics in Spain and Italy.

1530s

2922. NUÑEZ (SALACIENSE), Pedro. 1502–1578. (Astronomy 325) Studied medicine, mathematics, and astrology at the universities of Salamanca and Lisbon. In 1529 he was appointed royal cosmographer. From 1544 he was also professor of mathematics at Coimbra.

2923. MONARDES, Nicolás Bautista. ca. 1493–1588. (Natural history 1221) Graduated in medicine at the University of Alcalá in 1533 and became a successful physician in Seville. He was the author of numerous books on medicine from 1536 onward.

2924. ORTA, Garcia d'. ca. 1500–1568. (Natural history 1219) After studying at the universities of Salamanca and Alcalá he qualified as a medical practitioner in 1526. From 1534 until his death he was a physician in the Portugese colony of Goa, in India.

1560s

2925. ACOSTA, Cristóbal. ca. 1525–1594. (Natural history 1222) A Portugese physician. Travelled widely and spent four years in India where he met Garcia d'Orta.

1570s

2926. HERNÁNDEZ, Francisco. 1517–1587. (Natural history 1220) Physician to Philip II of Spain.

2927. ACOSTA, José de. 1539–1600. (Natural history 1223) A Spanish Jesuit who spent the period 1570–87 in Peru and Mexico.

1770s

2928. LISBON ACADEMY. Founded in 1779 as Academia das Sciencias de Lisboa.

1840s

2929. MADRID ACADEMY. Founded in 1847 as the Real Academia de Ciencias (Exactas, Fisicas y Naturales).

1870s

2930. GEOLOGICAL SURVEY OF SPAIN. Its publications began in 1873.

1880s

2931. RAMÓN Y CAJAL, Santiago. 1852-1934. (Anatomy 1646) M.D., Zara-goza, 1877. He subsequently trained himself to be a very skil-ful microscopist and histologist. In 1883 he was appointed professor of anatomy at Valencia, in 1887 professor of histology at Barcelona, and in 1892 professor of histology and patholog-ical anatomy at Madrid. He remained at Madrid for the rest of his life and built up a research school there. His achievements in neurohistology brought him many honours, both Spanish and foreign, and in 1906 he received the Nobel Prize (shared with his Italian rival, Golgi).

2.10   EASTERN EUROPE

1450s

2932. PEUERBACH, Georg. 1423-1461. (Astronomy 318) M.A., Vienna,
      1453. He was in Italy for a short time about 1450 and lectured
      on astronomy at Padua and Ferrara; he is said to have been
      offered positions at some of the Italian universities. After
      returning to Vienna he became court astrologer to the King of
      Hungary and later to the Emperor. He also lectured at Vienna
      University, mostly on classical literature, until his untimely
      death.

2933. REGIOMONTANUS (or MÜLLER), Johannes. 1436-1476. (Astronomy 319.
      Mathematics 68) M.A., Vienna, 1457. In the same year he was
      appointed to the staff of the University and became closely
      associated with Peuerbach. From 1461 to about 1467 he was in
      Italy where he met many notables and for a time lectured on
      astronomy at Padua. For a few years after his return he was
      court astrologer to the King of Hungary in Buda. In 1471 he
      settled in Nuremberg where he set up his own printing press
      and became the first publisher of astronomical and mathematical
      literature.

1510s

2934. COPERNICUS, Nicholas. 1473-1543. (Astronomy 330) Studied at
      the University of Cracow from 1491 and then spent the period
      1496-1503 mostly in Italy developing his astronomical and
      mathematical interests, and studying various subjects at the
      universities of Bologna, Padua, and Ferrara; from the last he
      obtained a doctorate of canon law. For the rest of his life
      he was a canon of the cathedral of Frauenburg (or Frombok--near
      the Baltic coast, east of Gdańsk), a post in which he was
      mainly occupied with administrative and economic matters. From
      the 1510s however he acquired a growing reputation as a mathe-
      matical astronomer.

1520s

2935. RUDOLFF, Christoff. Early 16th century. (Mathematics 83) Very
      little is known of his life. He was in Vienna in the early
      1520s and apparently studied at the University.

1580s

2936. BOODT, Anselmus Boetius de. ca. 1550-1632. (Mineralogy 1011)
      A native of Bruges. Studied law at Louvain, then medicine at
      Heidelberg and Padua, obtaining his M.D. at the latter. In
      1583 he migrated to Prague where he practised medicine, in 1604
      becoming physician to the Emperor Rudolph II who encouraged
      his study of minerals. On the death of the Emperor in 1612 he
      returned to Bruges.

1590s

2937. KEPLER, Johannes. 1571–1630. (Astronomy 343, 347, 349, 354.
      Physics 577, 580) A native of Württemburg. After receiving
      the M.A. at Tübingen (where he was introduced to Copernican
      astronomy by Mästlin) in 1591, he began the study of theology
      but gave it up to become a teacher of mathematics in the
      Lutheran school at Graz. In 1600 he joined Tycho Brahe's staff
      in Prague and when Brahe died the next year Kepler was appoint-
      ed his successor as mathematician to the Emperor Rudolph II.
      In 1612, in consequence of Rudolph's death, he left Prague and
      obtained an appointment as provincial mathematician in Linz
      where he remained until 1626. Caught up in the turmoil of the
      Thirty Years War, he spent his last years under the patronage
      of the famous general, Wallenstein.

2938. CROLL, Oswald. ca. 1560–1609. (Chemistry 781) A native of
      Germany. Studied at the universities of Marburg, Heidelberg,
      Strasbourg, and Geneva, and gained his M.D. about 1582. From
      1590 he was physician and councillor to several members of the
      high nobility in Eastern Europe and, after settling in Prague
      in 1602, to the Emperor Rudolph II.

2939. ZALUŽANSKÝ, Adam. ca. 1558–1613. (Botany 1302) Educated at
      the Charles University in Prague, then studied medicine at
      Helmstedt where he gained the M.D. in 1587. After teaching
      classics at the Charles University for a few years he became
      a medical practitioner in Prague.

1610s

2940. GULDIN, Paul. 1577–1643. (Mathematics 132) Jesuit. Teacher
      of mathematics at the Jesuit colleges in Rome and Graz, and
      later professor of mathematics at the University of Vienna.

1640s

2941. HEVELIUS, Johannes. 1611–1687. (Astronomy 361, 371, 377, 383)
      A native of Danzig (then an autonomous city). After studying
      law at Leiden in the early 1630s he held civic offices in his
      native city, eventually becoming a magistrate and city coun-
      cillor. From an early age he had been interested in astronomy
      and in the early 1640s he established a private observatory
      which, with the improvements he made to it, was for a time
      the best in Europe. He carried on a large correspondence with
      other astronomers and his observatory was visited by many of
      them, notably Halley.

1770s

2942. PRAGUE SOCIETY. Began in 1775 as a private society established
      by the prominent mineralogist, Count Ignaz von Born. In 1784
      it was transformed into a public body, the Böhmische Gesell-
      schaft für Wissenschaften, which acquired the prefix 'König-
      liche' six years later. In its early years the Society embarked
      on an ambitious survey of the natural resources of Bohemia but
      because of continuing financial difficulties the survey remained
      incomplete.

2943. SCHEMNITZ MINING ACADEMY. Founded in 1770 at Schemnitz (or Selmec) in Hungary by the government of the Austrian Empire. It was the second institution of its kind in Europe (the first having been established four years earlier at Freiberg in Saxony: see 2549).

2944. PROCHÁSKA, Georgius. 1749-1820. (Physiology 1737) M.D., Vienna, ca. 1776. He was professor of anatomy at Prague from 1778 and then at Vienna from 1791 almost until his death.

1800s

2945. BOLZANO, Bernard. 1781-1848. (Mathematics 211, 254) Graduated in philosophy, physics, and mathematics at Prague in 1800. Ordained a Catholic priest in 1804 and appointed to the new chair of the philosophy of religion at Prague in 1805. He was dismissed from the chair in 1819, during the era of reaction, but was able to continue his scholarly activities which embraced social, ethical, and theological questions as well as mathematics and logic. His influence on the development of mathematics was much less than it might have been because most of his mathematical work was not published.

2946. MOHS, Friedrich. 1773-1839. (Mineralogy 1095) A native of Saxony. Studied at the University of Halle and then at the Freiberg Mining Academy under Werner, graduating about 1800. After holding various positions in Austria relating to mining and mineralogy he was appointed curator of the mineral collection at the Johanneum museum in Graz in 1811. Subsequently he was professor of mineralogy at the Freiberg Academy (in succession to Werner) from 1818, and then at the University of Vienna from 1828.

2947. STERNBERG, Kaspar Maria von. 1761-1838. (Palaeobotany 1143) A member of the Bohemian landed aristocracy. Studied theology and became an ecclesiastic, holding various positions at the court of the Prince-Bishop of Regensburg from 1786 onward. He developed an interest in botany and when he accompanied the Prince-Bishop on a visit to Paris in 1805-06 his knowledge of the subject was much enlarged by his introduction to the French scientific scene. In 1808 he inherited the family estate near Prague where he lived for the rest of his life and where his botanical research was done. Around 1820 he was one of the chief founders of the Bohemian National Museum.

1820s

2948. BUDAPEST ACADEMY. Magyar Tudományos Akademia. (Ungarische Akademie der Wissenschaften) Begun by private initiative and finance in 1825 and approved by the King of Hungary (who was the Emperor of Austria) in 1831. It published books and periodicals in Hungarian and from 1882 issued the *Mathematische und naturwissenschaftliche Berichte aus Ungarn.*

2949. BOLYAI, János. 1802-1860. (Mathematics 215) Son of Farkas Bolyai from whom he derived both his precocity in mathematics and the concern with Euclid's fifth postulate that was to

bring him posthumous fame. He graduated at the Imperial
Engineering College in Vienna in 1822 and served as an army
engineer until 1833 when he was pensioned off as an invalid.
Largely because of the lack of recognition of his essay on
non-Euclidean geometry he published no other mathematical work.

..... PURKYNĚ, J.E. Graduated at Prague and was an assistant in anatomy there until 1823 when he gained a chair in Germany. He
returned to Prague in 1850 as professor of physiology. See 2609.

1830s

2950. WIENER MUSEUM DER NATURGESCHICHTE. Its publications began in
1835. It was later named the K.-K. Naturhistorisches Hofmuseum.

1840s

2951. VIENNA ACADEMY. Attempts to establish an academy in Vienna
originated with Leibniz, who was there in 1712-14. For different reasons at different times, so it appears, the successive
attempts did not meet with approval by the Imperial authorities
(though they did support academies in other parts of their
realms--in Prague, Brussels, and some of the cities of north
Italy). By the 1840s the situation had become embarrassing--
Vienna being the only capital in Europe (except Madrid) without
an academy.

The Kaiserliche Akademie der Wissenschaften was formally instituted in 1847 with two classes (mathematical-scien-
tific and historical-philological) and carefully contrived
arrangements concerning the politically sensitive issue of how
the membership was to be drawn from the several parts of the
Empire. Through the second half of the century the Academy
was very active and issued numerous periodicals, the chief of
which soon became very big.

2952. CONGRESS OF HUNGARIAN PHYSICIANS AND NATURALISTS. Magyar Orvosok
és Természetvizsgálók Nagygyülése. A national society of the
'advancement of science' type, presumably modelled on the
German society (2589). Its periodical began in 1841.

2953. DOPPLER, Johann Christian. 1803-1853. (Physics 678) Studied
at the Polytechnic Institute in Vienna, 1822-25. In 1841,
after some minor teaching posts, he became professor of mathematics at the Technical Academy in Prague. Two years later
he became a member of the Prague Society and he was made a
member of the Vienna Academy at its foundation in 1847. In
1850 he was appointed professor of experimental physics and
director of the new physics institute at the University of
Vienna.

2954. GEOLOGISCHE REICHSANSTALT, Vienna. Established in 1849 for the
geological survey of the Austro-Hungarian Empire. It became
an important centre for mineralogical and geological research.

2955. BRÜCKE, Ernst Wilhelm. 1819-1892. (Physiology 1760) A native
of Berlin. Graduated in medicine there in 1842 and became an
assistant to Müller and a member of the brilliant group associated with him. After several years of research at Berlin and

a brief stay at Königsberg he became professor of physiology at Vienna in 1849. He held the chair until his retirement in 1890 and built up an important research school which attracted students from several countries.

1850s

2956. STEFAN, Josef. 1835-1893. (Physics 732) D.Phil., Vienna, 1857. From 1858 he was a *Privatdozent* at Vienna, and in 1863 he became professor of physics. He was elected to the Vienna Academy in 1865 and was later prominent in its affairs. In his later years, as a successful teacher and experimenter, he was awarded various scholarly and national honours.

2957. SUESS, Eduard. 1831-1914. (Geology 1196, 1205) Because of his participation in the revolution of 1848 he was not allowed to continue his studies at the Polytechnic Institute in Vienna. However his ability in geology (mostly acquired through his own efforts) impressed Wilhelm Haidinger, the head of the Imperial Geological Survey, and in 1852 he gave Suess a post as assistant in the Imperial Geological Museum. After the publication of his early researches Haidinger was able to procure his appointment as an *extraordinarius* in geology at the University of Vienna in 1857, despite his lack of a degree. He became full professor five years later and held the post until his retirement in 1901.

     Suess was elected to the Vienna Academy in 1867 and later took a prominent part in its activities, becoming one of the statesmen of science in the German-speaking world. His concern for education and other social issues led him into politics from 1863 onward—first as a member of the Vienna City Council, then of the Diet of Lower Austria, and finally of the Reichsrat in which he sat for many years as the Liberal deputy from Vienna.

2958. MENDEL, Gregor. 1822-1884. (Heredity 1530) After a good education (including science) he entered the Augustinian monastery in Brünn in 1843 and was ordained a priest four years later. The atmosphere of the monastery was conducive to scientific research: there was a good library, several of the monks were teachers interested in science and agriculture, and one of them was already experimenting on hybridization of plants in a special garden for the purpose.

     In 1851 Mendel was sent by his abbot to the University of Vienna where he spent two years studying science, especially botany. After his return to Brünn he taught science in the local technical school until 1868 when he was elected abbot of the monastery, a position he held until his death.

1860s

2959. ROMANIAN ACADEMY. Established in 1866 on the model of the Institut de France.

2960. MACH, Ernst. 1838-1916. (Mechanics 563. Physiology 1781. Popularization 67) D.Phil., Vienna, 1860; then a *Privatdozent* in physics. In 1864 he became professor of mathematics at Graz

and in 1867 professor of physics at Prague; the latter position
he held until 1895 when he became professor of inductive philosophy at Vienna. He was a capable, if unconventional, physicist but is best known today for his positivist epistemology
and his views on the philosophy and history of science.

2961. BOLTZMANN, Ludwig. 1844-1906. (Physics 713, 716, 738) D.Phil.,
Vienna, 1867. From 1869 he held a succession of chairs, first
at Graz and Vienna, then (from 1890) at Munich and Leipzig,
and finally (from 1902) at Vienna again.

2962. TSCHERMAK, Gustav. 1836-1927. (Mineralogy 1102, 1105) After
studying at Vienna he took his doctorate at Tübingen in 1860.
He became a *Privatdozent* at Vienna in 1861 and in the following
year an assistant curator at the Hof-Mineralien-Cabinet. In
1868 he was appointed professor of mineralogy and petrography
at the University and concurrently director of the Hof-Mineralien-Cabinet. He gave up the latter post in 1877 when the
University provided him with an institute. Earlier, in 1871,
the growth of his research school had led him to begin the
journal *Mineralogische Mittheilungen* which soon became important (it was later known as *Tschermaks mineralogische und petro-
graphische Mittheilungen*).

Tschermak was elected to the Vienna Academy in 1875
and was one of the founders of the Austrian Mineralogical
Society. He was ennobled on his retirement in 1906 and received
many scholarly honours.

2963. ÖSTERREICHISCHE GESELLSCHAFT FÜR METEOROLOGIE. Founded in 1865.

2964. HANN, Julius Ferdinand. 1839-1921. (Meteorology 1069, 1075.
His important *Lehrbuch der Meteorologie* appeared in 1901)
D.Phil., Vienna, 1868. In the previous year he had been
appointed to the staff of the Zentralanstalt für Meteorologie
und Erdmagnetismus and, after an interlude as an *extraordinarius*
in physical geography at the University of Vienna from 1874,
he became director of the Zentralanstalt in 1877 (and concurrently professor of physics). He resigned the post in 1897 and,
after a brief sojourn at Graz, became professor of cosmic
physics at Vienna in 1900, holding the position until 1910.

In 1865 Hann had been one of the founders of the Österreichische Gesellschaft für Meteorologie and had thereafter
taken part in the editing of its *Zeitschrift*. The Austrian
and German meteorological societies combined their periodicals
in 1886 to form the *Meteorologische Zeitschrift* which Hann
continued to edit almost to the time of his death. It was the
world's leading meteorological journal and through it he exercised a great influence on the development of the science.

Hann became a member of the Vienna Academy in 1877 and
was later its secretary; he was also a foreign member of many
scientific bodies in other countries. He took a prominent part
in the First International Congress of Meteorologists in Leipzig
in 1872 and was a member of the International Meteorological
Committee from 1878 to 1898. He was ennobled in 1910.

..... HERING, E. Professor of physiology at Vienna, 1865-70, and at
Prague, 1870-95. See 2696.

1870s

2965. CRACOW ACADEMY. Akademia Umiejetności w Krakowie (Académie des Sciences et Lettres de Cracovie). Founded in 1873. It was preceded by the Towarzystwo Naukowe Krakowskie (Société des Sciences de Cracovie) whose origins went back to 1815, in close association with the University of Cracow.

2966. SOCIETY OF CZECH MATHEMATICIANS AND PHYSICISTS. Jednota Českých Matematiků a Fysiků. Its periodical began in 1872.

2967. EÖTVÖS, Roland *Baron* von. 1848-1919. (Physics 736) Came from a distinguished aristocratic family in Budapest. Studied at Heidelberg and Königsberg, taking his doctorate in physics at the former in 1870. After his return to Hungary he became professor of theoretical physics at Budapest in 1872, changing to the chair of experimental physics in 1878.

       Eötvös was for many years the president of the Hungarian Academy of Sciences, and in 1891 he was one of the founders of the Hungarian Mathematical and Physical Society, becoming its first president. He did much to promote the development of higher education in Hungary and in 1894-95 held the post of Minister of Public Instruction. In his later years he received many honours, Hungarian and foreign.

2968. SOCIETY OF CZECH CHEMISTS. Spolek Chemiků Českých. Its periodical began in 1875.

2969. HUNGARIAN GEOLOGICAL SOCIETY. Magyarhoni Földtani Társulat. Its periodical began in 1871.

1880s

..... PENCK, A. Professor of physical geography at Vienna, 1885-1906. See 2714.

2970. RABL, Carl. 1853-1917. (Cytology 1700) M.D., Vienna, 1882; then a prosector in the anatomical institute. He was professor of anatomy at the Ferdinand University in Prague from 1886, and at Leipzig from 1904. Most of his work was in morphology and comparative embryology but his outstanding contribution was in cytology.

1890s

2971. TSCHERMAK (VON SEYSENEGG), Erich. 1871-1962. (Heredity 1551) Studied at both the University and the Hochschule für Bodenkultur in Vienna, and later at the University of Halle where he gained a diploma in agriculture in 1895 and a D.Phil. in botany in the following year. After working at some agricultural stations in Germany he returned to Vienna in 1900 to join the staff of the Hochschule für Bodenkultur; he became an *extraordinarius* there in 1906 and full professor in 1909. For some years he was also director of the Royal Institute for Plant Breeding (later the Mendel Institute) in Moravia. In his later years he received many honours.

2972. ST. PETERSBURG ACADEMY. Initiated by the tsar-reformer, Peter
the Great, who in the early 1720s drew up the plans for it
(influenced to some extent by the ideas of Leibniz). He died
in 1725 but the Academy was formally established as the Academia Scientiarum Imperialis Petropolitana by his successor,
Catherine I, later in the same year.

      The early members of the Academy were necessarily all
non-Russians. They included several mathematicians, scientists,
and scholars who were soon to become outstanding (especially
Euler and Daniel Bernoulli) and the Academy's *Commentarii*,
begun in 1728, became highly respected in the West. The
original plans for the Academy included provisions for training
native Russians with the intention that in the course of time
the foreign members would be gradually replaced by Russians.
This scheme was unsuccessful however and until well into the
nineteenth century the Russian members were a small minority.
The first Russian, Lomonosov, was appointed in 1745.

      Despite all the trials it had to endure from court
favorites and reactionary bureaucrats, the eighteenth-century
Academy was remarkably successful. As well as participating
actively in the international science of the time it did an
immense amount of work in the scientific exploration of the
vast territories of the Russian Empire, and also performed a
valuable service in translating foreign books into Russian
and in various other ways disseminating a knowledge of science
in the country. For its later period see 2978.

2973. EULER, Leonhard. 1707-1783. (Mathematics 178, 181, 184, 190.
Mechanics 531, 535, 537, 539. Astronomy 393, 398, 409.
Physics 612, 622. Popularization 30) A native of Basel.
Graduated in philosophy at Basel University in 1724 (having
studied mathematics under Johann Bernoulli). He began publishing original work in mathematics in 1726 and, through his
friendship with members of the Bernoulli family, received an
invitation to join the new academy at St. Petersburg. He
arrived in St. Petersburg in 1727 (and never returned to Basel).

      The Academy provided a congenial setting for the
unfolding of Euler's genius and, reciprocally, the stream of
brilliant papers which he produced for its *Commentarii* was a
major factor in making the Academy known and respected in the
West, and consequently in giving it more significance in the
eyes of its Russian overlords. After the death of the Empress
Anne, however, the situation outside the Academy became threatening for foreigners and in 1741 Euler decided to accept an
attractive offer he had received from the King of Prussia.
For his subsequent career in Berlin see 2537.

..... BERNOULLI, D. Member of the St. Petersburg Academy, 1725-33.
      See 2895.

2974. DELISLE, Joseph Nicolas. 1688-1768. (For his early career in
      Paris see 2045) A foundation member of the St. Petersburg
      Academy, he arrived in the city in 1725 bringing with him the
      astronomical instruments that Peter the Great had bought in
      Paris a few years earlier. They were installed in the new
      observatory which had been built for the Academy and which,
      under Delisle's direction, soon became one of the best in
      Europe. There, as well as carrying on his own programme of
      observations, he trained the first generation of Russian
      astronomers. An important part of his students' activities
      was the determination of the co-ordinates of various localities
      in preparation for an accurate map of Russia.
            He returned to Paris in 1747; see 2045.

1740s

2975. LOMONOSOV, Mikhail Vasilievich. 1711-1765. (Not in Part 1)
      The first native Russian scientist of distinction. "It is
      his symbolic role more than his actual contributions that
      explains the continued growth of his reputation [in the twenti-
      eth century]. He represented a turning point in the history
      of science in Russia; more than any other pioneer, he helped
      to make science a vital part of Russian culture." (A. Vucinich)
            In 1736 Lomonosov was admitted as a student to the
      St. Petersburg Academy which sent him to the University of
      Marburg where he studied scientific and humanistic subjects
      for three years. After travelling in Germany and Holland he
      returned to St. Petersburg in 1741 to begin a very varied and
      sometimes stormy academic career. He was concerned with most
      of the physical sciences, but chiefly with physics and chemistry
      within the framework of corpuscularian natural philosophy. He
      published only a few papers in his lifetime, most of his work
      remaining in manuscript until it was rediscovered in the early
      twentieth century. Thus he had very little influence outside
      Russia (though enough to be made an honorary member of the
      academies of Stockholm and Bologna). Within Russia however
      he was a major figure in the Academy.
            His chief achievement was in the work he did, in his
      various roles, to build up the scientific (and literary) culture
      of the country, most notably in the leading part he played in
      the foundation of Moscow University in 1755.

1750s

..... AEPINUS, F.U.T. Member of the St. Petersburg Academy from 1757.
      See 2543.

1760s

2976. EULER, Leonhard. 1707-1783. (For entries in Part 1 see 2973.
      For his career in Berlin see 2537) On his return to St. Peters-
      burg in 1766 he was received by Catherine the Great as if he
      were of royal rank. It was an important event for the Academy
      whose prestige had been waning but was henceforth restored.

Euler continued to be its leading figure and he used his
prestige with the authorities to do what he could to strengthen
the Academy. During this last stage of his career his extraordinary productivity did not slacken, despite the blindness
that afflicted him soon after his return; indeed almost half
his works were produced after 1765. In the last year of his
life he made one of his most important discoveries in the
theory of numbers.

Though none of Euler's Russian disciples were particularly outstanding they laid the foundations of a mathematical
tradition in the country with their teaching and textbooks.

2977. PALLAS, Peter Simon. 1741-1811. (Natural history 1237. Geology
1118. Zoology 1464) Born and educated in Germany. Studied
medicine at Halle, Göttingen, and finally Leiden where he took
the M.D. in 1760. For the next six years he worked on invertebrate zoology in England and Holland, and became a member of
the Royal Society and the Leopoldina Academy. In 1767, already
an established scholar, he was appointed to the St. Petersburg
Academy in which he soon became a leading figure, especially
in connection with the Academy's ambitious projects for the
scientific exploration of the Russian Empire. He took a prominent part in the Academy's expedition of 1768-74, traversing
a large part of Russia and Siberia and returning with a vast
collection of specimens and data. His later work in the Academy, chiefly on zoology and the earth sciences, did much to
uphold its international prestige and to stimulate the embryonic scientific movement in Russia.

1800s

2978. ST. PETERSBURG ACADEMY. (For its early years see 2972) In 1803
the Academy was reorganized and granted a new charter which
made a number of improvements; it also changed the official
language from Latin to French, the Academy's title now becoming
Académie Impériale des Sciences de St. Pétersbourg and its
periodical the *Mémoires*.

As a result of the establishment of several universities, as well as numerous technical and medical schools, during
the course of the nineteenth century the number of native
Russian scientists increased greatly. Until late in the century
however the Academy remained dominated by non-Russians (nearly
all Germans) who deliberately kept the Russian members in a
minority (the exclusion of Mendeleev being an extreme case;
see 2991). This situation in the Academy was part of a larger
issue in Russian politics.

For later developments see 3010.

2979. SOCIÉTÉ IMPÉRIALE DES NATURALISTES DE MOSCOU. A natural history
society closely associated with Moscow University. Founded
with government approval in 1805. Its periodicals were published in French.

2980. SCHOOL OF MINES. Established in St. Petersburg, originally in
1783--presumably on the model of the French school (2091)
founded in that year. It was re-established in 1804 as the
Mining Institute and subsequently played an important part in

the development of geology, as well as mining technology, in Russia. From 1825 it published *Gornuy Zhurnal (Journal des Mines de Russie)*.

1810s

2981. LOBACHEVSKY, Nikolai Ivanovich. 1792-1856. (Mathematics 227) Graduated at the recently-established University of Kazan in 1812 (where he studied mathematics under J.M.C. Bartels, a friend of Gauss). He became a *Privatdozent* in physical and mathematical sciences at Kazan in 1814, an *extraordinarius* in 1816, and *ordinarius* in 1822. The rest of his life was spent at the University, and he was its rector from 1827 to 1846.
> Lobachevsky's great achievement in discovering non-Euclidean geometry was not recognized during his lifetime and his fame was posthumous (as was also the case with Bolyai).

2982. STRUVE, (F.G.) Wilhelm. 1793-1864. (Astronomy 436) Born and educated in Germany but left the country in 1808 to avoid being conscripted for the Napoleonic Wars. He went to Dorpat (now Tartu) in Estonia, within the Russian Empire, where there was a German-speaking university at which he studied. Becoming interested in astronomy, he was allowed to work in the university observatory; the observations he made there gained him the doctorate in 1813 and appointment as an *extraordinarius* in mathematics and astronomy. In 1820 he became full professor and director of the observatory which, through his efforts and the equipment he was able to obtain, became the best in the Russian Empire.
> Struve was elected to the St. Petersburg Academy in 1832 and two years later was appointed director of the big new observatory that the government had decided to establish at Pulkovo, near St. Petersburg. It was opened in 1839 and so well equipped and staffed that, under Struve's direction, it soon became one of the leading observatories of the world. When Struve retired in 1858 he was succeeded as director by his son, Otto Wilhelm Struve.

2983. IMPERIAL MINERALOGICAL SOCIETY. Imp. Mineralogicheskoe Obshchestvo. Founded in St. Petersburg in 1817; it was not very active until the 1840s. Its periodical was originally published in Russian, then from 1842 to 1866 in German, and thereafter in Russian. From 1866 the Society received a government grant to enable it to participate in the geological exploration of the country.

1820s

2984. OSTROGRADSKY, Mikhail Vasilievich. 1801-1862. (Mathematics 226) Graduated in mathematics and physics at Kharkov in 1820. From 1822 to 1827 he studied in Paris where he published some original work and won the commendation of Cauchy. On his return to Russia in 1828 he was elected to the St. Petersburg Academy and subsequently rejuvenated its mathematical activities which had been in a state of decline for many years.
> Though Ostrogradsky never taught at a university his energetic and successful teaching at several higher technical schools (military, naval, engineering, etc.) in St. Petersburg,

together with his textbooks, gave a powerful stimulus to the development of mathematics and related sciences in Russia. He thereby did much to set the scene for the creation of the St. Petersburg mathematical school by his younger contemporary, Chebyshev.

2985. HESS, Germain Henri (or GESS, German Ivanovich). 1802-1850. (Thermochemistry 889) Born in Geneva but brought up in Russia. He graduated in medicine at Dorpat in 1825 and then worked for a month in Stockholm with Berzelius who had a lasting influence on him. After returning to Russia his work in mineralogy led to his election to the St. Petersburg Academy in 1828. He remained an academician for the rest of his life but also taught chemistry at several institutions in St. Petersburg. His influence on the spread of chemistry in Russia was considerable, especially through his textbook which went through seven editions from its first publication in 1834.

..... PANDER, C.H. Member of the St. Petersburg Academy, 1821-27. See 2587.

1830s

2986. PULKOVO ASTRONOMICAL OBSERVATORY. Established at Pulkovo, near St. Petersburg, by the government of Nicholas I and officially opened in 1839. It was so well equipped and capably run by its director, Wilhelm Struve (2982), that it ranked as one of the best observatories in the world.

2987. LENZ, Emil. 1804-1865. (Physics 669) Born in Dorpat and educated at the university there. In 1828 he was appointed to a junior position in the St. Petersburg Academy and in 1834 became a full academician. He also taught physics at various technical colleges (military, engineering, etc.) and at the University of St. Petersburg where from 1840 he was dean of the mathematics and physics department. His teaching and textbooks had considerable influence and he took a prominent part in scientific affairs in Russia.

..... BAER, K.E. von. Member of the St. Petersburg Academy from 1834. See 2586.

1840s

2988. CHEBYSHEV, Pafnuty Lvovich. 1821-1894. (Mathematics 240. Mechanics 558) Graduated at Moscow in 1841 and took his master's degree in mathematics five years later. He was appointed assistant professor of mathematics at St. Petersburg University in 1847, associate professor in 1850, and full professor in 1860. In 1852 he spent several months in the West (visiting Paris, London, and Berlin) and in the following year was elected to the St. Petersburg Academy.

As a teacher Chebyshev was of major importance and was chiefly responsible for creating and guiding the distinctive Petersburg mathematical school (sometimes called the Chebyshev school) which became prominent on the international scene in the late nineteenth century. His students eventually occupied the mathematical chairs at St. Petersburg and most of the other Russian universities. In his later years he received many honours, Russian and foreign.

2989. RUSSIAN GEOGRAPHICAL SOCIETY. Imp. Russkoe Geograficheskoe
      Obshchestvo. Founded in St. Petersburg in 1845. It became a
      large and active society which sent out many expeditions and
      took a leading part in the development of the earth sciences
      generally (and also ethnography) in Russia. It also did much
      to disseminate knowledge of the earth sciences among the public.

1850s

2990. BUTLEROV, Aleksandr Mikhailovich. 1828-1886. (Chemistry 929)
      Graduated at Kazan in 1849 and then taught there, gaining his
      doctorate (at Moscow) in 1854 and becoming professor of chem-
      istry in 1857. During 1857-58 he visited chemistry schools
      in the West, spending much of his time in Paris where he worked
      in Wurtz' laboratory for some months. He met Couper there,
      and in Heidelberg he met Kekulé. After returning to Russia
      he made another brief visit to the West in 1861 to present his
      theory of chemical structure at a congress in Speyer.
          At Kazan Butlerov had several able students who later
      became important chemists, and when he became professor of
      chemistry at St. Petersburg in 1868 he created an active school
      there. He was appointed to the Academy in 1870 and was an
      energetic participant in the bitter disputes between the German
      and Russian factions within the Academy. He was also a promin-
      ent figure in the Russian Chemical Society and a foreign member
      of the main chemical societies of western Europe.

2991. MENDELEEV, Dmitry Ivanovich. 1834-1907. (Chemistry 942) Gradu-
      ated at St. Petersburg in 1856 and became a *Privatdozent*. In
      1859 he was sent to the West, chiefly Heidelberg and Paris,
      for two years of further study; while there he attended the
      Karlsruhe Congress (926) in 1860. He returned to his position
      on the staff of St. Petersburg University (with some short-term
      concurrent posts) and in 1867 was appointed professor of chem-
      istry. In the following year he was one of the founders of
      the Russian Chemical Society.
          The great fame which came to Mendeleev for his discovery
      of the periodic classification did not preserve him from some
      severe rebuffs. The refusal of the St. Petersburg Academy to
      elect him to membership in 1880 (as a result of the long-stand-
      ing rift between the German and Russian factions in the Academy)
      became a *cause célèbre*. Ten years later, as an indirect conse-
      quence of the political turmoil among the students at St. Peters-
      burg University, he was forced by the Ministry of Education to
      resign his chair. In the following year he was given a minor
      official position and in 1893 he was appointed director of the
      Bureau of Weights and Measures, a post he held until his death.

2992. BEILSTEIN, Konrad Friedrich. 1838-1906. (Chemistry 960)
      Born in St. Petersburg of German parents and at the age of
      fifteen sent to Germany to complete his education. His univ-
      ersity studies were done at Heidelberg, Munich, and Göttingen,
      and he took his doctorate at Göttingen in 1858. After further
      studies in Paris and Breslau he became a *Privatdozent* at Gött-
      ingen in 1860 and an *extraordinarius* five years later. In 1866
      he returned to St. Petersburg to become professor of chemistry
      at the Technical Institute where he remained for the rest of
      his life. He was elected to the Academy in 1881.

2993. RUSSIAN ENTOMOLOGICAL SOCIETY. Societas Entomologica Rossica. Founded in St. Petersburg in 1859. Its periodical, begun in 1861, was published in both Latin and Russian.

2994. SECHENOV, Ivan Mikhailovich. 1829-1905. (Not in Part 1. His researches were capable but not outstanding: his importance was as "the father of Russian physiology") Graduated in medicine at Moscow University in 1856. He then spent three years in Germany, working at Berlin (under Müller and others), Vienna (under Ludwig), and Heidelberg (under Helmholtz). On returning to Russia in 1860 he was appointed professor at the St. Petersburg Medical and Surgical Academy; during 1862 he obtained leave to go to Paris and work in Bernard's laboratory. He was later professor at the universities of Odessa (from 1871), St. Petersburg (from 1876), and Moscow (from 1891 to 1901).

     Sechenov was an able teacher and research director who modelled his academic role on that of Karl Ludwig whom he greatly admired. In the next generation most chairs in physiology in the Russian universities were held by his students. He also initiated a pronounced tradition of research in Russian physiology in the border region between neurophysiology (especially reflexes) and psychology, the most distinguished representative of the tradition being Pavlov.

1860s

2995. CONGRESS OF RUSSIAN SCIENTISTS AND PHYSICIANS. Siezd Russkikh Estestvoispuitatelei i Vrachei. A national society, doubtless based on the German model (2589), which held meetings every few years in different cities, beginning at St. Petersburg in 1867. It was very successful.

2996. MOSCOW MATHEMATICAL SOCIETY. Moskovoskoe Matematicheskoe Obshchestvo. Founded in 1864 and officially chartered in 1867. Its periodical, one of the first important scientific periodicals to be published in Russian, grew slowly at first but expanded rapidly near the end of the century.

2997. RUSSIAN CHEMICAL SOCIETY. Founded in 1868. Its periodical, published in Russian, began in the following year. In 1873 the Society merged with the recently-established Russian Physical Society to form the Russkoe Fiziko-Khimicheskoe Obshchestvo.

2998. MARKOVNIKOV, Vladimir Vasilevich. 1837-1904. (Chemistry 932) Studied under Butlerov at Kazan, and after graduating in 1860 became his assistant. During 1865-67 he studied in Germany, mainly at Heidelberg and Leipzig. After his return he succeeded Butlerov at Kazan but resigned in 1871 in protest at some actions of the University's administrators. In 1873 he was appointed to the chair of chemistry at Moscow where he created an active research school.

2999. VOEYKOV, Aleksandr Ivanovich. 1842-1916. (Meteorology 1076) Studied at St. Petersburg and then at several German universities, taking his doctorate in meteorology at Göttingen in 1865. After his return to Russia he became an active member of the Russian Geographical Society and took part in expeditions to various parts of the Russian Empire. During the

years 1872-76 he travelled to many countries around the world,
studying their geography and climate; he also established
contacts with numerous scientists and scientific institutions
in Europe and the United States. In 1885 he was appointed an
extraordinarius in geography at St. Petersburg and in 1887
ordinarius.

Voeykov was a leader in the organization of meteorological observations in Russia and served as editor of *Meteor-
ologicheskii Vestnik* (a periodical published by the Russian
Geographical Society) from its beginning in 1891 until his
death. Though he was the leading meteorologist in Russia and
a member of many scientific societies, Russian and foreign, he
was not elected to the St. Petersburg Academy--probably because
of his opposition to the non-Russian majority.

3000. KOVALEVSKY, Aleksandr Onufrievich. 1840-1901. (Embryology 1674)
Studied zoology and related subjects at St. Petersburg, Heidelberg, and Tübingen, and took his first degree at St. Petersburg
in 1862. In 1864 he began his career of research on comparative
embryology (working initially on marine organisms at the Naples
Zoological Station), his first results gaining him the D.Sc.
at St. Petersburg in 1866. In that year he was appointed to
the faculty at St. Petersburg; from 1868 he was at Kazan, from
1869 at Kiev, from 1873 at Odessa, and finally from 1891 at
St. Petersburg again. He was elected to the Academy in 1890
and, as the leading embryologist in Russia, received various
other honours.

1870s

3001. ZHUKOVSKY, Nikolai Egorovich. 1847-1921. (Mechanics 565)
Graduated at Moscow in 1868 and gained his master's degree in
1876 and his doctorate in 1882. From 1872 he taught mathematics and mechanics at Moscow Technical School, and in 1886 he
also became professor of mechanics at Moscow University. He
was a prominent member of the Moscow Mathematical Society and
was elected to the St. Petersburg Academy in 1900 but withdrew
from it because he was unwilling to leave Moscow and his
teaching posts.

Zhukovsky was an important teacher, especially (in the
twentieth century) of aerodynamics, of which he was one of the
pioneers. In 1918 he was made head of the new Central Aerodynamics Institute.

3002. DOKUCHAEV, Vasily Vasilievich. 1846-1903. (Soil science 1197)
Graduated at St. Petersburg in 1871 and in the following year
became curator of the University's geological collection.
After gaining his doctorate he was appointed professor of
geology at St. Petersburg in 1879; ill health forced his resig-
nation in 1897. In the 1890s his pioneering work on soil
science received considerable support from scientific and
agricultural organizations and from the Ministry of Agriculture.
In 1899 he founded the journal *Pochvovedenie* [Pedology].

3003. NENCKI, Marcelli. 1847-1901. (Biochemistry 978) A native of
Russian Poland. His participation in the Polish insurrection
of 1863/64 obliged him to emigrate to Germany. He studied
medicine at Berlin with an emphasis on what was then called

medical chemistry and obtained his M.D. in 1870. Soon afterwards he gained an appointment at the University of Bern where
a few years later he became professor of medical chemistry and
built up a flourishing research school. In 1890 he went to
St. Petersburg as head of the biochemistry section of the new
Institute of Experimental Medicine.

1880s

3004. MARKOV, Andrei Andreevich. 1856-1922. (Mathematics 291)
Graduated in 1878 at St. Petersburg (having studied under
Chebyshev by whom he was greatly influenced). After qualifying
as a *Privatdozent* in 1880 he gained his doctorate in 1884. He
was elected to the Academy in 1886 and in the same year was
appointed an *extraordinarius* at St. Petersburg University,
becoming full professor in 1893.

3005. LYAPUNOV, Aleksandr Mikhailovich. 1857-1918. (Mechanics 562)
Graduated in 1880 at St. Petersburg (having studied under
Chebyshev by whom he was greatly influenced). From 1881 onward
he was publishing original work and in 1885 he qualified as a
Privatdozent at Kharkov. He gained his doctorate in 1892 and
in the following year was appointed professor of mechanics at
Kharkov. He moved to St. Petersburg when he was elected to
the Academy in 1901 and thereafter was able to devote himself
entirely to research.

3006. FYODOROV, Evgraf Stepanovich. 1853-1919. (Crystallography 1107)
Originally a military engineer (partly, it seems, because of
his liking for mathematics). After a few years service he
resigned from the army and studied at the Mining Institute
from which he graduated in 1883. He then worked for the Department of Mines, taking part in expeditions to the Urals and
elsewhere.

3007. GEOLOGICAL SURVEY. Geologicheskago Komiteta (Comité Géologique).
Its first periodical began in 1882.

3008. PAVLOV, Ivan Petrovich. 1849-1936. (Physiology 1801) Studied
science at the University of St. Petersburg and then medicine
at the Military Medical Academy where he graduated in 1879.
After some early researches in the physiological laboratory
there he went to Germany in 1884 and worked with Ludwig at
Leipzig and Heidenhain at Breslau. Returning to St. Petersburg in 1886 he continued research at the Military Medical
Academy as a *Privatdozent*. In 1890 he was appointed professor
there, and director of the physiology section of the new Institute of Experimental Medicine, a post he held until 1925.
      Pavlov was awarded the Nobel Prize in 1904 for his
work on the physiology of digestion. In his later career he
organized a number of important research centres and created
a major school of research.

3009. VINOGRADSKY, Sergey Nikolaevich. 1856-1953. (Microbiology 1816)
Graduated in science at St. Petersburg in 1881 and began
research in microbiology. In 1885 he went to Strasbourg where
he worked under the eminent mycologist, A. De Bary, for three
years. From 1889 to 1890 he was at Zurich, working on bacterial

nitrification, and was then for a short time in Paris with
Pasteur. He returned to St. Petersburg in 1891 to become
director of the microbiological section of the new Institute
of Experimental Medicine; in 1905 he became director of the
whole Institute. He left Russia after the Revolution and from
1922 was director of the division of agricultural microbiology
at the Pasteur Institute.

1890s

3010. ST. PETERSBURG ACADEMY. During the 1890s the Academy was modern-
ized and strengthened, becoming much more active and much more
involved in the scientific life of the nation than it had been
previously. In contrast to the earlier situation (see 2978)
all the academicians elected after 1890 were Russians, and the
new members were, in general, more closely linked with the
universities and technical schools than many of their predecessors had been. In 1893 the government approved a reorganization of the Academy and considerably increased its finances.
As a result it was able to establish new research laboratories
and expand its publishing activities.

3011. RUSSIAN ASTRONOMICAL SOCIETY. Russkoe Astronomicheskoe Obshchestvo. Founded in St. Petersburg in 1890.

3012. LEBEDEV, Petr Nikolaevich. 1866-1912. (Physics 765) After
beginning the study of engineering in Moscow he went to Strasbourg in 1887 to study physics under Kundt. Returning to
Moscow four years later, he obtained a teaching position in
the University's department of physics. After gaining his
doctorate in 1900 he became professor of physics there. Despite
the shortness of his career he created an important school.
    In 1911 Lebedev was one of a number of professors at
Moscow University who resigned their chairs in protest at the
government's violation of the University's autonomy. The
trauma of the event contributed to his sudden death in the
following year.

3013. WALDEN, Paul. 1863-1957. (Chemistry 990. His extensive work
in physical chemistry was nearly all done after 1900) Born
and educated in Latvia, then part of the Russian Empire.
Graduated in 1889 at the (German-speaking) Polytechnic at Riga
where he worked first with Ostwald and then with C.A. Bischoff.
(It was in consequence of this dual influence that he later
became one of the pioneers of physical organic chemistry) He
was appointed to the staff of the Polytechnic and became professor of chemistry in 1899, after he had obtained a doctorate
at the University of St. Petersburg. In 1910 he was elected
to the St. Petersburg Academy and became active in the cause
of scientific education in Russia.
    In 1919, because of the political situation in Latvia,
Walden emigrated to Germany and became professor of chemistry
at Rostock. He held the post until his retirement in 1934.

3014. INSTITUTE OF EXPERIMENTAL MEDICINE. A well-endowed research
institute established by the government at St. Petersburg in
1891. Its physiological section was headed by Pavlov, its
biochemical section by Nencki, and its microbiological section
by Vinogradsky.

2.12   UNITED STATES

1710s

3015. LOGAN, James. 1674–1751. (Botany 1330) The most significant
scientist of the colonial period. A Quaker, born and educated
in Britain, he went to Pennsylvania with William Penn in 1699
and thereafter acted as Penn's agent in the colony. He built
up the best library in the country and through his own efforts
acquired a wide knowledge of science; on his two trips to
England (1708 and 1723–24) he made the acquaintance of a number
of English scientists. He is best known for his experiments
on sex in plants but he also published a work on optics.

1730s

..... COLLINSON, P. An "intelligencer" in London who was an important
contact for American scientists, including Franklin. See 2298
and 611.

1740s

3016. FRANKLIN, Benjamin. 1706–1790. (Physics 611. Earth sciences
1038) The famous statesman and the first American to receive
international recognition for scientific achievement. He was
apprenticed to a printer at the age of twelve and after various
adventures set up his own printing shop in Philadelphia in
1726. His business thrived and as well as taking a prominent
part in public affairs he became interested, from the early
1740s, in scientific experimentation. The publication of his
work on electricity in 1751–54 made him well known in the
scientific world: he was elected to the Royal Society of London
and awarded its Copley Medal in 1753 and he later became a
foreign member of the Paris Academy. His political career and
his inventions and other contributions to society are well known.

1760s

3017. AMERICAN PHILOSOPHY SOCIETY. Founded at Philadelphia in 1769
by a merger of two earlier groups, one of which had been originated by Benjamin Franklin; he was president of the new Society
until his death in 1790. The first volume of its *Transactions*
appeared in 1771 and attracted much interest in Europe but due
to the Revolution no more was published until 1789. Despite
its name the Society was a local, not a national, institution.
It was however the leading scientific institution in the
country until the middle of the nineteenth century.

1780s

3018. AMERICAN ACADEMY OF ARTS AND SCIENCES. Founded at Boston in
      1780; its *Memoirs* began in 1785. It was not, of course, an
      academy in the European style (the name 'Academy' is said to
      have been chosen because of the pro-French and anti-British
      feeling of the time) and, despite the adjective 'American',
      it was a local, not a national, body. The prime mover in its
      formation was John Adams.

1810s

3019. ACADEMY OF NATURAL SCIENCES, Philadelphia. Founded in 1812.
      A natural history society.

3020. SILLIMAN'S JOURNAL. Few, if any, non-institutional (or propri-
      etary) journals have played such a central role in the growth
      and development of a national scientific community as did the
      American Journal of Science, founded in 1818 by Benjamin
      Silliman (professor of chemistry, mineralogy, and geology at
      Yale College). As well as serving as the main vehicle of
      scientific publication in the country through most of the
      nineteenth century it also kept its readers informed about
      scientific developments in Europe. Though it covered all the
      sciences it had an emphasis on geology and related fields--an
      emphasis which reflected the main interests not only of Silliman
      and his editorial successors down to his grandson, E.S. Dana,
      but also of a large proportion of American scientists.
           In the late nineteenth century, when conditions became
      very difficult for non-institutional journals covering the
      whole of science, this incipient specialization became more
      pronounced and, as in the somewhat similar case of *The Philo-
      sophical Magazine* (2368), enabled the journal to survive.

1820s

3021. HENRY, Joseph. 1797-1878. (Physics 664) Educated at the Albany
      Academy and in 1826 became professor of mathematics and natural
      philosophy there. Six years later he was appointed professor
      of natural philosophy at the College of New Jersey (later
      Princeton University). In 1846 he became secretary of the
      newly-established Smithsonian Institution and thereafter strove
      to make it a source of support for research in the basic scien-
      ces and a centre for the interchange of ideas and the dissemin-
      ation of new findings. He held this central position in Amer-
      ican science until his death.

3022. STATE GEOLOGICAL SURVEYS. The first State surveys were those of
      North and South Carolina in 1823 and 1824. They were followed
      by Massachusetts which established the first major survey in
      1830. By 1850 twenty other States had followed suit and more
      did so later. The Federal survey was set up in 1879.

3023. REDFIELD, William. 1789-1857. (Meteorology 1048) Originally
      a saddle- and harness-maker in Connecticut. In 1824 he moved
      to New York where he was an engineer on steamboats and later
      railways. He became a prominent amateur scientist with inter-

ests in meteorology and palaeontology, and was a leader in the conversion of the Association of American Geologists and Naturalists (which had been founded in 1840) into the American Association for the Advancement of Science in 1848, serving as its first president.

3024. BACHE, Alexander Dallas. 1806-1867. (Not in Part 1. His importance was in his organizational role) Originally trained as an army engineer. From 1827 to 1836 he was professor of natural philosophy and chemistry at the University of Pennsylvania, and in the period 1836-42 he was involved in organizing college education in Philadelphia. During his time in Philadelphia he was developing his scientific interests, especially in terrestrial magnetism and meteorology, and in 1843 he was appointed head of the U.S. Coast Survey. He held the post until his death.

Bache was successful in obtaining government support for strengthening and enlarging the Coast Survey and by 1860 it had become a highly regarded organization and an outstanding example of the union of pure and applied science. Its activities extended into several fields, chiefly geodesy, geophysics, and astronomy, in addition to its practical tasks of surveying, mapping, etc.

"Bache is clearly one of the founders of the scientific community in the United States. His administration of the Coast Survey established a model for large-scale scientific organization that was followed either implicitly or explicitly by later groups. Bache and his close friend Joseph Henry [secretary of the Smithsonian Institution] established many of the patterns of interaction of science and the federal government." (*DSB*)

3025. LYCEUM OF NATURAL HISTORY, New York. Its periodical began in 1823. It became the New York Academy of Sciences in 1877.

1830s

3026. HALL, James. 1811-1898. (Geology 1171) Graduated at the Rensselaer Polytechnic at Troy, N.Y., in 1832, and soon afterwards became an instructor there. In 1837 he was appointed to the geological staff of the newly-established Natural History Survey of New York State. He became state palaeontologist in 1843 and subsequently the curator (in 1865) and then the director (in 1871) of the State Museum in Albany.

Hall was influential in the establishment and organization of geological surveys in several states, and from 1888 was the first president of the Geological Society of America. He was one of the foundation members of the National Academy of Sciences and received many honours, European as well as American.

3027. DANA, James Dwight. 1813-1895. (Mineralogy 1097. Geology 1172, 1182, 1187) Studied at Yale (natural history, geology, etc. under Benjamin Silliman, Snr) and graduated in 1833. After some travelling he became Silliman's assistant at Yale in 1836. Two years later he was selected to join the Wilkes Expedition to the Pacific area as geologist, thereby gaining four years of invaluable experience. After the return of the expedition

in 1842 he spent thirteen years on the massive official reports
of the geology and much of the zoology, a task that involved
much research (especially zoological).

In 1846 Dana succeeded Silliman as editor of the
American Journal of Science (see 3020) and for many years
this position enabled him to exert a strong influence on
American science in general and geology in particular. In
1856 Yale appointed him Silliman professor of natural history,
a post created for him (lest he be enticed away to Harvard)
and which he held until his death. He received many honours,
including medals from the Geological Society of London and the
Royal Society, and was a founding member of the National Academy of Sciences.

3028. BOSTON SOCIETY OF NATURAL HISTORY. Its publications began in 1834.

3029. GRAY, Asa. 1810-1888. (Botany 1364, 1371, 1379. Evolution 1539)
Graduated in medicine in 1831 but soon gave it up for botany.
After a succession of minor teaching positions, and a year in
Europe meeting botanists and visiting herbaria, he became
professor of natural history (actually botany) at Harvard in
1842. Though he had few advanced students he exercised a
major influence on the development of botany in the country
through his manuals and textbooks, and in various informal
ways. At Harvard he established a botanical garden and herbarium (later named the Gray Herbarium).

Gray became the leading taxonomic botanist in the
United States and his extensive studies of its flora did much
to unify and consolidate the taxonomy of North American plants.
In the 1860s and later he became known to the general public
for his defence of the Darwinian theory of evolution (in opposition to his Harvard colleague, Agassiz).

1840s

3030. SMITHSONIAN INSTITUTION. Established by Congress in 1846 as a
national institution "for the increase and diffusion of knowledge" in consequence of a bequest by the English scientist,
James Smithson. Its character was largely shaped by its first
secretary, Joseph Henry (3021).

3031. AMERICAN ASSOCIATION FOR THE ADVANCEMENT OF SCIENCE. Established
in 1848 on the model of the British Association (it incorporated the Association of American Geologists and Naturalists
which had been formed in 1840). It became the central scientific organization in the country and played a major role in
forming the American scientific community. In its early years
it was dominated by amateurs, typically concerned with natural
history, but control soon passed to the professionals.

3032. U.S. NAVAL OBSERVATORY. Its construction, in Washington, D.C.,
was authorized by Congress in 1842 and was completed two years
later. It was the first major American astronomical observatory.

3033. U.S. COAST SURVEY. Established in 1807 but reorganized in 1843
by its new head, A.D. Bache, who made it into an important
agency for pure and applied science; see 3024.

3034. MAURY, Matthew Fontaine. 1806-1873. (Oceanography 1061) Joined the navy in 1825 and served on some extensive voyages and expeditions. In 1842 he was appointed superintendant of the Depot of Charts and Instruments (out of which evolved the Naval Observatory and the Hydrographical Office). His work in hydrography, oceanography, and meteorology earned him an international reputation and he was the chief organizer of the first international conference on marine meteorology, held in Brussels in 1853.

3035. AGASSIZ, Louis. 1807-1873. (For entries in Part 1 and for his earlier career in Switzerland see 2910) Following his appointment as professor of zoology at Harvard in 1847 he built up his department into a major centre for teaching and research. In 1859 he succeeded in his cherished plan of establishing the Museum of Comparative Zoology, an institution which soon became well known. "The museum always bore the impress of Agassiz's conception of the relationship between graduate instruction, research, fieldwork, and publication, centered in an institution of higher learning and supported by private philanthropy and public funds." (*DSB*)

> Highly successful as a teacher (as well as an administrator, fund-raiser, and promoter), Agassiz had a far-reaching influence on the teaching of natural history in the United States. He was also one of the most prominent members of the American scientific community and a prime mover in the establishment of the National Academy of Sciences in 1863.

3036. BAIRD, Spencer Fullerton. 1823-1887. (Not in Part 1. Though he was an able systematic zoologist and a leading authority on North American birds his chief importance was as a scientific administrator and shaper of institutions) M.A., Dickinson College (Carlisle, Pa.), 1843. He was self-taught in zoology and received valuable stimulus from personal contacts with several prominent naturalists, including J.D. Dana and J.J. Audubon. In 1845 he became professor of natural history at Dickinson College, and five years later his publications (and influential friends) gained him the post of assistant secretary at the Smithsonian Institution. In 1878 he succeeded Joseph Henry as secretary.

> From the beginning of his career at the Smithsonian Baird steadily built up the zoological and anthropological collections of what was to become the U.S. National Museum, obtaining large amounts of material from the many government expeditions to the West (which he often helped to plan), and from numerous private collectors. At the same time he made the Museum into a leading scientific institution, serving as a centre for informal study and research.

> Baird's other main initiative was to persuade Congress to establish in 1871 the U.S. Fish Commission under his direction. He soon made this into an important body for research in marine biology as well as in the practical needs of the fishing industry. Like the National Museum, it too became a nursery for future naturalists and biologists.

1850s

3037. NEWCOMB, Simon. 1835-1909. (Astronomy 470) Had little formal
      education and was originally self-taught in mathematics and
      science. In 1857 he was appointed an assistant at the Nautical
      Almanac Office, then located in Cambridge, Mass.; he took the
      opportunity to attend a mathematics course at Harvard. Four
      years later he gained a post in the Naval Observatory, and in
      1877 he became superintendant of the Nautical Almanac Office,
      retaining the position until his retirement in 1897. From
      1884 to 1893 he was also professor of mathematics and astronomy
      at the Johns Hopkins University.
           In conjunction with A.M.W. Downing, superintendant of
      the British Nautical Almanac Office, Newcomb instigated an
      international movement to adopt a unified system of astronom-
      ical constants. At an international conference in 1896 it was
      agreed that all ephemerides would adopt a particular set of
      constants (which had mostly been formulated by Newcomb).
           In the late nineteenth century he was the most outstand-
      ing American astronomer and one of the best-known scientists
      in the country. He was one of the founders of the American
      Astronomical Society and its first president, an active member
      of the National Academy of Sciences, and for many years editor
      of the *American Journal of Mathematics*. He received many
      honours from both European and American institutions.

3038. DRAPER, Henry. 1837-1882. (Astronomy 461) His father, J.W.
      Draper, was a professor of chemistry and one of the pioneers
      of photography (including astronomical photography). The son
      graduated in medicine in 1858 and joined the staff of a New
      York hospital. At the same time he began his career as an
      amateur astronomer. In 1860 he was appointed professor of
      natural sciences at the University of the City of New York,
      becoming professor of physiology and dean of the medical
      faculty in 1866.
           For his photography of the transit of Venus in 1874
      Draper was awarded a special gold medal by Congress. He was
      elected to the Astronomische Gesellschaft in 1875 and to the
      National Academy of Sciences in 1877.

3039. FERREL, William. 1817-1891. (Earth sciences 1057, 1071)
      A schoolteacher who was largely self-taught in mathematics and
      science. His publications, beginning in 1853, led to his
      appointment to the staff of the Nautical Almanac Office in 1858.
      Ten years later he joined the Coast Survey and from 1882 until
      his retirement in 1886 he was with the Army's Signal Service
      (from which the Weather Bureau evolved). He became a member
      of the National Academy of Sciences in 1868 and was an honor-
      ary member of several European meteorological societies.

3040. MUSEUM OF COMPARATIVE ZOOLOGY, Harvard College. Founded in 1859
      by Agassiz: see 3035. Its *Bulletin* began in 1863.

1860s

3041. NATIONAL ACADEMY OF SCIENCES. Established almost inadvertently
      by Congress in 1863 (during the Civil War) which thereafter

took hardly any interest in it and gave it no funds. Until the First World War it remained little more than an honorific body.

3042. LANGLEY, Samuel Pierpont. 1834-1906. (Astronomy 475) Originally a civil engineer and architect. In 1865 he became an assistant at Harvard Observatory and two years later was appointed director of the Allegheny Observatory and professor of physics and astronomy at the University of Pittsburgh. In 1887 he accepted the post of secretary of the Smithsonian Institution (in succession to Baird) and three years later was successful in obtaining the necessary finance, from government and private sources, to establish the Smithsonian Astrophysical Observatory.

Langley was a member of the National Academy of Sciences and a foreign member of several of the leading European scientific bodies, and was highly respected in the American scientific community. To the general public he was best known for his experiments in aeronautics. In 1896 he was the first to construct an unmanned heavier-than-air flying machine.

3043. PICKERING, Edward Charles. 1846-1919. (Astronomy 474) Graduated at Harvard in 1865. After two years there in a junior teaching post he was appointed professor of physics at the new Massachusetts Institute of Technology where he introduced laboratory instruction in physics. In 1876 he became director of the Harvard Observatory and instituted research and cataloging programmes in astrophysics. Under his direction (which continued until his death) the Observatory became a major centre in the astronomical world.

3044. GIBBS, Josiah Willard. 1839-1903. (Physics 721. His important book on statistical mechanics appeared in 1902) Son of a Yale professor. In 1863 at Yale he received the first Ph.D. in engineering to be conferred in the United States (and one of the first in any field). From 1866 to 1869 he was in Europe --at Paris, Berlin, and Heidelberg--studying mathematics and physics intensively and assimilating the European scientific tradition. In 1871, two years after his return, he was appointed professor of mathematical physics at Yale, a post he held until his death (despite a tempting offer from the new Johns Hopkins University). He is generally regarded today as one of the greatest American scientists of the nineteenth century.

3045. CLARKE, Frank Wigglesworth. 1847-1931. (Geochemistry 965) Graduated in science at Harvard in 1867 and subsequently did research on mineral analysis. After holding some minor teaching posts he became professor of chemistry and physics at the University of Cincinnati in 1874. Nine years later he was appointed chief chemist to the U.S. Geological Survey, a post he held for forty years.

The American Chemical Society was first established as a small, local society in New York in 1876, but in the late 1880s Clarke was largely instrumental in converting it into a truly national society; he became its president in 1901. Around the turn of the century he was also a leader in the movement to set up the International Committee on Atomic Weights.

3046. U.S. DEPARTMENT OF AGRICULTURE. Established in 1862. In the late nineteenth century it grew to be the biggest organization for applied science in the world and the leading institution for government science in the United States. Its activities extended into many fields of science, pure as well as applied.

3047. AMERICAN MUSEUM OF NATURAL HISTORY, New York. Founded in 1869.

3048. AMERICAN ENTOMOLOGICAL SOCIETY. Founded in 1868.

3049. MARSH, Othniel Charles. 1831-1899. (Palaeontology 1190) Graduated at Yale in 1862 and then went to Europe for three years where he travelled widely and studied at Berlin, Heidelberg, and Breslau. In 1866, after his return, he was appointed professor of palaeontology at Yale, a post he held until his death. Concurrently he was director of the geological and palaeontological department of the Peabody Museum (founded by his uncle, the wealthy philanthropist George Peabody). From 1882 to 1892 he was also in charge of the U.S. Geological Survey's work in vertebrate palaeontology.

3050. COPE, Edward Drinker. 1840-1897. (Palaeontology 1189. Evolution 1533) Came from a wealthy family. From the late 1850s, after his secondary education, he taught himself natural history, spending much time in the chief American and European museums. His wealth enabled him to devote himself to research, and by the the mid-1860s he had begun the work on the vertebrate palaeontology of the western United States for which he became famous; much of the work was done in association with various government surveys of the western lands. In the 1880s he lost most of his money and in 1889 was obliged to take a teaching post in geology at the University of Pennsylvania. In 1895 he became professor of zoology and comparative anatomy.

3051. HYATT, Alpheus. 1838-1902. (Evolution 1532) Studied at Harvard under Agassiz, graduating in 1862. After serving in the Civil War he became curator of the Essex Institute at Salem, Mass., in 1867. From 1870 to 1888 he was professor of zoology and palaeontology at the Massachusetts Institute of Technology, and from 1877 he was also professor of biology at Boston University; he held the latter post until his death. He was also associated as a palaeontologist with the U.S. Geological Survey and the Harvard Museum of Comparative Zoology.

　　　In 1868 Hyatt was one of the founders, and for a few years the editor, of the *American Naturalist*, the first American journal devoted to the biological sciences. He became a prominent figure in the Boston Society of Natural History and was elected to the National Academy of Sciences in 1875. In 1883 he was the chief founder and first president of the American Society of Naturalists. He also took a leading part in 1881 in establishing the marine biological laboratory that is now at Woods Hole, Mass.

1870s

..... SYLVESTER, J.J. Professor of mathematics at Johns Hopkins University, 1876-83. See 2369.

3052. ROWLAND, Henry Augustus. 1848-1901. (Physics 724, 735) Gradu-

ated in civil engineering at Rensselaer Polytechnic Institute
(at Troy, N.Y.) in 1870 and two years later became a teacher
of physics there. In 1875 he was appointed professor of physics
at the new Johns Hopkins University and before taking up the
post travelled in Europe for a year, meeting many physicists,
inspecting laboratories, and buying apparatus for his new
department. His subsequent researches in experimental physics
brought him wide recognition and election to the National
Academy of Sciences and, as a foreign member, to the Royal
Society of London and the Paris Academy. In 1899 he was one
of the founders and first president of the American Physical
Society.

3053. MICHELSON, Albert Abraham. 1852-1931. (Physics 741. Some of
his most important work was done after 1900) Graduated at the
U.S. Naval Academy in 1873 and two years later was appointed
an instructor in science there. In 1878 he began work on his
lifelong interest, the precise measurement of the speed of
light. Realizing the need for a deeper scientific training
he went to Europe for two years, studying at Berlin, Heidelberg,
and Paris. It was while he was in Berlin (in Helmholtz' laboratory) that he conceived the idea for his famous ether-drift
experiment and made the first attempt at it.

In 1883, soon after his return, Michelson was appointed
professor of physics at the Case School of Applied Science in
Cleveland, Ohio; he moved in 1889 to the Clark University at
Worcester, Mass., and finally in 1892 to the University of
Chicago. In his later years he received many honours from
scientific societies, American and European, and in 1907 became
the first American scientist to win the Nobel Prize.

3054. AMERICAN CHEMICAL SOCIETY. Founded in 1876. See 3045.

3055. REMSEN, Ira. 1846-1927. (Not in Part 1. He was a capable
chemist but his chief importance was in his institutional role)
After graduating in medicine at Columbia University in 1867 he
decided to go to Germany to study chemistry. He spent a year
at Munich and then went to Göttingen where he worked under
Fittig, gaining his D.Phil. in 1870. In that year Fittig moved
to Tübingen and Remsen went with him as his assistant, a post
he held for two years. Soon after his return to the United
States in 1872 he became professor of chemistry at Williams
College (Williamstown, Mass.) where he continued his research
and wrote the first of his influential textbooks.

In 1876 Remsen was appointed professor of chemistry at
the new Johns Hopkins University in Baltimore. In accord with
the University's innovative educational policy, he introduced
some of the main features of the organization and educational
methods of the German universities--above all, the integration
of research and teaching at the graduate level. (It was through
Remsen and his like-minded colleagues that Johns Hopkins became
known as "Göttingen in Baltimore") The large research school
that he built up along these lines, together with his several
textbooks, and his establishment in 1879 of the *American Chemical Journal* (the first significant chemical journal in America), had a far-reaching effect on the development of chemistry
in the United States. From 1901 until his retirement in 1913
Remsen was president of Johns Hopkins.

3056. U.S. GEOLOGICAL SURVEY. Established by the Federal Government
      in 1879. (Many of the individual States had established their
      own surveys from the 1820s onwards--see 3022) From 1881 it
      was directed very ably by J.W. Powell.

3057. WRIGHT, George Frederick. 1838-1921. (Geology 1208) Graduated
      at Oberlin College (Ohio) in 1859 and after theological studies
      became a pastor of the Congregational Church in 1862. He had
      long been interested in geology and when, in 1872, he moved to
      Andover (Mass.) he developed contacts with scientists at Harvard,
      notably Asa Gray whom he persuaded to publish his *Darwiniana*
      (1539). Subsequently Wright's published work established him
      in the American geological community. In 1881 he took part in
      the Second Geological Survey of Pennsylvania and later in the
      same year he was appointed to a post at Oberlin College where
      he taught both geology and theology. He was active in several
      scientific societies and was one of the founders of the Geolog-
      ical Society of America in 1889.

3058. DAVIS, William Morris. 1850-1934. (Geomorphology 1209) Gradu-
      ated at Harvard in 1870. In 1877, after working as a meteor-
      ologist and a field geologist, he was appointed assistant to
      the professor of geology at Harvard. He was promoted through
      several grades to become professor of physical geography in
      1890 and professor of geology in 1898. He was one of the
      founders of the Geological Society of America in 1889 and
      became a member of the National Academy of Sciences. In his
      later years he received many honours and awards.

3059. U.S. NATIONAL MUSEUM. Its publications began in 1875 but it had
      evolved earlier as a branch of the Smithsonian Institution,
      largely through the work of S.F. Baird (3036).

3060. AMERICAN SOCIETY OF MICROSCOPISTS. Founded in 1878.

3061. WHITMAN, Charles Otis. 1842-1910. (Zoology 1521) Initially a
      schoolteacher. He had long been interested in natural history
      and in the early 1870s he happened to encounter Louis Agassiz
      who had a big influence on him. As a result he went to Europe
      in 1875 to study zoology, first with Dohrn at the Naples Zoolog-
      ical Station and then with Leuckart at Leipzig where he gained
      the D.Phil. in 1878. For the next two years he was professor
      of zoology at Tokyo University and helped to introduce the
      science into Japan.
            After another stay at Naples in 1881-82 Whitman became
      assistant at the Museum of Comparative Zoology at Harvard.
      In 1887 he founded the *Journal of Morphology*, the first Amer-
      ican journal of its kind. Following some short-term appoint-
      ments he became professor of zoology at the new University of
      Chicago in 1892. He was a leading figure in the establishment
      of the Marine Biological Laboratory at Woods Hole, Mass., in
      1888 and served as its first director. He was also one of the
      main movers in the foundation of the American Morphological
      Society (later the American Society of Zoologists) in 1890.

1880s

3062. NEW YORK MATHEMATICAL SOCIETY. Founded in 1888. In 1894 it
      became the American Mathematical Society.

3063. LICK OBSERVATORY. Completed in 1888 and donated to the University of California by the philanthropist, James Lick. Its 26-inch refracting telescope was the largest yet constructed and it was one of the first major observatories to be built at a high altitude (on Mount Wilson).

3064. ASTRONOMICAL SOCIETY OF THE PACIFIC, San Francisco. Established in 1889, the main movers being the astronomers of Lick Observatory.

3065. BARNARD, Edward Emerson. 1857-1923. (Astronomy 473, 484) Came from a very poor family and had little formal schooling. He worked as a photographer and in his spare time taught himself astronomy so effectively that in 1883 he was appointed to a post at the observatory of Vanderbilt University. He took the opportunity to study at the University but was not eligible for a degree. His remarkable achievements as an observer, however, led to his appointment in 1888 to the staff of the new Lick Observatory. In 1895 he transferred to the Yerkes Observatory (then being constructed) where he remained until his death.

      One of the greatest observational astronomers of the time (partly because of his extraordinary eyesight), Barnard received many honours and awards from American and European scientific bodies (including three gold medals from the Paris Academy) and was a prominent member of the chief American societies.

3066. KEELER, James Edward. 1857-1900. (Astronomy 486) Graduated at Johns Hopkins University in 1881. Until 1886 he was assistant to S.P. Langley at the Allegheny Observatory but was able to take a year off in 1883 to study in Europe (at Heidelberg and Berlin). From 1886 he was involved in the construction of Lick Observatory and from its completion in 1888 was a member of its staff. In 1891 he became director of the Allegheny Observatory (in succession to Langley) and in 1898 returned to Lick as its director. He was co-editor with G.E. Hale of the *Astrophysical Journal* from its inception in 1895. Despite his relatively short career he received a number of honours.

3067. GEOLOGICAL SOCIETY OF AMERICA. Founded in 1889.

3068. MARINE BIOLOGICAL LABORATORY, Woods Hole, Mass. Founded in 1888.

3069. WILSON, Edmund Beecher. 1856-1939. (Embryology 1685. Cytology 1704. His important researches on heredity were done from 1903 onward) After taking his bachelor's degree at Yale in 1878 he was a graduate research student at the new Johns Hopkins University until 1881 when he received the Ph.D. He then spent two years in Europe--at Cambridge, Leipzig, and the Naples Zoological Station. After his return he held some short-term appointments and then became head of the biology department at Bryn Mawr College in 1885. In 1891-92 he spent another year in Europe (at Munich with Boveri and at the Naples station) before taking up his appointment as professor of zoology at Columbia University where he remained for the rest of his career. He was a highly respected and successful teacher and his department at Columbia became an important research centre.

3070. AMERICAN PHYSIOLOGICAL SOCIETY. Founded in 1887.

3071. LOEB, Jacques. 1859-1924. (Physiology 1793) Born and educated
      in Germany. M.D., Strasbourg, 1884. He began physiological
      research at Würzburg in 1886, as assistant to Fick, and contin-
      ued at the Naples Zoological Station from 1889. In 1891 he
      settled in the United States, becoming a professor at the
      universities of Chicago (from 1892) and California (from 1902).
      From 1910 until his death he was a member of the Rockefeller
      Institute in New York. Much of his experimental work was done
      at marine biological laboratories--at Pacific Grove, California,
      and then at Woods Hole, Massachusetts.

1890s

3072. YERKES OBSERVATORY. Constructed at Williams Bay, Wisconsin,
      for the University of Chicago, and completed in 1897. In
      addition to its 40-inch refracting telescope, the largest ever
      made, it was well equipped for astrophysical investigations.

3073. AMERICAN ASTRONOMICAL SOCIETY. Founded in 1899, the chief movers
      being S. Newcomb and G.E. Hale. The original name was the
      American Astronomical and Astrophysical Society, it being
      thought desireable at the time to give some emphasis to the
      still new field of astrophysics.

3074. AMERICAN PHYSICAL SOCIETY. Founded in 1899. The first president
      was H.A. Rowland. The Society took over *The Physical Review*
      which had been begun at Cornell University in 1893.

3075. BOTANICAL SOCIETY OF AMERICA. Founded in 1894.

3076. AMERICAN MORPHOLOGICAL SOCIETY. Founded in 1890. It was later
      renamed the American Society of Zoologists.

SYNCHRONIC SUMMARY OF PART 2

The first column of figures gives the number of entries for persons in each country (or region) in the decade, and the second column the number of entries for institutions. The third column gives the entry numbers (the first for the decade when there are more than one).

1450s

| Italy | 2 | – | 1820+ | Eastern Europe | 2 | – | 2932+ |

Totals: persons 4; institutions 0

1460s

| Italy | 1 | – | 1822 |

Totals: persons 1; institutions 0

1470s

| Italy | 1 | – | 1823 |

Totals: persons 1; institutions 0

1480s

| Italy | 2 | – | 1824+ | Germany | 1 | – | 2475 |
| France | 1 | – | 1963 | | | | |

Totals: persons 4; institutions 0

1490s

| Italy | 1 | – | 1826 | Germany | 1 | – | 2476 |

Totals: persons 2; institutions 0

1500s

| Italy | 1 | – | 1827 | Germany | 1 | – | 2477 |
| France | 2 | – | 1964+ | Spain | 1 | – | 2920 |

Totals: persons 5; institutions 0

1510s

| Britain | 1 | – | 2215 | Spain | 1 | – | 2921 |
| Germany | 3 | – | 2478+ | Eastern Europe | 1 | – | 2934 |

Totals: persons 6; institutions 0

1520s

Italy	3	–	1828+	Switzerland	1	–	2883
France	1	–	1966	Eastern Europe	1	–	2935
Germany	4	–	2481+			

Totals: persons 10; institutions 0

1530s

Italy	3	–	1831+	North Netherlands	1	–	2758
France	1	1	1967+	South Netherlands	2	–	2805+
Britain	2	–	2216+	Spain	1	–	2923
Germany	7	–	2485+	Portugal	2	–	2922+

Totals: persons 19; institutions 1

1540s

Italy	8	1	1834+	Germany	1	–	2492
France	6	–	1969+	South Netherlands	1	–	2807
Britain	1	–	2218	Switzerland	1	–	2884

Totals: persons 18; institutions 1

1550s

Italy	10	–	1843+	Britain	2	–	2219+
France	4	–	1975+			

Totals: persons 16; institutions 0

1560s

Italy	5	1	1853+	Switzerland	3	–	2885+
France	2	–	1979+	Portugal	1	–	2925
Germany	5	–	2493+			

Totals: persons 16; institutions 1

1570s

Italy	3	–	1859+	Germany	2	–	2498+
France	2	–	1981+	Scandinavia	2	–	2823+
Britain	4	–	2221+	Spain	2	–	2926+

Totals: persons 15; institutions 0

1580s

Italy	6	–	1862+	South Netherlands	1	–	2808
France	2	–	1983+	Scandinavia	1	–	2825
Britain	1	–	2225	Switzerland	1	1	2888+
Germany	1	1	2500+	Eastern Europe	1	–	2936
North Netherlands	2	1	2759+			

Totals: persons 16; institutions 3

1590s

Italy	4	–	1868+	Germany	3	–	2502+
France	–	1	1985	North Netherlands	1	–	2762
Britain	2	1	2226+	Eastern Europe	3	–	2937+

Totals: persons 13; institutions 2

1600s

Italy	3	1	1872+	Germany	6	—	2505+
France	1	—	1986	North Netherlands	1	—	2763
Britain	4	—	2229+			

Totals: persons 15; institutions 1

1610s

Italy	4	—	1876+	North Netherlands	1	—	2764
France	2	—	1987+	South Netherlands	2	—	2809+
Britain	1	—	2233	Scandinavia	1	—	2826
Germany	3	—	2511+	Eastern Europe	1	—	2940

Totals: persons 15; institutions 0

1620s

Italy	3	—	1880+	Germany	1	—	2514
France	6	—	1989+	North Netherlands	1	—	2765
Britain	2	1	2234+	South Netherlands	1	—	2811

Totals: persons 14; institutions 1

1630s

Italy	3	—	1883+	Germany	—	1	2515
France	4	1	1995+	North Netherlands	1	1	2766+
Britain	4	—	2237+			

Totals: persons 12; institutions 3

1640s

Italy	5	—	1886+	North Netherlands	4	—	2768+
France	6	—	2000+	South Netherlands	1	—	2812
Britain	6	—	2241+	Scandinavia	1	—	2827
Germany	1	—	2516	Eastern Europe	1	—	2941

Totals: persons 25; institutions 0

1650s

Italy	6	1	1891+	North Netherlands	3	—	2772+
France	4	—	2006+	South Netherlands	1	—	2813
Britain	10	—	2247+	Scandinavia	2	—	2828+
Germany	1	1	2517+			

Totals: persons 27; institutions 2

1660s

Italy	4	—	1898+	North Netherlands	4	—	2775+
France	9	2	2010+	Scandinavia	—	1	2830
Britain	12	2	2257+	Switzerland	1	—	2890
Germany	3	—	2519+			

Totals: persons 33; institutions 5

1670s

Italy	1	—	1902	Germany	1	—	2522

1670s (cont'd)

| France | 5 | 1 | 2021+ | North Netherlands | 2 | – | 2779+ |
| Britain | 5 | 2 | 2271+ | | | | |

Totals: persons 14; institutions 3

1680s

Italy	1	–	1903	North Netherlands	1	–	2781
France	7	–	2027+	Scandinavia	–	1	2831
Britain	5	–	2278+	Switzerland	1	–	2891
Germany	1	1	2523+				

Totals: persons 16; institutions 2

1690s

Italy	3	1	1904+	Germany	5	–	2525+
France	2	1	2034+	Switzerland	1	–	2892
Britain	5	–	2283+				

Totals: persons 16; institutions 2

1700s

Italy	3	–	1908+	Germany	1	1	2530+
France	5	2	2037+	North Netherlands	1	–	2782
Britain	3	–	2288+	Switzerland	1	–	2893

Totals: persons 14; institutions 3

1710s

Italy	4	1	1911+	North Netherlands	1	–	2783
France	2	1	2044+	Scandinavia	–	1	2832
Britain	6	–	2291+	United States	1	–	3015

Totals: persons 14; institutions 3

1720s

France	5	–	2047+	Scandinavia	1	–	2833
Britain	3	–	2297+	Switzerland	2	–	2894+
Germany	1	–	2532	Russia	2	1	2972+
North Netherlands	2	–	2784+				

Totals: persons 16; institutions 1

1730s

France	5	1	2052+	Scandinavia	3	1	2834+
Germany	3	–	2533+	Switzerland	1	–	2896
North Netherlands	1	–	2786				

Totals: persons 13; institutions 2

1740s

Italy	5	–	1916+	Scandinavia	3	1	2838+
France	6	–	2058+	Switzerland	1	1	2897+
Britain	2	–	2300+	Russia	1	–	2975
Germany	2	1	2536+	United States	1	–	3016
North Netherlands	1	–	2787				

Totals: persons 22; institutions 3

1750s

Italy	2	–	1921+	North Netherlands	1	1	2788+
France	6	–	2064+	Scandinavia	1	–	2842
Britain	2	1	2302+	Switzerland	–	1	2899
Germany	5	3	2539+				

Totals: persons 17; institutions 6

1760s

Italy	3	2	1923+	Scandinavia	2	1	2843+
France	3	–	2070+	Russia	2	–	2976+
Britain	7	1	2305+	United States	–	1	3017
Germany	4	2	2547+				

Totals: persons 21; institutions 7

1770s

Italy	3	1	1928+	Scandinavia	3	–	2846+
France	10	1	2073+	Switzerland	2	–	2900+
Britain	2	3	2313+	Portugal	–	1	2928
Germany	5	1	2553+	Eastern Europe	1	2	2942+
South Netherlands	–	1	2814				

Totals: persons 26; institutions 10

1780s

Italy	1	2	1932+	Germany	4	–	2559+
France	10	2	2084+	United States	–	1	3018
Britain	7	4	2318+				

Totals; persons 22; institutions 9

1790s

France	8	4	2096+	Scandinavia	2	–	2849+
Britain	4	1	2329+	Switzerland	2	–	2902+
Germany	5	1	2563+				

Totals: persons 21; institutions 6

1800s

Italy	3	–	1935+	South Netherlands	1	–	2815
France	10	3	2109+	Scandinavia	2	–	2851+
Britain	9	1	2334+	Eastern Europe	3	–	2945+
Germany	7	1	2569+	Russia	–	3	2978+

Totals: persons 35; institutions 8

1810s

Italy	1	–	1938	Scandinavia	2	–	2853+
France	8	1	2122+	Switzerland	1	1	2904+
Britain	9	–	2344+	Russia	2	1	2981+
Germany	12	–	2577+	United States	–	2	3019+

Totals: persons 35; institutions 5

1820s

Italy	1	–	1939	Switzerland	3	1	2906+
France	14	1	2131+	Eastern Europe	1	1	2948+
Britain	11	2	2353+	Russia	2	–	2984+
Germany	23	1	2589+	United States	3	2	3021+
Scandinavia	3	–	2855+				

Totals: persons 61; institutions 8

1830s

Italy	1	1	1940+	Scandinavia	3	1	2858+
France	10	4	2146+	Switzerland	1	–	2910
Britain	9	7	2366+	Eastern Europe	–	1	2950
Germany	10	1	2613+	Russia	1	1	2986+
South Netherlands	1	–	2816	United States	3	1	3026+

Totals: persons 39; institutions 17

1840s

Italy	4	–	1942+	Scandinavia	4	1	2862+
France	12	1	2160+	Spain	–	1	2929
Britain	19	5	2382+	Eastern Europe	2	3	2951+
Germany	23	3	2624+	Russia	1	1	2988+
North Netherlands	–	1	2790	United States	3	4	3030+

Totals: persons 68; institutions 20

1850s

Italy	2	–	1946+	South Netherlands	–	1	2817
France	8	3	2173+	Scandinavia	–	1	2867
Britain	16	3	2406+	Eastern Europe	3	–	2956+
Germany	20	1	2650+	Russia	4	1	2990+
North Netherlands	–	1	2791	United States	3	1	3037+

Totals: persons 56; institutions 12

1860s

Italy	2	2	1948+	Scandinavia	3	1	2868+
France	6	2	2184+	Switzerland	–	1	2911
Britain	6	1	2425+	Eastern Europe	4	2	2959+
Germany	24	3	2671+	Russia	3	3	2995+
North Netherlands	–	1	2792	United States	7	4	3041+
South Netherlands	–	2	2818+				

Totals: persons 55; institutions 22

1870s

Italy	1	1	1952+	Scandinavia	3	2	2872+
France	9	5	2192+	Switzerland	3	2	2912+
Britain	20	5	2432+	Spain	–	1	2930
Germany	24	5	2698+	Eastern Europe	1	4	2965+
North Netherlands	6	1	2793+	Russia	3	–	3001+
South Netherlands	1	1	2820+	United States	6	4	3052+

Totals: persons 77; institutions 31

1880s

Italy	4	2	1954+	Switzerland	–	1	2917+
France	5	2	2206+	Spain	1	–	2931
Britain	11	4	2457+	Eastern Europe	1	–	2970
Germany	15	7	2727+	Russia	5	1	3004+
North Netherlands	1	2	2800+	United States	4	6	3062+
Scandinavia	3	2	2877+				

Totals: persons 50; institutions 27

1890s

Italy	–	3	1960+	Scandinavia	–	1	2882
France	1	1	2213+	Switzerland	1	1	2918+
Britain	1	2	2472+	Eastern Europe	1	–	2971
Germany	5	4	2749+	Russia	2	3	3010+
North Netherlands	1	1	2803+	United States	–	5	3072+
South Netherlands	1	–	2822				

Totals: persons 13; institutions 21

Notes: (a) The sharp decrease in the number of persons in the 1890s
is due to the fact that few who began their careers in
that decade made major contributions before 1900 (the
terminating date for the present chronology). The number
in the 1880s may be likewise affected but to a much
smaller extent.

(b) The number of persons included in Part 2 is 1000 but the
sum of the decadal totals recorded above is 1008. This
discrepancy is due to the fact that a few persons did
important work in two different countries (three countries
in the case of Lagrange) and so were counted more than
once.

APPENDIX TO PART 2

TEACHING INSTITUTIONS

Most of the following institutions do not have separate entries in
Part 2 because they encompassed much more than science. They are
listed here to indicate their relative importance for science at
different periods; the entry numbers given refer to persons who held
teaching (or administrative) positions in them. (Many persons are
cited more than once because they moved from one institution to
another in the course of their careers or--especially in France--
because they held two or more posts concurrently) Some small and
less important institutions have been omitted.

ITALY

Universities

Bologna. *15th cent.* 1826. *16th cent.* 1827, 1829, 1831, 1846, 1851,
1858, 1859. *17th cent.* 1880, 1886, 1894, 1897, 1898, 1903. *18th
cent.* 1906, 1908. *19th cent.* 1941, 1946, 1948.

Padua. *15th cent.* 1825. *16th cent.* 1833, 1850, 1853, 1855, 1857,
1863, 1865, 1866. *17th cent.* 1867, 1878, 1882, 1893, 1898. *18th
cent.* 1903, 1905, 1910, 1922. *19th cent.* 1952.

Pavia. *15th cent.* 1821. *16th cent.* 1831. *17th cent.* 19. *18th
cent.* 1916, 1925, 1929, 1930 (see note in entry 1925). *19th cent.*
1942, 1948, 1951.

Ferrara. *15th cent.* 1820, 1822. *16th cent.* 1841.

Pisa. *16th cent.* 1839, 1840, 1847, 1863. *17th cent.* 1876, 1885,
1888, 1897, 1901. *18th cent.* 1926, 1931. *19th cent.* 1941, 1943,
1948, 1956.

Rome. (a) Archiginnasio della Sapienza (the university of the
        Renaissance period). *16th cent.* 1838, 1840, 1847,
        1849, 1852, 1862. *17th cent.* 1876, 1907.
    (b) Collegio Romano (the Jesuits' university). *16th cent.*
        1854. *17th cent.* 1259, 1883. *18th cent.* 1916.
        19th cent. 1944.
    (c) The modern university (founded in 1871). 1945, 1946,
        1948, 1956.

Turin. *18th cent.* 1918. *19th cent.* 1935, 1955, 1956, 1957.

Modena. *18th cent.* 1925, 1930, 1934. *19th cent.* 1959.

Technical Schools

Bologna, Institute (can be regarded as an early technical school--
    see 1911; later in the eighteenth century it acquired some medical

chairs). *18th cent.* 1908, 1927.

Milan, Technical Institute. *late 19th cent.* 1942, 1946.

Rome, Polytechnic. *late 19th cent.* 1946.

FRANCE

Before the Revolution

University of Paris. *16th cent.* 1965, 1968, 1969, 1973, 2920. *17th cent.* 1988, 2029. *18th cent.* 2036, 2040, 2053, 2054.

University of Montpellier. *16th cent.* 1971, 1984, 1985. *17th cent.* 2019.

Collège Royal (1967, 2037). *16th cent.* 1839, 1966, 1969. *17th cent.* 1988, 1990, 1995, 2021. *18th cent.* 2029, 2030, 2040, 2045, 2062, 2066.

Jardin du Roi (1998, 2056). *17th cent.* 1999, 2008, 2017, 2026, 2033. *18th cent.* 2036, 2040, 2041, 2046, 2047, 2051, 2057, 2059, 2060, 2062, 2068, 2080, 2081, 2082, 2083, 2090, 2095.

Military and Naval Establishments. *18th cent.* 2030, 2054, 2064, 2074, 2075, 2076, 2086.

After the Revolution

Collège de France. 2086, 2094, 2106, 2114, 2115, 2119, 2130, 2137, 2138, 2144, 2149, 2152, 2170, 2171, 2176, 2185, 2190, 2200, 2205.

Muséum d'Histoire Naturelle (2097). 2062, 2080, 2081, 2082, 2090, 2092, 2095, 2102, 2103, 2105, 2106, 2107, 2115, 2116, 2120, 2127, 2139, 2140, 2142, 2157, 2167, 2182, 2191, 2198, 2213.

Ecole Polytechnique (2098). 2074, 2075, 2078, 2079, 2085, 2086, 2088, 2090, 2111, 2112, 2113, 2114, 2117, 2119, 2120, 2123, 2124, 2125, 2132, 2133, 2136, 2148, 2149, 2151, 2152, 2154, 2160, 2185, 2198.

Ecole des Ponts et Chaussées. 2088, 2113, 2124, 2135.

Ecole des Mines (2091). 2092, 2093, 2102, 2138, 2167, 2178, 2200, 2202.

Ecole de Pharmacie (2118). 2102, 2128, 2176, 2209.

Ecole Normale Supérieure (2184). 2166, 2172, 2186, 2195.

Faculties of Science (2109)
Paris (at the Sorbonne). 2086, 2092, 2104, 2107, 2111, 2119, 2120, 2123, 2125, 2131, 2132, 2133, 2136, 2137, 2141, 2142, 2148, 2149, 2151, 2157, 2160, 2165, 2166, 2170, 2178, 2186, 2187, 2191, 2194, 2195, 2206.

Montpellier. 2161, 2164, 2903.   Besançon. 2166.

Lyons. 2141, 2154.   Lille. 2172.

Bordeaux. 2153, 2191, 2206.   Grenoble. 2177.

Rennes. 2158.   Toulouse. 2195, 2210, 2793.

Strasbourg. 2164, 2167, 2172.

Faculty of Medicine, Paris. 2090, 2102, 2136, 2165, 2183, 2204.

Minor Technical Schools. 2124, 2136, 2157; 2131; 2141, 2198; 2208.

BRITAIN

Before 1800

University of Oxford. *17th cent.* 2228, 2242, 2243, 2253, 2256, 2263, 2274, 2279. *18th cent.* 2284, 2296, 2297, 2313.

University of Cambridge. *17th cent.* 2239, 2254, 2262, 2265. *18th cent.* 2287, 2288, 2294.

London

    Gresham College (2226). *17th cent.* 2228, 2233, 2235, 2263, 2266, 2269, 2286.

    Medical and Surgical Institutions. *17th cent.* 2232, 2269, 2281.

Scottish Universities

    Edinburgh. *17th cent.* 2279. *18th cent.* 2292, 2299, 2301, 2302, 2304, 2325.

    St. Andrews. *17th cent.* 2264.

    Glasgow. *18th cent.* 2302.

After 1800

University of Oxford. 2348, 2369, 2407, 2414, 2449, 2458.

University of Cambridge. 2323, 2344, 2345, 2349, 2357, 2384, 2385, 2387, 2413, 2428, 2431, 2433, 2437, 2451, 2454, 2456, 2459, 2460, 2468, 2469.

London

    Royal Institution (2329). 2339, 2358, 2411, 2414, 2428.

    University College. 2354, 2369, 2373, 2392, 2431, 2442, 2449, 2455, 2467, 2468.

    King's College. 456a, 1757, 2347, 2370, 2395, 2413, 2453.

    Royal College of Chemistry (2391). 2637.

    (Royal) School of Mines (2419). 2382, 2393, 2403.

    Normal School of Science (2457) (later Royal College of Science). 2403, 2426, 2441.

    Medical and Surgical Institutions. 1269, 2365, 2393, 2414, 2471.

Provincial Colleges

    Owens College, Manchester. 2393, 2396, 2412, 2417, 2427, 2434, 2461, 2707.

    Mason College, Birmingham. 2444.

    Liverpool University College. 2440, 2471.

    Bristol University College. 2442, 2466.

Scottish Universities

  Edinburgh. 2335, 2343, 2371, 2378, 2382, 2408, 2420, 2422, 2424,
    2428, 2429, 2455.

  Glasgow. 2337, 2342, 2389, 2410.

  St. Andrews. 2335.

  Aberdeen. 2413, 2422.

Irish Colleges

  Trinity College, Dublin. 2353, 2439, 2473.

  Queen's College, Belfast. 2374, 2408, 2422.

  Queen's College, Cork. 2383.

  Queen's College, Galway. 2460.

GERMANY

Before 1800

Universities

  Giessen. *17th cent*. 2529.

  Göttingen. *18th cent*. 2538, 2557, 2896.

  Halle. *17th cent*. 2525, 2526. *18th cent*. 2532.

  Heidelberg. *16th cent*. 2500, 2506.

  Ingolstadt. *16th cent*. 2481. *17th cent*. 2512, 2513.

  Jena. *16th cent*. 2485. *17th cent*. 2514.

  Königsberg. *16th cent*. 2485.

  Leipzig. *16th cent*. 2487. *17th cent*. 2527. *18th cent*. 2556.

  Mainz. *17th cent*. 2518. *18th cent*. 2558.

  Marburg. *16th cent*. 2483, 2491, 2504.

  Tübingen. *15th cent*. 2475. *16th cent*. 2490, 2500. *17th cent*.
    2528. *18th cent*. 2550.

  Wittenberg. *16th cent*. 2486, 2499. *17th cent*. 2509.

  Würzburg. *16th cent*. 2808. *17th cent*. 2518.

Freiberg Mining Academy (2549). *18th cent*. 2554.

After 1800

Universities

  Berlin. 2561, 2572, 2579, 2590, 2594, 2601, 2605, 2610, 2612, 2614,
    2615, 2625, 2626, 2627, 2631, 2633, 2637, 2641, 2646, 2648, 2657,
    2658, 2676, 2683, 2686, 2687, 2697, 2700, 2708, 2720, 2725, 2726,
    2731, 2739.

  Bonn. 2582, 2592, 2596, 2610, 2631, 2632, 2654, 2666, 2668, 2695,
    2705, 2717, 2730.

Breslau.   2609, 2614, 2633, 2649, 2655, 2685, 2687, 2722.

Erlangen.   2591, 2611, 2619, 2671, 2701, 2708, 2709.

Freiberg i.B.   2619, 2640, 2659, 2693, 2719.

Giessen.   2603, 2644, 2653, 2671, 2672, 2703.

Göttingen.   2570, 2594, 2600, 2602, 2611, 2620, 2650, 2653, 2701, 2705, 2727, 2737, 2741.

Halle.   2576, 2659, 2666, 2673, 2722.

Heidelberg.   2573, 2581, 2606, 2616, 2620, 2621, 2624, 2631, 2633, 2635, 2642, 2664, 2670, 2706, 2712, 2713, 2723.

Jena.   2575, 2622, 2641, 2664, 2680, 2690, 2695, 2699, 2720, 2721, 2757.

Kiel.   2663, 2681, 2686, 2687, 2692, 2694, 2707, 2731, 2732.

Königsberg.   2578, 2585, 2586, 2593, 2599, 2607, 2624, 2631, 2727, 2741.

Leipzig.   2577, 2588, 2600, 2604, 2638, 2644, 2647, 2665, 2679, 2681, 2688, 2696, 2701, 2709, 2710, 2733, 2738, 2868, 2961.

Marburg.   2616, 2638, 2725, 2739.

Munich.   2575, 2597, 2603, 2619, 2640, 2656, 2658, 2669, 2682, 2702, 2711, 2721, 2753, 2961.

Strasbourg.   2651, 2656, 2657, 2659, 2676, 2677, 2711, 2713.

Tübingen.   2608, 2642, 2655, 2657, 2688, 2755.

Würzburg.   2645, 2660, 2667, 2677, 2679, 2703, 2708, 2734, 2736, 2748.

Freiberg Mining Academy.   933, 974, 2583, 2604, 2617, 2946.

Technical Schools (mostly known originally as Polytechnische Schulen or Polytechniken; they grew from small beginnings to become Technische Hochschulen by the end of the century)

Aachen.   2683, 2707, 2734.

Berlin.   2602, 2605, 2625, 2651, 2656, 2751.

Brunswick.   2652.

Darmstadt.   2677.

Hanover.   2735.

Karlsruhe.   2653, 2655, 2680, 2682, 2730.

Kassel.   2602.

Munich.   2624, 2680, 2701.

Nuremberg.   2581, 2597.

NORTHERN NETHERLANDS (HOLLAND)

Universities

Leiden.   *16th cent.* 1978.   *17th cent.* 2763, 2768, 2770, 2781.   *18th cent.* 2782, 2783, 2784, 2785.   *19th cent.* 2795, 2801.

Groningen.   *17th cent.* 2892.   *18th cent.* 2789.   *19th cent.* 2793.

Franeker.   *18th cent.* 2789.

Utrecht.   *18th cent.* 2784.   *19th cent.* 2697.

Amsterdam. *19th cent.* 2670, 2794, 2796, 2797, 2801, 2804.

Medical and Surgical Institutions. *17th cent.* 2774, 2776. *18th cent.* 2789.

Delft Polytechnic. *late 19th cent.* 2799.

· SOUTHERN NETHERLANDS (BELGIUM)

Universities

Louvain. *16th cent.* 2805, 2807, 2808. *19th cent.* 2623, 2822.

Liège. *19th cent.* 2623, 2821.

Ghent. *19th cent.* 2654.

Ecole Royale Militaire. *19th cent.* 2816.

SCANDINAVIA

Before 1800

Universities

Copenhagen. *16th cent.* 2825. *17th cent.* 2023, 2248, 2826, 2827, 2828.

Uppsala. *17th cent.* 2829. *18th cent.* 2835, 2836, 2840, 2844, 2847.

Kiel. *18th cent.* 2848, 2850.

Military Academy, Stockholm. *18th cent.* 2842.

After 1800

Universities

Copenhagen. 2851, 2865, 2866, 2874.

Uppsala. 2854, 2863, 2878.

Lund. 2853, 2854, 2859.

Christiania (Oslo). 2860, 2868, 2869, 2870.

Medical College, Stockholm (Karolinska Institutet). 2852, 2856, 2857, 2875.

Polytechnic Institute, Copenhagen. 2851, 2864.

Institute of Technology, Stockholm (later University of Stockholm). 2872, 2874, 2878, 2879.

SWITZERLAND

Before 1800

Universities

Basel. *16th cent.* 2885, 2887, 2889. *17th cent.* 2890, 2891. *18th cent.* 2892, 2895.

Geneva.* *18th cent.* 2894, 2900.

After 1800

Universities

Basel. 2658, 2665, 2688, 2717.

Bern.* 2907, 3003.

Geneva.* 2903, 2908.

Lausanne.* 2914.

Zurich. 2575, 2620, 2640, 2645, 2647, 2667, 2679, 2728, 2913, 2918.

  *Originally called Academy, later University.

Polytechnikum (later Technische Hochschule), Zurich. 2632, 2651, 2652, 2700, 2706, 2736, 2913.

SPAIN AND PORTUGAL

Universities

Zaragoza. *16th cent.* 2920.   Barcelona. *19th cent.* 2931.

Coimbra. *16th cent.* 2922.   Madrid. *19th cent.* 2931.

Valencia. *19th cent.* 2931.

EASTERN EUROPE

Before 1800

Universities

Vienna. *15th cent.* 2932, 2933. *17th cent.* 2940. *18th cent.* 2944.

Prague. *16th cent.* 2939. *18th cent.* 2944.

After 1800

Universities

Vienna. 2714, 2946, 2953, 2955, 2956, 2957, 2960, 2961, 2962, 2964.

Prague. 2609, 2696, 2945, 2960, 2970.

Innsbruck. 2722.

Graz. 2960, 2961.

Budapest. 2967.

Technical School, Prague. 2953.

Medical Schools, Vienna. 2647, 2696.

RUSSIA

After 1800

Universities

Moscow. 2994, 2998, 3001, 3012.   Kiev. 3000.

Dorpat. 2607, 2982.   Kazan. 2981, 2990, 2998.

UNITED STATES

After 1800

INDEX

References are to entry numbers (page numbers are not used, except in some of the notes). Main entries for persons in Part 1 are signified by italic type, and main entries for persons and institutions in Part 2 by bold type; all other references are in roman. The alphabetical arrangement is on the letter-by-letter principle, with a few slight modifications.

Titles of books and periodicals are not indexed, except in a few special cases. Principles, laws, and effects, etc., named after individual persons are generally not indexed but may be found in the entries for the persons concerned in Part 1. Many teaching institutions are not listed here but are included in the Appendix to Part 2.